Controlled and Conditioned Invariants in Linear System Theory

Giuseppe Basile
Department of Electronics, Computers, and Systems
University of Bologna, Italy
     and
University of Florida

Giovanni Marro
Department of Electronics, Computers, and Systems
University of Bologna, Italy

 Prentice Hall, Englewood Cliffs, New Jersey 07632

Library of Congress Cataloging-in-Publication Data

Basile, G.
   Controlled and conditioned invariants in linear system theory / G. Basile and G. Marro.
   p. cm.
   Includes index.
   ISBN 0-13-172974-8
   1. System analysis. 2. Linear systems. I. Marro, G. II. Title.
QA402.B375   1991   91-14177
003—dc20   CIP

Editorial/production: bookworks   Prepress buyer: Kelly Behr
Acquisition editor: Karen Gettman Manufacturing buyer: Susan Brunke
Cover design: Lundgren Graphics   The diskette is used with permission of The MathWorks, Inc.

© 1992 by Prentice-Hall, Inc.
A Simon & Schuster Company
Englewood Cliffs, New Jersey 07632

The publisher offers discounts on this book when ordered in bulk quantities. For more information, write:

   Special Sales/Professional Marketing
   Prentice Hall
   Professional & Technical Reference Division
   Englewood Cliffs, NJ 07632

If your diskette is defective or damaged in transit, return it directly to Prentice Hall at the address below for a no-charge replacement within 90 days of purchase. Mail the defective diskette together with your name and address to:
   Prentice Hall
   Attention: Ryan Colby
   College Operations
   Englewood Cliffs, NJ 07632

The author and publisher of this book have used their best efforts in preparing this book and software. These efforts include the development, research, and testing of the procedures and programs to test their effectiveness. The author and publisher make no warranty of any kind, expressed or implied, with regard to these programs or the documentation contained in this book. The author and publisher shall not be liable in any event for incidental or consequential damages in connection with, or arising out of, the furnishing, performance, and use of these programs.

All rights reserved. No part of this book may be reproduced, in any form or by any means, without permission in writing from the publisher.

Printed in the United States of America

10 9 8 7 6 5 4 3 2 1

ISBN 0-13-172974-8

Prentice-Hall International (UK) Limited, *London*
Prentice-Hall of Australia Pty. Limited, *Sydney*
Prentice-Hall Canada Inc., *Toronto*
Prentice-Hall Hispanoamericana, S.A., *Mexico*
Prentice-Hall of India Private Limited, *New Delhi*
Prentice-Hall of Japan, Inc., *Tokyo*
Simon & Schuster Asia Pte. Ltd., *Singapore*
Editora Prentice-Hall do Brasil, Ltda., *Rio de Janeiro*

CONTENTS

	PREFACE	vii
	SYMBOLS	xi
CHAPTER 1	**INTRODUCTION TO SYSTEMS**	1

    1.1 Basic Concepts and Terms, *1*
    1.2 Some Examples of Dynamic Systems, *4*
    1.3 General Definitions and Properties, *11*
    1.4 Controlling and Observing the State, *21*
    1.5 Interconnecting Systems, *25*
          1.5.1 Graphic Representations of Interconnected Systems, 25
          1.5.2 Cascade, Parallel, and Feedback Interconnections, 29
    1.6 A Review of System and Control Theory Problems, *32*
    1.7 Finite-State Systems, *36*
          1.7.1 Controllability, 40
          1.7.2 Reduction to the Minimal Form, 43
          1.7.3 Diagnosis and State Observation, 47
          1.7.4 Homing and State Reconstruction, 49
          1.7.5 Finite-Memory Systems, 52

CHAPTER 2	**GENERAL PROPERTIES OF LINEAR SYSTEMS**	56

    2.1 The Free State Evolution of Linear Systems, *56*
          2.1.1 Linear Time-Varying Continuous Systems, 56
          2.1.2 Linear Time-Varying Discrete Systems, 60
          2.1.3 Function of a Matrix, 62
          2.1.4 Linear Time-Invariant Continuous Systems, 65
          2.1.5 Linear Time-Invariant Discrete Systems, 72
    2.2 The Forced State Evolution of Linear Systems, *75*
          2.2.1 Linear Time-Varying Continuous Systems, 76
          2.2.2 Linear Time-Varying Discrete Systems, 78

2.2.3 Linear Time-Invariant Systems, 79
2.2.4 Computation of the Matrix Exponential Integral, 84
2.2.5 Approximating Continuous with Discrete, 87
2.3 IO Representations of Linear Constant Systems, 89
2.4 Relations Between IO and ISO Representations, 92
2.4.1 The Realization Problem, 94
2.5 Stability, 101
2.5.1 Linear Time-Varying Systems, 101
2.5.2 Linear Time-Invariant Systems, 104
2.5.3 The Liapunov and Sylvester Equations, 107
2.6 Controllability and Observability, 111
2.6.1 Linear Time-Varying Systems, 111
2.6.2 Linear Time-Invariant Systems, 119

CHAPTER 3 THE GEOMETRIC APPROACH: CLASSIC FOUNDATIONS 126

3.1 Introduction, 126
3.1.1 Some Subspace Algebra, 126
3.2 Invariants, 129
3.2.1 Invariants and Changes of Basis, 129
3.2.2 Lattices of Invariants and Related Algorithms, 131
3.2.3 Invariants and System Structure, 132
3.2.4 Invariants and State Trajectories, 134
3.2.5 Stability and Complementability, 136
3.3 Controllability and Observability, 140
3.3.1 The Kalman Canonical Decomposition, 143
3.3.2 Referring to the Jordan Form, 148
3.3.3 SISO Canonical Forms and Realizations, 150
3.3.4 Structural Indices and MIMO Canonical Forms, 155
3.4 State Feedback and Output Injection, 160
3.4.1 Asymptotic State Observers, 168
3.4.2 The Separation Property, 172
3.5 Some Geometric Aspects of Optimal Control, 176
3.5.1 Convex Sets and Convex Functions, 178
3.5.2 The Pontryagin Maximum Principle, 182
3.5.3 The Linear-Quadratic Regulator, 193
3.5.4 The Time-Invariant LQR Problem, 195

CHAPTER 4 THE GEOMETRIC APPROACH: ANALYSIS 204

4.1 Controlled and Conditioned Invariants, 204
4.1.1 Some Specific Computational Algorithms, 209
4.1.2 Self-Bounded Controlled Invariants and their Duals, 210
4.1.3 Constrained Controllability and Observability, 215
4.1.4 Stabilizability and Complementability, 217
4.2 Disturbance Localization and Unknown-Input State Estimation, 223

- 4.3 Unknown-Input Reconstructability, Invertibility, and Functional Controllability, 230
  - 4.3.1 A General Unknown-Input Reconstructor, 232
  - 4.3.2 System Invertibility and Functional Controllability, 236
- 4.4 Invariant Zeros and the Invariant Zero Structure, 238
  - 4.4.1 The Generalized Frequency Response, 239
  - 4.4.2 The Role of Zeros in Feedback Systems, 243
- 4.5 Extensions to Quadruples, 245
  - 4.5.1 On Zero Assignment, 248

CHAPTER 5 THE GEOMETRIC APPROACH: SYNTHESIS 252

- 5.1 The Five-Map System, 252
  - 5.1.1 Some Properties of the Extended State Space, 255
  - 5.1.2 Some Computational Aspects, 260
  - 5.1.3 The Dual-Lattice Structures, 266
- 5.2 The Dynamic Disturbance Localization and the Regulator Problem, 273
  - 5.2.1 Proof of the Nonconstructive Conditions, 276
  - 5.2.2 Proof of the Constructive Conditions, 279
  - 5.2.3 General Remarks and Computational Recipes, 286
  - 5.2.4 Sufficient Conditions in Terms of Zeros, 290
- 5.3 Reduced-Order Devices, 291
  - 5.3.1 Reduced-Order Observers, 295
  - 5.3.2 Reduced-Order Compensators and Regulators, 296
- 5.4 Accessible Disturbance Localization and Model-Following Control, 299
- 5.5 Noninteracting Controllers, 303

CHAPTER 6 THE ROBUST REGULATOR 310

- 6.1 The Single-Variable Feedback Regulation Scheme, 310
- 6.2 The Autonomous Regulator: A General Synthesis Procedure, 315
  - 6.2.1 On the Separation Property of Regulation, 324
  - 6.2.2 The Internal Model Principle, 326
- 6.3 The Robust Regulator: Some Synthesis Procedures, 328
- 6.4 The Minimal-Order Robust Regulator, 338
- 6.5 The Robust Controlled Invariant, 344
  - 6.5.1 The Hyper-Robust Disturbance Localization Problem, 350
  - 6.5.2 Some Remarks on Hyper-Robust Regulation, 351

APPENDIX A MATHEMATICAL BACKGROUND 355

- A.1 Sets, Relations, Functions, 355
  - A.1.1 Equivalence Relations and Partitions, 364
  - A.1.2 Partial Orderings and Lattices, 365
- A.2 Fields, Vector Spaces, Linear Functions, 369
  - A.2.1 Bases, Isomorphisms, Linearity, 373

- A.2.2 *Projections, Matrices, Similarity, 378*
- A.2.3 *A Brief Survey of Matrix Algebra, 382*
- A.3 Inner Product, Orthogonality, *386*
  - A.3.1 *Orthogonal Projections, Pseudoinverse of a Linear Map, 391*
- A.4 Eigenvalues, Eigenvectors, *394*
  - A.4.1 *The Schur Decomposition, 398*
  - A.4.2 *The Jordan Canonical Form. Part I, 399*
  - A.4.3 *Some Properties of Polynomials, 403*
  - A.4.4 *Cyclic Invariant Subspaces, Minimal Polynomial, 405*
  - A.4.5 *The Jordan Canonical Form. Part II, 408*
  - A.4.6 *The Real Jordan Form, 409*
  - A.4.7 *Computation of the Characteristic and Minimal Polynomial, 410*
- A.5 Hermitian Matrices, Quadratic Forms, *414*
- A.6 Metric and Normed Spaces, Norms, *417*
  - A.6.1 *Matrix Norms, 422*
  - A.6.2 *Banach and Hilbert Spaces, 426*
  - A.6.3 *The Main Existence and Uniqueness Theorem, 429*

APPENDIX B COMPUTATIONAL BACKGROUND   434

- B.1 The Gauss-Jordan Elimination Method and the LU Factorization, *434*
- B.2 The Gram-Schmidt Orthonormalization Process and the QR Factorization, *438*
  - B.2.1 *The QR Factorization for Singular Matrices, 441*
- B.3 The Singular Value Decomposition, *442*
- B.4 Some Matlab Subroutines for Geometric Approach Computations, *444*

INDEX   455

PREFACE

This book is based on material developed by the authors for an introductory course in System Theory and an advanced course on Multivariable Control Systems at the University of Bologna, Italy and the University of Florida, Gainesville, Florida. A characterizing feature of the graduate-level course is the use of new geometric-type techniques in dealing with linear systems, both from the analysis and synthesis viewpoints. A primary advantage of these techniques is the formulation of the results in terms of very simple concepts that give the feeling of problems not masked by heavy, misleading mathematics. To achieve this, fundamental tools known as "controlled invariants" and their duals, "conditioned invariants" (hence the title of the volume) have been developed with a great deal of effort during the last twenty years by numerous researchers in system and control theory. Among them, we would like to mention W.M. Wonham, A.S Morse, J.B. Pearson, B.A. Francis, J.C. Willems, F. Hamano, H. Akashi, B.P. Molinari, J.M.H. Schumacher, S.P. Bhattacharyya, C. Commault, all of whose works have greatly contributed to setting up and augmenting the foundations and applications of this geometric approach.

The presentation is organized as follows. Chapter 1 familiarizes the reader with the basic definitions, properties, and typical problems of general dynamic systems. As an application, finite state systems are investigated in light of these concepts. Chapter 2 deals with linear system analysis: it is shown that the linear structure allows the results to be carried forward in a simpler form and easier computational procedures to be developed. Basic topics, such as stability, controllability, and observability, are presented and discussed. Both chapters are supported by the mathematical background given in Appendix A. The material presented up to this point meets the needs of an introductory-level course in system theory. Topics in Appendix A may be used in part or entirely, as required by the reader's previous educational curriculum.

The remainder of the book addresses an advanced linear system audience and stresses the geometric concepts. Chapter 3 establishes a connection between basic concepts of linear algebra (like invariants, complementability, changes of basis) and properties of linear time-invariant dynamic systems. Controllability and observability are revisited in this light and the most important canonical forms and realization procedures are briefly presented. Then, elementary synthesis problems such as pole assignment, asymptotic observer theory, state feedback, and output injection are discussed. The section at the end of the chapter covers the geometric aspects of optimization and frames the classical linear-quadratic regulator problem in this context. Chapter 4 first introduces the most specific tools of the geometric approach, then investigates other linear time-invariant system properties, like constrained and functional controllability and observability, system invertibility, and invariant zeros. Controlled and conditioned invariants are widely used to treat all these topics.

Chapter 5 presents the most general linear time-invariant systems synthesis problems, such as regulator and compensator design based on output dynamic feedback. Complete constructive solutions of these problems, including the reduced-order cases, are presented, again using geometric tools and the concept of invariant zero. Chapter 6 presents methods for extending the geometric techniques to the case where some parameters of the controlled system are subject to variation and the overall control scheme has to be "robust" against this, a case which is very important in practice. Finally, Appendix B provides the computational bases and some software to support problems and exercises.

In courses that are more oriented to practice of regulation rather than rigorous, unified mathematical description, most of Chapter 5 may be omitted. In fact Chapter 6, on robust regulation, which extends to the multivariable case some classic automatic control design techniques, includes a completely self-contained simplified statement of the regulator problem.

A few words on notation: numbering of formulas and bibliographic references starts from 1 at the beginning of each section and chapter respectively: when numbers are cited, if pertaining to a different section or chapter, they are prefixed by the section or chapter number (e.g., referring to formula (15) of Section 5.1 is shown as (5.1.15) and to reference [13] of Chapter 3 as [3.13]).

This material has been developed and brought to its final form with the assistance of many people to whom we wish to express our sincere appreciation. Among those to whom we owe a particular debt of gratitude are Dr. A. Piazzi, who made a substantial contribution to our research in the field of geometric approach in recent years and in establishing most of the new material published here, and Professor M. Tibaldi, who made helpful suggestions on treatment of optimal control in a geometric context.

We also wish to acknowledge the continuous and diligent assistance of

Mrs. M. Losito of the Department of Electronics, Computer Sciences and Systems of the University of Bologna for her precision in technically correcting the manuscript and preparing the software relative to specific algorithms and CAD procedures, and Mrs. T. Muratori, of the same department, for her artistic touch in preparing the layout of the text and the figures.

G. Basile and G. Marro
Bologna, Italy

SYMBOLS

Standard symbols and abbreviations

$\forall$	for all
$\ni$	such that
$\exists$	there exists
$\Rightarrow$	implies
$\Leftrightarrow$	implies and is implied by
$:=$	equal by definition
$\mathcal{A}, \mathcal{X}$	sets or vector spaces
a, x	elements of sets or vectors
$\emptyset$	the empty set
$\{x_i\}$	the set whose elements are x_i
$\mathcal{A}_f, \mathcal{X}_f$	function spaces
$\in$	belonging to
$\subset$	contained in
$\subseteq$	contained in or equal to
$\supset$	containing
$\supseteq$	containing or equal to
$\cup$	union
$\uplus$	aggregation (union with repetition count)
$\cap$	intersection
$\dot{-}$	difference of sets with repetition count
$\times$	cartesian product
$\oplus$	direct sum
B	the set of the binary symbols 0 and 1
N	the set of all natural integers
Z	the set of all integer numbers
R	the set of all real numbers
C	the set of all complex numbers
$\mathbf{R}^n$	the set of all n-tuples of real numbers
$[t_0, t_1]$	a closed interval
$[t_0, t_1)$	a right-open interval
$f(\cdot)$	a time function
$\dot{f}(\cdot)$	the first derivative of function $f(\cdot)$

$f(t)$	the value of $f(\cdot)$ at t
$f\|_{[t_0,t_1]}$	a segment of $f(\cdot)$
j	the imaginary unit
z^*	the conjugate of complex number z
sign x	the signum[1] function (x real)
$\|z\|$	the absolute value[1] of complex number z
arg z	the argument[1] of complex number z
$\|x\|$	the norm of vector x
$\langle x,y \rangle$	the inner or scalar product of vectors x and y
grad f	the gradient of function $f(x)$
sp $\{x_i\}$	the span of vectors $\{x_i\}$
dim $\mathcal{X}$	the dimension of subspace $\mathcal{X}$
$\mathcal{X}^\perp$	the orthogonal complement of subspace $\mathcal{X}$
$\mathcal{O}(x,\epsilon)$	the ϵ-neighborhood of x
int $\mathcal{X}$	the interior of set $\mathcal{X}$
clo $\mathcal{X}$	the closure of set $\mathcal{X}$
A, X	matrices or linear transformations
O	a null matrix
I	an identity matrix
I_n	the $n \times n$ identity matrix
A^T	the transpose of A
A^*	the conjugate transpose of A
A^{-1}	the inverse of A (A square nonsingular)
A^+	the pseudoinverse of A (A nonsquare or singular)
adj A	the adjoint of A
det A	the determinant of A
tr A	the trace of A
$\rho(A)$	the rank of A
im A	the image of A

---

[1] Let x be a real number; the signum function of x is defined as

$$\text{sign } x := \begin{cases} 1 & \text{for } x \geq 0 \\ -1 & \text{for } x < 0 \end{cases}$$

and can be used, for instance, for a correct computation of the argument of the complex number $z = u + jv$:

$$|z| = \sqrt{u^2 + v^2}$$

$$\arg z = \arcsin \frac{v}{|z|} \text{sign } u + \frac{\pi}{2}(1 - \text{sign } u)\text{sign } v$$

where the co-domain of function arcsin has been assumed to be $(-\pi/2, \pi/2]$.

Symbols

$\nu(A)$	the nullity of A
ker A	the kernel of A
$\|A\|$	the norm of A
$A\|_{\mathcal{J}}$	the restriction of linear map A to the A-invariant $\mathcal{J}$
$A\|_{\mathcal{X}/\mathcal{J}}$	the linear map induced by A on the quotient space $\mathcal{X}/\mathcal{J}$
□	end of discussion

Specific symbols and abbreviations

$\mathcal{J}$	a generic invariant
$\mathcal{V}$	a generic controlled invariant
$\mathcal{S}$	a generic conditioned invariant
$\max \mathcal{J}(A,\mathcal{C})$	the maximal A-invariant contained in $\mathcal{C}$
$\min \mathcal{J}(A,\mathcal{B})$	the minimal A-invariant containing $\mathcal{B}$
$\max \mathcal{V}(A,\mathcal{B},\mathcal{E})$	the maximal $(A,\mathcal{B})$-controlled invariant contained in $\mathcal{E}$
$\min \mathcal{S}(A,\mathcal{C},\mathcal{D})$	the minimal $(A,\mathcal{C})$-conditioned invariant containing $\mathcal{D}$
$\max \mathcal{V}_R(A(p),\mathcal{B}(p),\mathcal{E})$	the maximal robust $(A(p),\mathcal{B}(p))$-controlled invariant contained in $\mathcal{E}$
$\mathcal{R}$	the reachable set of pair (A,B): $\mathcal{R}=\min\mathcal{J}(A,\mathcal{B})$, $\mathcal{B}:=\mathrm{im}B$
$\mathcal{Q}$	the unobservable set of pair (A,C): $\mathcal{Q}=\max\mathcal{J}(A,\mathcal{C})$, $\mathcal{C}:=\ker C$
$\mathcal{R}_{\mathcal{E}}$	the reachable set on $\mathcal{E}$: $\mathcal{R}_{\mathcal{E}}=\mathcal{V}^*\cap\min\mathcal{S}(A,\mathcal{E},\mathcal{B})$, $\mathcal{V}^*:=\max\mathcal{V}(A,\mathcal{B},\mathcal{E})$, $\mathcal{E}:=\ker E$
$\mathcal{Q}_{\mathcal{D}}$	the unobservable set containing $\mathcal{D}$: $\mathcal{Q}_{\mathcal{D}}=\mathcal{S}^*+\max\mathcal{V}(A,\mathcal{D},\mathcal{C})$, $\mathcal{S}^*:=\min\mathcal{S}(A,\mathcal{C},\mathcal{D})$, $\mathcal{D}:=\mathrm{im}D$
$\Phi(\mathcal{B}+\mathcal{D},\mathcal{E})$	the lattice of all $(A,\mathcal{B}+\mathcal{D})$-controlled invariants self-bounded with respect to $\mathcal{E}$ and containing $\mathcal{D}$: $\Phi(\mathcal{B}+\mathcal{D},\mathcal{E}):=\{\mathcal{V}: A\mathcal{V}\subseteq\mathcal{V}+\mathcal{B},\ \mathcal{D}\subseteq\mathcal{V}\subseteq\mathcal{E},\ \mathcal{V}\supseteq\mathcal{V}^*\cap\mathcal{B}\}$
$\Psi(\mathcal{C}\cap\mathcal{E},\mathcal{D})$	the lattice of all $(A,\mathcal{C})$-conditioned invariants self-hidden with respect to $\mathcal{D}$ and contained in $\mathcal{E}$: $\Psi(\mathcal{C}\cap\mathcal{E},\mathcal{D}):=\{\mathcal{S}: A(\mathcal{S}\cap\mathcal{C})\subseteq\mathcal{S},\ \mathcal{D}\subseteq\mathcal{S}\subseteq\mathcal{E},\ \mathcal{S}\subseteq\mathcal{S}^*+\mathcal{C}\}$
$\mathcal{V}_m$	the infimum of $\Phi(\mathcal{B}+\mathcal{D},\mathcal{E})$: $\mathcal{V}_m=\mathcal{V}^*\cap\mathcal{S}_1^*$, $\mathcal{S}_1^*:=\min\mathcal{S}(A,\mathcal{E},\mathcal{B}+\mathcal{D})$
$\mathcal{S}_M$	the supremum of $\Psi(\mathcal{C}\cap\mathcal{E},\mathcal{D})$: $\mathcal{S}_M=\mathcal{S}^*+\mathcal{V}_1^*$, $\mathcal{V}_1^*:=\max\mathcal{V}(A,\mathcal{D},\mathcal{C}\cap\mathcal{E})$
$\mathcal{V}_M$	a special element of $\Phi(\mathcal{B}+\mathcal{D},\mathcal{E})$, defined as $\mathcal{V}_M:=\mathcal{V}^*\cap(\mathcal{V}_1^*+\mathcal{S}_1^*)$
$\mathcal{S}_m$	a special element di $\Psi(\mathcal{C}\cap\mathcal{E},\mathcal{D})$, defined as $\mathcal{S}_m:=\mathcal{S}^*+\mathcal{V}_1^*\cap\mathcal{S}_1^*$

1
INTRODUCTION TO SYSTEMS

1.1 BASIC CONCEPTS AND TERMS

In this chapter standard system theory terminology is introduced and explained in terms that are as simple and self-contained as possible, with some representative examples. Then, the basic properties of systems are analyzed, and concepts such as state, linearity, time-invariance, minimality, equilibrium, controllability, and observability are briefly discussed. Finally, as a first application, finite-state systems are presented.

Terms like "system," "system theory," "system science," and "system engineering" have come into common use in the last three decades from various fields (process control, data processing, biology, ecology, economics, traffic-planning, electricity systems, management, etc.), so that they have now come to assume various shades of meaning. Therefore, before beginning our treatment of systems, we shall try to exactly define the object of our study and outline the class of problems, relatively restricted, to which we shall refer in this book.

The word *system* denotes an object, device, or phenomenon whose time evolution appears through the variation of a certain number of measurable attributes as with, for example, a machine tool, an electric motor, a computer, an artificial satellite, the economy of a nation.

A *measurable attribute* is a characteristic that can be correlated with one or more numbers, either integer, real or complex, or simply a set of symbols. Examples include the rotation of a shaft (a real number), the voltage or impedance between two given points of an electric circuit (a real or complex number), any color belonging to a set of eight well-defined colors (an element of a set of eight symbols; for instance, digits ranging from 1 to 8 or letters from a to h), the position of a push button (a symbol equal to 0 or 1, depending on whether it is released or pressed). In dealing with distributed-parameter systems, attributes can be represented by real or complex-valued functions of space coordinates. Examples include the temperature along a continuous furnace (a real function of space), the voltage of a given frequency along a transmission line (a complex function of space coordinates).

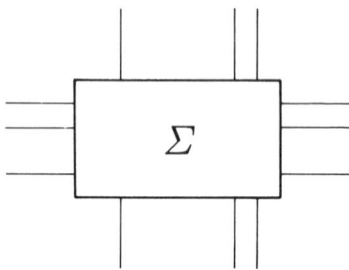

Figure 1-1 Schematic representation of a system.

In order to reproduce and analyze the behavior of a system, it is necessary to refer to a *mathematical model* which, generally with a certain approximation, represents the links existing between the various measurable attributes or *variables* of the system. The same system can be related to several mathematical models, each of which may correspond to a different compromise between precision and simplicity, and may also depend on the particular problem.

Since mathematical models are themselves systems, although abstract, it is customary to denote both the object of the study and its mathematical model by the word "system." The discipline called *system theory* pertains to the derivation of mathematical models for systems, their classification, investigation of their properties, and their use for the solution of engineering problems.

A system can be represented as a block and its variables as connections with the *environment* or other systems, as shown by the simple diagram of Fig. 1-1.

As a rule, in order to represent a system with a mathematical model, it is first necessary to divide its variables into *causes* or *inputs* and *effects* or *outputs*. Inputs correspond to independent and outputs to dependent variables. A system whose variables are so divided is called an *oriented system* and can be represented as shown in Fig. 1-2, with the connections oriented by means of arrows.

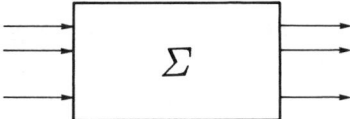

Figure 1-2 Schematic representation of an oriented system.

It is worth noting that the distinction between causes and effects appears quite natural, so it is often tacitly assumed in studying physical systems; nevertheless in some cases it is anything but immediate. Consider, for instance, the simple electric circuit shown in Fig. 1-3(a), whose variables are v and i. It can be oriented as in Fig. 1-3(b), i.e., with v as input and i as output: this is the most natural choice if the circuit is supplied by a voltage generator. But the same system may be supplied by a current generator, in which case i would be the cause and v the effect and the corresponding oriented block diagram would be as shown in Fig. 1-3(c).

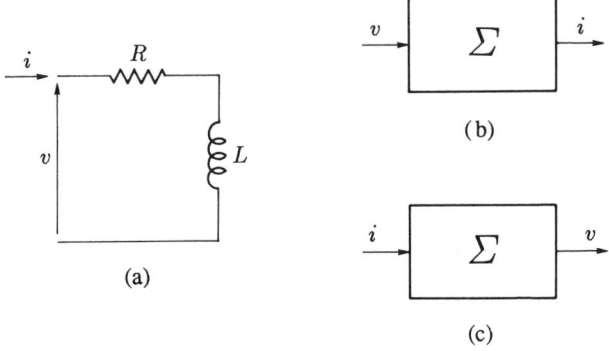

Figure 1-3 An electric system with two possible orientations.

Systems can be divided into two main classes: *memoryless* or *purely algebraic systems*, in which the values of the outputs at any instant of time depend only on the values of the inputs at the same time, and *systems with memory* or *dynamic systems*, in which the values of the outputs depend also on the past time evolution of the inputs.

In dynamic systems the concept of *state* plays a fundamental role: in intuitive terms, the state of a system is the information that is needed at every instant of time, in order to be able to predict the effect of the past history of the system on its future behavior. The state consists of a set of variables or, in distributed-parameter systems, of one or more functions of space coordinates, and is subject to variation in time depending on the time evolution of the inputs.

The terms "input," "state," and "output" of a system usually refer to all its input, state, and output variables as a whole, whereas the terms *input function*, *output function*, and *motion* refer to the time evolution of such variables. In particular, input and output functions are often called input and output *signals*; the terms *stimulus* and *response* are also used.

A system that is not connected to the environment by any input is called a *free* or *autonomous system*; if, on the contrary, there exist any such inputs that represent stimuli from the environment, they are called *exogenous* (variables or signals) and it is said to be a *forced system*. In control problems, it is natural to divide inputs into *manipulable variables* and *nonmanipulable variables*. The former are those whose values can be imposed at every instant of time in order to achieve a given control goal. The latter are those that cannot be arbitrarily varied; if unpredictable, they are more precisely called *disturbances*.

1.2 SOME EXAMPLES OF DYNAMIC SYSTEMS

This section presents some examples of dynamic systems and their mathematical models, with the aim of investigating their common features.

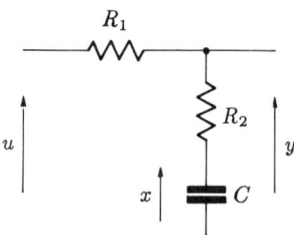

Figure 1-4 A simple electric circuit.

Example 1.2-1. A Simple Electric Circuit

Consider the electric circuit shown in Fig. 1-4. It is described by the equations, one differential and one algebraic,

$$\dot{x}(t) = a\, x(t) + b\, u(t) \qquad (1)$$

$$y(t) = c\, x(t) + d\, u(t) \qquad (2)$$

where the functions on the right side are respectively called *state velocity function* and *output function*; u and y denote the input and output voltages, x the voltage across the capacitor, which can be assumed as the (only) state variable, and $\dot{x}$ the time derivative dx/dt. Constants a, b, c, and d are related to the electric parameters shown in the figure by the following easy-to-derive relations:

$$a := -\frac{1}{C(R_1 + R_2)} \qquad b := \frac{1}{C(R_1 + R_2)}$$

Sect. 1.2 Some Examples of Dynamic Systems

$$c := \frac{R_1}{R_1 + R_2} \qquad d := \frac{R_2}{R_1 + R_2} \tag{3}$$

The differential equation (1) is easily solvable. Let t_0 and t_1 ($t_1 > t_0$) be two given instants of time, x_0 the initial state, i.e., the state at t_0 and $u(\cdot)$ a given piecewise continuous function of time whose domain is assumed to contain the time interval $[t_0, t_1]$. The solution of (1) for $t \in [t_0, t_1]$ is expressed by

$$x(t) = x_0 \, e^{a(t-t_0)} + \int_{t_0}^{t} e^{a(t-\tau)} b \, u(\tau) \, d\tau \tag{4}$$

as can be easily checked by direct substitution.[1] Function (4) is called *state transition function*: it provides the state $x(t)$ as a function of t, t_0, x_0, and $u[t_0, t]$. By substituting (4) into (2) we obtain the so-called *response function*

$$y(t) = c \left(x_0 \, e^{a(t-t_0)} + \int_{t_0}^{t} e^{a(t-\tau)} b \, u(\tau) \, d\tau \right) + d \, u(t) \tag{5}$$

Example 1.2-2. An Electromechanical System

Let us now consider the slightly more complicated electromechanical system shown in Fig. 1-5, i.e., an armature-controlled d.c. electric motor. Its behavior is described by the following set of two differential equations, which express respectively the equilibrium of the voltages along the electric mesh and that of the torques acting on the shaft:

$$v_a(t) = R_a \, i_a(t) + L_a \frac{di_a}{dt}(t) + v_c(t) \tag{6}$$

$$c_m(t) = B \, \omega(t) + J \frac{d\omega}{dt}(t) + c_r(t) \tag{7}$$

In (6) v_a is the applied voltage, R_a and L_a the armature resistance and inductance, i_a and v_c the armature current and counter emf, while in (7) c_m is the motor torque, B, J, and ω the viscous friction coefficient, the moment of inertia, and the angular velocity of the shaft, and c_r an externally applied load torque. If the excitation voltage v_e is assumed to be constant, the following two additional relations hold:

$$v_c(t) = k_1 \, \omega(t) \qquad c_m(t) = k_2 \, i_a(t) \tag{8}$$

---

[1] Recall the rule for the computation of the derivative of an integral depending on a parameter:

$$\frac{d}{dt} \int_{a(t)}^{b(t)} f(x, t) \, dx = f\big(b(t), t\big) \dot{b} - f\big(a(t), t\big) \dot{a} + \int_{a(t)}^{b(t)} \dot{f}(x, t) \, dx$$

where

$$\dot{f}(x, t) := \frac{\partial}{\partial t} f(x, t)$$

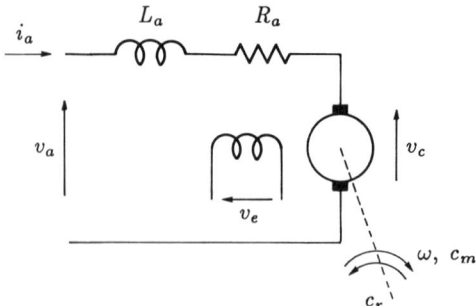

Figure 1-5 An electric motor.

where k_1 and k_2 denote constant coefficients, which are numerically equal to each other if the adopted units are coherent (volt and amp for voltages and currents, Nm and rad/sec for torques and angular velocities). Orient the system assuming as input variables $u_1 := v_a$, $u_2 := c_r$ and as output variable $y := \theta$, the angular position of the shaft, which is related to ω by the simple equation

$$\frac{d\theta}{dt}(t) = \omega(t) \tag{9}$$

Then assume as state variables the armature current, the angular velocity, and the angular position of the shaft, i.e., $x_1 := i_a$, $x_2 := \omega$, $x_3 := \theta$. Equations (6–9) can be written in compact form (using matrices) as

$$\dot{x}(t) = A\,x(t) + B\,u(t) \tag{10}$$
$$y(t) = C\,x(t) + D\,u(t) \tag{11}$$

where[2] $x := (x_1, x_2, x_3)$, $u := (u_1, u_2)$ and

$$A := \begin{bmatrix} -R_a/L_a & -k_1/L_a & 0 \\ k_2/J & -B/J & 0 \\ 0 & 1 & 0 \end{bmatrix} \quad B := \begin{bmatrix} 1 & 0 \\ 0 & -1 \\ 0 & 0 \end{bmatrix}$$

$$C := [0\ \ 0\ \ 1] \quad D := [0\ \ 0] \tag{12}$$

Note that the mathematical model of the electric motor has the same structure as that of the simple electric circuit considered before, but with the constants replaced by matrices. It is worth pointing out that such a structure is common to all lumped-parameter linear time-invariant dynamic systems, which are the most important in connection with control problems, and will also be the protagonists in this book. A further remark: the last term in equation (11) can be deleted, D being a null matrix. In fact, in this case the input does not influence the output directly, but only through the state. Systems with this property are very common and are called *purely dynamic systems*.

---

[2] Here and in the following the same symbol is used for a vector belonging to $\mathbf{R}^n$ and a $n \times 1$ matrix.

Sect. 1.2 Some Examples of Dynamic Systems

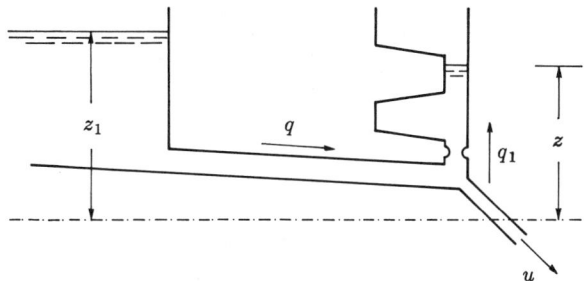

Figure 1-6 A surge tank installation.

Example 1.2-3. A Hydraulic System
A standard installation for a hydroelectric plant can be represented as in Fig. 1-6: it consists of a reservoir, a conduit connecting it to a surge tank, which in turn is connected to the turbines by means of a penstock. At the bottom of the surge tank there is a throttle, built in order to damp the water level oscillations. Let z_1 be the total elevation of water level in the reservoir, z that in the surge tank, $F(z)$ the cross-sectional area of the surge tank, which is assumed to be variable, z_2 the static head at the end of the conduit, q the flow per second in the conduit, q_1 that into the surge tank, and u that in the penstock. Neglecting water inertia in the surge tank, it is possible to set up the equations

$$k_2 \left(z_1(t) - z_2(t) \right) = k_1 q(t) |q(t)| + \dot{q}(t) \tag{13}$$

$$z_2(t) - z(t) = k_3 q_1(t) |q_1(t)| \tag{14}$$

$$\dot{z}(t) = F(z) q_1(t) \tag{15}$$

$$q_1(t) = q(t) - u(t) \tag{16}$$

which can be referred to, respectively, as the conduit equation, the throttle equation, the surge tank equation, and the flow continuity equation; k_1, k_2, and k_3 denote constants. By substituting for z_2 and q_1, the first-order differential equations

$$\dot{q}(t) = -k_1 q(t) |q(t)| + k_2 \left(z_1(t) - z(t) - k_3 \left(q(t) - u(t) \right) |q(t) - u(t)| \right) \tag{17}$$

$$\dot{z}(t) = F(z) \left(q(t) - u(t) \right) \tag{18}$$

are obtained. Let z_2 be assumed as the output variable: this choice is consistent since z_2 is the variable most directly related to the power-delivering capability of the plant: it can be expressed by the further equation

$$z_2(t) = z(t) + k_3 \left(q(t) - u(t) \right) |q(t) - u(t)| \tag{19}$$

If the water level elevation in the reservoir is assumed to be constant, the only input is u, which is typically a manipulable variable. We choose as state variables the flow per second in the conduit and the water level elevation in the surge tank, i.e., $x_1 := q$, $x_2 := z$, and, as the only output variable the static head at the penstock, i.e., $y := z_2$. Equations (17–19) can be written in the more compact form

$$\dot{x}(t) = f(x(t), u(t)) \qquad (20)$$

$$y(t) = g(x(t), u(t)) \qquad (21)$$

where $x := (x_1, x_2)$ and f, g are nonlinear continuous functions.

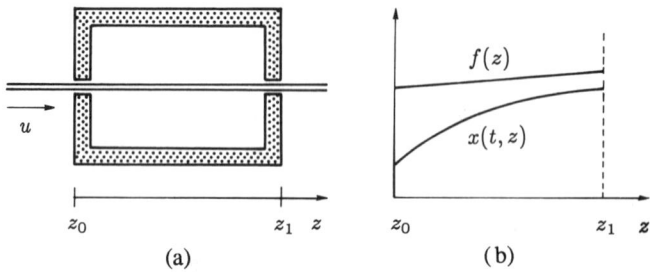

Figure 1-7 A continuous furnace and related temperature distributions.

Example 1.2-4. A Distributed-Parameter System

As an example of a distributed-parameter system, we consider the continuous furnace represented in Fig. 1-7(a): a strip of homogeneous material having constant cross-sectional area is transported with adjustable speed u through a furnace. Both the temperature distributions in the furnace and in the strip are assumed to be variable in the direction of movement z and uniform within sections orthogonal to this direction. Denote by $f(z)$ the temperature along the furnace, which is assumed to be constant in time, and by $x(t, z)$ that along the strip, which is a function of both time and space. The system is described by the following one-dimensional heat diffusion equation:

$$\frac{\partial x(t, z)}{\partial t} = k_1 \frac{\partial^2 x(t, x)}{\partial z^2} + u(t) \frac{\partial x(t, z)}{\partial z} + k_2 \left(x(t, z) - f(z)\right) \qquad (22)$$

where k_1 and k_2 are constants related respectively to the internal and surface thermal conductivity of the strip. We assume the speed u as the input variable and the temperature of the strip at the exit of the furnace as the output variable, i.e.

$$y(t) = x(t, z_1) \qquad (23)$$

The function $x(t, \cdot)$ represents the state at time t; the partial differential equation (22) can be solved if the initial state $x(t_0, \cdot)$ (initial condition), the strip temperature before heating $x(\cdot, z_0)$ (boundary condition), usually constant, and the input function $u(\cdot)$, are given.

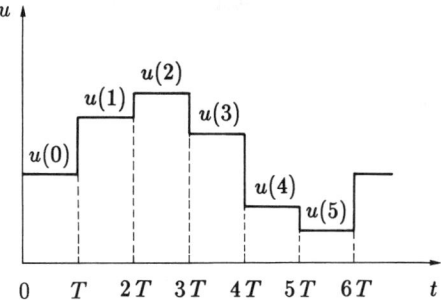

Figure 1-8 Piecewise constant function.

Example 1.2-5. A Discrete-Time System

We refer again to the electric circuit shown in Fig. 1-4 and assume that its input variable u is changed in time by steps, as shown in Fig. 1-8, and the output variable is detected only at the time instants $T, 2T, \ldots$. Such a situation occurs when a continuous-time system is controlled by means of a digital processor, whose inputs and outputs are *sampled data*.

Denote by $u(i)$ the input value in the time interval $[iT, (i+1)T]$ and by $y(i)$ the output value at the time iT; the system is easily shown as being described by a difference equation and an algebraic equation, i.e.,

$$x(i+1) = a_d\, x(i) + b_d\, y(i) \tag{24}$$

$$y(i) = c\, x(i) + d\, u(i) \tag{25}$$

where coefficients c and d are the same as in equation (2), whereas a_d, b_d are related to a, b and the sampling period T by

$$a_d = e^{aT} \qquad b_d = b \int_0^T e^{a(T-\tau)}\, d\tau = \frac{b}{a}\left(e^{aT} - 1\right) \tag{26}$$

Subscript d in the coefficients stands for "discrete." In discrete-time systems, time is an integer variable instead of a real variable and time evolutions of the system variables are represented by sequences instead of continuous or piecewise continuous functions of time. Let j, i ($i > j$) be any two (integer) instants of time, x_0 the initial state, i.e., the state at time j, and $u(\cdot)$ the input sequence in any time interval containing $[j, i]$.[3] The state transition function is obtained by means of a recursive application of (24) and is expressed by

$$x(i) = a_d^{i-j}\, x_0 + \sum_{k=1}^{i-1} a_d^{i-k-1}\, b_d\, u(k) \tag{27}$$

---

[3] For the sake of precision, note that this time interval is necessary in connection with the response function, but can be restricted to $[j, i-1]$ for the state transition function.

The response function is obtained by substituting (27) into (25) as

$$y(i) = c\left(a_d^{i-j} x_0 + \sum_{k=1}^{i-1} a_d^{i-k-1} b_d u(k)\right) + d u(i) \tag{28}$$

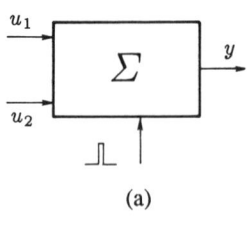

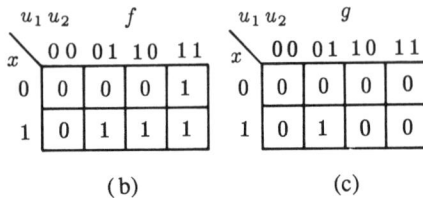

Figure 1-9 A finite-state system and its characterizing functions.

Example 1.2-6. A Finite-State System

The finite-state system is represented as a block in Fig. 1-9(a): the input variables u_1, u_2 are the positions of two push buttons and the output variable y is the lighting of a lamp. The value of each variable is represented by one of two symbols, for instance 1 or 0 according to whether the push buttons are pressed or released and whether the lamp is lighted or not. The input data are assumed to be sampled, i.e., they are accepted when a clock pulse is received by the system; also the possible output variable changes occur at clock pulses, so that their time evolution is inherently discrete. The system behavior is described in words as follows: "lamp lights up if the current input symbol is 01 and, between symbols 00 and 11, 11 previously appeared as the latter." A mathematical model that fits this behavior is

$$x(i+1) = f\big(x(i), u(i)\big) \tag{29}$$

$$y(i) = g\big(x(i), u(i)\big) \tag{30}$$

where f, g are the so-called *next-state function* and *output function*; x is a binary state variable whose value (which is also restricted to 0 or 1) changes only when the current sampled input is 00 or 11: hence x implements the "system memory." If functions f and g are those defined in the tables shown in Fig. 1-9(b) and in Fig. 1-9(c), it is evident that the output variable changes in time according to the previous word description.

Representative examples of finite-state systems are digital processors, i.e., the most widespread electronic systems of technology today. Their state can be represented by a finite number, although very large, of binary symbols, each corresponding to a bit of memory; their input is a keyboard, hence it is similar to that of the above simple example; their output is a sequence of symbols which is translated into a string of characters by a display, a monitor, or a printer. Their time evolution, at least in principle, can be represented by a mathematical model consisting of a next-state and an output function. The former is ruled by a high-frequency clock and is such that an input symbol is accepted (i.e., influences system behavior) only when it is changed with respect to the previous one.

1.3 GENERAL DEFINITIONS AND PROPERTIES

Referring to the examples presented in the previous section, which, although very simple, are representative of the most important classes of dynamic systems, let us now state general definitions and properties in which the most basic connections between system theory and mathematics are shown.

First, consider the sets to which the variables and functions involved in the system mathematical model must belong. In general it is necessary to specify

1. a *time set* $\mathcal{T}$
2. an *input set* $\mathcal{U}$
3. an *input function set* $\mathcal{U}_f$
4. a *state set* $\mathcal{X}$
5. an *output set* $\mathcal{Y}$

The values and functions belonging to the above sets are called *admissible*. Only two possibilities will be considered for the time set: $\mathcal{T} = \mathbf{R}$ (time is measured by a real number) and $\mathcal{T} = \mathbf{Z}$ (time is measured by an integer number). It is worth noting that the properties required for a set to be a time set from a strict mathematical viewpoint are fewer than the properties of either $\mathbf{R}$ or $\mathbf{Z}$; for instance, multiplication does not need to be defined in a time set. Nevertheless, since the familiar $\mathbf{R}$ and $\mathbf{Z}$ fit our needs, it is convenient to adopt them as the only possible time sets and avoid any subtle investigation in order to find out what is strictly required for a set to be a time set. On the basis of this decision, the following definitions are given.

Definition 1.3-1. Continuous-Time System
Systems are said to be *continuous-time* if $\mathcal{T} = \mathbf{R}$, *discrete-time* if $\mathcal{T} = \mathbf{Z}$.

Definition 1.3-2. Purely Algebraic System
A *memoryless* or *purely algebraic system* is composed of sets $\mathcal{T}, \mathcal{U}, \mathcal{Y}$, and an *input-output function* or *input-output map*:

$$y(t) = g\big(u(t), t\big) \qquad (1)$$

Definition 1.3-3. Dynamic Continuous-Time System
A *dynamic continuous-time system* is composed of sets T (= **R**), $\mathcal{U}$, $\mathcal{U}_f$, $\mathcal{X}$, $\mathcal{Y}$ of a *state velocity function*

$$\dot{x}(t) = f\big(x(t), u(t), t\big) \qquad (2)$$

having a unique solution for any admissible initial state and input function and of an *output function* or *output map*

$$y(t) = g\big(x(t), u(t), t\big) \qquad (3)$$

Definition 1.3-4. Dynamic Discrete-Time System
A *dynamic discrete-time system* is composed of sets T (= **Z**), $\mathcal{U}$, $\mathcal{U}_f$, $\mathcal{X}$, $\mathcal{Y}$ of a *next-state function* [1]

$$x(i+1) = f\big(x(i), u(i), i\big) \qquad (4)$$

and of an *output function* or *output map*

$$y(i) = g\big(x(i), u(i), i\big) \qquad (5)$$

The following definition refers to a specialization of dynamic systems that occurs very frequently in practice.

Definition 1.3-5. Purely Dynamic System
A *purely dynamic system* is one in which the output map reduces to

$$y(t) = g\big(x(t), t\big) \qquad (6)$$

Therefore, a purely dynamic system is such that input does not affect the output directly, but only through the state; thus, in continuous-time purely dynamic systems the output is a continuous function of time and in discrete-time purely dynamic systems the output is delayed by at least one sampling period with respect to the input. Any dynamic system can be considered as composed of a purely dynamic system and a purely algebraic one, interconnected as shown in Fig. 1-10. Most system theory problems are approached by referring to purely dynamic systems: since the mathematical model of a purely algebraic system is very simple (it reduces to a function), the extension of the theory to the general case is usually straightforward.

---

[1] In the specific case of discrete-time systems, symbols i or k instead of t are used to denote time. However, in general definitions reported in this chapter, which refer both to the continuous and the discrete-time case, the symbol t is used to denote a real as well as an integer variable.

Sect. 1.3 General Definitions and Properties

Definition 1.3-6. Time-Invariant (Time-Varying) System
A system is called *time-invariant* or *constant* if time is not an explicit argument of the functions of its mathematical model; otherwise, it is called *time-varying*.

Functions referred to in the above statement are those on the right of equations (1–6). For the sake of generality they have been written for time-varying systems: it is sufficient to omit time as the last argument in brackets in order to obtain the corresponding equations for time-invariant systems.

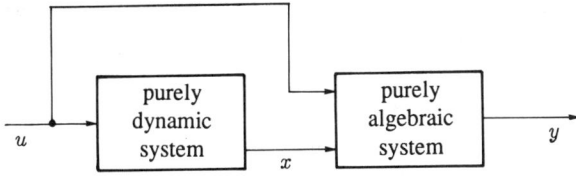

Figure 1-10 Decomposition of a general dynamic system.

The next concept to introduce is that of linearity, which is of paramount importance in system theory because it allows numerous properties to be derived and many rigorous synthesis procedures to be sketched. By studying linear systems the designer is provided with a rich store of experience that is also very useful in approaching the most general nonlinear problems.

Definition 1.3-7. Linear System
A system is *linear* if the sets $\mathcal{U}$, $\mathcal{U}_f$, $\mathcal{X}$, $\mathcal{Y}$ are vector spaces (all over the same field $\mathcal{F}$) and the functions that compose its mathematical model are linear with respect to x, u for all admissible t. A dynamic system that is not linear is called *nonlinear*.

As a consequence of the above definition, in the case of purely algebraic linear systems instead of (1) we will consider the equation

$$y(t) = C(t)\,u(t) \tag{7}$$

whereas in the case of continuous-time linear dynamic systems, instead of (2,3) we refer more specifically to the equations

$$\dot{x}(t) = A(t)\,x(t) + B(t)\,u(t) \tag{8}$$
$$y(t) = C(t)\,x(t) + D(t)\,u(t) \tag{9}$$

In the above equations $A(t)$, $B(t)$, $C(t)$, and $D(t)$ denote linear maps depending on time which, in particular, are constant if systems are time-invariant.

Similarly, for discrete-time linear dynamic systems instead of (4,5) we refer to the equations

$$x(i+1) = A_d(i)\,x(i) + B_d(i)\,u(i) \tag{10}$$
$$y(i) = C_d(i)\,x(i) + D_d(i)\,u(i) \tag{11}$$

where $A_d(i)$, $B_d(i)$, $C_d(i)$ and $D_d(i)$ are also linear maps that are constant if systems are time-invariant.

If, in particular, $\mathcal{U}:=\mathbf{R}^p$, $\mathcal{X}:=\mathbf{R}^n$, $\mathcal{Y}:=\mathbf{R}^q$, i.e., input, state, and output are respectively represented by a p-tuple, an n-tuple, and a q-tuple of real numbers, $A(t)$, $B(t)$, $C(t)$, $D(t)$, and the corresponding symbols for the discrete-time case can be considered to denote real matrices of proper dimensions, which are functions of time if the system is time-varying and constant if the system is time-invariant.

In light of the definitions just stated, let us again consider the six examples presented in the previous section. The systems in Examples 1.2-1–1.2-4 are continuous-time, whereas those in Examples 1.2-5 and 1.2-6 are discrete-time. All of them are time-invariant, but may be time-varying if some of the parameters that have been assumed to be constant are allowed to vary as given functions of time: for instance, the elevation z_1 of the water level in the reservoir of the installation shown in Fig. 1-9 may be subject to daily oscillations (depending on possible oscillations of power request) or yearly oscillations according to water inlet dependence on seasons. As far as linearity is concerned, the systems considered in Examples 1.2-1, 1.2-2, and 1.2-5 are linear, whereas all the others are nonlinear.

The input sets are $\mathbf{R}$, $\mathbf{R}^2$, $\mathbf{R}$, $\mathbf{R}$, $\mathbf{R}$, $\mathbf{B}$ respectively, the state sets are $\mathbf{R}$, $\mathbf{R}^3$, $\mathbf{R}^2$, $\mathbf{R}_f$, $\mathbf{R}$, $\mathbf{B}$, and the output sets are $\mathbf{R}$, $\mathbf{R}$, $\mathbf{R}$, $\mathbf{R}$, $\mathbf{R}$, $\mathbf{B}$. $\mathbf{R}_f$ denotes a vector space of functions with values in $\mathbf{R}$. Also the input function set $\mathcal{U}_f$ must be specified, particularly for continuous-time systems in order to guarantee that the solutions of differential equation (2) have standard smoothness and uniqueness properties. In general $\mathcal{U}_f$ is assumed to be the set of all piecewise continuous functions with values in $\mathcal{U}$, but in some special cases, it could be different: if, for instance, the input of a dynamic system is connected to the output of a purely dynamic system, input functions of the former are restricted to being continuous. In discrete-time systems in general $\mathcal{U}_f$ is a sequence with values in $\mathcal{U}$ without any special restriction.

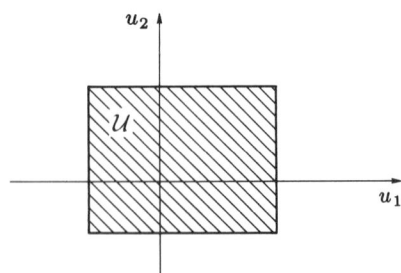

Figure 1-11 A possible input set contained in $\mathbf{R}^2$.

A proper choice of the input set can be used in order to take into account

Sect. 1.3 General Definitions and Properties

bounds for the values of input variables. For instance, it is possible to model independent bounds of each of two input variables by assuming the subset of $\mathbf{R}^2$ shown in Fig. 1-11 as the input set. Such bounds may correspond to safety limits for control action so that the controlled device is not damaged and/or to limits that cannot be exceeded because of sharp physical constraints. Note that such a limitation of the input set causes nonlinearity.

Examples. It is reasonable to specify a bound V_a for the absolute value of the applied voltage v_a to the electric motor considered in Example 1.2-2, in order to avoid damage due to overheating: $-V_a \leq v_a(t) \leq V_a$. It is physically impossible for flow u in the hydraulic installation considered in Example 1.2-3 to be negative and exceed an upper bound U depending on the diameter of the nozzle in the turbine and the maximum static head at the output of penstock: $0 \leq u(t) \leq U$.

In Definitions 1.3-3 and 1.3-4 dynamic systems are simply introduced by characterizing two possible classes of mathematical models for them. Note that, although the concepts of input and output are primitive, being related to the connections of the system to the environment, the concept of state has been introduced as a part of the mathematical model, not necessarily related to the presence of corresponding internal physical variables. Indeed, the state is necessary in order to pursue the natural way of thinking of systems as objects basically ruled by the relationship of cause and effect. The following property is a formalization of this concept of state.

Property 1.3-1. Concept of State

The *state* of a dynamic system is an element (of a set called *state set*) subject to variation in time and such that its value $x(t_0)$ at a given instant of time t_0, together with an input function segment $u|_{[t_0,t_1]}$, univocally determines the output function segment $y|_{[t_0,t_1]}$.

Property 1.3-1 implies the property of causality: all dynamic systems which are considered in this book are *causal* or *nonanticipative*, i.e., their output at any instant of time t does not depend on the values of input at instants of time greater than t.

The nature of the state variables deeply characterizes dynamic systems, so that it can be assumed as a classification for them, according to the following definition.

Definition 1.3-8. Finite-State (Finite-Dimensional) System

A dynamic system is called *finite-state*, *finite-dimensional*, *infinite-dimensional* if its state set is respectively a finite set, a finite-dimensional vector space, or an infinite-dimensional vector space.

A more compact mathematical description of dynamic system behavior is obtained as follows: equation (2) – by assumption – and equation (4) – by inherent property – have a unique solution that can be expressed as a function of the initial instant of time t_0, the initial state $x_0 := x(t_0)$, and the input function $u(\cdot)$, that is:

$$x(t) = \varphi(t, t_0, x_0, u(\cdot)) \qquad (12)$$

Function φ is called *state transition function*. Being the solution of a differential or a difference equation, it has some special features, such as:

1. *time orientation*: it is defined for $t \geq t_0$, but not necessarily for $t < t_0$;

2. *causality*: its dependence on the input function is restricted to the time interval $[t_0, t]$:

$$\varphi(t, t_0, x_0, u_1(\cdot)) = \varphi(t, t_0, x_0, u_2(\cdot)) \quad \text{if } u_1|_{[t_0,t]} = u_2|_{[t_0,t]}$$

3. *consistency*:

$$x = \varphi(t, t, x, u(\cdot))$$

4. *composition*: consecutive state transitions are congruent. i.e.,

$$\varphi(t, t_0, x_0, u(\cdot)) = \varphi(t, t_1, x_1, u(\cdot))$$

provided that

$$x_1 := \varphi(t_1, t_0, x_0, u(\cdot)) , \quad t_0 \leq t_1 \leq t$$

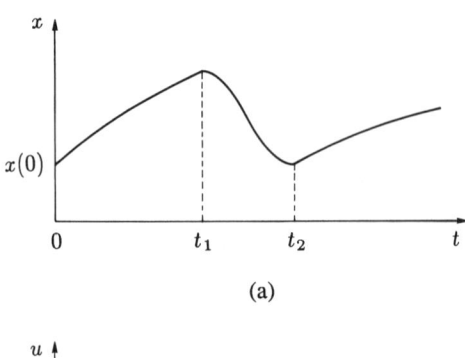

(a)

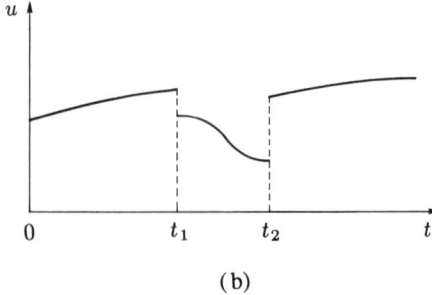

(b)

Figure 1-12 A possible motion and the corresponding input function.

Sect. 1.3 General Definitions and Properties

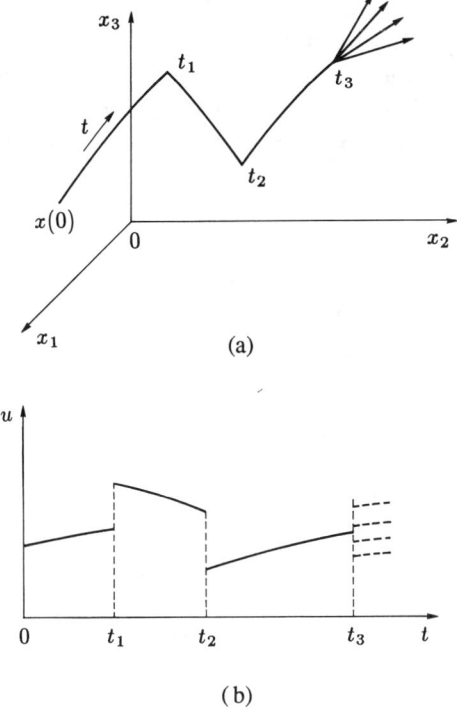

Figure 1-13 A possible trajectory and the corresponding input function.

The pair $(t, x(t)) \in T \times \mathcal{X}$ is called an *event*: when the initial event $(t_0, x(t_0))$ and the input function $u(\cdot)$ are known, the state transition function provides a set of events, namely a function $x(\cdot) : T \to \mathcal{X}$, which is called *motion*.

To be precise, motion in the time interval $[t_0, t_1]$ is the set

$$\{(t, x(t)) : x(t) = \varphi(t, t_0, x(t_0), u(\cdot)), \ t \in [t_0, t_1]\} \tag{13}$$

The image of motion in the state set, i.e., the set

$$\{x(t) : x(t) = \varphi(t, t_0, x(t_0), u(\cdot)), \ t \in [t_0, t_1]\} \tag{14}$$

of all the state values in the time interval $[t_0, t_1]$ is called *trajectory* (of the state in $[t_0, t_1]$).

When the state set $\mathcal{X}$ coincides with a finite-dimensional vector space $\mathbf{R}^n$, motion can be represented as a line in the event space $T \times \mathcal{X}$ and trajectory as a line in the state space $\mathcal{X}$, graduated versus time. The representation in the event space of a motion of the electric circuit described in Example 1.2-1 and the corresponding input function are shown in Fig. 1-12, while the representation in the state space of a possible trajectory of the electromechanical system described in Example 1.2-2 and the corresponding input function are shown in Fig. 1-13. For any given initial state different input functions cause different trajectories, all initiating at the same point of the state space; selecting input at a particular instant of time (for instance, t_3) allows different orientations in space of the tangent to the trajectory at t_3, namely of the state velocity $(\dot{x}_1, \dot{x}_2, \dot{x}_3)$.

The analysis of dynamic system behavior mainly consists of studying trajectories and the possibility of influencing them through the input. Then the geometric representation of trajectories is an interesting visualization of state transition function features and limits. In particular, it clarifies state trajectory dependence on input.

Substituting (12) into (3) or (5) yields

$$y(t) = \gamma\big(t, t_0, x_0, u|_{[t_0,t]}\big) \quad t \geq t_0 \qquad (15)$$

Function γ is called the *response function* and provides the system output at generic time t as a function of the initial instant of time, the initial state, and a proper input function segment. Therefore (15) represents the relationship of cause and effect which characterizes the time behavior of a dynamic system and can be considered as an extension of the cause and effect relationship expressed by (1) for memoryless systems. Equation (15) expresses a line in the output space, which is called *output trajectory*.

A very basic concept in system theory is that of the minimality of an input-state-output mathematical representation.

Definition 1.3-9. Indistinguishable States

Consider a dynamic system. Two states $x_1, x_2 \in \mathcal{X}$ are called *indistinguishable* in $[t_0, t_1]$ if

$$\gamma\big(t, t_0, x_1, u(\cdot)\big) = \gamma\big(t, t_0, x_2, u(\cdot)\big) \quad \forall \, t \in [t_0, t_1], \, \forall \, u(\cdot) \in \mathcal{U}_f \qquad (16)$$

Definition 1.3-10. Equivalent States

Consider a dynamic system. Two states $x_1, x_2 \in \mathcal{X}$ that are indistinguishable in $[t_0, t_1] \, \forall \, t_0, t_1 \in \mathcal{T}, t_1 > t_0$ are called *equivalent*.

Definition 1.3-11. Minimal System

A dynamic system that has no equivalent states is said to be in *minimal form* or, simply, *minimal*.

Any nonminimal dynamic system can be made minimal by defining a new state set in which every new state corresponds to a class of equivalent old states, and by redefining the system functions accordingly.

Sect. 1.3 General Definitions and Properties

Definition 1.3-12. Equivalent Systems

Two dynamic systems Σ_1, Σ_2 are said to be *equivalent* if $\mathcal{U}_1 = \mathcal{U}_2$, $\mathcal{U}_{f1} = \mathcal{U}_{f2} = \mathcal{U}_f$, $\mathcal{Y}_1 = \mathcal{Y}_2$ and to any state $x_1 \in \mathcal{X}_1$ of Σ_1 it is possible to associate a state $x_2 \in \mathcal{X}_2$ of Σ_2, and vice versa, such that [2]

$$\gamma_1(t, t_0, x_1, u(\cdot)) = \gamma_2(t, t_0, x_2, u(\cdot)) \quad \forall t_0, \ \forall t \geq t_0, \ \forall u(\cdot) \in \mathcal{U}_f \tag{17}$$

In some control problems it is necessary to stop the time evolution of the state of a dynamic system at a particular value. This is possible only if such a value corresponds to an equilibrium state according to the following definition.

Definition 1.3-13. (Temporary) Equilibrium State

In a dynamic system any state $x \in \mathcal{X}$ is a *temporary equilibrium state* in $[t_0, t_1]$ if there exists an admissible input function $u(\cdot) \in \mathcal{U}_f$ such that

$$x = \varphi(t, t_0, x, u(\cdot)) \quad \forall t \in [t_0, t_1] \tag{18}$$

The state x is called simply an *equilibrium state* if it is a temporary equilibrium state in $[t_0, t_1]$ for all pairs $t_0, t_1 \in T$, $t_1 > t_0$.

Note that, by virtue of the property of time-shifting of causes and effects, in time-invariant systems all temporary equilibrium states in any finite time interval are simply equilibrium states. Referring to the geometric representation of the state evolution as a state space trajectory, equilibrium states are often also called *equilibrium points*.

When the corresponding dynamic system is either time-invariant or linear, functions φ and γ have special properties, which will now be investigated.

Property 1.3-2. Time-Shifting of Causes and Effects

Let us consider a time-invariant system and for any $\tau \in T$ and all input functions $u(\cdot) \in \mathcal{U}_f$ define the *shifted input function* as

$$u_\Delta(t+\tau) := u(t) \quad \forall t \in T \tag{19}$$

assume that $u_\Delta(\cdot) \in \mathcal{U}_f$ for all $u(\cdot) \in \mathcal{U}_f$, i.e., that the input function set is closed with respect to the shift operation. The state transition function and the response function satisfy the following relationships:

$$x(t) = \varphi(t, t_0, x_0, u(\cdot)) \quad \Leftrightarrow \quad x(t+\tau) = \varphi(t+\tau, t_0+\tau, x_0, u_\Delta(\cdot)) \tag{20}$$

$$y(t) = \gamma(t, t_0, x_0, u(\cdot)) \quad \Leftrightarrow \quad y(t+\tau) = \gamma(t+\tau, t_0+\tau, x_0, u_\Delta(\cdot)) \tag{21}$$

Proof. We refer to system equations (2,3) or (4,5) and assume that the system is time-invariant, so that functions on the right are independent of time. Property is

---

[2] If Σ_1 e Σ_2 are both in minimal form, this correspondence between initial states is clearly one-to-one.

a consequence of the fact that shifting any function of time implies also shifting its derivative (in the case of continuous-time systems) or all future values (in the case of discrete-time systems), so that equations are still satisfied if all the involved functions are shifted. □

Assuming in (20) and (21) $\tau := -t_0$, we obtain

$$x(t) = \varphi(t, t_0, x_0, u(\cdot)) \quad \Leftrightarrow \quad x(t-t_0) = \varphi(t-t_0, 0, x_0, u_\Delta(\cdot))$$
$$y(t) = \gamma(t, t_0, x_0, u(\cdot)) \quad \Leftrightarrow \quad y(t-t_0) = \gamma(t-t_0, 0, x_0, u_\Delta(\cdot))$$

from which it can be inferred that

1. when the system referred to is time-invariant, the initial instant of time can be assumed to be zero without any loss of generality;
2. state transition and response functions of time-invariant systems are actually dependent on the difference $t-t_0$ instead of t and t_0 separately.

Property 1.3-3. Linearity of State Transition and Response Functions

Let us consider a linear dynamic system and denote by α, β any two elements of the corresponding field $\mathcal{F}$, by x_{01}, x_{02} any two admissible initial states and by $u_1(\cdot)$, $u_2(\cdot)$ any two admissible input function segments. State transition and response functions satisfy the following relationships:

$$\varphi(t, t_0, \alpha x_{01} + \beta x_{02}, \alpha u_1(\cdot) + \beta u_2(\cdot)) =$$
$$\alpha \varphi(t, t_0, x_{01}, u_1(\cdot)) + \beta \varphi(t, t_0, x_{02}, u_2(\cdot)) \tag{22}$$

$$\gamma(t, t_0, \alpha x_{01} + \beta x_{02}, \alpha u_1(\cdot) + \beta u_2(\cdot)) =$$
$$\alpha \gamma(t, t_0, x_{01}, u_1(\cdot)) + \beta \varphi(t, t_0, x_{02}, u_2(\cdot)) \tag{23}$$

which express the linearity of φ and γ with respect to the initial state and input function.

Proof. We refer to equation (8) or (10) and consider its solutions corresponding to the different pairs of initial state and input function segments x_{01}, $u_1(\cdot)$ and x_2, $u_2(\cdot)$, which can be expressed as

$$x_1(t) = \varphi(t, t_0, x_{01}, u_1(\cdot))$$
$$x_2(t) = \varphi(t, t_0, x_{02}, u_2(\cdot))$$

By substituting on the right of (8) or (10) $\alpha x_1(t) + \beta x_2(t)$, $\alpha u_1(t) + \beta u_2(t)$ in place of $x(t)$, $u(t)$ and using linearity, we obtain on the left the quantity $\alpha \dot{x}_1(t) + \beta \dot{x}_2(t)$ in the case of (8), or $\alpha x_1(t+1) + \beta x_2(t+1)$ in the case of (10). Therefore, $\alpha x_1(t) + \beta x_2(t)$ is a solution of the differential equation (8) or difference equation (10); hence (22) holds. As a consequence, (23) also holds, provided that γ is the composite function of two linear functions. □

In the particular case $\alpha = \beta = 1$, equations (22, 23) correspond to the so-called *property of superposition of the effects*.

Property 1.3-4. Decomposability of State Transition and Response Functions
In linear systems the state transition (response) function corresponding to the initial state x_0 and the input function $u(\cdot)$ can be expressed as the sum of the *zero-input* state transition (response) function corresponding to the initial state x_0 and the *zero-state* state transition (response) function corresponding to the input function $u(\cdot)$.

Proof. Given any admissible initial state x_0 and any admissible input function $u|_{[t_0,t]}$, assume in (22,23), $\alpha := 1$, $\beta := 1$, $x_{01} := x_0$, $x_{02} := 0$, $u_1(\cdot) := 0$, $u_2(\cdot) := u(\cdot)$; it follows that

$$\varphi(t, t_0, x_0, u(\cdot)) = \varphi(t, t_0, x_0, 0) + \varphi(t, t_0, 0, u(\cdot)) \tag{24}$$

$$\gamma(t, t_0, x_0, u(\cdot)) = \gamma(t, t_0, x_0, 0) + \gamma(t, t_0, 0, u(\cdot)) \quad \square \tag{25}$$

The former term of the above decomposition is usually referred to as the *free motion* (*free response*), the latter as the *forced motion* (*forced response*). The following properties are immediate consequences of the response decomposition property.

Property 1.3-5.
Two states of a linear system are indistinguishable in $[t_0, t_1]$ if and only if they generate the same free response in $[t_0, t_1]$.

Property 1.3-6.
A linear system is in minimal form if and only if for any initial instant of time t_0 no different states generate the same free response.

1.4 CONTROLLING AND OBSERVING THE STATE

The term *controllability* denotes the possibility of influencing the motion $x(\cdot)$ or the response $y(\cdot)$ of a dynamical system Σ by means of the input function (or control function) $u(\cdot) \in \mathcal{U}_f$.

In particular, one may be required to steer a system from a state x_0 to x_1 or from an event (t_0, x_0) to (t_1, x_1): if this is possible, the system is said to be *controllable* from x_0 to x_1 or from (t_0, x_0) to (t_1, x_1). Equivalent statements are: "the state x_0 (or event (t_0, x_0)) is controllable to x_1 (or to (t_1, x_1))" and "the state x_1 (or the event (t_1, x_1)) is reachable from x_0 (or from (t_0, x_0))."

Example. Suppose the electric motor in Fig. 1-5 is in a given state x_0 at $t=0$: a typical controllability problem is to reach the zero state (i.e., to null armature current, angular velocity, and angular position) at a time instant t_1, (which may be specified in advance or not), by an appropriate choice of the input function segment $u|_{[0,t_1]}$; if this problem has a solution, state x_0 is said to be controllable to the zero state (in the time interval $[t_0, t_1]$).

Controllability analysis is strictly connected to the definition of particular subsets of the state space $\mathcal{X}$, that is:

1. the *reachable set at final time* $t=t_1$ *from event* (t_0, x_0)

$$\mathcal{R}^+(t_0, t_1, x_0) := \{x_1 : x_1 = \varphi(t_1, t_0, x_0, u(\cdot)), \ u(\cdot) \in \mathcal{U}_f\} \quad (1)$$

2. the *reachable set at any time in* $[t_0, t_1]$ *from event* (t_0, x_0)

$$\mathcal{W}^+(t_0, t_1, x_0) := \{x_1 : x_1 = \varphi(\tau, t_0, x_0, u(\cdot)), \ \tau \in [t_0, t_1], \ u(\cdot) \in \mathcal{U}_f\} \quad (2)$$

3. the *controllable set to event* (t_1, x_1) *from initial time* t_0

$$\mathcal{R}^-(t_0, t_1, x_1) := \{x_0 : x_1 = \varphi(t_1, t_0, x_0, u(\cdot)), \ u(\cdot) \in \mathcal{U}_f\} \quad (3)$$

4. the *controllable set to event* (t_1, x_1) *from any time in* $[t_0, t_1]$

$$\mathcal{W}^-(t_0, t_1, x_1) := \{x_0 : x_1 = \varphi(t_1, \tau, x_0, u(\cdot)), \ \tau \in [t_0, t_1], \ u(\cdot) \in \mathcal{U}_f\} \quad (4)$$

In the previous definitions the ordering relation $t_0 \leq t_1$ is always tacitly assumed. Clearly

$$\mathcal{R}^+(t_0, t_1, x) \subseteq \mathcal{W}^+(t_0, t_1, x) \quad \forall x \in \mathcal{X} \quad (5)$$
$$\mathcal{R}^-(t_0, t_1, x) \subseteq \mathcal{W}^-(t_0, t_1, x) \quad \forall x \in \mathcal{X} \quad (6)$$

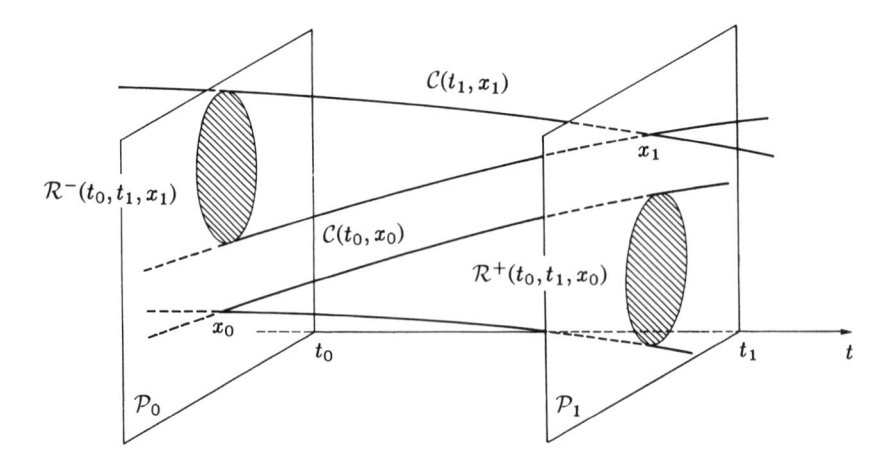

Figure 1-14 Sets of reachable and controllable states.

The geometric meaning of the above definitions is clarified by Fig. 1-14, which refers to the particular case $\mathcal{X} = \mathbf{R}^2$: in the event space, $\mathcal{R}^+(t_0, t_1, x_0)$ is obtained by intersecting set $\mathcal{C}(t_0, x_0)$ of all admissible motions which include the event (t_0, x_0) with the hyperplane $\mathcal{P}_1 := \{(t, x) : t = t_1\}$; $\mathcal{W}^+(t_0, t_1, x_0)$ is obtained by projecting set $\mathcal{C}(t_0, x_0) \cap \mathcal{M}$, where $\mathcal{M} := \{(t, x) : t \in [t_0, t_1]\}$, on $\mathcal{P}_1 := \{(t, x) : t = t_1\}$ along the t axis. $\mathcal{R}^-(t_0, t_1, x_0)$ and $\mathcal{W}^-(t_0, t_1, x_0)$ are derived in a similar way.

Sect. 1.4 Controlling and Observing the State

Definition 1.4-1. Reachability from (Controllability to) an Event
The state of a dynamic system Σ or, by extension, system Σ itself, is said to be *completely reachable from event* (t_0, x) *in time interval* $[t_0, t_1]$ if $W^+(t_0, t_1, x) = \mathcal{X}$, *completely controllable to event* (t_1, x) *in time interval* $[t_0, t_1]$ if $W^-(t_0, t_1, x) = \mathcal{X}$.

In time-invariant systems, $\mathcal{R}^+(t_0, t_1, x)$, $W^+(t_0, t_1, x)$, $\mathcal{R}^-(t_0, t_1, x)$, $W^-(t_0, t_1, x)$ do not depend on t_0, t_1 in a general way, but only on the difference $t_1 - t_0$, so that the assumption $t_0 = 0$ can be introduced without any loss of generality and notation is simplified as:

1. $\mathcal{R}^+_{t_1}(x)$: the reachable set at $t = t_1$ from event $(0, x)$;
2. $W^+_{t_1}(x)$: the reachable set at any time in $[0, t_1]$ from event $(0, x)$;
3. $\mathcal{R}^-_{t_1}(x)$: the controllable set to x at $t = t_1$ from initial time 0;
4. $W^-_{t_1}(x)$: the controllable set to x at any time in $[0, t_1]$ from initial time 0.

Given any two instants of time t_1, t_2 satisfying $t_1 \leq t_2$, the following hold:

$$W^+_{t_1}(x) \subseteq W^+_{t_2}(x) \quad \forall x \in \mathcal{X} \tag{7}$$

$$W^-_{t_1}(x) \subseteq W^-_{t_2}(x) \quad \forall x \in \mathcal{X} \tag{8}$$

Notations $W^+(x)$, $W^-(x)$ refer to the limits

$$W^+(x) := \lim_{t \to \infty} W^+_t(x) \qquad W^-(x) := \lim_{t \to \infty} W^-_t(x)$$

i.e., denote the *reachable set from* x and the *controllable set to* x in an arbitrarily large interval of time.

Definition 1.4-2. Completely Controllable System
A time-invariant system is said to be *completely controllable* or *connected* if it is possible to reach any state from any other state (so that $W^+(x) = W^-(x) = \mathcal{X}$ for all $x \in \mathcal{X}$).

Consider now the state observation. The term *observability* denotes generically the possibility of deriving initial state $x(t_0)$ or final state $x(t_1)$ of a dynamic system Σ when time evolutions of input and output in time interval $[t_0, t_1]$ are known. Final state observability is denoted also with the term *reconstructability*. The state observation and reconstruction problems may not always admit a solution: this happens, in particular, for observation when the initial state belongs to a class whose elements are indistinguishable in $[t_0, t_1]$.

Like controllability, observability is also analyzed by considering proper subsets of the state set $\mathcal{X}$, which characterize dynamic systems regarding the possibility of deriving state from input and output evolutions, i.e.:

1. the set of all initial states consistent with functions $u(\cdot)$, $y(\cdot)$ in the time interval $[t_0, t_1]$

$$\mathcal{Q}^-\big(t_0, t_1, u(\cdot), y(\cdot)\big) := \big\{x_0 : y(\tau) = \gamma\big(\tau, t_0, x_0, u(\cdot)\big),\ \tau \in [t_0, t_1]\big\} \tag{9}$$

2. the set of all final states consistent with functions $u(\cdot)$, $y(\cdot)$ in the time interval $[t_0, t_1]$

$$\mathcal{Q}^+(t_0, t_1, u(\cdot), y(\cdot)) \\ := \{x_1 : x_1 = \varphi(t_1, t_0, x_0, u(\cdot)), \ x_0 \in \mathcal{Q}^-(t_0, t_1, u(\cdot), y(\cdot))\} \quad (10)$$

It is clear that in (9) and (10) $y(\cdot)$ is not arbitrary, but constrained to belong to the set of all output functions admissible with respect to the initial state and the input function. This set is defined by

$$\mathcal{Y}_f(t_0, u(\cdot)) := \{y(\cdot) : y(t) = \gamma(t, t_0, x_0, u(\cdot)), \ t \geq t_0, \ x_0 \in \mathcal{X}\} \quad (11)$$

Definition 1.4-3.
The state of a dynamic system Σ or, by extension, system Σ itself, is said to be *observable in* $[t_0, t_1]$ *by a suitable experiment* (called *diagnosis*) if there exists at least one input function $u(\cdot) \in \mathcal{U}_f$ such that (9) reduces to a single element for all $y(\cdot) \in \mathcal{Y}_f(t_0, u(\cdot))$; it is said to be *reconstructable in* $[t_0, t_1]$ *by a suitable experiment* (called *homing*) if there exists at least one input function $u(\cdot) \in \mathcal{U}_f$ such that (10) reduces to a single element for all $y(\cdot) \in \mathcal{Y}_f(t_0, u(\cdot))$.

A dynamic system without any indistinguishable states in $[t_0, t_1]$ is not necessarily observable in $[t_0, t_1]$ by a diagnosis experiment since different input functions may be required to distinguish different pairs of initial states. This is typical in finite-state systems and quite common in general nonlinear systems.

Definition 1.4-4. Completely Observable (Reconstructable) System
The state of a dynamic system Σ or, by extension, system Σ itself, is said to be *completely observable in* $[t_0, t_1]$ if for all input functions $u(\cdot) \in \mathcal{U}_f$ and for all output functions $y(\cdot) \in \mathcal{Y}_f(t_0, u(\cdot))$ set (9) reduces to a single element; it is said to be *completely reconstructable in* $[t_0, t_1]$ if for all input functions $u(\cdot) \in \mathcal{U}_f$ and for all output functions $y(\cdot) \in \mathcal{Y}_f(t_0, u(\cdot))$ set (10) reduces to a single element.

Since the final state is a function of the initial state and input, clearly every system that is observable by a suitable experiment is also reconstructable by the same experiment and every completely observable system is also completely reconstructable.

In time-invariant systems $\mathcal{Q}^-(t_0, t_1, u(\cdot), y(\cdot))$, $\mathcal{Q}^+(t_0, t_1, u(\cdot), y(\cdot))$ do not depend on t_0 and t_1 in a general way, but only on the difference $t_1 - t_0$, so that, as in the case of controllability, the assumption $t_0 = 0$ can be introduced without any loss of generality. In this case the simplified notations $\mathcal{Q}^-_{t_1}(u(\cdot), y(\cdot))$, $\mathcal{Q}^+_{t_1}(u(\cdot), y(\cdot))$ will be used.

The above sets are often considered in solving problems related to system control and observation. The most significant of these problems are:

1. *Control between two given states*: given two states x_0 and x_1 and two instants of time t_0 and t_1, determine an input function $u(\cdot)$ such that $x_1 = \varphi(t_1, t_0, x_0, u(\cdot))$.

2. *Control to a given output*: given an initial state x_0, an output value y_1 and two instants of time t_0, t_1, $t_1 > t_0$, determine an input $u(\cdot)$ such that $y_1 = \gamma(t_1, t_0, x_0, u(\cdot))$.

3. *Control for a given output function*: given an initial state x_0, an admissible output function $y(\cdot)$ and two instants of time t_0, t_1, $t_1 > t_0$, determine an input $u(\cdot)$ such that $y(t) = \gamma(t, t_0, x_0, u(\cdot))$ for all $t \in [t_0, t_1]$.

4. *State observation*: given corresponding input and output functions $u(\cdot), y(\cdot)$ and two instants of time t_0, t_1, $t_1 > t_0$, determine an initial state x_0 (or the whole set of initial states) consistent with them, i.e., such that $y(t) = \gamma(t, t_0, x_0, u(\cdot))$ for all $t \in [t_0, t_1]$.

5. *State reconstruction*: given corresponding input and output functions $u(\cdot)$, $y(\cdot)$ and two instants of time t_0, t_1, $t_1 > t_0$, determine a final state x_1 (or the whole set of final states) consistent with them, i.e., corresponding to an initial state x_0 such that $x_1 = \varphi(t_1, t_0, x_0, u(\cdot))$, $y(t) = \gamma(t, t_0, x_0, u(\cdot))$ for all $t \in [t_0, t_1]$.

6. *Diagnosis*: like 4, except that the solution also includes the choice of a suitable input function.

7. *Homing*: like 5, except that the solution also includes the choice of a suitable input function.

Moreover, problems often arise where observation and control are simultaneously required. For instance, problems 1 and 2 would be of this type if initial state x_0 was not given.

1.5 INTERCONNECTING SYSTEMS

Decomposing complex systems into simpler interconnected subsystems makes their analysis easier. It is useful because many properties of the overall system are often determined by analyzing corresponding properties of subsystems. Furthermore, it is convenient to keep different types of devices distinct, for instance those whose behavior can be influenced by a suitable design (like controllers and, more generally, signal processors), and those that, on the contrary, cannot be affected in any way.

1.5.1 Graphic Representations of Interconnected Systems

Complex systems consisting of numerous interconnected parts are generally represented in drawings by means of *block diagrams* and *signal-flow graphs*. They will be adopted here too, so, although they are very well known, it seems convenient to briefly recall their distinguishing features and interpretative conventions.

Block diagrams. Block diagrams are a convenient representation for systems that consist of numerous interconnected parts. In this book they will be used in a rather informal way: they will be referred without any graphic difference to the single-variable as well as to the multivariable case, and the mathematical model of the subsystem represented with a single block will be reported inside the block not in a unified way, but in the form that is most consistent with the text.

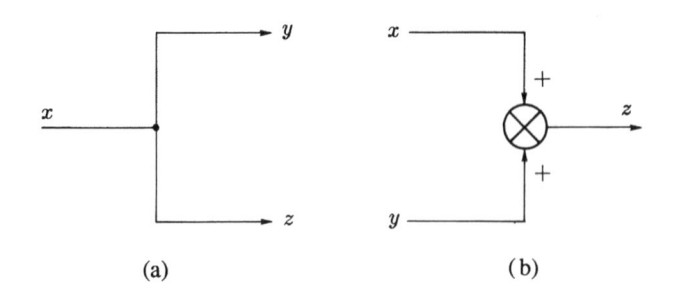

Figure 1-15 Branching point and summing junction.

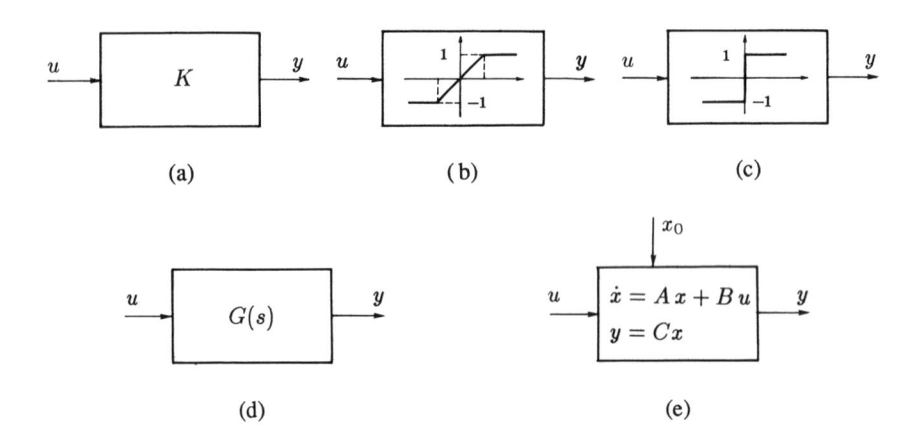

Figure 1-16 Some types of blocks.

The main linkage elements between blocks are the *branching point*, represented in Fig. 1-15(a) and the *summing junction*, represented in Fig. 1-15(b). They are described respectively by the simple relations

$$y(t) = x(t)$$
$$z(t) = x(t)$$

and

$$z(t) = x(t) + y(t)$$

Some types of blocks are shown in Fig. 1-16(a–e): block (a) represents the linear purely algebraic constant input-output relation $y = K\,u$, where K is a real constant or a real matrix; (b) and (c) represent nonlinear purely algebraic constant input-output relations, specifically a saturation or a block of saturations and an ideal relay or block of ideal relays (or signum functions): if referred to multivariable cases, they are understood to have the same number of outputs as inputs; (d) represents a dynamic link specified by means of a transfer function in the single-variable case or a transfer matrix in the multivariable case; (e) a dynamic link specified by an ISO description: note, in this case, the possible presence of a further input denoting the initial state. In some block diagrams, as for instance those shown in Figs. 1-10 and 1-25 to 1-27, no mathematical model is specified inside the blocks, but simply a description in words of the corresponding subsystem. When, on the other hand, a precise mathematical description is given for each block of a diagram, the complete diagram is equivalent to a set of equations for the overall system, in which interconnection equations are those of branching points and summing junctions.

Signal-flow graphs. Signal-flow graphs are preferred to block diagrams to represent complex structures consisting of several elementary (single-input single-output) parts, each described by a transfer constant or a transfer function. Their use is restricted to show the internal structure of some linear systems which, although possibly of the multivariable type, can be represented as a connection of single-variable elements. The major advantage of signal-flow graphs over block diagrams is that the transfer constant or the transfer function relating any input to any output can be derived directly from a simple analysis of the topological structure of the graph.

A signal-flow graph is composed of *branches* and *nodes*. Every branch joins two nodes in a given direction denoted by an arrow, i.e., is *oriented* and characterized by a coefficient or transfer function, called *transmittance* or *gain*. Every node represents a *signal*, which by convention is expressed by a linear combination of the signals from whose nodes there exist branches directed to it, with the transmittances of these branches as coefficients. A node that has no entering branches is called an *independent* or *input node*, while the other nodes are called *dependent nodes*: clearly, every dependent node represents a linear equation, so that the graph is equivalent to as many linear equations in as many unknowns as there are dependent nodes.

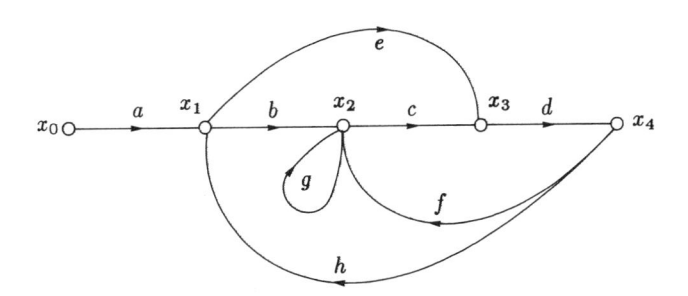

Figure 1-17 A signal-flow graph.

As an example, consider the simple signal-flow graph represented in Fig. 1-17: transmittances are denoted by a, b, c, d, e, f, g, h and can be real constants or transfer functions. The graph corresponds to the set of linear equations

$$x_1 = a\,x_0 + h\,x_4 \tag{1}$$
$$x_2 = b\,x_1 + g\,x_2 + f\,x_4 \tag{2}$$
$$x_3 = e\,x_1 + c\,x_2 \tag{3}$$
$$x_4 = d\,x_3 \tag{4}$$

in the unknowns x_1, x_2, x_3, x_4.

Choose x_4 as the output signal: we can derive a gain that relates x_4 to x_0 by solving equations (1–4). Otherwise, we can take advantage of the relative sparseness of the signal-flow graph (in the sense that nodes are not connected by branches in all possible ways) to use a topological analysis method which in most practical cases turns out to be very convenient.

For this, some further definitions are needed. A *path* joining two given nodes is a sequence of adjacent branches that originates in the first node and terminates in the second passing through any node only once. The *transmittance of a path* P is the product of the transmittances of all the branches in the path. A *loop* is a closed path. The *transmittance of a loop* L is the product of the transmittances of all the branches in the loop. A loop consisting of a single branch and a single node, as for instance loop g in Fig. 1-17, is called a *self-loop*. Two paths or two loops, or a path and a loop are said to be *nontouching* if they do not have any common node.

The *Mason formula* allows determination of the gain relating any dependent node to any source node of a signal-flow graph as a function of the transmittances of all paths joining the considered nodes and all loops in the graph. In order to express the formula, a little topological analysis is needed: denote by P_i, $i \in \mathcal{P}$, where $\mathcal{J}_p$ is a set of indexes, the transmittances of all different paths joining the considered nodes; by L_j, $j \in \mathcal{J}_1$, those of all different loops in the graph; by $\mathcal{J}_2$ the set of the pairs of indices corresponding to nontouching loops; by $\mathcal{J}_3$ that of the triples of indices corresponding to nontouching loops by three, and so on; furthermore, let $\mathcal{J}_{1,i}$ be the set of the indices of all loops not touching path P_i; $\mathcal{J}_{2,i}$ that of the indices of all pairs of nontouching loops not touching path P_i, and so on. When a set of indexes is empty, so are all subsequent ones.

The Mason formula for the transmittance coefficient relating the considered nodes is

$$T = \frac{1}{\Delta} \sum_{i \in \mathcal{J}_p} P_i\, \Delta_i \tag{5}$$

where

$$\Delta := 1 - \sum_{i \in \mathcal{J}_1} L_i + \sum_{(i,j) \in \mathcal{J}_2} L_i L_j - \sum_{(i,j,k) \in \mathcal{J}_3} L_i L_j L_k + \ldots \tag{6}$$

$$\Delta_i := 1 - \sum_{i \in \mathcal{J}_{1,i}} L_i + \sum_{(i,j) \in \mathcal{J}_{2,i}} L_i L_j - \sum_{(i,j,k) \in \mathcal{J}_{3,i}} L_i L_j L_k + \ldots \tag{7}$$

Δ is called the *determinant of the graph*, whereas Δ_i denotes the determinant of the partial graph obtained by *deleting* path P_i, i.e., by deleting all nodes belonging to P_i and all pertinent branches.

Going back to the example of Fig. 1-17, in order to derive the gain T relating x_4 to x_0, first identify all paths and loops and determine their transmittances:

$$P_1 = abcd, \quad P_2 = aed, \quad P_3 = abf$$

$$L_1 = edh, \quad L_2 = bcdh, \quad L_3 = bfh, \quad L_4 = g$$

then consider the corresponding nonempty index sets:

$$\mathcal{J}_p = \{1,2,3\}, \quad \mathcal{J}_1 = \{1,2,3,4\}, \quad \mathcal{J}_2 = \{(1,4)\}, \quad \mathcal{J}_{12} = \{4\}$$

The Mason formula immediately yields

$$T = \frac{abcd + aed(1-g) + abf}{1 - edh - bcdh - bfh - g + edhg}$$

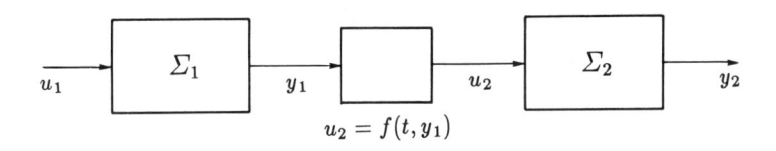

Figure 1-18 Cascaded systems.

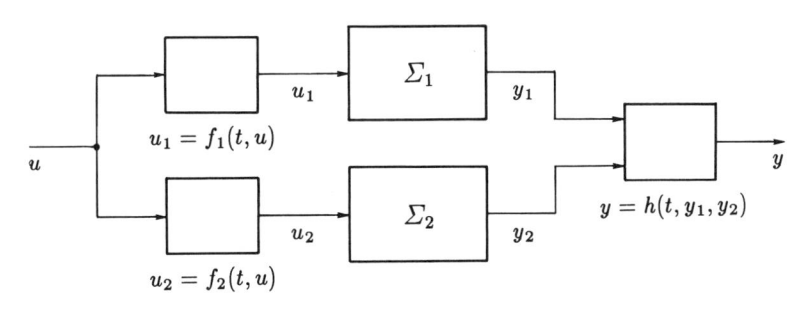

Figure 1-19 Parallel systems.

1.5.2 Cascade, Parallel, and Feedback Interconnections

It is useful to define three basic interconnections of systems, which are often referred to when considering decomposition problems.

1. *Cascade*. Two dynamic systems Σ_1 and Σ_2 are said to be connected in cascade (or, briefly, cascaded) if, at any instant of time, the input of Σ_2 is a function of the output of Σ_1, as shown in Fig. 1-18. For the cascade connection to be possible, condition $\mathcal{T}_1 = \mathcal{T}_2$ is necessary. The input set of the overall system is $\mathcal{U} = \mathcal{U}_1$, whereas the output set is $\mathcal{Y} = \mathcal{Y}_2$ and the state set is $\mathcal{X} = \mathcal{X}_1 \times \mathcal{X}_2$.

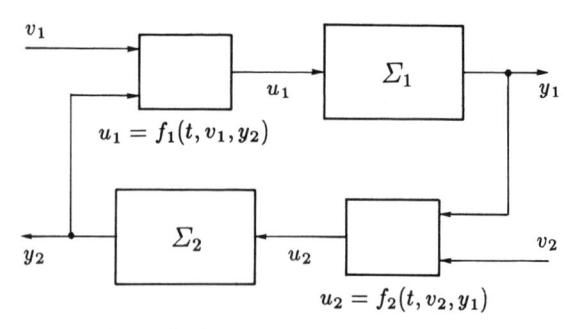

Figure 1-20 Feedback-connected systems.

2. *Parallel.* Two dynamic systems Σ_1 and Σ_2 are said to be connected in parallel if, at any instant of time, their inputs are functions of a single variable $u \in \mathcal{U}$, which is the input of the overall system, while the output $y \in \mathcal{Y}$ of the overall systems is, at any instant of time, a function of both outputs y_1 and y_2. Condition $T_1 = T_2$ is also required in this case; the state set of the overall system is $\mathcal{X} = \mathcal{X}_1 \times \mathcal{X}_2$. The parallel connection is shown in Fig. 1-19.

3. *Feedback.* Two dynamic systems Σ_1 and Σ_2 are said to be connected in mutual feedback if, at any instant of time, their inputs u_1 and u_2 are functions of y_2 and y_1 and of two further variables, $v_1 \in \mathcal{V}_1$ and $v_2 \in \mathcal{V}_2$ respectively. The input, output, and state sets of the overall system are $\mathcal{U} = \mathcal{V}_1 \times \mathcal{V}_2$, $\mathcal{Y} = \mathcal{Y}_1 \times \mathcal{Y}_2$, $\mathcal{X} = \mathcal{X}_1 \times \mathcal{X}_2$. Condition $T_1 = T_2$ is also required in this case. The feedback connection is shown in Fig. 1-20.

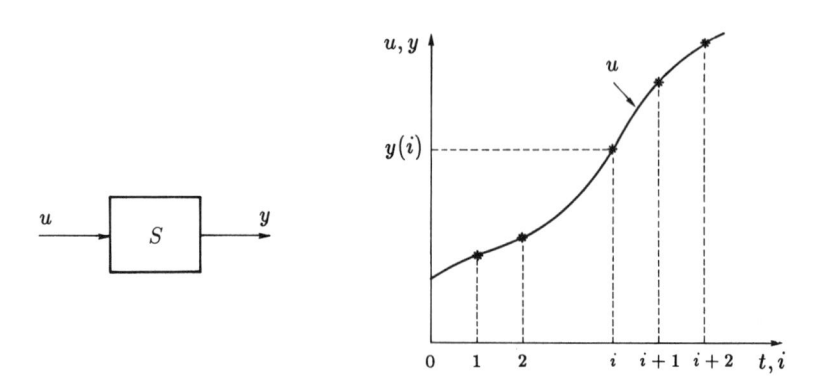

Figure 1-21 Sampler.

When two (or more) systems are connected to each other, common signals must be congruent with regard both to the sets over which variables are defined and to time, which must be real or integer for all considered systems. However, connection is also possible in the absence of these congruences provided that suitable signal converters are used.

Sect. 1.5 Interconnecting Systems

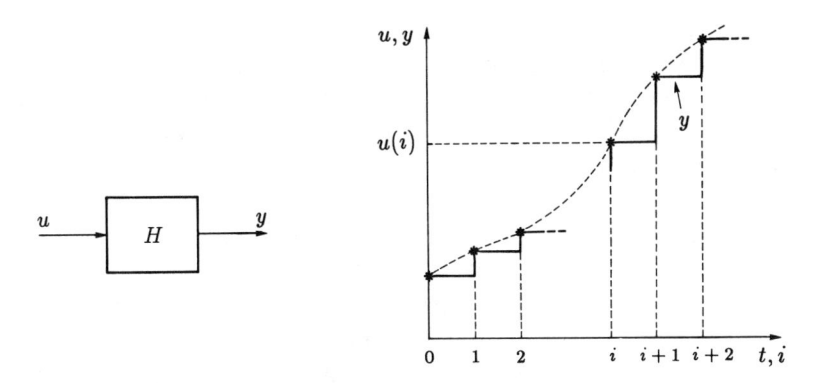

Figure 1-22 Hold device.

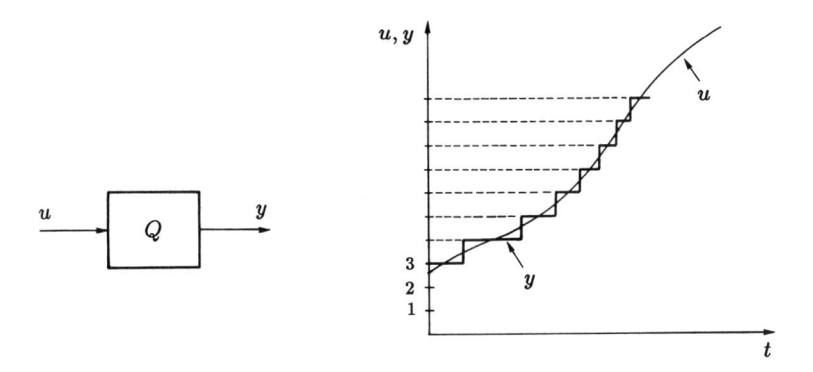

Figure 1-23 Quantizer.

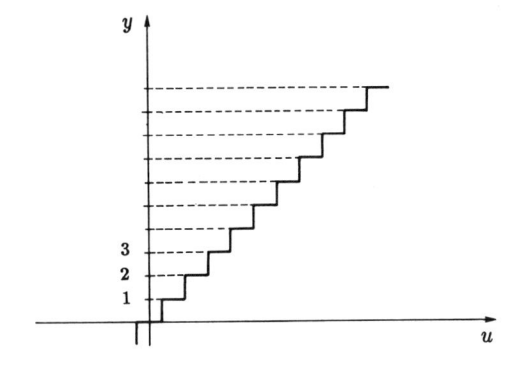

Figure 1-24 Input-output characteristic function of a quantizer.

An output signal from a continuous system can be converted into an input signal to a discrete system by means of a device, called a *sampler*, which performs the processing represented in Fig. 1-21. This consists of taking samples of the continuous signal at given instants of time. The reverse conversion is achieved by using a *hold device*, which maintains its output at the value corresponding to the last received sample, as shown in Fig. 1-22.

In order to obtain congruence of values, devices called *quantizers* are used. For instance, see in Fig. 1-23 the processing that transforms a real-valued function of time into an integer-valued function of time; the input-output characteristic function of the device is shown in Fig. 1-24.

When a continuous system is connected to a discrete system whose variables have a finite number of values (for instance, a digital processor), both a sampler and a quantizer, connected to each other in cascade, are required.

1.6 A REVIEW OF SYSTEM AND CONTROL THEORY PROBLEMS

In this section the most important problems pertinent to the system and control theory area are briefly presented. The solution of some of them by means of state-space techniques is the aim of this book, so they will be the object of more ample consideration and discussion in subsequent chapters.

System theory problems can be divided into two major classes: *analysis* and *synthesis* problems, the former referring to the investigation of inherent properties of systems, usually stated and approached through mathematical models, the latter to the derivation of mathematical tools or artificial subsystems, called *controllers* or *regulators*, to influence system behavior properly. Of course, there is a strict connection between analysis and synthesis since most general properties of systems are investigated to achieve precise information on the solvability of certain synthesis problems or on the feasibility of some synthesis-oriented procedures.

The most important analysis problems are the following:

1. *modeling*: to derive a suitable mathematical model for a system;

2. *motion* and *response analysis*: to determine the state time evolution (motion) and the output time evolution (response), given the initial state and the input function;

3. *stability analysis*: in plain terms, stability is the property of a system to react with bounded variations of state and output functions to bounded variations of the input function;

4. *controllability analysis*: to investigate the possibility of reaching given values of state or output vectors or obtaining particular types of state or output evolutions by means of admissible input functions;

5. *observability analysis*: to investigate the possibility of achieving knowledge of the state from complete or partial knowledge of the input and output

functions;

6. *identifiability analysis*: to investigate the possibility of deriving an input-output model or some of its parameters from complete or partial knowledge of the input and output functions.

Some related synthesis problems are:

1. *control input synthesis*: to determine an input function that, from a given initial state, causes system evolution to meet a specified control task;

2. *control input and initial state and time synthesis*: same as above, but the initial state and, in the time-varying case, the initial time have to be determined besides the input function, for a specified control task;

3. *synthesis of a state observer*: to determine a procedure or a device to derive the state of a system from a finite record of input and output functions;

4. *synthesis of an identifier*: to determine a procedure or a device that derives a model of a system or some parameters of it from a finite record of input and output functions;

5. *synthesis of an automatic control apparatus*: to design a processor that, taking into account measurements of some of the output variables, automatically sets the manipulable variables to achieve a given control task.

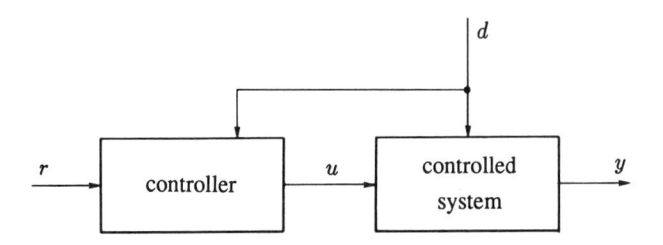

Figure 1-25 Open-loop control connection.

Problems 1 and 2 above are typical *open-loop* or *feedforward* control problems, since the controller works without any information on the actual system time evolution, i.e., without being able to check whether the control objective is being reached. The possibility of implementing a completely open-loop control strategy largely depends on the precision of the available mathematical model of the controlled system. The corresponding connection is shown in Fig. 1-25, where r denotes the reference input for the controller, which is here understood in a broad sense as the complete amount of information needed to specify the control task, u the manipulable input, d the nonmanipulable input or disturbance, y the output.

The "control task" is different from case to case; for instance, it may consist of reproducing a given state or output trajectory with a minimum error, or in reaching a given final state from a given initial state in an optimal

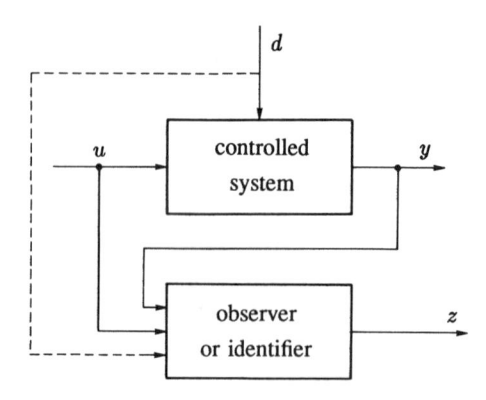

Figure 1-26 Connection for observation or identification.

way; optimality is usually expressed in mathematical terms by a *performance index*, which is usually a given functional of the system time evolution, i.e., of input, state and output functions, to be minimized. The block diagram in Fig. 1-25 shows a logical cause-effect connection rather than an actual physical connection: in fact, computation of an optimal open-loop control law is not necessarily performed in real time.

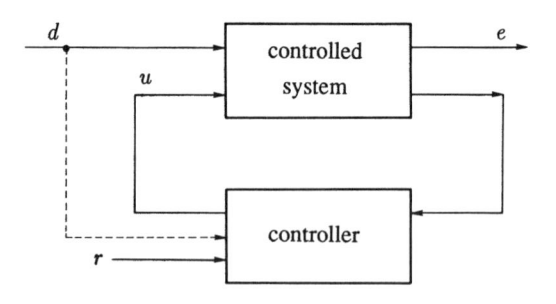

Figure 1-27 Closed-loop control connection.

Problems 3 and 4 are represented by the block diagram shown in Fig. 1-26, where the manipulable input u is assumed to be completely known, disturbance input d may be inaccessible or not for measurement, and z indicates the information provided by the device (estimate of state or parameters of an input-output model). Again observation and identification, being open-loop operations, are not necessarily performed in real time by means of a device continuously connected to the system like the one shown in the figure, but can also be viewed as off-line processing of recorded data.

When, on the other hand, continuous monitoring of the controlled system is needed to obtain the desired control task, it is necessary to use the *closed-loop* or *feedback* connection shown in Fig. 1-27 as a block diagram, in which the controller receives information both from the environment, through input r, and the controlled system, through connection y. In the controlled system, d still denotes the disturbance input, e the *regulated output*, possibly different from the *informative output* y. The feedback connection is the most interesting for on-line control purposes since it allows any type of error made in achieving the control task to be compensated by performing corrections over the manipulable variables related to measurements of the most representative quantities of the controlled system. Thus, lack of accuracy in the mathematical model or in the manipulable variables actuation is automatically adjusted.

Example 1.6-1. Optimal Control

Suppose a guided missile has to be controlled to reach an earth satellite. The manipulable variables are the intensity and direction of engine thrust. A preliminary step for control is to determine an optimum policy, i.e., the initial time, the initial situation if the launching pad is not fixed, and the control function according to a given performance index; this could, for instance, be a minimum overall fuel consumption or the minimum time for a given total amount of fuel. The solution of this problem is usually achieved before the launch, hence it is completely open-loop.

Example 1.6-2. Tracking Control

After launch of the aforesaid missile, we are faced with a completely different control problem: to ensure that the previously determined trajectory is actually followed despite possible, relatively small imperfections of the mathematical model used and the thrust controlling apparatus. To this end, the actual trajectory is continuously monitored and corrections are made should it tend to diverge from the desired one; this kind of control is a typical example of a closed-loop action and is called *tracking*. Contrary to the open-loop, it can automatically compensate for unknown and unpredictable disturbances or simply for precision limits of the mathematical model.

Example 1.6-3. Adaptive Control

A robot arm must handle several types of objects whose mass is not known in advance, and deposit them in specified positions. This, too, can be considered a tracking problem, but the system does not have a completely known dynamic behavior because the mass is not known in advance. The controller in this case must be *adaptive* or *self-tuning*, i.e., must vary its dynamics to achieve satisfactory performance in all circumstances (good speed and absence of oscillations). Control action and parametric identification are obtained together, with interacting policies. In fact, by means of movement a twofold effect is obtained: to position objects and evaluate their masses by measuring the corresponding effort, so that a motion trajectory which fits both these requirements must be followed.

The above examples, although referred to particular cases, point out some general aspects of control problems. Complicated control policies, such as those relating to overall optimization of multivariable plants, are usually computed off-line, hence open-loop, but implemented by means of suitable closed-loop automatic tracking apparatus. In some cases, an on-line identification process coordinated with the control task is required to improve performance.

1.7 FINITE-STATE SYSTEMS

A *finite-state system*, or *finite-state machine*, or *automaton* is a discrete-time system whose input, state, and output sets are finite. Finite-state systems are used as models of numerous physical and nonphysical objects, like computers, automatic machine-tools, computer programs, telephone switching apparatus.

In the framework of system theory, they are quite interesting because, although they require a very simple mathematical background (the most elementary concepts of algebra), finite-state systems are generally nonlinear. For this reason, simple examples taken from finite-state system theory can be used, for instance, to clarify how restrictive linearity assumption is with respect to general system behavior. Furthermore, the algorithms that solve control and observation problems of finite-state systems help to gain insight into the meaning and way of operation of the most relevant algorithms of the geometric approach to linear systems, which have a very similar structure.

Finite-state systems, being particular discrete-time systems, satisfy Definition 1.3-4. The input set is $\mathcal{U} := \{u_1, \ldots, u_p\}$, the input function set $\mathcal{U}_f$ is the set of all sequences $u(\cdot) : \mathcal{T} \to \mathcal{U}$, the state set is $\mathcal{X} := \{x_1, \ldots, x_n\}$, and the output set is $\mathcal{Y} := \{y_1, \ldots, y_q\}$. Discrete time will be herein denoted by i so that the next-state function and the output function can be written as

$$x(i+1) = f(x(i), u(i)) \tag{1}$$
$$y(i) = g(x(i), u(i)) \tag{2}$$

Transitions occur when a suitable *synchronizing event* or *clock signal* (for instance, a sequence of impulses) is applied to a clock input. The synchronizing events can be generated independently of the other inputs or related to them; for instance, when a keyboard is used as the input device, an impulse is generated when any key is pressed, causing the corresponding symbols to be accepted. Hence, the synchronizing events do not necessarily need to be uniformly spaced in time.

In the usual automata terminology the mathematical model expressed by (1,2), referring to a nonpurely dynamic system, is called a *Mealy model*, whereas that expressed by the purely dynamic system

$$x(i+1) = f(x(i), u(i)) \tag{3}$$
$$y(i) = g(x(i)) \tag{4}$$

is called a *Moore model*.

A particular finite-state purely dynamic system is the *unit delay* (see Fig. 1-28) described by the equations

$$x(i+1) = u(i) \tag{5}$$
$$y(i) = x(i) \tag{6}$$

Sect. 1.7 Finite-State Systems

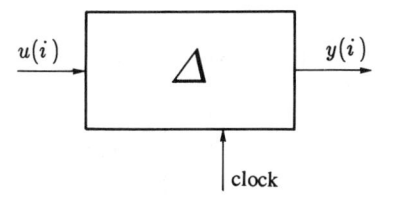

Figure 1-28 A unit delay.

or by the sole input-output equation

$$y(i+1) = u(i) \qquad (7)$$

As in the general case, a *memoryless*, or *purely algebraic*, or *purely combinatorial* finite-state system is one whose mathematical model reduces to the sole algebraic relation

$$y(i) = g(u(i)) \qquad (8)$$

Property 1.7-1.
Any finite-state system can be realized by interconnecting a purely combinatorial system and a unit delay.

Proof. Consider the connection shown in Fig. 1-29, which is clearly described by (1,2) and is obtained by connecting a memoryless system (pointed out by dashed lines) and a unit delay. □

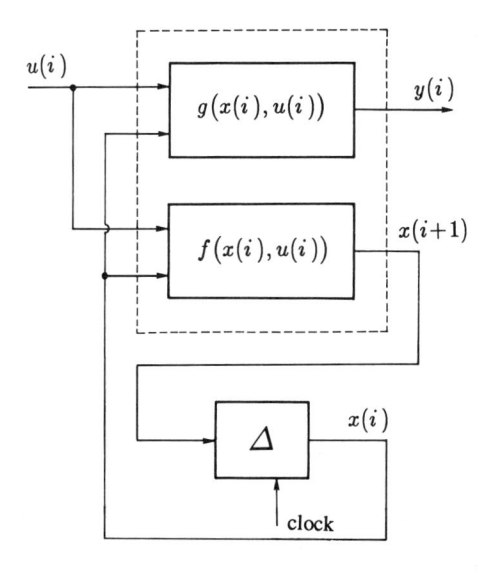

Figure 1-29 Realization of a finite-state system.

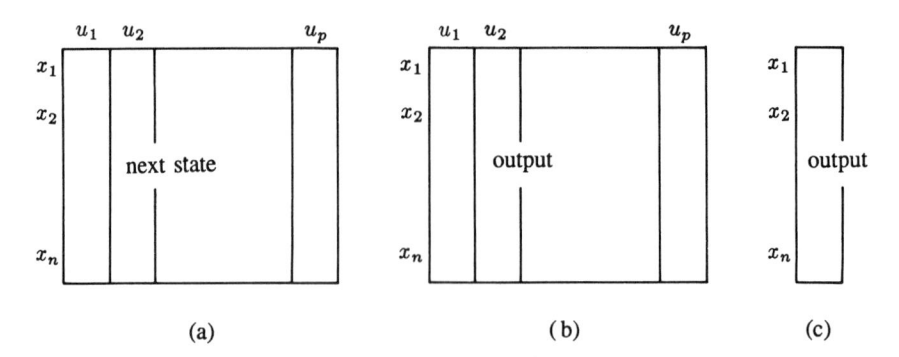

Figure 1-30 Transition and output tables.

Functions f and g which, according to (1,2), describe the behavior of finite-state systems, can be specified by means of two tables or a graph, as follows.

1. *Transition and output tables*: In the most general case (Mealy model) transition and output tables are shown in Fig. 1-30(a,b). They have n rows and p columns, labeled with the state and input symbols respectively: the intersection of row x_i and column u_j shows the value of next-state function $f(x_i, u_j)$ and output function $g(x_i, u_j)$. In the case of a purely dynamic system (Moore model) the output table is simplified as shown in Fig. 1-30(c).

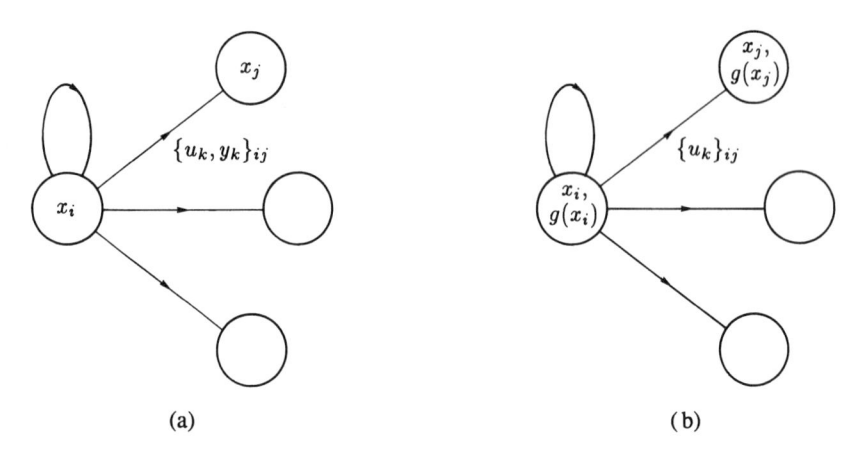

Figure 1-31 Transition graphs.

2. *Transition graphs*: The transition graph has n nodes, which represent the n states of the system. Referring to Fig. 1-31(a), consider the generic node x_i and denote by $\mathcal{U}_{ij}$ the set of all input symbols such that $f(x_i, \mathcal{U}_{ij}) = \{x_j\}$, i.e., of symbols which cause transition from state x_i to x_j. If $\mathcal{U}_{ij}$ is nonempty, the graph has an oriented branch joining nodes x_i and x_j, which is labeled

Sect. 1.7 Finite-State Systems

with the set of symbols $\{u_k/y_k\}$, where k is any subscript such that $u_k \in \mathcal{U}_{ij}$ and $u_k := g(x_i, u_j)$. In other words, for every node x_i there are as many outgoing branches as there are possible transitions, i.e., as there are possible future states, which in the graph appear as the terminal nodes of these branches. Each branch is labeled with symbols of the type u_k/y_k, as many as the input symbols causing the considered transition. This type of graph refers to the Mealy model; the graph for the Moore model is shown in Fig. 1-31(b): outputs are related to nodes instead of to each different transition in the branches.

	A	R	E	σ	␣
L	M	L	L	L	L
M	M	N	L	L	L
N	M	L	O	L	L
O	M	L	L	L	L

	A	R	E	σ	␣
L	0	0	0	0	0
M	0	0	0	0	0
N	0	0	0	0	0
O	0	0	0	0	1

(a)

(b)

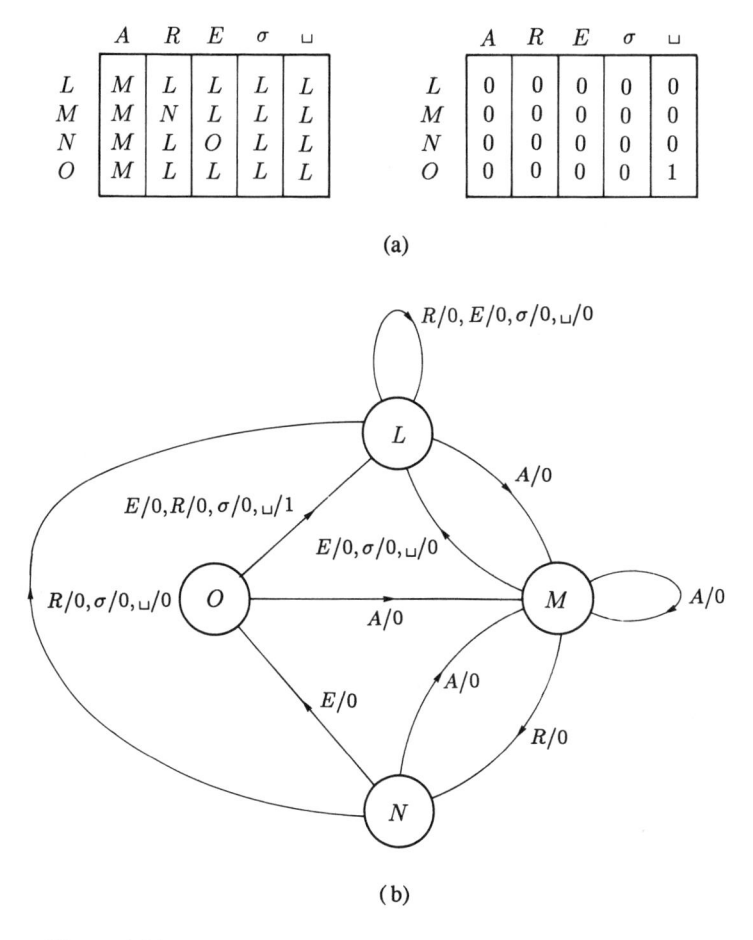

Figure 1-32 Transition and output table and transition graph of a sequence detector.

Example 1.7-1. A Sequence Detector

The input set is the set of characters of a keyboard and the device is required to detect the sequence $ARE_\sqcup$, where ␣ denotes space; the output set is $\{0,1\}$ and the output is required to assume value 1 every time the sequence is detected. The transition table, output table, and transition graph corresponding to this description in words are shown in Fig. 1-32: for the sake of simplicity, input σ is used for any character different from $ARE_\sqcup$.

Let us now examine the most important characterizing features of finite-state systems related to control and observation.

1.7.1 Controllability

Following notation introduced in Section 1.3 let

$$\varphi(k, 0, x(0), u(\cdot)) \qquad (9)$$
$$\gamma(k, 0, x(0), u(\cdot)) \qquad (10)$$

be the transition and response functions of a finite-state system. Clearly (9, 10) are implicitly defined by recursion formulae (1, 2) or (3, 4).

Given any two states $x_i, x_j \in \mathcal{X}$, x_i is said to be controllable to x_j in k steps or x_j is said to be reachable from x_i in k steps if there exists an input sequence $u|_{[0, k-1]}$ such that

$$x_j = \varphi(k, 0, x_i, u(\cdot)) \qquad (11)$$

Given any two states $x_i, x_j \in \mathcal{X}$, x_i is said to be controllable to x_j or x_j is said to be reachable from x_i if there exists an input sequence $u|_{[0, k-1]}$ such that (11) holds.

Given any state $x \in \mathcal{X}$, with $\mathcal{R}_k^+(x)$ we shall denote the set of all reachable states, or briefly, the reachable set, from x in k steps, with $\mathcal{W}_k^+(x)$ the reachable set from x in any number of steps not greater than k. Clearly, $\mathcal{R}_k^+(x) \subseteq \mathcal{W}_k^+(x)$. Similarly, with $\mathcal{R}_k^-(x)$ and $\mathcal{W}_k^-(x)$ we will refer to the set of all states controllable to x in k steps or in any number of steps not greater than k. Clearly, $\mathcal{R}_k^-(x) \subseteq \mathcal{W}_k^-(x)$.

Algorithm 1.7-1.

Sets $\mathcal{R}_k^+(x)$ and $\mathcal{R}_k^-(x)$ are provided by the recursion equations

$$\mathcal{R}_0^+(x) = \{x\}$$
$$\mathcal{R}_i^+(x) = \bigcup_{j=1}^{p} f(\mathcal{R}_{i-1}^+(x), u_j) \quad (i = 1, \ldots, k) \qquad (12)$$

$$\mathcal{R}_0^-(x) = \{x\}$$
$$\mathcal{R}_i^-(x) = \bigcup_{j=1}^{p} f^{-1}(\mathcal{R}_{i-1}^-(x), u_j) \quad (i = 1, \ldots, k) \qquad (13)$$

Sect. 1.7 Finite-State Systems

where $f(\mathcal{R}_{i-1}^+(x), u_j)$ and $f^{-1}(\mathcal{R}_{i-1}^-(x), u_j)$ denote respectively the image of $\mathcal{R}_{i-1}^+(x)$ and the inverse image of $\mathcal{R}_{i-1}^-(x)$ in the function $f(x, u_j) : \mathcal{X} \to \mathcal{X}$ (for any given u_j).

Proof. The meaning of the recursion relation (12) is clear: the set of reachable states from x in i steps is the union of sets that are obtained by transforming, with respect to f, the set of reachable states from x in $i-1$ steps for all input symbols. A similar argument applies to (13). □

Algorithm 1.7-2.
Sets $\mathcal{W}_k^+(x)$ and $\mathcal{W}_k^-(x)$ are provided by the recursion equations

$$\mathcal{W}_0^+(x) = \{x\}$$

$$\mathcal{W}_i^+(x) = \mathcal{W}_0^+(x) \cup \left(\bigcup_{j=1}^p f\left(\mathcal{W}_{i-1}^+(x), u_j\right) \right) \quad (i=1,\ldots,k) \tag{14}$$

$$\mathcal{W}_0^-(x) = \{x\}$$

$$\mathcal{W}_i^-(x) = \mathcal{W}_0^-(x) \cup \left(\bigcup_{j=1}^p f^{-1}\left(\mathcal{W}_{i-1}^-(x), u_j\right) \right) \quad (i=1,\ldots,k) \tag{15}$$

Proof. First, we consider relations (14) and prove that the recursion formula is equivalent to

$$\mathcal{W}_i^+(x) = \mathcal{W}_{i-1}^+(x) \cup \left(\bigcup_{j=1}^p f\left(\mathcal{W}_{i-1}^+(x), u_j\right) \right) \quad (i=1,\ldots,k) \tag{16}$$

which, in turn, simply expresses the definition of $\mathcal{W}_i^+(x)$. Clearly, in (16) it is $\mathcal{W}_i^+(x) \supseteq \mathcal{W}_{i-1}^+(x)$ $(i=1,\ldots,k)$, so that

$$\bigcup_{j=1}^p f\left(\mathcal{W}_{i-1}^+(x), u_j\right) \supseteq \bigcup_{j=1}^p f\left(\mathcal{W}_{i-2}^+(x), u_j\right)$$

Substitute in the i-th relation of (16) the first term on the right, provided by the previous relation (the $i-1$-th). It follows that

$$\mathcal{W}_i^+(x) = \mathcal{W}_{i-2}^+(x) \cup \left(\bigcup_{j=1}^p f\left(\mathcal{W}_{i-2}^+(x), u_j\right) \right) \cup \left(\bigcup_{j=1}^p f\left(\mathcal{W}_{i-1}^+(x), u_j\right) \right)$$

$$= \mathcal{W}_{i-2}^+(x) \cup \left(\bigcup_{j=1}^p f\left(\mathcal{W}_{i-1}^+(x), u_j\right) \right) \tag{17}$$

In a similar way it is possible to prove that in (17) $\mathcal{W}_{i-2}^+(x)$ can be substituted by $\mathcal{W}_{i-3}^+(x)$, and so on until (14) is obtained. Relations (15) are proved by a similar argument. □

By $W^+(x)$ we shall denote the *set of states reachable from* x (with sequences of any length) and by $W^-(x)$ the *set of states controllable to* x. The following holds.

Theorem 1.7-1.
Sets $W^+(x)$ and $W^-(x)$ can be determined respectively with the recursion formulae (14) and (15), stopping at the first value of i such that $W_{i+1}^+(x) = W_i^+(x)$ or $W_{i+1}^-(x) = W_i^-(x)$.

Proof. If, for a value of i, say k, $W_{k+1}^+(x) = W_k^+(x)$, sequence (14) for all values of i greater than k provides the same set, since at each additional step the same formula is applied to the same set. This argument also holds for sequence (15). □

Corollary 1.7-1.
Consider a finite-state system having n states. If state x_j is reachable from x_i, transition can be obtained in, at most, $n-1$ steps.

Proof. It has been remarked in the previous proof that the number of elements of sets $W_i^+(x)$ ($i = 0, 1, 2, \ldots$) strictly increases until the condition $W_{i+1}^+(x) = W_i^+(x)$ is met. Since $W_0(0)$ has at least one element, the total number of transitions cannot be greater than $n-1$. □

Theorem 1.7-2.
A finite-state system is completely controllable or strongly connected if and only if $W^+(x) = W^-(x) = \mathcal{X}$ for all $x \in \mathcal{X}$.

Proof. Only if. In a strongly connected system both the transition from x to any other state and the inverse transition must be possible.

If. For any two states $x_i, x_j \in \mathcal{X}$, since $x_i \in W^-(x)$, $x_j \in W^+(x)$, it is possible to reach x from x_i and x_j from x. □

Algorithms 1.7-1 and 1.7-2 are the basic tools for solving control problems of finite-state systems. The most common of these problems are the following.

Problem 1.7-1. Control Between Two Given States
Given any two states x_i, x_j, find a minimal-length input sequence that causes transition from x_i to x_j.

Solution. For the problem to admit a solution, one of the following relations must clearly be satisfied:

$$x_j \in W^+(x_i) \tag{18}$$

or

$$x_i \in W^-(x_j) \tag{19}$$

Refer, for instance, to (19): let k be such that $x_i \in W_k^-(x_j)$, $x_i \notin W_{k-1}^-(x_j)$; by definition, there exists an input $u(0)$ such that $x(1) = f(x_i, u(0)) \in W_{k-1}^-(x_j)$, an input $u(1)$ such that $x(2) = f(x(1), u(1)) \in W_{k-2}^-(x_j)$, and so on. Thus, an input sequence $u|_{[0, k-1]}$ exists that tranfers the state from x_i to x_j. □

Problem 1.7-2. Control to a Given Output

Given an initial state x_i and an output value y_j, find a minimal-length input sequence that, starting from x_i, produces y_j as the last output symbol.

Solution. The set of all states that, by an appropriate choice of the input, can produce the output y_j, is

$$\mathcal{X}_j := \bigcup_{r=1}^{p} g^{-1}(\{y_j\}, u_r) \tag{20}$$

By applying Algorithm 1.7-2 with $W_0^-(\mathcal{X}_j) := \mathcal{X}_j$ an integer k is determined such that $x_i \in W_k^-(\mathcal{X}_j)$, $x_i \notin W_{k-1}^-(\mathcal{X}_j)$, then it is possible to proceed as in the previous problem in order to derive the input sequence $u|_{[0,k-1]}$. Let $x_k \in \mathcal{X}_j$ be the state that can be reached by applying this sequence: by definition, there exists an input $u(k)$ such that $y_j = g(x(k), u(k))$; this completes the sequence $u|_{[0,k]}$, which solves the problem. □

Problem 1.7-3. Control for a Given Output Sequence

Find, if possible, an input sequence $u|_{[0,k]}$ that produces a given output sequence $y|_{[0,k]}$ starting at a given initial state x_i.

Solution. The set of all initial states compatible with the given output sequence is provided by the recursion algorithm

$$\mathcal{X}_k = \bigcup_{r=1}^{p} g^{-1}(\{y(k)\}, u_r)$$

$$\mathcal{X}_{k-i} = \bigcup_{r=1}^{p} \left(g^{-1}(\{y(k-i)\}, u_r) \cap f^{-1}(\mathcal{X}_{k-i+1}, u_r) \right) \quad (i=1,\ldots,k) \tag{21}$$

which can be explained in the following terms: $\mathcal{X}_k$ is the set of all states from which, by a suitable input, it is possible to obtain the output $y(k)$, while $\mathcal{X}_{k-1}$ is the similar set which allows the output $y(k-1)$ and transition to a state belonging to $\mathcal{X}_k$ to be obtained, and so on. For the problem to admit a solution, it is clearly necessary that $x(0) \in \mathcal{X}_0$: the input sequence $u|_{[0,k]}$ can be determined as for Problem 1.7-1. □

1.7.2 Reduction to the Minimal Form

In the particular case of finite-state systems, Definition 1.3-9 can be stated in the following terms: two states $x_i, x_j \in \mathcal{X}$ are said to be *indistinguishable in k steps* or *k-indistinguishable* if

$$\gamma(r, 0, x_i, u(\cdot)) = \gamma(r, 0, x_j, u(\cdot)) \quad \forall r \in [0, k], \; \forall u(\cdot)$$

or, in words, if, for any input sequence of length $k+1$, the same output sequence is obtained starting either at x_i or x_j. Note that k-indistinguishability, being clearly reflexive, symmetric, and transitive, is an equivalence relation. The induced state partition will be denoted by P_k. Since the set of all partitions of a finite set $\mathcal{X}$ is a lattice, it is possible to define in the set of all state partitions a partial ordering relation, addition and multiplication, a supremum (the maximal partition P_M, with a unique block), and an infimum (the minimal partition P_m, with as many blocks as there are elements in $\mathcal{X}$). Furthermore, given a function $f : \mathcal{X} \to \mathcal{Y}$ and any partition P of $\mathcal{Y}$, the inverse image of P in f is defined as the partition of $\mathcal{X}$ whose blocks are the inverse images in f of the blocks of P. In light of these additional definitions, the following algorithm can be set.

Algorithm 1.7-3.

The k-indistinguishability partition P_k is provided by the recursion equations

$$P_0 = \prod_{j=1}^{p} g^{-1}(P_m, u_j)$$

$$P_i = P_0 \cdot \left(\prod_{j=1}^{p} f^{-1}(P_{i-1}, u_j) \right) \quad (i=1,\ldots,k) \tag{22}$$

Proof. The first of (22) provides the 0-indistinguishability partition, i.e., the partition of states whose blocks in relation (2) provide the same output for all inputs. Consider, instead of the next recursion relations (22),

$$P_i = P_{i-1} \cdot \left(\prod_{j=1}^{p} f^{-1}(P_{i-1}, u_j) \right) \quad (i=1,\ldots,k) \tag{23}$$

which will be proved to be equivalent to them. The i-th of (23) expresses that, for any two states to be i-indistinguishable, they must be $(i-1)$-indistinguishable and any transition from them has to occur toward $(i-1)$-indistinguishable states. In fact, if there should exist an input corresponding to a transition toward $(i-1)$-distinguishable states, an input sequence $u|_{[0,k]}$ with this input as the first element would allow us to distinguish the two states referred to.

We shall prove now that in (22) the recursion formula is equivalent to (23): note that in sequence (23) $P_i \leq P_{i-1}$ ($i=1,\ldots,k$), so that

$$\prod_{j=1}^{p} f^{-1}(P_{i-1}, u_j) \leq \prod_{j=1}^{p} f^{-1}(P_{i-2}, u_j) \tag{24}$$

Then, substitute in the i-th relation of (23) the first term on the right, P_{i-1}, provided by the previous relation. It follows that

$$P_i = P_{i-2} \cdot \left(\prod_{j=1}^{p} f^{-1}(P_{i-1}, u_j) \right) \cdot \left(\prod_{j=1}^{p} f^{-1}(P_{i-2}, u_j) \right)$$

$$= P_{i-2} \cdot \left(\prod_{j=1}^{p} f^{-1}(P_{i-1}, u_j) \right) \tag{25}$$

Sect. 1.7 Finite-State Systems

In a similar way, it is possible to prove that in (25) P_{i-2} can be substituted by P_{i-3}, and so on, until (22) is obtained. □

In the particular case of finite-state systems, Definition 1.3-10 can be stated in the following terms: two states x_i and x_j are said to be *equivalent* if they are k-indistinguishable for any k. The corresponding state partition $P = \{p_1, \ldots, p_s\}$ is called *equivalence partition*.

Theorem 1.7-3.
The equivalence partition P can be determined with the recursion relations (22), stopping at the first value of i such that $P_{i+1} = P_i$.

Proof. If, for a value of i, say k, $P_{k+1} = P_k$, sequence (21) for all values of i greater than k provides the same partition, since at each additional step the same formula is applied to the same partition. □

Corollary 1.7-2.
Consider a finite-state system having n states. Any two $(n-2)$-indistinguishable states are equivalent.

Proof. Partition P_0 has at least two blocks and the subsequent P_i $(i = 1, 2, \ldots)$ have a number of blocks strictly increasing until the condition $P_{k+1} = P_k$ is obtained for a certain k. Hence, the value of k cannot be greater than $n-2$. □

According to Definition 1.3-11, a finite-state system is said to be *in minimal form* or *minimal* if it has no equivalent states, i.e., if $P = P_m$. According to Definition 1.3-12, two finite-state systems Σ_1 and Σ_2 are said to be *equivalent* if $\mathcal{U}_1 = \mathcal{U}_2$, $\mathcal{Y}_1 = \mathcal{Y}_2$, and if for any state $x_1 \in \mathcal{X}_1$ $(x_2 \in \mathcal{X}_2)$ of one of them there exists a state $x_2 \in \mathcal{X}_2$ $(x_1 \in \mathcal{X}_1)$ of the other such that

$$\gamma_1(k, 0, x_1, u(\cdot)) = \gamma_2(k, 0, x_2, u(\cdot)) \quad \forall k \geq 0, \; \forall u(\cdot) \qquad (26)$$

Theorem 1.7-4.
Two finite-state systems Σ_1 and Σ_2 are equivalent if and only if the composite system Σ defined by

1. $\mathcal{U} = \mathcal{U}_1 = \mathcal{U}_2$
2. $\mathcal{Y} = \mathcal{Y}_1 = \mathcal{Y}_2$
3. $\mathcal{X} = \mathcal{X}_1 \cup \mathcal{X}_2$
4. $f(x, u) = \begin{cases} f_1(x, u) & \text{if } x \in \mathcal{X}_1 \\ f_2(x, u) & \text{if } x \in \mathcal{X}_2 \end{cases}$
5. $g(x, u) = \begin{cases} g_1(x, u) & \text{if } x \in \mathcal{X}_1 \\ g_2(x, u) & \text{if } x \in \mathcal{X}_2 \end{cases}$

has states of both Σ_1 and Σ_2 in every block of the equivalence partition.

Proof. A characterizing feature of system Σ is that all transitions occur between states of $\mathcal{X}_1$ if the initial state belongs to $\mathcal{X}_1$ and between states of $\mathcal{X}_2$ if the initial state belongs to X_2. In order to prove the theorem, simply note that relation (26), which states the equivalence of Σ_1 and Σ_2, implies the equivalence of x_1 and x_2, considered as states of Σ. □

Definition 1.7-1. Minimal Form of a Finite-State System

Let Σ be a finite-state system not in minimal form: the minimal form of Σ is the equivalent system Σ', defined by

1. $U' = U$
2. $\mathcal{Y}' = \mathcal{Y}$
3. $X' = P = \{p_1, \ldots, p_s\}$
4. $f'(p_i, u) = p_j$ if $f(x_i, u) = x_j$, $x_i \in p_i$, $x_j \in p_j$
5. $g'(p_i, u) = g(x_i, u)$ if $x_i \in p_i$

Concepts of k-indistinguishability and equivalence partition are used in order to solve observation and reconstruction problems. One of the most important of these is the "pairwise diagnosis experiment," which is formulated as follows.

Problem 1.7-4. Pairwise Diagnosis Experiment

Let x_i and x_j be the unique admissible initial states of a minimal finite-state system Σ: determine an input sequence having minimum length that allows us to determine which of them is the actual initial state from the output function.

Solution. Let k be such that x_i and x_j belong to different blocks of P_k but to the same block of P_{k-1}, i.e., such that they are k-distinguishable and $(k-1)$-indistinguishable. Then, there exists an input $u(0)$ which produces transitions from x_i and x_j toward $(k-1)$-distinguishable but $(k-2)$-indistinguishable states, an input $u(1)$ which produces transitions from such states toward $(k-2)$-distinguishable but $(k-3)$-indistinguishable states, and so on: the input $u(k-1)$ produces transitions toward 0-distinguishable states, so that a suitable input $u(k)$ causes outputs to be different. The determined input sequence $u|_{[0,k]}$ allows the two initial states to be distinguished by considering the last element of the corresponding output sequence $y|_{[0,k]}$. Note that, in any case, $k \leq n-2$. □

Example. An experiment to distinguish the initial states L and M in the finite-state system shown in Fig. 1-32 can be determined as follows: first, subsequent k-indistinguishability partitions P_k ($k=0,1,\ldots$) are determined until L and M are in different blocks. In the particular case referred to we get

$$P_0 = \{L, M, N; O\}$$
$$P_1 = \{L, M; N; O\}$$
$$P_2 = \{L; M; N; O\}$$

States L and M are 2-distinguishable: in fact, input R causes transitions to states L and N, then E to L and O, which are 0-distinguishable, since ⊔ causes the output to be 0 and 1 respectively. Then the problem is solved by input sequence $\{R, E, ⊔\}$, which produces either output sequence $\{0,0,0\}$ if the initial state is L or $\{0,0,1\}$ if it is M.

1.7.3 Diagnosis and State Observation

According to the definition stated in Section 1.4, the *diagnosis* problem of a finite-state system Σ, assumed to be minimal, is the following: given an *admissible initial state set* $\mathcal{X}_A \subseteq \mathcal{X}$, determine the actual initial state by applying a suitable input sequence and considering the corresponding output sequence; in other words, an input sequence has to be determined such that the corresponding output sequence is different for every $x(0) \in \mathcal{X}_A$.

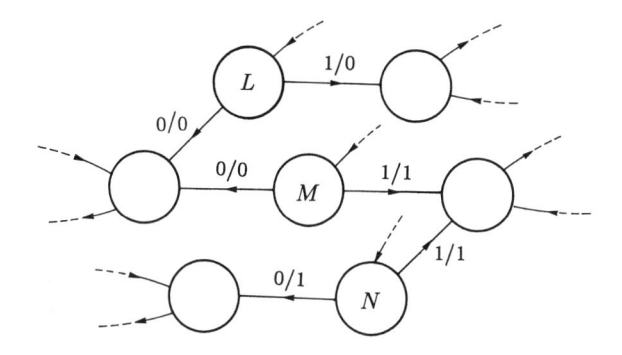

Figure 1-33 A case where diagnosis is not possible.

It has been shown in the previous section (Problem 1.7-4) that the diagnosis problem always admits a solution if $\mathcal{X}_A$ has only two elements. If, on the other hand, elements of $\mathcal{X}_A$ are more numerous, the problem may not admit a solution. As an example of such a case, consider the partial graph shown in Fig. 1-33 concerning a system with $\{0,1\}$ as input and output set: an input sequence beginning with 0 destroys all chances of distinguishing L and M, while an input sequence beginning with 1 destroys all chances of distinguishing M and N. Therefore, diagnosis with $\mathcal{X}_A := \{L, M, N\}$ is not possible, at least with a *simple experiment*, i.e., with a single trial.

If the system can be "reset" after every trial (i.e., brought again to the unknown initial state), a *multiple experiment* can be performed: it will be shown (Theorem 1.7-5) that this kind of experiment always solves the diagnosis problem.

The diagnosis experiments, simple or multiple, can be *preset* or *adaptive*: in a preset experiment the input sequence is determined in advance, while in an adaptive sequence it depends on the output values as they arrive.

In some instances the diagnosis problem is solved with a simple, adaptive experiment, but not with a simple, preset experiment. As an example for this, consider the partial graph shown in Fig. 1-34, again referring to a system with $\{0,1\}$ as input and output set. Let $\mathcal{X}_A := \{O, P, Q, R\}$: applying the input sequence $\{0,1\}$ destroys all chances of distinguishing Q and R, while $\{0,0\}$

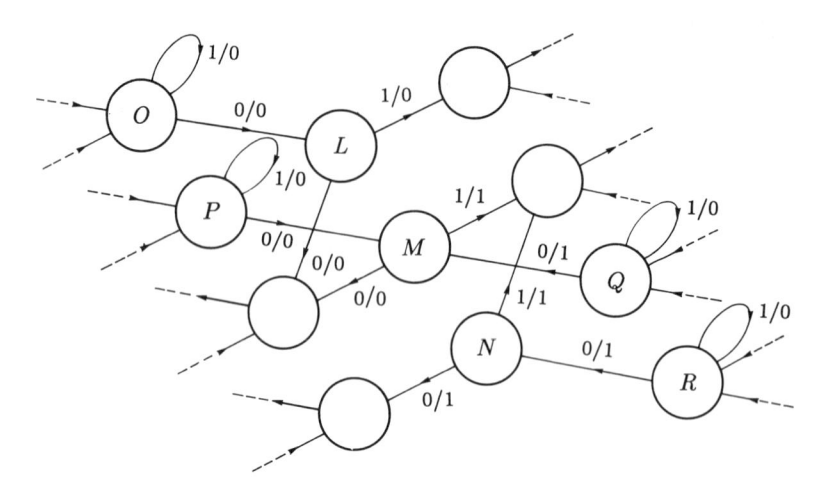

Figure 1-34 A case where adaptive diagnosis is possible and preset diagnosis is not.

destroys those of distinguishing O and P. However, it is possible to follow another policy: apply input 0 and observe the output; if it is 0, Q and R are excluded as initial states and subsequent application of input 1 allows us to distinguish between O and P, while if it is 1, O and P are excluded as initial states and subsequent application of input 0 allows us to distinguish between Q and R.

A simple diagnosis experiment, although not generally providing the initial state in $\mathcal{X}_A$ (if elements of $\mathcal{X}_A$ are more than two), allows the exclusion of a certain number of elements of $\mathcal{X}_A$ as initial states (at least one). In fact, there exists an input sequence that allows any pair of states x_i, x_j in $\mathcal{X}_A$ to be distinguished (i.e., a pairwise diagnosis sequence for them) and corresponds to a partition P_0 of initial states with at least two blocks, since x_i and x_j clearly must belong to different blocks. Hence the following result.

Theorem 1.7-5.
The diagnosis problem always admits a solution with a multiple experiment, which can be preset or adaptive.

Proof. Applying an input sequence that allows two states of $\mathcal{X}_A$ to be distinguished (pairwise diagnosis experiment) induces a partition of $\mathcal{X}_A$ with a different output sequence associated to each block. This procedure can be repeated for each block (multiple preset experiment) or for the particular block to which the initial state actually turns out to belong (multiple adaptive experiment), until complete information on the initial state is obtained. □

The initial state partition induced by the knowledge of the output sequence is provided by the following simple algorithm, which refers to the most general case $\mathcal{X}_A = \mathcal{X}$.

Algorithm 1.7-4.

Knowledge of the output sequence $y|_{[0,k]}$ corresponding to a given input sequence $u|_{[0,k]}$ allows the establishment of a set of admissible initial states, which coincides with one of the blocks of the partition P_0 provided by the recursion equation

$$P_k = g^{-1}\left(P_m, u(k)\right)$$
$$P_{k-i} = g^{-1}\left(P_m, u(k-i)\right) \cdot f^{-1}\left(P_{k-i+1}, u(k-i)\right) \quad (i=1,\ldots,k) \qquad (27)$$

Proof. P_k is the partition of final states $x(k)$ induced by the property of producing the same output $y(k)$ under the input $u(k)$ (which is clearly an equivalence relation); P_{k-1} is induced by the property of causing the same partial output sequence $\{y(k-1), y(k)\}$ under the partial input sequence $\{u(k-1), u(k)\}$ (also an equivalence relation): in other words, states $x(k-1)$ belonging to a block of P_{k-1} produce the same output under the input $u(k-1)$ and with this input are transformed into future states that produce the same output with input $u(k)$. A similar argument applies for the generic P_{k-i}. □

The *observation problem* (i.e., to derive information on the initial state from the knowledge of corresponding input and output sequences) can be solved as stated in the following problem, by an iterative procedure very similar to Algorithm 1.7-4.

Problem 1.7-5. State Observation

Determine the initial state set in $\mathcal{X}_A$ compatible with given sequences of input and output $u|_{[0,k]}$, $y|_{[0,k]}$.

Solution. The recursion relation

$$\mathcal{X}_k = g^{-1}\left(\{y(k)\}, u(k)\right)$$
$$\mathcal{X}_{k-i} = g^{-1}\left(\{y(k-i)\}, u(k-i)\right) \cap f^{-1}\left(\mathcal{X}_{k-i+1}, u(k-i)\right) \quad (i=1,\ldots,k) \qquad (28)$$

provides, as the last term, the set of initial states compatible with the given sequences, i.e., $\mathcal{E}_k^-(u(\cdot), y(\cdot))$. The solution of the stated problem is clearly $\mathcal{X}_A \cap \mathcal{X}_0$. □

1.7.4 Homing and State Reconstruction

According to the definition stated in Section 1.4, the *homing* problem of a finite-state system Σ, which is assumed to be minimal, consists of the following problem: given an *admissible initial state set* $\mathcal{X}_A \subseteq \mathcal{X}$, determine the final (or current) state by applying a proper input sequence and considering the corresponding output sequence.

Like diagnosis experiments, homing experiments can be *preset* or *adaptive*, according to whether the input sequence is determined in advance or made to depend on the output values as they arrive. Contrary to diagnosis, homing always admits a solution with a simple experiment.

To present the algorithm that solves the homing problem, it is convenient to introduce the concept of *partialization* of a set $\mathcal{X}$: a partialization of $\mathcal{X}$ is a collection Q of subsets of $\mathcal{X}$ (which, similar to those of partition, will be called "blocks" herein) such that
1. the same element can belong to several blocks of Q;
2. the sum of all elements in Q (possibly repeated) is not greater than n.

It is easy to check that, given both a function $f : \mathcal{X} \to \mathcal{X}$ and a partialization Q of $\mathcal{X}$, the set $Q' := f(Q)$ whose elements are the images of all blocks of Q with respect to f is also a partialization of $\mathcal{X}$. In particular, transforming with respect to f a partition of $\mathcal{X}$, which is also a partialization, provides a partialization. The product of a partition P by a partialization Q is the partialization whose blocks are obtained by intersecting blocks of P and Q in all possible ways.

Algorithm 1.7-5.

Knowledge of the output sequence $y|_{[0,k]}$ corresponding to a given input sequence $u|_{[0,k]}$ allows the establishment of a set of admissible future states $x(k+1)$, which coincides with one of the blocks of the partialization Q_{k+1} and a set of admissible current states $x(k)$, which coincides with one of the blocks of the partialization Q'_k, provided by the recursion relations

$$Q_0 = \{\mathcal{X}_A\}$$
$$Q_i = f\left(Q'_{i-1}, u(i-1)\right) \quad (i=1,\ldots,k+1) \tag{29}$$

where

$$Q'_{i-1} = Q_{i-1} \cdot g^{-1}\left(P_m, u(i-1)\right) \tag{30}$$

Proof. First, we consider sequence (29): Q_1 clearly is the partialization of states $x(1)$ induced by the property of deriving with a transition corresponding to $u(0)$ from states $x(0)$ belonging to $\mathcal{X}_A$ and producing the same output $y(0)$ under the input $u(0)$, Q_2 is the partialization of states induced by the property of deriving with a transition corresponding to $u(1)$ from states $u(1)$ of a single block of Q_1 producing the same output $y(1)$ under the input $u(1)$; in other words, every block of Q_2 collects all states $x(2)$ which correspond to the same output sequence $\{y(0), y(1)\}$ under the input sequence $\{u(0), u(1)\}$, provided the initial state belongs to $\mathcal{X}_A$. In a similar way, by induction, the expression of the generic Q_i is derived. The meaning of the right side member of (30), which is a part of the right side member of (29), is implied by the previous argument. □

Theorem 1.7-6.

Homing of a minimal finite-state system can always be performed with a single preset experiment.

Proof. Note that in the iterative procedure set by Algorithm 1.7-5 every block of Q_{i-1} may be partitioned (by intersection with $g^{-1}(P_m, u(i-1))$) and every part is transformed to a block of Q_i: hence, the number of elements of every block of Q_i is not greater than that of the corresponding blocks of Q_{i-1}. Thus, an input sequence

Sect. 1.7 Finite-State Systems

that is not favorable (i.e., that does not improve the knowledge of state) in the worst case leaves the maximum number of elements per block in Q_i and Q'_i unchanged. If, on the other hand, at a certain time i any two states, say x_i and x_j, belong to the same block, it is sufficient to apply a pairwise diagnosis sequence for them from that time on to be sure to improve current state knowledge. By joining sequences that pairwise separate all states, an input sequence $u|_{[0,k]}$ is obtained which solves the homing problem, because it corresponds to a partialization Q_{k+1} whose blocks all contain a single state. □

In the homing problem, as in the diagnosis problem, an adaptive experiment is, in general, shorter than a preset one, because the future input sequence is determined during the experiment referring to only one block of Q_i, singled out by examining the previously occurred output sequence. The *reconstruction problem* (i.e., to derive information on the final or current state from the knowledge of corresponding input and output sequences) can be solved by the following iterative procedure very similar to Algorithm 1.7-5.

Problem 1.7-6. State Reconstruction

Given the set of admissible initial states $\mathcal{X}_A$, determine the sets of states $x(k+1)$ or $x(k)$ compatible with given input and output sequences $u|_{[0,k]}$ and $y|_{[0,k]}$.

Solution. By an argument similar to the one developed to prove Algorithm 1.7-5, the following recursion relations are derived:

$$\mathcal{X}_0 = \{\mathcal{X}_A\}$$
$$\mathcal{X}_i = f\big(\mathcal{X}'_{i-1}, u(i-1)\big) \quad (i=1,\ldots,k+1) \tag{31}$$

where

$$\mathcal{X}'_{i-1} = \mathcal{X}_{i-1} \cap g^{-1}\big(y(i-1), u(i-1)\big) \tag{32}$$

which provide the sets $\mathcal{X}_{k+1}$ of future states $x(k+1)$ and $\mathcal{X}'_k$ of current states $x(k)$ compatible with the given input and output sequences and with the given initial state set $\mathcal{X}_A$. □

When $\mathcal{X}_A = \mathcal{X}$, the described procedure provides the set $\mathcal{E}^+_k(u(\cdot), y(\cdot))$ defined in Section 1.4 as $\mathcal{X}'_k$.

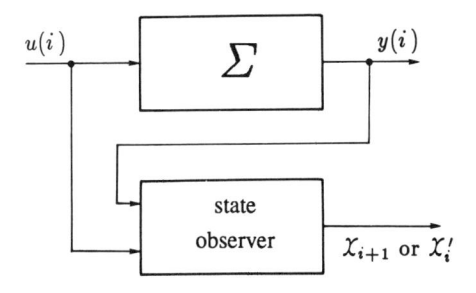

Figure 1-35 Connection of a state observer.

Relations (31,32) allow us to define a *state observer*, i.e., a finite-state system which, connected to Σ as shown in Fig. 1-35, continuously provides the set of future states $\mathcal{X}_{i+1}$ or the set of current states $\mathcal{X}'_i$. The input set of the observer is $\mathcal{U} \times \mathcal{Y}$, both state and output sets coincide with the set of all subsets of $\mathcal{X}$, while next-state and output function are easily derivable from (31,32). If the initial state of the observer is the single-element set corresponding to the actual state of the observed system, its output is the single-element set containing future state $x(i+1)$ or current state $x(i)$; if, on the other hand, the initial state is different, but congruent (i.e., a set containing the actual state of the observed system, possibly the whole state set $\mathcal{X}$), the observer provides as output the maximum information on future and current state of the observed system; in order to "synchronize" the observer for complete information, it is sufficient to apply an input sequence corresponding to a preset homing experiment, which always exists by virtue of Theorem 1.7-6.

1.7.5 Finite-Memory Systems

Definition 1.7-2. Finite-Memory System

A finite-state system Σ is said to be *finite-memory* if it can be represented by an input-output model of the type

$$y(i) = g'\big(u(i), u(i-1), \ldots, u(i-\mu), y(i-1), y(i-2), \ldots, y(i-\mu)\big) \tag{33}$$

where g' denotes a function from $\mathcal{U}^{\mu+1} \times \mathcal{Y}^\mu$ to $\mathcal{Y}$.

The minimal value of the integer μ is called the *memory* of Σ. A finite-memory finite-state system can be realized according to the interconnection scheme shown in Fig. 1-36, which refers to the input-output model (33) instead of that shown in Fig. 1-29, which refers to the input-state-output model (1,2).

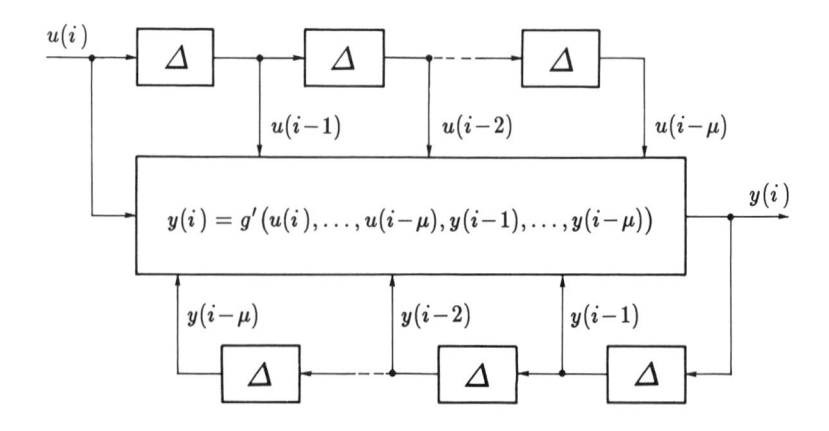

Figure 1-36 Realization of a finite-memory system.

Sect. 1.7 Finite-State Systems

The most common finite-state systems are generally finite-memory; for instance, the sequence detector shown in Fig. 1-32 is finite-memory. On the other hand, some finite-state systems, although very simple, are infinite-memory. As an example, consider the system shown in Fig. 1-37, where $\mathcal{U} = \mathcal{Y} = \{0, 1\}$. If an arbitrarily long input sequence consisting of all 0 is applied, the system remains in any one of the two states with output 0 while, when input is 1, output depends on the state. Therefore, the output in any instant of time cannot be expressed as a function of previous input and output sequences and current input.

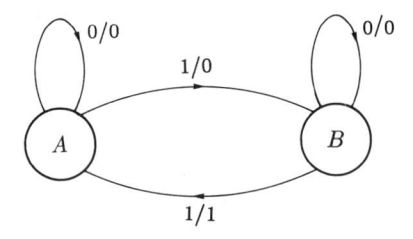

Figure 1-37 An example of infinite-memory system.

Theorem 1.7-7.

If system $(1, 2)$ is minimal and finite-memory, its state can be expressed as a function of previous input and output sequences having length μ, i.e.,

$$x(i) = f'\big(u(i-1), u(i-2), \ldots, u(i-\mu), y(i-1), y(i-2), \ldots, y(i-\mu)\big) \qquad (34)$$

where f' denotes a function from $\mathcal{U}^\mu \times \mathcal{Y}^\mu$ to $\mathcal{X}$.

Proof. If (34) does not hold, the same input sequence $u|_{[i-\mu, i-1]}$ can take the system into two distinguishable states also causing the same output sequence $y|_{[i-\mu, i-1]}$. These states being distinguishable, there exists an input sequence $u|_{[i, i+r]}$ ($r \geq 0$) such that the corresponding output sequences $y|_{[i, i+r]}$ are different, so that (33) cannot hold. □

As a consequence of Theorem 1.7-7, any input sequence $u|_{[0, \mu-1]}$ having length μ can be used as a homing sequence for a minimal finite-state system with finite memory μ. In other words, such a system is always reconstructable. On the other hand, if the system is not finite-memory, by definition there always exists at least one input sequence $u|_{[0,r]}$, with r arbitrarily large, which does not solve the homing problem.

Going a step further with this argument, it is possible to derive a procedure to establish whether a minimal finite-state system is finite-memory or not. To achieve this it is necessary to introduce the concept of *cover* of a set $\mathcal{X}$: a cover C of $\mathcal{X}$ is a collection of subsets of $\mathcal{X}$ (again called "blocks") with the following properties:

1. the same element can belong to several blocks of C;
2. the union of all blocks coincides with X;
3. no block is contained in another block.

Note that a partition is a particular cover. It can be easily proved that the set of all covers of X is a lattice with the partial order relation: $C_i \leq C_j$ if every block of C_i is contained in C_j. Denote by $R(\mathcal{P})$ the *reduction* operation which, for any set $\mathcal{P}$ of subsets of X, consists of the elimination of every subset contained in another set. Addition and multiplication of two covers are defined as: $C_1 + C_2 := R(\mathcal{A})$, where $\mathcal{A}$ is the union of the blocks of C_1 and C_2; $C_1 \cdot C_2 := R(\mathcal{B})$, where $\mathcal{B}$ is the set whose elements are obtained by intersecting blocks of C_1 and C_2 in all possible ways. Minimal cover C_m and maximal cover C_M of a given set X are equal respectively to the minimal partition P_m and to the maximal partition P_M.

Algorithm 1.7-6.
Let Σ be a minimal, strictly connected, finite-state system: if the sequence of covers of X provided by the recursion relation

$$C_0 = C_M$$
$$C_i = R\left(\bigcup_{j=1}^{p} f\left(C_{i-1} \cdot g^{-1}(P_m, u_j), u_j\right)\right) \quad (i=1,2,\ldots) \tag{35}$$

is such that $C_k = C_m$, Σ is finite-memory with memory $\mu = k$. On the other hand, if $C_i = C_{i-1} \neq C_m$ for a certain i, Σ is not finite-memory.

Proof. Since Σ is strictly connected, every state has at least a predecessor, so that sets C_i on the left of (35) are covers of X. Furthermore, $C_i \leq C_{i-1}$: in fact from $C_1 \leq C_0$ it follows that

$$\bigcup_{j=1}^{p} f\left(C_1 \cdot g^{-1}(P_m, u_j), u_j\right) \leq \bigcup_{j=1}^{p} f\left(C_0 \cdot g^{-1}(P_m, u_j), u_j\right)$$

hence, $C_2 \leq C_1$, and so on.

If, for a value of i, say k, $C_{k+1} = C_k$, sequence (35) for all values of i greater than k provides the same cover, since at each additional step the same formula is applied to the same cover.

Note that all blocks of partializations Q_i ($i=1,\ldots,k+1$) provided by Algorithm 1.7-5 for a given input sequence $u|_{[0,k]}$ and for $X_A := X$, are contained in blocks of C_i. Hence, if there exists an input sequence $u|_{[0,r-1]}$, for r large at will, which does not allow the final state to be determined, the equality $C_r = C_m$ is not possible.

On the other hand, since every block of C_i has a corresponding sequence $u|_{[0,i-1]}$ such that this block belongs to Q_i, if any sequence $u|_{[0,\mu-1]}$ allows the determination of the final state, condition $C_\mu = C_m$ is clearly necessary. □

REFERENCES

1. BOOTH, T.L., *Sequential Machines and Automata Theory*, Wiley, New York, 1967.
2. GILL, A., *Introduction to the Theory of Finite-State Machines*, McGraw-Hill, New York, 1962.
3. GINSBURG, S., *An Introduction to Mathematical Machine Theory*, Addison-Wesley, Reading, Mass., 1962.
4. HARTMANIS, J., and STEARN, R.E., *Algebraic Structure Theory of Sequential Machines*, Prentice Hall, Englewood Cliffs, N.J., 1966.
5. KALMAN, R.E., FALB, P.L., and ARBIB, M.A., *Topics in Mathematical System Theory*, McGraw-Hill, New York, 1969.
6. KLIR, G.J., *An Approach to General System Theory*, Van Nostrand Reinhold, New York, 1969.
7. LIU, C.L., "Determination of the final state of an automaton whose initial state is unknown," *IEEE Trans. Electron. Computers*, vol. EC-12, pp. 918–921, 1963.
8. LUENBERGER, D.G., *Introduction to Dynamic Systems: Theory, Models and Applications*, Wiley, New York, 1979.
9. MASSEY, J.L., "Notes on finite-memory sequential machines," *IEEE Trans. Electron. Computers*, vol. EC-15, pp. 658–659, 1966.
10. PADULO, L., and ARBIB, M.A., *System Theory*, W.B. Saunders, Philadelphia, 1974.
11. RINALDI, S., *Teoria dei Sistemi*, Clup and Hoepli, Milan, Italy, 1973.
12. SAIN, M.K., *Introduction to Algebraic System Theory*, Academic, New York, 1981.
13. VON BERTALANFFY, L., *General System Theory*, Braziller, New York, 1970.
14. WANG, P.C.K., "Control of distributed parameter systems," *Advances in Control Systems - I*, edited by Leondes, C.T., Academic, New York, 1964.
15. WINDEKNECHT, T.G., *General Dynamical Processes: a Mathematical Introduction*, Academic, New York and London, 1971.
16. WOOD, P.E. JR., *Switching Theory*, McGraw-Hill, New York, 1968.
17. ZADEH, L., and DESOER, C.A., *Linear System Theory: the State Space Approach*, McGraw-Hill, New York, 1963.
18. ZADEH, L.A., and POLAK, E., *System Theory*, McGraw-Hill, New York, 1969.

2

GENERAL PROPERTIES OF LINEAR SYSTEMS

2.1 THE FREE STATE EVOLUTION OF LINEAR SYSTEMS

When dealing with multivariable dynamic systems, in most cases mathematical models are used that consist of vector (i.e., with vectors as variables) differential or difference equations. Although these equations are generally nonhomogeneous because of the inputs, in the fundamental case of linear systems the basic features of their solutions are closely related to those of the corresponding homogeneous equations, which describe the free evolution of the state. Hence, it is helpful to study and classify the types of solutions of homogeneous linear equations and their connections with the general properties of linear maps; in this framework a fundamental tool is the state transition matrix, which is herein defined and analyzed.

2.1.1 Linear Time-Varying Continuous Systems

Consider the linear time-varying system

$$\dot{x}(t) = A(t)\, x(t) \qquad (1)$$

where $x \in \mathcal{F}^n$ ($\mathcal{F} := \mathbf{R}$ or $\mathcal{F} := \mathbf{C}$) and $A(t)$ is an $n \times n$ matrix of piecewise continuous functions of time with values in $\mathcal{F}$. On the assumption that the real-valued function $k(t) := \|A(t)\|$ is bounded and piecewise continuous, Theorem A.6-4 ensures the existence of a unique solution of (1) such that $x(t_0) = x_0$ for all $x_0 \in \mathcal{F}^n$ and all $t \in \mathbf{R}$.

Sect. 2.1 The Free State Evolution of Linear Systems

Note that:

1. the set of all solutions of (1) is a vector space: in fact, given any two solutions $x_1(\cdot)$, $x_2(\cdot)$ of (1), $\alpha_1 x_1(\cdot) + \alpha_2 x_2(\cdot)$ is also a solution of (1) for all $\alpha_1, \alpha_2 \in \mathcal{F}$;

2. the zero function $x(\cdot) = 0$ is a solution of (1): it is called the *trivial solution*; due to uniqueness, no other solution can vanish at any instant of time.

The state transition matrix is a fundamental tool used to achieve a better insight into the structure of the vector space of the solutions of (1).

Definition 2.1-1. State Transition Matrix

Let $\varphi_i(\cdot)$ $(i=1,\ldots,n)$ be the n solutions of (1) with initial conditions $\varphi_i(t_0) = e_i$ $(i=1,\ldots,n)$, where e_i denotes the i-th vector of the main basis of $\mathcal{F}^n$, i.e., the i-th column of the $n \times n$ identity matrix. The matrix $\Phi(\cdot, t_0)$ having the functions $\varphi_i(\cdot)$ as columns is called the *state transition matrix* of system (1).

In other words, the state transition matrix is the solution of the matrix differential equation [1]

$$\dot{X}(t) = A(t) X(t) \qquad (2)$$

with initial condition $X(t_0) = I$. In (2) $X(t)$ denotes an $n \times n$ matrix with elements in $\mathcal{F}$.

A basic property of the state transition matrix is set in the following theorem.

Theorem 2.1-1.

State transition matrix $\Phi(t, t_0)$ is nonsingular for all $t, t_0 \in \mathbf{R}$, $t \geq t_0$.[2]

Proof. Denote, as before, by $\varphi_i(t)$ $(i = 1,\ldots,n)$ the n columns of the state transition matrix. By contradiction, assume that for a particular t the equality

$$\alpha_1 \varphi_1(t) + \ldots + \alpha_n \varphi_n(t) = 0$$

holds with the α_i $(i=1,\ldots,n)$ not all zero; since the right side is a solution of (1), due to uniqueness it must be the trivial solution, so that

$$\alpha_1 \varphi_1(t_0) + \ldots + \alpha_n \varphi_n(t_0) = \alpha_1 e_1 + \ldots + \alpha_n e_n = 0$$

which is impossible, since vectors e_i $(i=1,\ldots,n)$ are a linearly independent set. □

---

[1] Time derivatives or time integrals of any matrix $X(t)$ of functions of time are the matrices whose elements are the time derivatives or time integrals of the elements of $X(t)$.

[2] The statement also holds for $t < t_0$; this extension of the state transition matrix will be discussed a little later.

Corollary 2.1-1.
The set of all solutions of (1) is an n-dimensional vector space over $\mathcal{F}$.

Proof. Consider any solution $x(\cdot)$ of (1) and denote by x_1 its value at any instant of time $t_1 \geq t_0$. Since $\Phi(t_1, t_0)$ is nonsingular, there exists a vector $a \in \mathcal{F}^n$ such that
$$x_1 = \Phi(t_1, t_0) a \qquad (3)$$
hence, due to uniqueness
$$x(\cdot) = \Phi(\cdot, t_0) a$$
This means that the n columns of $\Phi(\cdot, t_0)$ are a basis for the vector space of all solutions of (1). □

Another basic property of the state transition matrix can be derived from the previous argument. Since $\Phi(t_0, t_0) = I$, relation (3) shows that a is the value $x(t_0)$ of function $x(\cdot)$ at t_0. Hence (3) can be rewritten as
$$x(t_1) = \Phi(t_1, t_0) x(t_0) \qquad (4)$$
or
$$x(t_0) = \Phi^{-1}(t_1, t_0) x(t_1)$$
Clearly, state transition matrix $\Phi(t_1, t_0)$ represents the transformation of the initial state $x(t_0)$ at time t_0 into the state $x(t_1)$ at time t_1 performed by the differential equation (1). Being nonsingular, it can be also used to solve the inverse problem, i.e., to derive the state at time t_0 to which a given state at t_1 corresponds in the relative solution of (1). In other words (4) is also consistent when $t_1 < t_0$.

The state transition matrix satisfies:
1. inversion:
$$\Phi(t, t_0) = \Phi^{-1}(t_0, t) \qquad (5)$$

2. composition:
$$\Phi(t, t_0) = \Phi(t, t_1) \Phi(t_1, t_0) \qquad (6)$$

3. separation:
$$\Phi(t, t_0) = \Theta(t) \Theta^{-1}(t_0) \qquad (7)$$

4. time evolution of the determinant:
$$\det \Phi(t, t_0) = e^{\int_{t_0}^{t} \operatorname{tr} A(\tau) \, d\tau} \qquad (8)$$

where $\operatorname{tr} A$ denotes the trace of matrix A (the sum of all the elements on the main diagonal).

Note that (7) can be obtained from (5) by setting for instance $\Theta(t) := \Phi(t, 0)$ or $\Theta(t) := \Phi(t, t_0)$ for any t_0;

Sect. 2.1 The Free State Evolution of Linear Systems

Proof of 4. The time derivative of $\det \Phi(t, t_0)$ is the sum of the determinants of all the matrices obtained by substituting the elements of a row (column) of $\Phi(t, t_0)$ with their time derivatives. For instance, the first element of this sum is the determinant of

$$\begin{bmatrix} \dot{\varphi}_{11}(t) & \dot{\varphi}_{12}(t) & \cdots & \dot{\varphi}_{1n}(t) \\ \varphi_{21}(t) & \varphi_{22}(t) & \cdots & \varphi_{2n}(t) \\ \vdots & \vdots & \ddots & \vdots \\ \varphi_{n1}(t) & \varphi_{n2}(t) & \cdots & \varphi_{nn}(t) \end{bmatrix} \quad (9)$$

Since the state transition matrix satisfies equation (2), it follows that

$$\dot{\varphi}_{1i}(t) = \sum_{j=1}^{n} a_{1j}(t) \varphi_{ji}(t) \quad (i = 1, \ldots, n) \quad (10)$$

By replacing the first row of (9) with (10) and recalling some properties of the determinants, such as linearity with respect to any row and vanishing when any row is a linear combination of other rows, it is easily seen that the determinant of (9) is $a_{11}(t) \det \Phi(t, t_0)$. By taking into account all terms of the sum, it follows that

$$\frac{d}{dt} \det \Phi(t, t_0) = \operatorname{tr} A(t) \det \Phi(t, t_0)$$

This scalar differential equation together with the initial condition $\det \Phi(t_0, t_0) = 1$ clearly implies (8). □

The following properties are consequences of the argument that proves Theorem A.6-4; in particular, Property 2.1-2 directly follows from Corollary A.6-1.

Property 2.1-1.

The Peano-Baker sequence

$$\Phi_o(t, t_0) = I \; ,$$
$$\Phi_i(t, t_0) = I + \int_{t_0}^{t} A(\tau) \Phi_{i-1}(\tau, t_0) \, d\tau \quad (i = 1, 2, \ldots) \quad (11)$$

converges uniformly to the state transition matrix $\Phi(t, t_0)$.

Property 2.1-2.

The elements of state transition matrix $\Phi(t, t_0)$ are continuous functions of time.

Property 2.1-1 suggests an iterative procedure to compute the state transition matrix of time-varying systems.

Definition 2.1-2. (Continuous-Time) Adjoint System

The linear time-varying system

$$\dot{p}(t) = -A^T(t) p(t) \quad (A(\cdot) \text{ real}) \quad (12)$$

or

$$\dot{p}(t) = -A^*(t) p(t) \quad (A(\cdot) \text{ complex}) \quad (13)$$

is called the *adjoint system* of system (1).

Property 2.1-3.
The inner product of a solution $x(t)$ of (1) and a solution $p(t)$ of (12) or (13) is a constant.

Proof. Consider the case of $A(\cdot)$ being real and set the equalities

$$\frac{d}{dt}\langle x(t), p(t)\rangle = \langle \dot{x}(t), p(t)\rangle + \langle x(t), \dot{p}(t)\rangle$$
$$= \langle A(t)x(t), p(t)\rangle + \langle x(t), -A^T(t)p(t)\rangle$$
$$= \langle A(t)x(t), p(t)\rangle - \langle A(t)x(t), p(t)\rangle = 0$$

Since the components of $x(t)$ and $p(t)$ are continuous functions, it follows that $\langle x(t), p(t)\rangle$ is a constant. The same argument with obvious changes applies when $A(\cdot)$ is complex. □

Property 2.1-4.
Let $\Phi(t,\tau)$ be the state transition matrix of system (1), and $\Psi(t,\tau)$ that of the adjoint system (12) or (13). Then

$$\Psi^T(t,\tau)\Phi(t,\tau) = I \quad (A(\cdot) \text{ real}) \tag{14}$$

or

$$\Psi^*(t,\tau)\Phi(t,\tau) = I \quad (A(\cdot) \text{ complex}) \tag{15}$$

Proof. Let $A(\cdot)$ be real. Note that for any τ all elements of matrix $\Psi^T(t,\tau)\Phi(t,\tau)$ are left inner products of a solution of (1) by a solution of (12); hence the matrix is constant. On the other hand, $\Psi^T(\tau,\tau) = \Phi(\tau,\tau) = I$; hence this constant matrix is the identity matrix I. The conjugate transposes substitute the transpose matrices if $A(\cdot)$ is complex. □

2.1.2 Linear Time-Varying Discrete Systems

The previous arguments will now be extended to the case of discrete-time systems. Consider the linear time-varying homogeneous difference system

$$x(i+1) = A_d(i)\, x(i) \tag{16}$$

and apply a procedure similar to the aforementioned in order to derive the state transition matrix. Instead of vector equation (16), refer to the corresponding matrix equation

$$X(i+1) = A_d(i)\, X(i) \tag{17}$$

where matrices of sequence $X(i)$ are assumed to be square. Solution of (17) with initial condition $X(j) = I$ is the state transition matrix $\Phi(i,j)$ of system (16).

Unlike continuous-time systems, state transition matrix $\Phi(i,j)$ may be singular in this case: this happens if and only if $A_d(k)$ is singular for at least one value of k such that $j \leq k \leq i-1$. However, if equation (16) has been obtained from (1) by means of a sampling process, the corresponding state transition matrix is nonsingular for all i, j.

The discrete-time state transition matrix satisfies the following properties:
1. inversion:
$$\Phi(i,j) = \Phi^{-1}(j,i) \tag{18}$$
2. composition:
$$\Phi(i,j) = \Phi(i,k)\Phi(k,j) \tag{19}$$
3. separation (if for at least one k, $\Theta(j) := \Phi(k,j)$ is nonsingular for all j):
$$\Phi(i,j) = \Theta(i)\Theta^{-1}(j) \tag{20}$$
4. time evolution of the determinant:
$$\det \Phi(i,j) = \prod_{k=j}^{i-1} \det A_d(k) \tag{21}$$

Also, the adjoint system concept and related properties can easily be extended as follows. Proofs are omitted, since they are trivial extensions of those for the continuous-time case.

Definition 2.1-3. (Discrete-Time) Adjoint System
The linear time-varying system
$$p(i-1) = -A_d^T(i-1)p(i) \quad (A(\cdot) \text{ real}) \tag{22}$$
or
$$p(i-1) = -A_d^*(i-1)p(i) \quad (A(\cdot) \text{ complex}) \tag{23}$$
if $A(\cdot)$ is complex, is called the *adjoint system* of system (16).

Property 2.1-5.
The inner product of a solution $x(i)$ of (16) and a solution $p(i)$ of (22) or (23) is a constant.

Property 2.1-6.
Let $\Phi(i,j)$ be the state transition matrix of system (1) and $\Psi(i,j)$ that of the adjoint system (22) or (23). Then
$$\Psi^T(i,j)\Phi(i,j) = I \quad (A(\cdot) \text{ real}) \tag{24}$$
or
$$\Psi^*(i,j)\Phi(i,j) = I \quad (A(\cdot) \text{ complex}) \tag{25}$$

2.1.3 Function of a Matrix

Consider a function $f : \mathcal{F} \to \mathcal{F}$ ($\mathcal{F} := \mathbf{R}$ or $\mathcal{F} := \mathbf{C}$) that can be expressed as an infinite series of powers, i.e.,

$$f(x) = \sum_{i=0}^{\infty} c_i x^i \qquad (26)$$

The argument of such a function can be extended to become a matrix instead of a scalar through the following definition.

Definition 2.1-4. Function of a Matrix

Let A be an $n \times n$ matrix with elements in $\mathcal{F}$ and f a function that can be expressed by power series (26); *function $f(A)$ of matrix A* is defined by

$$f(A) := \sum_{i=0}^{\infty} c_i A^i \qquad (27)$$

Note that, according to this definition, A and $f(A)$ commute, i.e., $A f(A) = f(A) A$. The infinite series (27) can be expressed in finite terms by applying one of the following procedures.

The Interpolating Polynomial Method. Let $m(\lambda)$ be the minimal polynomial (monic) of A and $\alpha_0, \ldots, \alpha_{m-1}$ its coefficients, so that

$$A^{m+k} = -\left(\alpha_{m-1} A^{m+k-1} + \alpha_{m-2} A^{m+k-2} + \ldots + \alpha_0 A^k\right) \quad (k = 0, 1, \ldots)$$

hence it is possible to express any power of A equal to or higher than m in the right side of (27) as a linear combination of lower powers of A, so that by collection of the common factors,

$$f(A) = \sum_{i=0}^{m-1} \gamma_i A^i \qquad (28)$$

This means that any function of a matrix $f(A)$ can be expressed as a polynomial with degree not greater than that of the minimal polynomial of A. Let $\varphi(\lambda)$ and $\psi(\lambda)$ be any pair of polynomials such that $f(A) = \varphi(A) = \psi(A)$ or $\varphi(A) - \psi(A) = O$. Thus, the minimal polynomial $m(\lambda)$ divides $\varphi(\lambda) - \psi(\lambda)$, i.e., there exists a polynomial $q(\lambda)$ such that

$$\varphi(\lambda) - \psi(\lambda) = m(\lambda) q(\lambda)$$

Consider the eigenvalues of A ($\lambda_1, \ldots, \lambda_h$), which are roots of the minimal polynomial, and denote by $m_1, \ldots, m_h$ their multiplicities in $m(\lambda)$. Since

$$m(\lambda_i) = m'(\lambda_i) = \ldots = m^{(m_i-1)}(\lambda_i) = 0 \quad (i = 1, \ldots, h)$$

Sect. 2.1 The Free State Evolution of Linear Systems

it follows that
$$\varphi(\lambda_i) = \psi(\lambda_i)$$
$$\varphi'(\lambda_i) = \psi'(\lambda_i)$$
$$\cdots\cdots\cdots$$
$$\varphi^{(m_i-1)}(\lambda_i) = \psi^{(m_i-1)}(\lambda_i) \quad (i = 1,\ldots, h)$$

We conclude that all the polynomials equal to $f(A)$ and their derivatives up to the (m_i-1)-th assume the same values over spectrum $\{\lambda_i\ (i=1,\ldots,n)\}$ of A. Let

$$\varphi(\lambda) := \sum_{i=0}^{m-1} \gamma_i \lambda^i \tag{29}$$

be the polynomial at the right side of (28). Since the minimal polynomial and its derivatives are zero at λ_i ($i = 1,\ldots,h$), by the same argument used to derive (28) from (27) (i.e., by direct substitution into the infinite series (26) and its derivatives) it follows that

$$f(\lambda_i) = \sum_{k=0}^{m-1} \gamma_k \lambda_i^k = \varphi(\lambda_i) \quad (i = 1,\ldots,h)$$
$$f'(\lambda_i) = \varphi'(\lambda_i) \quad (i=1,\ldots,h)$$
$$\cdots\cdots\cdots$$
$$f^{(m_i-1)}(\lambda_i) = \varphi^{(m_i-1)}(\lambda_i) \quad (i = 1,\ldots,h) \tag{30}$$

Coefficients γ_i ($i = 1,\ldots,m-1$) can easily be obtained from (30). In fact, by substituting (29) into (30) we get a set of m linear equations that can be written in compact form as

$$V\gamma = v \tag{31}$$

V denotes an $m \times m$ matrix that can be partitioned by rows into $m_i \times m$ matrices V_i ($i=1,\ldots,h$) defined as

$$V_i := \begin{bmatrix} 1 & \lambda_i & \lambda_i^2 & \cdots & \lambda_i^{m-1} \\ 0 & 1 & 2\lambda_i & \cdots & (m-1)\lambda_i^{m-2} \\ 0 & 0 & 2 & \cdots & \frac{(m-1)!}{(m-3)!}\lambda_i^{m-3} \\ \vdots & \vdots & \vdots & \ddots & \vdots \\ 0 & 0 & 0 & \cdots & \frac{(m-1)!}{(m-m_i)!}\lambda_i^{m-m_i} \end{bmatrix}$$

In (31) $\gamma \in \mathbf{R}^m$ denotes the vector having coefficients γ_i ($i = 0,\ldots,m-1$) as components and v is defined as

$$v := \Big(f(\lambda_1),\ f'(\lambda_1),\ \ldots,\ f^{(m_1-1)}(\lambda_1),\ \ldots,$$
$$f(\lambda_h),\ f'(\lambda_h),\ \ldots,\ f^{(m_h-1)}(\lambda_h)\Big) \tag{32}$$

Each row of V_i is the derivative of the previous one with respect to λ_i. The element in row i and column j has coefficient zero for $j < i$ and $(j-1)!/(j-i)!$ for $j \geq i$. Matrix V is nonsingular; in fact, if during computation of each submatrix V_i we divide the general row i ($i > 1$) by i before differentiating it, we obtain a matrix V' such that

$$\det V = k \det V'$$

where k is the product of the integers by which the rows of V have been divided. V' has a structure similar to V; it consists of blocks like

$$V_i' = \begin{bmatrix} 1 & \lambda_i & \lambda_i^2 & \cdots & \lambda_i^{m-1} \\ 0 & 1 & 2\lambda_i & \cdots & (m-1)\lambda_i^{m-2} \\ 0 & 0 & 1 & \cdots & \frac{(m-1)!}{2(m-3)!}\lambda_i^{m-3} \\ \vdots & \vdots & \vdots & \ddots & \vdots \\ 0 & 0 & 0 & \cdots & \frac{(m-1)!}{(m_i-1)!(m-m_i)!}\lambda_i^{m-m_i} \end{bmatrix}$$

where the general coefficient belonging to row i and column j is zero for $j < i$ and $(j-1)!/((i-1)!(j-i)!)$ for $j \geq i$; this particular structure corresponds to the transpose of a generalized Vandermonde matrix, whose determinant is given by

$$\det V' = \prod_{1 \leq i < j \leq h} (\lambda_j - \lambda_i)^{m_j m_i}$$

hence it is different from zero, as the eigenvalues λ_i ($i = 1, \ldots, h$) are assumed to be noncoincident. V being nonsingular, vector γ is finally derived as $\gamma = V^{-1} v$. Note that any element of $f(A)$ is a linear function of v, i.e.

$$\big(f(A)\big)_{ij} = \langle k_{ij}(A), v \rangle$$

where vectors $k_{ij} \in \mathbf{C}^m$ ($i, j = 1, \ldots, m$) depend only on A, while v depends both on A and f.

The Maclaurin Expansion and the Jordan Form. Let function f be analytic at the origin. Consider, as a particular case of (27), the Maclaurin expansion

$$f(A) := \sum_{i=0}^{\infty} \frac{f^{(i)}(x)}{i!}\bigg|_{x=0} A^i \tag{33}$$

Denote by B the Jordan form of A, expressed by (A.4.11). From

$$B = T^{-1} A T$$

it follows that

$$B^i = T^{-1} A^i T \quad \forall i \in \mathbf{N}$$

Sect. 2.1 The Free State Evolution of Linear Systems

hence

$$f(B) = T^{-1} f(A) T \quad \text{or} \quad f(A) = T f(B) T^{-1}$$

Function $f(B)$ is easily derived, because of the particular structure of the Jordan form, shown in (A.4.9). In fact

$$f(B) = \begin{bmatrix} f(B_{11}) & O & \cdots & O & \cdots & O \\ O & f(B_{12}) & \cdots & O & \cdots & O \\ \vdots & \vdots & \ddots & \vdots & \ddots & \vdots \\ O & O & \cdots & f(B_{1,k_1}) & \cdots & 0 \\ \vdots & \vdots & \ddots & \vdots & \ddots & \vdots \\ O & O & \cdots & O & \cdots & f(B_{h,k_h}) \end{bmatrix} \quad (34)$$

while the function of a single $\ell \times \ell$ Jordan block is obtained from series (33) as

$$f(B_{ij}) = \begin{bmatrix} f(\lambda_i) & f'(\lambda_i) & \tfrac{1}{2} f''(\lambda_i) & \cdots & \tfrac{1}{(\ell-1)!} f^{(\ell-1)}(\lambda_i) \\ 0 & f(\lambda_i) & f'(\lambda_i) & \cdots & \tfrac{1}{(\ell-2)!} f^{(\ell-2)}(\lambda_i) \\ 0 & 0 & f(\lambda_i) & \cdots & \tfrac{1}{(\ell-3)!} f^{(\ell-3)}(\lambda_i) \\ \vdots & \vdots & \vdots & \ddots & \vdots \\ 0 & 0 & 0 & \cdots & f(\lambda_i) \end{bmatrix} \quad (35)$$

where $f^{(k)}(\lambda_i)$ denotes the k-th derivative of $f(x)$ at $x = \lambda_i$.

2.1.4 Linear Time-Invariant Continuous Systems

For stationary systems the concepts just presented assume a simpler form. Furthermore, the computational support needed for their use in engineering design framework is quite standard, well checked, and reliable.

Consider the time-invariant or constant linear homogeneous system

$$\dot{x}(t) = A\, x(t) \quad (36)$$

where A denotes a real or complex $n \times n$ matrix. Since A is constant, it is customary in this case to assume $t_0 = 0$. As before, denote the state transition matrix by $\Phi(t, 0)$ and consider the successive approximations method (11):

$$\Phi_0(t, 0) = I$$

$$\Phi_i(t, 0) = I + At + \frac{A^2 t^2}{2} + \ldots + \frac{A^i t^i}{i!} \quad (i = 1, 2, \ldots)$$

from which it follows that

$$\Phi(t, 0) = \lim_{i \to \infty} \Phi_i(t, 0) = \sum_{i=0}^{\infty} \frac{A^i t^i}{i!} = e^{At} \quad (37)$$

where the last equality is a consequence of Definition 2.1-4 of a function of a matrix.

Therefore, the state transition matrix of a constant system is the *matrix exponential*. As in the scalar case, the matrix exponential satisfies

$$e^{A(t+\tau)} = e^{At} e^{A\tau} \tag{38}$$

which is an immediate consequence of the composition property of the state transition matrix. On the other hand, in general

$$e^{(A+B)t} \neq e^{At} e^{Bt} \tag{39}$$

In fact, consider the expansions

$$e^{(A+B)t} = I + (A+B)t + \frac{(A+B)^2 t^2}{2} + \dots$$

and

$$e^{At} e^{Bt} = \left(I + At + \frac{A^2 t^2}{2} + \dots\right)\left(I + Bt + \frac{B^2 t^2}{2} + \dots\right)$$
$$= I + (A+B)t + \frac{A^2 t^2}{2} + ABt^2 + \frac{B^2 t^2}{2} + \dots$$

By subtraction we derive

$$e^{(A+B)t} - e^{At} e^{Bt} = (BA - AB)\frac{t^2}{2} + \dots$$

By continuing this expansion, it is easily seen that the right side vanishes if and only if $AB = BA$, i.e., (39) holds with the equality sign if and only if A and B commute.

Some methods for computation of the matrix exponential are now presented.

The Power Series Expansion. The most natural way to compute the matrix exponential is to use the definition formula

$$e^{At} = \sum_{i=0}^{\infty} \frac{A^i t^i}{i!} \tag{40}$$

The series on the right side of (40) for any finite t converges to a matrix having finite norm. In fact, let $m := \|A\|$; since $\|A^i\| \leq m^i$ for all $i \in \mathbb{N}$, it follows that

$$\left\|\sum_{i=0}^{\infty} \frac{A^i t^i}{i!}\right\| \leq \sum_{i=0}^{\infty} \frac{\|A^i\| |t^i|}{i!} \leq \sum_{i=0}^{\infty} \frac{m^i |t^i|}{i!} = e^{m|t|}$$

Sect. 2.1 The Free State Evolution of Linear Systems

Hence (40) can be used for computational purposes; however, it requires many multiplications and involves a truncation error that greatly depends on the properties of matrix A. A preliminary scaling is often introduced to improve accuracy. For instance, repeatedly divide matrix A by 2 (say q times) until condition $\|At\| < 1/2$ is met. Then apply expansion (40) until the difference in norm of two consecutive partial sums is equal to a machine zero and perform the inverse scaling by squaring q times the obtained result.[3]

Use of the Jordan Form. In this case matrix (34) becomes

$$e^{Bt} = \begin{bmatrix} e^{B_{11}t} & O & \cdots & O & \cdots & O \\ O & e^{B_{12}t} & \cdots & O & \cdots & O \\ \vdots & \vdots & \ddots & \vdots & \ddots & \vdots \\ O & O & \cdots & e^{B_{1,k_1}t} & \cdots & 0 \\ \vdots & \vdots & \ddots & \vdots & \ddots & \vdots \\ O & O & \cdots & O & \cdots & e^{B_{h,k_h}t} \end{bmatrix} \quad (41)$$

while the exponential of a single $\ell \times \ell$ Jordan block is the following particularization of (35):

$$e^{B_{ij}t} = \begin{bmatrix} e^{\lambda_i t} & t e^{\lambda_i t} & \frac{t^2}{2} e^{\lambda_i t} & \cdots & \frac{t^{\ell-1}}{(\ell-1)!} e^{\lambda_i t} \\ 0 & e^{\lambda_i t} & t e^{\lambda_i t} & \cdots & \frac{t^{\ell-2}}{(\ell-2)!} e^{\lambda_i t} \\ 0 & 0 & e^{\lambda_i t} & \cdots & \frac{t^{\ell-3}}{(\ell-3)!} e^{\lambda_i t} \\ \vdots & \vdots & \vdots & \ddots & \vdots \\ 0 & 0 & 0 & \cdots & e^{\lambda_i t} \end{bmatrix} \quad (42)$$

The main drawback of this procedure is that the derivation of transformation T is quite laborious and subject to ill-conditioning effects.

The Interpolating Polynomial Method. By using the procedure considered in Subsection 2.1.3, the general element of the exponential matrix can be derived as

$$\left(e^{At}\right)_{ij} = \langle k_{ij}(A), v \rangle \quad (43)$$

where both $k_{ij}(A)$ and v belong to $\mathbb{C}^m$ (m denotes the degree of the minimal polynomial of A). Vector v is

$$v := \left(e^{\lambda_1 t}, t e^{\lambda_1 t}, \ldots, t^{m_1 - 1} e^{\lambda_1 t}, \ldots, e^{\lambda_h t}, t e^{\lambda_h t}, \ldots, t^{m_h - 1} e^{\lambda_h t}\right) \quad (44)$$

Use of the Schur Form. It is shown in Section A.4 that for any real or complex $n \times n$ matrix A there exists a unitary similarity transformation U such that

$$B = U^* A U = \begin{bmatrix} \lambda_1 & b_{12} & \cdots & b_{1n} \\ 0 & \lambda_2 & \cdots & b_{2n} \\ \vdots & \vdots & \ddots & \vdots \\ 0 & 0 & \cdots & \lambda_n \end{bmatrix} \quad (45)$$

---

[3] See Golub and Van Loan [B.3], p. 558.

Matrix exponential e^{Bt} can be computed column by column as the solution of the n differential equations

$$\dot{z}_i(t) = B\, z_i(t)\,, \quad z_i(0) = e_i \qquad (i = 1,\ldots,n) \tag{46}$$

where e_i denotes the i-th vector of the main basis of $\mathbf{C}^n$.

Note that, in particular, all the components of solution $z_i(t)$ with an index greater than i are equal to zero, while the i-th component is the complex exponential $e^{\lambda_i t}$. Due to the particular structure of B, solutions of (46) are easily obtained by substitution of the components of each vector $z_i(t)$, starting from the i-th, into the scalar differential equations corresponding to the previous components.

It will be shown that every nonzero component of $z_i(t)$ is a linear combination of exponential terms like

$$f_{rj}(t) = t^r\, e^{\lambda_j t} \tag{47}$$

where exponent r at most equals $m_i - 1$, the multiplicity of eigenvalue λ_i less one. Since superposition holds, the problem reduces to the solution of some scalar differential equations of the general type

$$\dot{z}(t) = \lambda_i\, z(t) + b\, f_{rj}(t)\,, \quad z(0) = 0 \tag{48}$$

Two cases are possible.

1. $\lambda_i = \lambda_j$. The solution of (48) is

$$z(t) = \frac{b\, t^{r+1}}{r+1}\, e^{\lambda_i t} \tag{49}$$

since

$$z(t) = \int_0^t e^{\lambda_i(t-\tau)} b\, f_{rj}(\tau)\, d\tau = b\, e^{\lambda_i t} \int_0^t \tau^r\, d\tau = \frac{b\, t^{r+1}}{r+1}\, e^{\lambda_i t}$$

2. $\lambda_i \neq \lambda_j$. In this case $z(t)$ is computed by considering the sequence of functions $w_\ell(t)$ $(\ell=0,\ldots,r)$ defined below, which are the solutions of (48) with forcing terms $f_{\ell j}(t)$ $(\ell=0,\ldots,r)$. The solution is $z(t)=w_r(t)$, with

$$w_0(t) = b\, \frac{e^{\lambda_i t} - e^{\lambda_j t}}{\lambda_i - \lambda_j}$$

$$w_\ell(t) = \frac{1}{\lambda_i - \lambda_j}\left(\ell\, w_{\ell-1}(t) - b\, t^\ell\, e^{\lambda_j t}\right) \quad (\ell=1,\ldots,r) \tag{50}$$

The first of (50) is derived immediately. In order to prove the subsequent ones, consider

$$w_\ell(t) = \int_0^t e^{\lambda_i(t-\tau)} b\, \tau^\ell\, e^{\lambda_j \tau}\, d\tau = b\, e^{\lambda_i t} \int_0^t e^{(\lambda_j - \lambda_i)\tau}\, \tau^\ell\, d\tau$$

Integrating by parts with τ^ℓ as finite factor and $e^{(\lambda_j - \lambda_i)\tau}\, d\tau$ as differential factor

$$w_\ell(t) = b\, e^{\lambda_i t} \left(\frac{\tau^\ell e^{(\lambda_j - \lambda_i)\tau}}{\lambda_j - \lambda_i} \bigg|_0^t - \frac{\ell}{\lambda_j - \lambda_i} \int_0^t \tau^{\ell-1}\, e^{(\lambda_j - \lambda_i)\tau}\, d\tau \right)$$

$$= \frac{1}{\lambda_j - \lambda_i} \left(b\, t^\ell\, e^{\lambda_j t} - \ell \int_0^t e^{\lambda_i(t-\tau)}\, b\, \tau^{\ell-1}\, e^{\lambda_j \tau}\, d\tau \right)$$

Sect. 2.1 The Free State Evolution of Linear Systems

The previous computational methods point out an important property: all elements of the matrix exponential e^{At} are linear combinations with constant complex coefficients of the time functions which appear as the components of vector v in (44). These functions, whose general expressions are also given at the right side of (47), are called *modes* of system (36).

Modes expressed by $e^{\lambda t}$ with real λ are shown in Fig. 2-1(a–c): the three cases correspond respectively to λ positive, negative, and zero; modes expressed by $t^r e^{\lambda t}$, again with real λ positive, negative, and zero, are shown in Fig. 2-1(d–f).

If A is real it is convenient to sum the modes corresponding to pairs of complex conjugate eigenvalues that are linearly combined with complex conjugate coefficients in every element of the matrix exponential. A real function is obtained by means of the following standard procedure. Consider the sum

$$S := h\, t^r\, e^{\lambda t} + h^*\, t^r\, e^{\lambda^* t} \tag{51}$$

and denote by u, v, and σ, ω the real and imaginary parts of h and λ respectively, so that

$$S = t^r \left((u + jv) e^{(\sigma + j\omega)t} + (u - jv) e^{(\sigma - j\omega)t} \right)$$

By defining

$$m := 2|h|, \quad \varphi := \arg h$$

it follows that

$$S = \frac{m}{2} t^r e^{\sigma t} \left(e^{j(\omega t + \varphi)} + e^{-j(\omega t + \varphi)} \right)$$
$$= m\, t^r e^{\sigma t} \cos(\omega t + \varphi)$$
$$= m\, t^r e^{\sigma t} \sin\left(\omega t + \varphi + \frac{\pi}{2}\right)$$

When r is zero the resulting time function is typically one of those shown in Fig. 2-2(a–c) (respectively for λ positive, negative, and zero), while for r different from zero, for instance $r=2$, it is one of those shown in Fig. 2-2(d–f).

The same result could have been obtained by considering the "real" Jordan form whose general block is of type (A.4.26): denote by σ, ω the real and imaginary part of λ and define

$$C := \begin{bmatrix} \sigma & \omega \\ -\omega & \sigma \end{bmatrix} \tag{52}$$

so that

$$e^{Ct} = \begin{bmatrix} e^{\sigma t} \cos \omega t & e^{\sigma t} \sin \omega t \\ -e^{\sigma t} \sin \omega t & e^{\sigma t} \cos \omega t \end{bmatrix} \tag{53}$$

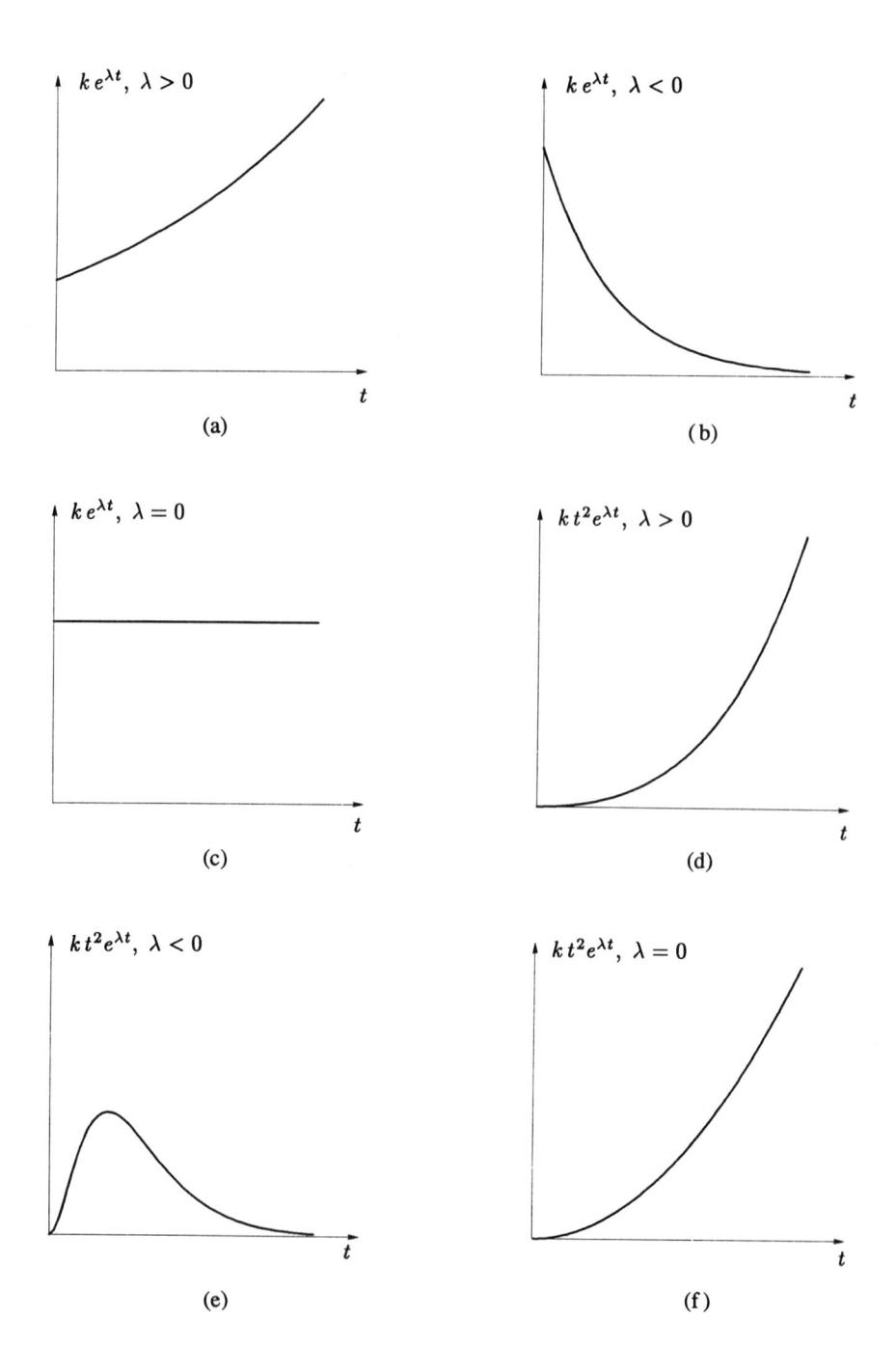

Figure 2-1 Modes corresponding to real eigenvalues.

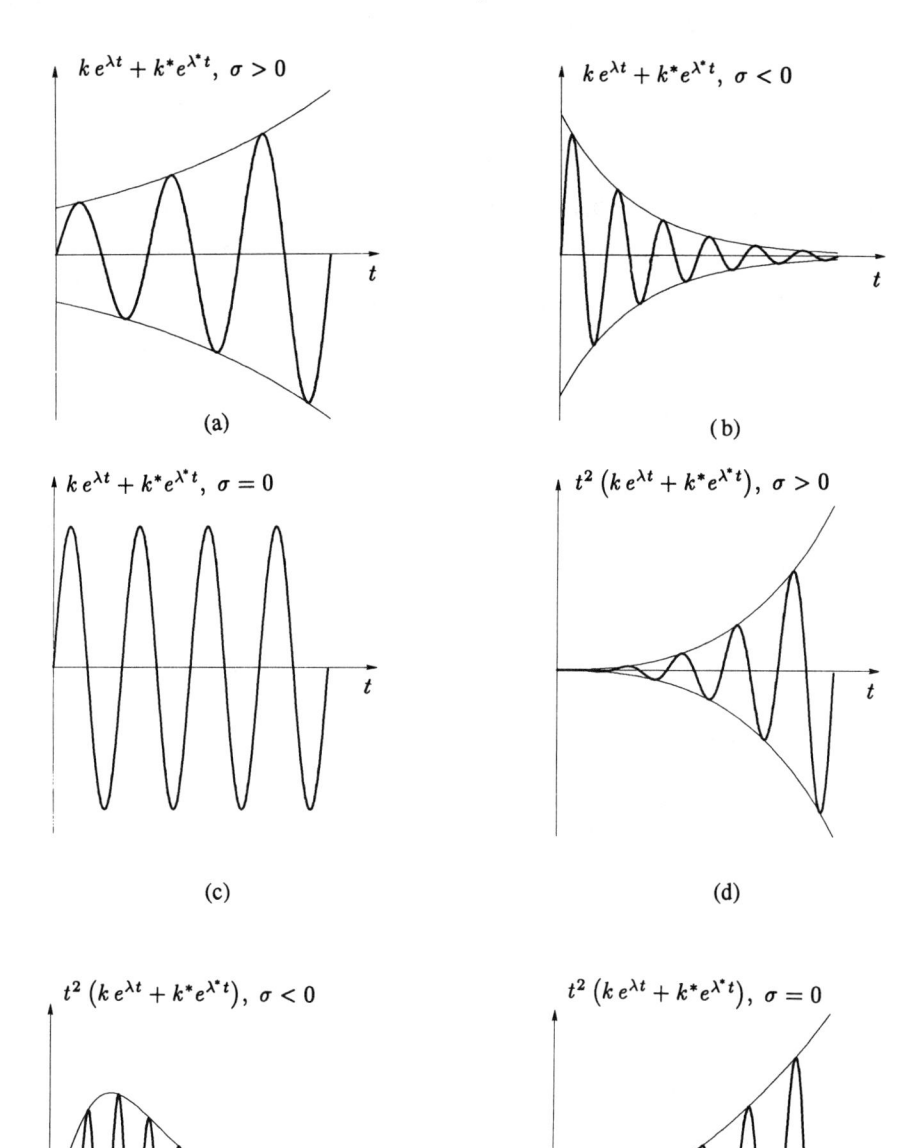

Sect. 2.1 The Free State Evolution of Linear Systems

(a) $ke^{\lambda t} + k^*e^{\lambda^* t}$, $\sigma > 0$

(b) $ke^{\lambda t} + k^*e^{\lambda^* t}$, $\sigma < 0$

(c) $ke^{\lambda t} + k^*e^{\lambda^* t}$, $\sigma = 0$

(d) $t^2\left(ke^{\lambda t} + k^*e^{\lambda^* t}\right)$, $\sigma > 0$

(e) $t^2\left(ke^{\lambda t} + k^*e^{\lambda^* t}\right)$, $\sigma < 0$

(f) $t^2\left(ke^{\lambda t} + k^*e^{\lambda^* t}\right)$, $\sigma = 0$

Figure 2-2 Modes corresponding to pairs of complex conjugate eigenvalues.

as can be checked by applying the identity

$$\frac{d}{dt}e^{Ct} = C\, e^{Ct}$$

It is easily seen that the exponential of a "real" Jordan block of type (A.4.26) has the same structure as matrix (42), but with 2×2 matrices $e^{C_i t}$ instead of scalars $e^{\lambda_i t}$ in the terms on and above the main diagonal, 2×2 null matrices instead of the scalar 0's below the main diagonal.

It is now possible to draw some interesting conclusions: the general mode (47) corresponds to a function of time whose behavior as time approaches infinity is one of the following:

1. It converges to zero if the real part of the corresponding eigenvalue is negative;
2. It remains bounded if r is zero and the real part of the corresponding eigenvalue is zero;
3. It diverges if the real part of the corresponding eigenvalue is positive or if r is different from zero and the real part of the eigenvalue is zero.

The above modes are called respectively *asymptotically* or *strictly stable*, (merely) *stable* and *unstable*.

2.1.5 Linear Time-Invariant Discrete Systems

All the previous arguments will be briefly extended to discrete-time systems. Consider the discrete-time constant homogeneous system

$$x(i+1) = A_d\, x(i) \qquad (54)$$

whose state transition matrix is clearly

$$\Phi(i,j) = A_d^{i-j} \qquad (55)$$

i.e., reduces to the *power of a matrix*.

The four procedures to compute the state transition matrix of continuous-time systems can also be profitably used in the discrete-time case.

The Direct Method. Repeating the matrix multiplication many times may involve significant errors due to truncation and great computation time. To minimize these drawbacks, it is convenient to use the *binary powering method*:[4] expand exponent k in binary form as

$$k = \sum_{i=0}^{n} \beta_i\, 2^i$$

---

[4] See Golub and Van Loan [B.3], p. 552.

Sect. 2.1 The Free State Evolution of Linear Systems

then initialize $i \leftarrow 0$, $Z \leftarrow A_d$, $B \leftarrow I$ if $\beta_0 = 0$ or $B \leftarrow A_d$ if $\beta_0 = 1$ and, until $i = n$, compute $i \leftarrow i+1$, $Z \leftarrow Z^2$, $B \leftarrow B$ if $\beta_i = 0$ or $B \leftarrow ZB$ if $\beta_0 = 1$. At the end, the result $B = A_d^k$ is obtained. Coefficients β_i ($i = 0, 1, \ldots, n$) can be obtained at each step as the remainders of repeated divisions of k by 2 in the set of integers, until the quotient is zero.

Use of the Jordan form. From

$$B = T^{-1} A_d T$$

it follows that

$$B^k = T^{-1} A_d^k T, \quad \text{hence} \quad A_d^k = T B^k T^{-1}$$

Consider (A.4.11). Clearly

$$B^k = \begin{bmatrix} B_{11}^k & O & \cdots & O & \cdots & O \\ O & B_{12}^k & \cdots & O & \cdots & O \\ \vdots & \vdots & \ddots & \vdots & \ddots & \vdots \\ O & O & \cdots & B_{1,k_1}^k & \cdots & 0 \\ \vdots & \vdots & \ddots & \vdots & \ddots & \vdots \\ O & O & \cdots & O & \cdots & B_{h,k_h}^k \end{bmatrix} \quad (56)$$

while the k-th power of a single $\ell \times \ell$ Jordan block is easily obtained by direct computation as

$$B_{ij}^k = \begin{bmatrix} \lambda_i^k & k\lambda_i^{k-1} & \binom{k}{2}\lambda_i^{k-2} & \cdots & \binom{k}{\ell-1}\lambda_i^{k-\ell+1} \\ 0 & \lambda_i^k & k\lambda_i^{k-1} & \cdots & \binom{k}{\ell-2}\lambda_i^{k-\ell+2} \\ 0 & 0 & \lambda_i^k & \cdots & \binom{k}{\ell-3}\lambda_i^{k-\ell+3} \\ \vdots & \vdots & \vdots & \ddots & \vdots \\ 0 & 0 & 0 & \cdots & \lambda_i^k \end{bmatrix} \quad (57)$$

where

$$\binom{k}{h} := \begin{cases} \dfrac{k(k-1)\ldots(k-h+1)}{h!} & \text{for } k > h \\ 1 & \text{for } k = h \\ 1 & \text{for } h = 0 \\ 0 & \text{for } k < h \end{cases}$$

The Interpolating Polynomial Method. By means of the procedure described in Subsection 2.1.3, the general element of the k-th power of matrix A_d is obtained as

$$\left(A_d^k\right)_{ij} = \langle k_{ij}(A_d), v \rangle \quad (58)$$

where both $k_{ij}(A_d)$ and v belong to $\mathbf{C}^m$ and m denotes the degree of the minimal polynomial of A_d. Vector v is defined as

$$v := \left(\lambda_1^k, \ k\lambda_1^{k-1}, \ \ldots, \ k(k-1)\ldots(k-m_1+2)\lambda_1^{k-m_1+1}, \ \ldots, \right.$$

$$\left. \lambda_h^k, \ k\lambda_h^{k-1}, \ \ldots, \ k(k-1)\ldots(k-m_h+2)\lambda_h^{k-m_h+1} \right) \quad (59)$$

Relation (59) only makes sense when $k \geq m$. However, the following equivalent expression for v holds without any constraint on the value of k.

$$v = \left(\lambda_1^k, k\lambda_1^{k-1}, 2!\binom{k}{2}\lambda_1^{k-2}, \ldots, (m_1-1)!\binom{k}{m_1-1}\lambda_1^{k-m_1+1}, \ldots,\right.$$
$$\left.\lambda_h^k, k\lambda_h^{k-1}, 2!\binom{k}{2}\lambda_h^{k-2}, \ldots, (m_h-1)!\binom{k}{m_h-1}\lambda_h^{k-m_h+1}\right) \qquad (60)$$

Use of the Schur form. By means of a procedure similar to the one presented for computation of the matrix exponential, it is possible to reduce the computation of A_d^k to the solution of scalar difference equations of the general type

$$z(k+1) = \lambda_i\, z(k) + b\, f_{rj}(k), \qquad z(0) = 0 \qquad (61)$$

where

$$f_{rj}(k) = r!\binom{k}{r}\lambda_j^{k-r} \qquad (62)$$

In the solution of (61) two cases are possible.
1. $\lambda_i = \lambda_j$. Solution of (61) is

$$z(k) = \frac{b}{r+1}(r+1)!\binom{k}{r+1}\lambda_i^{k-r-1} \qquad (63)$$

2. $\lambda_i \neq \lambda_j$. In this case $z(k)$ is computed by considering the sequence of functions $w_\ell(k)$ $(\ell=0,\ldots,r)$ defined below, which are the solutions of (61) with forcing terms $f_{\ell j}(k)$ $(\ell=0,\ldots,r)$. The solution is $z(k)=w_r(k)$, with

$$w_0(k) = b\,\frac{\lambda_i^k - \lambda_j^k}{\lambda_i - \lambda_j}$$
$$w_\ell(k) = \frac{1}{\lambda_i - \lambda_j}\left(\ell w_{\ell-1}(k) - b\ell!\binom{k}{\ell}\lambda_j^{k-\ell}\right) \quad (\ell=1,\ldots,r) \qquad (64)$$

Also in the case of discrete-time systems the components of vector v in (59) or (60) are called *modes*. Modes corresponding to a pair of complex conjugate eigenvalues appear with complex conjugate coefficients. The sum (51) in this case is changed into

$$S := \binom{k}{r}\left(h\lambda^{k-r} + h^*\lambda^{*\,k-r}\right) \qquad (65)$$

or, by using the previously introduced notation

$$S = \binom{k}{r}\left((u+jv)(\sigma+j\omega)^{k-r} + (u-jv)(\sigma-j\omega)^{k-r}\right)$$

By setting
$$\rho := |\lambda|, \quad \vartheta := \arg \lambda$$
$$m := 2|h|, \quad \varphi := \arg h$$
it follows that
$$S = \frac{m}{2} \binom{k}{r} \rho^k \left(e^{j((k-r)\vartheta+\varphi)} + e^{-j((k-r)\vartheta+\varphi)} \right)$$
$$= m \binom{k}{r} \rho^k \cos((k-r)\vartheta + \varphi)$$
$$= m \binom{k}{r} \rho^k \sin\left((k-r)\vartheta + \varphi + \frac{\pi}{2}\right)$$

Furthermore, instead of (52,53) the following matrices are easily derived:
$$C = \begin{bmatrix} \rho \cos \vartheta & \rho \sin \vartheta \\ -\rho \sin \vartheta & \rho \cos \vartheta \end{bmatrix} \tag{66}$$

$$C^k = \begin{bmatrix} \rho^k \cos k\vartheta & \rho^k \sin k\vartheta \\ -\rho^k \sin k\vartheta & \rho^k \cos k\vartheta \end{bmatrix} \tag{67}$$

The k-th power of a "real" Jordan block of type (A.4.26) has the same structure as matrix (57) but, on and above the main diagonal, functions $\lambda_i^{k-\ell+1}$ are replaced by 2×2 matrices $C_i^{k-\ell+1}$ and, below the main diagonal, zeros are replaced by 2×2 null matrices.

The general mode (62) has one of the following behaviors as k approaches infinity:

1. It converges to zero if the magnitude of the corresponding eigenvalue is less than one;

2. It remains bounded if r is zero and the magnitude of the corresponding eigenvalue is equal to one;

3. It diverges if the magnitude of the corresponding eigenvalue is greater than one or if r is different from zero and the magnitude of the eigenvalue is equal to one.

These modes are called respectively *asymptotically* or *strictly stable*, (merely) *stable*, and *unstable*.

2.2 THE FORCED STATE EVOLUTION OF LINEAR SYSTEMS

Let us now examine specifically linear systems whose evolution in time is described in Section 1.3 by equations (1.3.8,9) in the continuous-time case and (1.3.10,11) in the discrete-time case. In order to derive solutions for these equations, we will refer to the mathematical background presented in the previous section, in particular to the concepts of the state transition matrix and the function of a matrix.

2.2.1 Linear Time-Varying Continuous Systems

First refer to differential equation (1.3.8) and denote by $\Phi(t, t_0)$ the state transition matrix of the related homogeneous equation

$$\dot{x}(t) = A(t) x(t) \tag{1}$$

An expression for the state transition function $\varphi(t, t_0, x_0, u|_{[t_0, t]})$ is derived as follows.

Theorem 2.2-1.
The solution of differential equation (1.3.8), with initial condition $x(t_0) = x_0$ and piecewise continuous input function $u(\cdot)$, is

$$x(t) = \Phi(t, t_0) x(t_0) + \int_{t_0}^{t} \Phi(t, \tau) B(\tau) u(\tau) \, d\tau \tag{2}$$

Proof. First, clearly (2) satisfies the initial condition, since $\Phi(t_0, t_0) = I$. Then by differentiating both members of (2) and taking into account the rule for computation of the derivative of an integral depending on a parameter (see footnote 1 in Section 1.2) one obtains

$$\dot{x}(t) = \dot{\Phi}(t, t_0) x(t_0) + \Phi(t, t) B(t) u(t) + \int_{t_0}^{t} \dot{\Phi}(t, \tau) B(\tau) u(\tau) \, d\tau$$

$$= A(t) \left(\Phi(t, t_0) x_0 + \int_{t_0}^{t} \Phi(t, \tau) B(\tau) u(\tau) \, d\tau \right) + B(t) u(t)$$

$$= A(t) x(t) + B(t) u(t) \quad \square$$

Since Theorem 2.2-1 is of basic importance in linear system theory, we present another simple proof of it.

Another Proof of Theorem 2.2-1. For any identically nonsingular and differentiable matrix function of time $X(t)$, consider the relation

$$\frac{d}{dt} X^{-1}(t) = -X^{-1}(t) \dot{X}(t) X^{-1}(t)$$

which is obtained by differentiating the identity $X^{-1}(t) X(t) = I$. Replacing $X(t)$ with $\Phi(t, t_0)$ yields

$$\frac{d}{dt} \left(\Phi^{-1}(t, t_0) x(t) \right) = -\Phi^{-1}(t, t_0) \dot{\Phi}(t, t_0) \Phi^{-1}(t, t_0) x(t) + \Phi^{-1}(t, t_0) \dot{x}(t)$$

From $\dot{\Phi}(t, t_0) = A(t) \Phi(t, t_0)$ and differential equation (1.3.8) it follows that

$$\frac{d}{dt} \left(\Phi^{-1}(t, t_0) x(t) \right) = \Phi^{-1}(t, t_0) \left(\dot{x}(t) - A(t) x(t) \right)$$

$$= \Phi^{-1}(t, t_0) B(t) u(t)$$

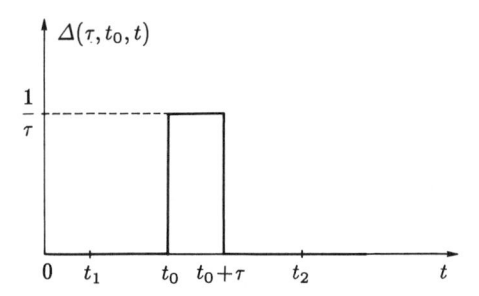

Sect. 2.2 The Forced State Evolution of Linear Systems

Figure 2-3 An impulse.

By integrating both members, we finally obtain

$$\Phi^{-1}(t, t_0)\, x(t) = c + \int_{t_0}^{t} \Phi^{-1}(\tau, t_0)\, B(\tau)\, u(\tau)\, d\tau$$

where c denotes a constant vector depending on the initial condition. By using inversion and composition properties of the state transition matrix, it is immediately shown that the above formula is equivalent to (2). □

Substitution of (2) into (1.3.9) immediately provides the following expression for the response function $\gamma(t, t_0, x_0, u|_{[t_0, t]})$:

$$y(t) = C(t)\, \Phi(t, t_0)\, x_0 + C(t) \int_{t_0}^{t} \Phi(t, \tau)\, B(\tau)\, u(\tau)\, d\tau + D(t)\, u(t)$$

$$= C(t)\, \Phi(t, t_0)\, x_0 + \int_{t_0}^{t} C(t)\, \Phi(t, \tau)\, B(\tau)\, u(\tau)\, d\tau + D(t)\, u(t) \qquad (3)$$

The integrals on the right of (2, 3) are *convolution integrals* whose *kernels* are the functions

$$V(t, \tau) := \Phi(t, \tau)\, B(\tau) \qquad (4)$$

$$W(t, \tau) := C(t)\, \Phi(t, \tau)\, B(\tau) \qquad (5)$$

Matrix $W(t, \tau)$ is called the *impulse response* of the system, and a reason for this denomination is explained as follows. Consider the piecewise continuous function $\Delta(\tau, t_0, \cdot)$ represented in Fig. 2-3 and suppose that parameter τ tends to zero: at the limit, we obtain a *Dirac impulse* having infinite amplitude and unitary area. The Dirac impulse at $t = 0$ will be denoted by $\delta(t)$ and at $t = t_0$ by $\delta(t - t_0)$. Still referring to Fig. 2-3, we define

$$\int_{t_1}^{t_2} \delta(t - t_0)\, dt := \lim_{\tau \to 0} \int_{t_1}^{t_2} \Delta(\tau, t_0, t)\, dt = 1$$

Similarly, for any continuous function of time $f(\cdot)$, we define
$$\int_{t_1}^{t_2} f(t)\,\delta(t-t_0)\,dt := \lim_{\tau \to 0} \int_{t_1}^{t_2} f(t)\,\Delta(\tau, t_0, t)\,dt = f(t_0)$$

Referring now to a purely dynamic system [$D(t) = O$ in (1.3.9)] with zero initial state, apply the impulse represented in Fig. 2-3 to its i-th input, the other inputs being set equal to zero. At the limit for τ approaching zero we obtain
$$y_i(t) = \int_{t_0}^{t} W(t,\tau)\,e_i\,\delta(t-t_0)\,d\tau = W(t,t_0)\,e_i \quad (i=1,\ldots,p)$$
where e_i denotes the i-th column of the identity matrix I_p. Hence each single column of $W(t, t_0)$ represents the system response to a Dirac impulse applied at t_0 to each single input.

Relation (2) for a purely dynamic system with zero initial state becomes
$$y(t) = \int_{t_0}^{t} W(t,\tau)\,u(\tau)\,d\tau \tag{6}$$

This means that the zero-state response of a purely dynamic linear system depends only on its impulse response function. Equation (6) is a typical *input-output model* or *IO model* of a linear system, while equations (1.3.8, 9) represent an *input-state-output model* or *ISO model*.

2.2.2 Linear Time-Varying Discrete Systems

The previous considerations can easily be extended to the discrete-time case, corresponding to system equations (1.3.10, 11). Often discrete-time systems derive from continuous ones subject to piecewise constant input functions of the type shown in Fig. 1-8 and with output accordingly sampled. Denote by $t_0 + iT$ ($i = 0, 1, \ldots$) the times corresponding to input changes and output sampling. Matrices of the discrete-time model are related to those of the continuous-time one by:

$$A_d(i) = \Phi\big(t_0 + iT, t_0 + (i+1)T\big) \tag{7}$$
$$B_d(i) = \int_{t_0+iT}^{t_0+(i+1)T} \Phi\big(t_0 + (i+1)T, \tau\big)\,B(\tau)\,d\tau \tag{8}$$
$$C_d(i) = C(t_0 + iT) \tag{9}$$
$$D_d(i) = D(t_0 + iT) \tag{10}$$

State transition matrix $\Phi(i, j)$ in this case is derived in connection with the homogeneous difference equation
$$x(i+1) = A_d(i)\,x(i) \tag{11}$$

The following theorem is the discrete counterpart of Theorem 2.2-1 and can easily be proved by direct check.

Theorem 2.2-2.

The solution of difference equation (1.3.10) with initial condition $x(j)=x_0$ is

$$x(i) = \Phi(i,j)\,x_0 + \sum_{k=j}^{i-1} \Phi(i,k+1)\,B_d(k)\,u(k) \qquad (12)$$

Substitution of (12) into (1.3.10) yields

$$y(i) = C_d(i)\,\Phi(i,j)\,x_0 + C_d(i)\sum_{k=j}^{i-1} \Phi(i,k+1)\,B_d(k)\,u(k) + D_d(i)\,u(i) \qquad (13)$$

Right side members of (12, 13) represent respectively the state transition and response function of the discrete-time case. The matrix

$$W(i,j) := C_d(i)\,\Phi(i,j+1)\,B_d(j) \qquad (14)$$

is called the *impulse response* of the considered discrete-time system. Its meaning is analogous to the continuous-time case: if the system is purely dynamic (i.e., if $D_d(i)=O$), its k-th column represents the zero-state response to an input identically zero, except for the k-th component, which is equal to one at time j.

2.2.3 Linear Time-Invariant Systems

The class of systems under consideration will now be further restricted to time-invariant ones which, in the continuous-time case, are described by

$$\dot{x}(t) = A\,x(t) + B\,u(t) \qquad (15)$$
$$y(t) = C\,x(t) + D\,u(t) \qquad (16)$$

and in the discrete-time case by

$$x(i+1) = A_d\,x(i) + B_d\,u(i) \qquad (17)$$
$$y(i) = C_d\,x(i) + D_d\,u(i) \qquad (18)$$

Time-invariance implies the following very important features:
1. Better computability of the state transition matrix and easier solvability of the nonhomogeneous differential equation which describes the state evolution of the system subject to control. In fact, while in the time-varying case it is necessary to use numerical integration procedures (such as Runge-Kutta methods), for time-invariant systems it is possible to express the state transition

matrix in finite terms, for instance with the interpolating polynomial method presented in the previous section.

2. A more straightforward and deeper insight into the mathematical essence of the structural constraints which may condition the action on the state through the input and/or the knowledge of the state through the output: such an insight will be particularly stressed with the geometric techniques presented in Chapters 3 and 4.

3. The possibility of relating state-space with polynomial matrix models which, although very restricted in use (since they are not at all extensible to the time-varying and nonlinear cases), allow a more satisfactory approach to some structure-independent problems such as identification. This alternative modeling of linear constant systems will be briefly reviewed in the next two sections.

In the continuous-time case the state transition function on the right of (2) is expressed in terms of the matrix exponential as follows:

$$x(t) = e^{At}x_0 + \int_0^t e^{A(t-\tau)} B\, u(\tau)\, d\tau \tag{19}$$

while in the discrete-time case the state transition function on the right of (12) is expressed in terms of the matrix power as

$$x(i) = A_d^{\,i} x_0 + \sum_{j=0}^{i-1} A_d^{\,i-j-1} B_d\, u(j) \tag{20}$$

Note that when dealing with time-invariant systems the initial time is usually assumed to be zero without any loss of generality because of the time-shifting property. The two terms on the right of (19) and (20) represent the zero-input and zero-state transition function.

We now consider the computation of the right side member of (19). The first term, i.e., the zero-input state transition function, is expressed in terms of a simple matrix exponential, while a mathematical description of the input function $u(\cdot)$ is needed to compute the second term, i.e., the convolution integral. Two important cases will be considered:

1. The input function is available as a sequence of samples;
2. The input function is available as a solution of a known homogeneous time-invariant differential equation.

Input Available as a Sequence of Samples. This case is easily transformed into a discrete-time case: denote by $u(t_i)$ ($i=0,1,\ldots$) the input samples and suppose that the actual input function is constant between samples, i.e., that $u(t) = u(t_i)$, $t_i \leq t < t_{i+1}$. This assumption is technically sound, since in many cases input to a continuous-time system is provided by a digital processor with a "hold" output

Sect. 2.2 The Forced State Evolution of Linear Systems

circuitry. In other cases, it is possible to use a discretization fine enough to have good reproduction of the actual input, without any other approximation in solving the system differential equation. In order to improve computational precision, it is also possible to introduce a linear interpolation between samples by means of an artifice, as will be shown in Subsection 2.2.5.

Consider a general instant of time t and denote by t_i, t_{i+1} two subsequent sampling instants such that $t_i \leq t < t_{i+1}$; equation (19) can be written as

$$x(t) = e^{At}x_0 + \sum_{k=0}^{i-1}\left(\int_{t_k}^{t_{k+1}} e^{A(t-\tau)}d\tau\right) B\,u(t_k) + \left(\int_{t_i}^{t} e^{A(t-\tau)}d\tau\right) B\,u(t_i)$$

$$= e^{At}x_0 + \sum_{k=0}^{i-1} e^{A(t-t_{k+1})} f\big(A, (t_{k+1}-t_k)\big)\, B\,u(t_k) + f(A,t)\,B\,u(t_i) \qquad (21)$$

where $f(A,t)$ denotes the matrix exponential integral, defined as

$$f(A,t) := \int_0^t e^{A\tau}d\tau \qquad (22)$$

whose computation will be considered in the next subsection. In the previous derivation the following identity has been used:

$$\int_{t_0}^{t_1} e^{A(t_1-\tau)}d\tau = -\int_{t_1-t_0}^{0} e^{Ax}dx = \int_0^{t_1-t_0} e^{A\tau}d\tau$$

In conclusion, the state transition function of a constant continuous-time system with a piecewise constant input is expressed as a finite sum whose terms are easily computable using matrix exponential and matrix exponential integral. If sampling is uniform with period T and the output is also synchronously sampled, the system is described by the discrete-time model (17, 18) where matrices A_d, B_d, C_d, D_d are related to the corresponding ones of the continuous-time case by relations

$$A_d := e^{AT} \qquad (23)$$
$$B_d := f(A,T)\,B \qquad (24)$$
$$C_d := C \qquad (25)$$
$$D_d := D \qquad (26)$$

which particularize (7–10). In (24), $f(A,T)$ still denotes the function defined by (22). In conclusion, computation is performed by using (20), which only requires a computational algorithm for the matrix power.

Input Provided by an Exosystem. We shall now consider the other remarkable case in which the convolution integral in (19) is easily computable. The input is assumed to be a linear function of the solution of a homogeneous linear differential equation, i.e., to be provided as the output of the time-invariant free system

$$\dot{v}(t) = W\,v(t)\,, \quad v(0) = v_0 \tag{27}$$
$$u(t) = L\,v(t) \tag{28}$$

where W and L denote properly dimensioned real matrices. In explicit form, we have

$$u(t) = L\,e^{Wt}\,v_0 \tag{29}$$

This case has considerable practical importance: in fact, it allows the reproduction of "test signals" such as steps, ramps, sine and cosine functions, which are widely used for comparison and classification of dynamic system features. Since input in this case is a linear combination of modes, in order to distinguish the outside modes from those inherent in the system itself, we will call the former *exogenous modes* and the latter *internal modes*. System (27, 28), which generates exogenous modes, is called an *exosystem*. Design techniques based on using simple test signals as inputs, which are very easily manipulable in computations, are widely used in regulator synthesis.

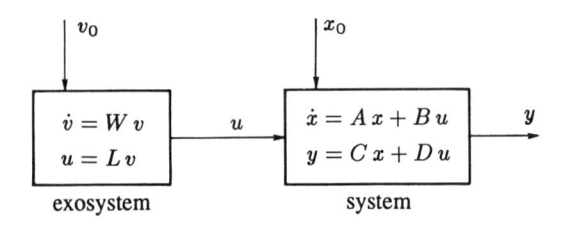

Figure 2-4 A system connected to an exosystem.

Referring to the connection shown in Fig. 2-4, we introduce the *extended state*

$$\hat{x} := \begin{bmatrix} x \\ v \end{bmatrix} \quad \text{with} \quad \hat{x}_0 := \begin{bmatrix} x_0 \\ v_0 \end{bmatrix}$$

and the corresponding extended matrices

$$\hat{A} := \begin{bmatrix} A & BL \\ O & W \end{bmatrix}\,, \quad \hat{C} := [\,C \quad DL\,]$$

The time evolution of the extended state and, in particular, of the system state that is a part of it, is obtained as the solution of the homogeneous linear matrix differential equation

$$\dot{\hat{x}}(t) = \hat{A}\,\hat{x}(t)\,, \quad \hat{x}(0) = \hat{x}_0 \tag{30}$$

Thus

$$\hat{x}(t) = e^{\hat{A}t}\,\hat{x}_0$$

Sect. 2.2 The Forced State Evolution of Linear Systems

and, consequently

$$y(t) = \hat{C}\,\hat{x}(t) = \hat{C}\,e^{\hat{A}t}\,\hat{x}_0$$

i.e., the response function is determined by means of a matrix exponential computation.

To show how this technique is used, let us report some examples. Consider the single input system

$$\dot{x}(t) = A\,x(t) + b\,u(t) \qquad (31)$$

$$y(t) = C\,x(t) + d\,u(t) \qquad (32)$$

where b and d denote row matrices, and set the problem to determine its zero-state responses to the test signals shown in Fig. 2-5.

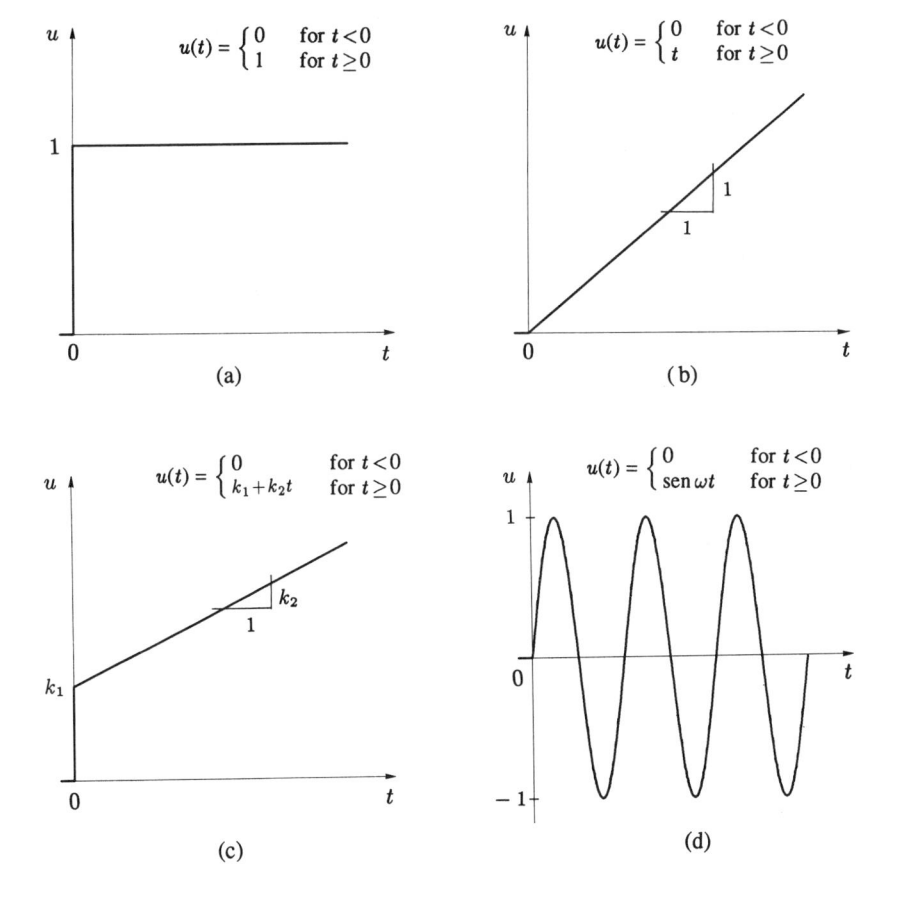

(a) $u(t) = \begin{cases} 0 & \text{for } t < 0 \\ 1 & \text{for } t \geq 0 \end{cases}$

(b) $u(t) = \begin{cases} 0 & \text{for } t < 0 \\ t & \text{for } t \geq 0 \end{cases}$

(c) $u(t) = \begin{cases} 0 & \text{for } t < 0 \\ k_1 + k_2 t & \text{for } t \geq 0 \end{cases}$

(d) $u(t) = \begin{cases} 0 & \text{for } t < 0 \\ \operatorname{sen} \omega t & \text{for } t \geq 0 \end{cases}$

Figure 2-5 Some test signals.

The signal shown in Fig. 2-5(a) is called a *unit step*: the zero-state response of system (31, 32) to it is determined by extending the state with a scalar extra state coordinate v and considering the time evolution of the free system

$$\dot{\hat{x}}(t) = \hat{A}\,\hat{x}(t), \quad \hat{x}(0) = \hat{x}_0 \tag{33}$$
$$y(t) = \hat{C}\,\hat{x}(t) \tag{34}$$

where

$$\hat{A} := \begin{bmatrix} A & b \\ O & 0 \end{bmatrix}, \quad \hat{x}_0 := \begin{bmatrix} O \\ 1 \end{bmatrix}, \quad \hat{C} := [C \ d]$$

A similar procedure can be applied for the signal shown in Fig. 2-5(b), called a *unit ramp*, for that shown in Fig. 2-5(c), consisting of the sum of a step and a ramp, and for the *sinusoid* shown in Fig. 2-5(d). In these cases two scalar extra state coordinates v_1 and v_2 are needed and matrices in (33, 34) are defined as

$$\hat{A} := \begin{bmatrix} A & b & O \\ O & 0 & 1 \\ O & 0 & 0 \end{bmatrix}, \quad \hat{x}_0 := \begin{bmatrix} O \\ 0 \\ 1 \end{bmatrix}, \quad \hat{C} := [C \ d \ 0]$$

$$\hat{A} := \begin{bmatrix} A & b & O \\ O & 0 & 1 \\ O & 0 & 0 \end{bmatrix}, \quad \hat{x}_0 := \begin{bmatrix} O \\ k_1 \\ k_2 \end{bmatrix}, \quad \hat{C} := [C \ d \ 0]$$

$$\hat{A} := \begin{bmatrix} A & b & O \\ O & 0 & \omega \\ O & -\omega & 0 \end{bmatrix}, \quad \hat{x}_0 := \begin{bmatrix} O \\ 0 \\ 1 \end{bmatrix}, \quad \hat{C} := [C \ d \ 0]$$

2.2.4 Computation of the Matrix Exponential Integral

Consider the matrix exponential integral, i.e., function $f(A, t)$ defined by (22), which is used in expression (24) for the input distribution matrix of the discrete system corresponding to a uniformly sampled continuous system. If A is nonsingular, it can easily be expressed in terms of the matrix exponential by means of the relation

$$f(A, t) = A^{-1}\left(e^{At} - I\right) \tag{35}$$

which follows from the term-by-term integration of the infinite series

$$e^{At} = I + At + \frac{A^2 t^2}{2} + \ldots + \frac{A^i t^i}{i!} + \ldots \tag{36}$$

which yields

$$f(A, t) = It + \frac{A t^2}{2} + \frac{A^2 t^3}{6} + \ldots + \frac{A^i t^{i+1}}{(i+1)!} + \ldots \tag{37}$$

clearly equivalent to (35). If A is singular, it is possible to use one of the following two computational methods.

Sect. 2.2 The Forced State Evolution of Linear Systems

The Interpolating Polynomial Method. Consider the finite sum

$$f(A,t) = \sum_{i=0}^{m-1} \eta_i(t) A^i \tag{38}$$

Define

$$w(t) := \left(\int_0^t e^{\lambda_1 \tau} d\tau, \ \int_0^t \tau e^{\lambda_1 \tau} d\tau, \ \ldots, \ \int_0^t \tau^{m_1-1} e^{\lambda_1 \tau} d\tau, \ \ldots, \right.$$

$$\left. \int_0^t e^{\lambda_h \tau} d\tau, \ \int_0^t \tau e^{\lambda_h \tau} d\tau, \ \ldots, \ \int_0^t \tau^{m_h-1} e^{\lambda_h \tau} d\tau \right) \tag{39}$$

and compute the vector $\eta(t) \in \mathbf{R}^m$ having coefficients $\eta_i(t)$ ($i = 0, \ldots, m-1$) as components, by means of the relation

$$\eta(t) = V^{-1} w(t) \tag{40}$$

where V is the same nonsingular matrix introduced in Subsection 2.1.3 to compute a general function of a matrix. The components of vector $w(t)$ are integrals of the general type

$$I_k(t) := \int_0^t \tau^k e^{\lambda \tau} d\tau \tag{41}$$

which, if $\lambda \neq 0$, can be computed by means of the recursion formula

$$I_k(t) = \left. \frac{\tau^k e^{\lambda \tau}}{\lambda} \right|_{\tau=0}^{\tau=t} - \frac{k}{\lambda} I_{k-1}(t) \tag{42}$$

This immediately follows from the well-known integration by parts

$$\int_{t_0}^{t_1} f(\tau) \dot{g}(\tau) d\tau = \left. f(\tau) g(\tau) \right|_{\tau=t_0}^{\tau=t_1} - \int_{t_0}^{t_1} \dot{f}(\tau) g(\tau) d\tau$$

with the assumptions

$$f(\tau) := \tau^k \quad \text{hence} \quad \dot{f}(\tau) = k \tau^{k-1}$$

$$\dot{g}(\tau) d\tau := e^{\lambda \tau} d\tau \quad \text{hence} \quad g(\tau) = \frac{1}{\lambda} e^{\lambda \tau}$$

On the other hand, for $\lambda = 0$ it is possible to obtain directly

$$I_k(t) = \frac{t^{k+1}}{k+1}$$

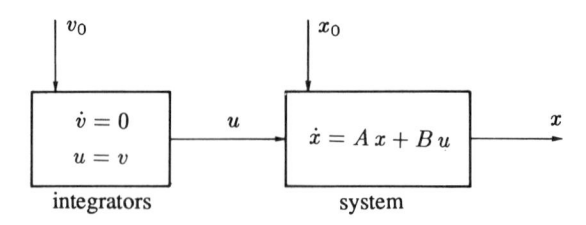

Figure 2-6 Representation of extended system (44, 45).

Summing up:

$$I_k(t) = \begin{cases} \frac{e^{\lambda t}}{\lambda}\left(t^k - \frac{k}{\lambda}t^{k-1} + \frac{k(k-1)}{\lambda^2}t^{k-2} - \right.\\ \qquad \left.\ldots + (-1)^k \frac{k!}{\lambda^k}\right) - (-1)^k \frac{k!}{\lambda^{k+1}} & \text{for } \lambda \neq 0 \\ \frac{t^{k+1}}{k+1} & \text{for } \lambda = 0 \end{cases} \quad (43)$$

A Submatrix of an Extended Matrix Exponential. Function $f(A,t)$ can be computed also as a submatrix of the matrix exponential of a properly extended system. We shall present this method for the direct computation of matrix B_d defined in (24): of course, it will provide $f(A,t)$ in the particular case $B_d := I_n$. Consider the continuous-time extended free system

$$\dot{\hat{x}}(t) = \hat{A}\,\hat{x}(t), \quad \hat{x}(0) = \hat{x}_0 \quad (44)$$

with

$$\hat{x} := \begin{bmatrix} x \\ v \end{bmatrix}, \quad \hat{x}_0 := \begin{bmatrix} x_0 \\ v_0 \end{bmatrix}, \quad \hat{A} := \begin{bmatrix} A & B \\ O & O \end{bmatrix} \quad (45)$$

which represents the interconnection shown in Fig. 2-6. Clearly $v(t) = v_0$ in time interval $[0,T]$ for any T, so that, from

$$\hat{x}(T) = e^{\hat{A}T}\hat{x}_0 = \begin{bmatrix} e^{AT}x_0 + f(A,T)Bv_0 \\ v_0 \end{bmatrix}$$

it follows that

$$e^{\hat{A}T} = \begin{bmatrix} e^{AT} & f(A,T)B \\ O & I_p \end{bmatrix} = \begin{bmatrix} A_d & B_d \\ O & I_p \end{bmatrix} \quad (46)$$

i.e., A_d, B_d are easily derived as submatrices of the above extended matrix exponential.

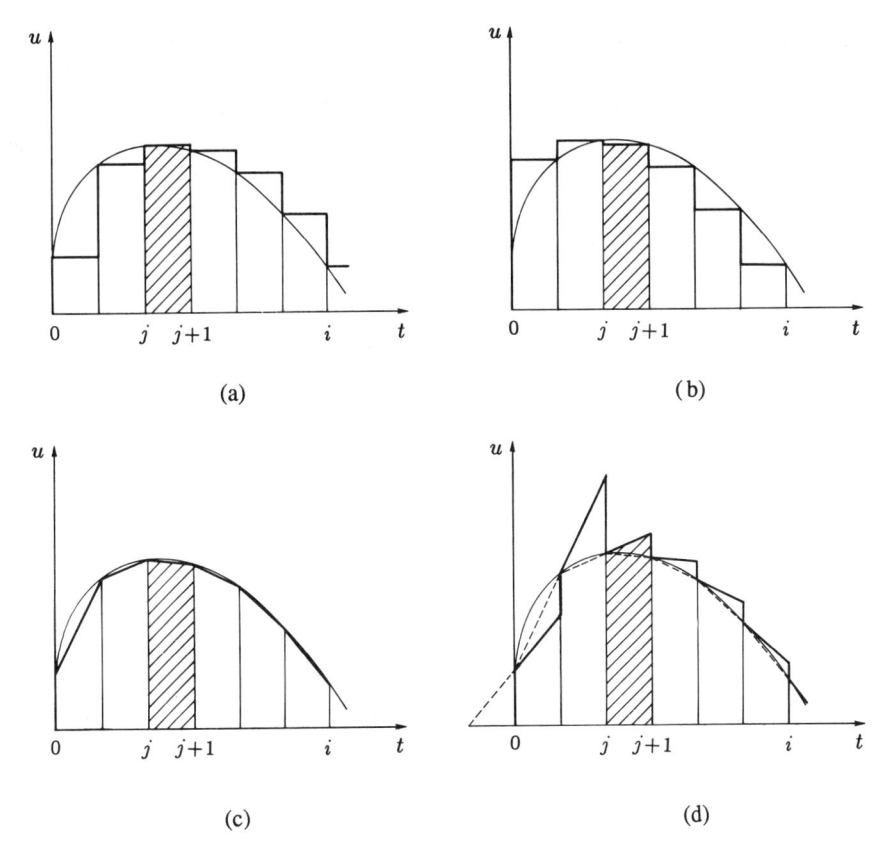

Sect. 2.2 The Forced State Evolution of Linear Systems

Figure 2-7 Several ways to reconstruct a continuous function from samples.

2.2.5 Approximating Continuous with Discrete

Discretizing the input function was presented in Subsection 2.2.3 as a method to compute the convolution integral on the right of (19): the input was approximated with a piecewise constant function, so that the convolution integral was transformed into a finite sum with terms easily computable as functions of the "matrix exponential" and "matrix exponential integral" types.

Such a reconstruction of a continuous function from a finite sequence of samples is not unique: other means to reach the same goal are represented in Fig. 2-7. Two of them, those shown in Figs. 2-7(a) and 2-7(d), can also be used to convert, in real time, a sequence of samples into a function of time, so that they correspond to actual *hold devices*, while those shown in Figs. 2-7(b) and 2-7(c) cannot be implemented in real time since they require knowledge of the next sample at any instant of time; nevertheless, they can be used to compute the convolution integral, because in this case the sequence of all samples can be assumed to be a priori known.

The four approximations are called, respectively: *backward rectangular* or *zero-order hold*, *forward rectangular*, *trapezoidal*, and *first-order hold*. The first-order hold approximation consists of maintaining between the sample at iT and that at $(i+1)T$, the constant slope corresponding to linear interpolation between samples at $(i-1)T$ and iT.

Recall definitions (23, 24) for A_d, B_d: the convolution integral in backward rectangular and forward rectangular approximations is computed by expressing the effect at time k of the pulse at generic time j (emphasized by hatching in figure) and summing over j. The following expressions are obtained in the two cases:

$$\sum_{j=0}^{k-1} A_d^{k-j-1} B_d\, u(j) \qquad (47)$$

$$\sum_{j=0}^{k-1} A_d^{k-j-1} B_d\, u(j+1) \qquad (48)$$

In the cases of trapezoidal and first-order hold approximations, the generic impulse is trapezoidal instead of rectangular and can be expressed as the sum of a rectangle and a triangle. The effect of the triangular pulse can be computed by means of an artifice, by considering that the triangular pulse with amplitude Δu can be obtained from an auxiliary dynamic system that integrates over T a rectangular pulse with amplitude $\Delta u/T$ with zero initial condition. Consider the extended system

$$\begin{bmatrix} \dot{x}(t) \\ \dot{z}(t) \end{bmatrix} = \begin{bmatrix} A & B \\ O & O \end{bmatrix} \begin{bmatrix} x(t) \\ z(t) \end{bmatrix} + \begin{bmatrix} O \\ I_p \end{bmatrix} u(t)$$

which, in fact, represents the original system with an integrator on each input. Denote by A_1 and B_1 the corresponding matrices and set

$$C_1 := [\, I_n \quad O\,]$$

The effect at time T of a triangular pulse with amplitude Δu applied between 0 and T is

$$\frac{1}{T} C_1 f(A_1, T) B_1\, \Delta u$$

Let

$$B_t := \frac{1}{T} C_1 f(A_1, T) B_1$$

The convolution integral is approximated in the two cases by the sums

$$\sum_{j=0}^{k-1} A_d^{k-j-1}\Big(B_d u(j) + B_t\big(u(j+1) - u(j)\big)\Big) \quad (49)$$

$$\sum_{j=0}^{k-1} A_d^{k-j-1}\Big(B_d u(j) + B_t\big(u(j) - u(j-1)\big)\Big), \quad u(-1):=0 \quad (50)$$

Matrices A_d, B_d, B_t can be computed as submatrices of the exponential of a properly extended matrix. In fact, it is easily shown that

$$e^{\hat{A}T} = \begin{bmatrix} A_d & B_d & TB_t \\ O & I_p & TI_p \\ O & O & I_p \end{bmatrix} \quad \text{if} \quad \hat{A} := \begin{bmatrix} A & B & O \\ O & O & I_p \\ O & O & O \end{bmatrix} \quad (51)$$

2.3 IO REPRESENTATIONS OF LINEAR CONSTANT SYSTEMS

Consider a time-invariant continuous-time linear system Σ with input $u \in \mathbf{R}^p$ and output $y \in \mathbf{R}^q$. A typical *input-output representation* (or briefly *IO representation*) of Σ is a differential equation of the type [1]

$$\sum_{k=0}^{\mu} Q_k \frac{d^k}{dt^k} y(t) = \sum_{k=0}^{\mu} P_k \frac{d^k}{dt^k} u(t) \quad (1)$$

where P_k and Q_k ($k=1,\ldots,\mu$) denote real matrices with dimensions $q \times p$ and $q \times q$ respectively; in particular, Q_μ is assumed to be nonsingular. The integer μ is called the *order* of the representation.

For a simpler notation and more straightforward algebraic handling, it is customary to represent with a polynomial any differential operator consisting

---

[1] Differential equation (1) has a meaning only if input and output functions are differentiable at least μ times. This assumption is very restrictive in system theory, where it is often necessary to refer to piecewise continuous functions, which are not differentiable, at least at discontinuity points. For a rigorous approach (1) should be interpreted in the light of distribution theory, i.e., in the framework of a suitable extension of the concept of function. This is not at all worth doing for the particular case at hand, since the drawback can be overcome in a simple and realistic way by considering (1) merely as a conventional representation of the integral equation obtained by integrating both its members μ times, and introducing each time in one of its members a proper integration constant whose value is related to the system initial condition.

of a linear combination of the derivatives of a given time function, like both members of (1). Let, for any function of time $x(\cdot)$:[2]

$$s\, x(t) := \frac{d}{dt} x(t)\ ,\quad s^2\, x(t) := \frac{d^2}{dt^2} x(t)\ ,\ \ldots$$

and, accordingly, write (1) as

$$\sum_{k=0}^{\mu} Q_k\, s^k\, y(t) = \sum_{k=0}^{\mu} P_k\, s^k\, u(t)$$

or simply

$$Q(s)\, y(t) = P(s)\, u(t) \tag{2}$$

where $P(s)$ and $Q(s)$ denote polynomial matrices defined by

$$P(s) := \sum_{k=0}^{\mu} P_k\, s^k\ ,\quad Q(s) := \sum_{k=0}^{\mu} Q_k\, s^k \tag{3}$$

The discrete-time case can be handled in a very similar way. Consider a time-invariant discrete-time linear system Σ with input $u \in \mathbf{R}^p$ and output $y \in \mathbf{R}^q$. A typical input-output representation of Σ is a difference equation of the type

$$\sum_{k=0}^{\mu} Q_k\, y(i+k) = \sum_{k=0}^{\mu} P_k\, u(i+k) \tag{4}$$

where P_k and Q_k ($k=1,\ldots,\mu$) denote real matrices with dimensions $q \times p$ and $q \times q$ respectively; in particular, Q_μ is assumed to be nonsingular. The integer μ is called the *order* of the representation. For a simpler notation, referring to any sequence $x(i)$ ($i=1,2,\ldots$) define

$$z\, x(i) := x(i+1)\ ,\quad z^2\, x(i) := x(i+2)\ ,\ \ldots$$

and write (4) accordingly as

$$Q(z)\, y(i) = P(z)\, x(i) \tag{5}$$

---

[2] It is worth noting at this point that the introduction of the Laplace transform and related mathematical background is very far beyond the scope of this book; here symbol s and polynomials in s are simply short notations for the linear operator "first derivative" and a linear combination of a function and its subsequent order derivatives.

Sect. 2.3 IO Representations of Linear Constant Systems

where $P(z)$ and $Q(z)$ are the polynomial matrices

$$P(z) := \sum_{k=0}^{\mu} P_k z^k, \quad Q(z) := \sum_{k=0}^{\mu} Q_k z^k \qquad (6)$$

In the cases of single-input single-output systems, i.e., when $p = q = 1$, equations (2) and (5), referring respectively to the continuous- and the discrete-time cases, can be written simply as

$$y(t) = \frac{P(s)}{Q(s)} u(t), \quad y(i) = \frac{P(z)}{Q(z)} u(i)$$

In this way the product of a time function or a time sequence by a ratio of polynomials in the variable s (the derivative operator) or z (the shift operator) is given a well-defined conventional meaning. Such ratios are called *transfer functions* of the systems referred to and are a complete representation of their zero-state dynamic behavior; their usefulness is particularly clear in dealing with complex systems consisting of numerous interconnected parts. In fact, the rules for the reduction of block diagrams and signal-flow graphs, which are briefly presented as complementary material of this chapter, can be applied to transfer functions as well as to real transmittance coefficients, and lead to overall transfer functions which in the previous convention represent the actual differential or difference equation relating the particular input and output referred to in the reduction. This expedient leads to significantly simpler notation and implementation of mathematical passages: in fact, all operations on polynomials performed in the course of the reduction like, for instance, products and sums, correspond to similar operations on the represented differential operators. In block diagrams a dynamic subsystem can be simply represented as a single block, as shown in Fig. 2-8(a) (continuous-time case) or in Fig. 2-8(b) (discrete-time case).

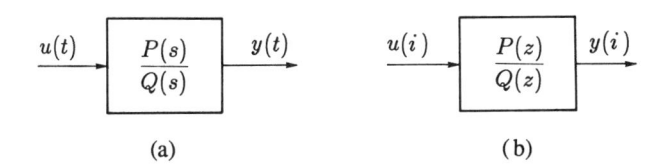

Figure 2-8 Block representations of single-input single-output linear systems.

Note that in transfer functions the degree of the polynomial at the numerator is not greater than that of the polynomial at the denominator, because of a well-known *physical realizability condition*, which excludes terms directly related to the time derivative of the input signal from the system response. Should this condition not be satisfied, in the response to a sine wave there would be terms with amplitude increasing when input frequency increases, which is a physical nonsense.

The roots of polynomial equations

$$P(s) = 0 \quad \text{or} \quad P(z) = 0$$

and

$$Q(s) = 0 \quad \text{or} \quad Q(z) = 0$$

are called respectively *zeros* and *poles* of the transfer function $P(s)/Q(s)$ or $P(z)/Q(z)$.

Similar arguments can be developed for multi-input multi-output systems; in fact (2) and (5) can also be written as

$$y(t) = Q^{-1}(s) P(s) u(t) \quad \text{with} \quad Q^{-1}(s) = \frac{\text{adj} Q(s)}{\det Q(s)}$$

and

$$y(i) = Q^{-1}(z) P(z) u(i) \quad \text{with} \quad Q^{-1}(z) = \frac{\text{adj} Q(z)}{\det Q(z)}$$

Each element of the *transfer matrices*

$$G(s) := Q^{-1}(s) P(s) \quad \text{and} \quad G(z) := Q^{-1}(z) P(z)$$

is a polynomial ratio, and represents the transfer function that relates the input and output corresponding to its column and its row respectively.

In the multivariable case *poles* are the roots of the polynomial equations $\det Q(s) = 0$ or $\det Q(z) = 0$, while the extension of the concept of zero leads to the so-called *invariant zeros*, which will be defined in Section 4.4 in the framework of the geometric approach.

2.4 RELATIONS BETWEEN IO AND ISO REPRESENTATIONS

It is quite evident that the IO (input-output) and the ISO (input-state-output) representations of linear dynamic constant systems are equivalent and related to each other. This section considers the problem of deriving any one of them from the other.

The IO description appears to be more compact and more direct, particularly when a mathematical model of a real system is derived from records of the input and output functions by means of an identification procedure; however, in engineering practice the state-space approach is more convenient for the following reasons:

1. the relative ease with which some nonlinearities and physical bounds, such as saturations, are taken into account;
2. a more direct dependence of the model coefficients on the actual values of physical parameters, which facilitates the study of sensitivity and robustness.

These reasons are important enough to adopt the ISO representation and restrict the IO one to the domain of mathematical inquisitiveness, provided the widespread belief that IO models are necessary to solve such problems as regulation, noninteraction, and stabilization, is absolutely unfounded. In fact, in this book such problems will be solved referring exclusively to ISO models: the passage from one to the other of the representations will be considered in this section, merely as an interesting exercise.

We will consider first the passage from ISO to IO representation, for which a constructive procedure is suggested in the proof of the following theorem.

Theorem 2.4-1.

System (2.2.15, 16) admits an IO representation of the type (2.3.2, 3), in which μ is the degree of the minimal polynomial of A.

Proof. Consider the subsequent time derivatives of both members of (16) and take into account (15):

$$y(t) = C\,x(t) + D\,u(t)$$
$$s\,y(t) = C\,A\,x(t) + C\,B\,u(t) + D\,s\,u(t)$$
$$s^2\,y(t) = C\,A^2\,x(t) + C\,A\,B\,u(t) + C\,B\,s\,u(t) + D\,s^2\,u(t)$$
$$\ldots\ldots$$
$$s^\mu\,y(t) = C\,A^\mu\,x(t) + \sum_{j=0}^{\mu-1} C\,A^j\,B\,s^{\mu-j-1}\,u(t) + D\,s^\mu\,u(t)$$

Let $\lambda^\mu + q_1\lambda^{\mu-1} + \ldots + q_\mu$ be the minimal polynomial of A. Multiply the first of the above relations by q_μ, the second by $q_{\mu-1}$, and so on, so that the last but one is multiplied by q_1, and sum all of them. It follows that

$$s^\mu\,y(t) + \sum_{i=1}^{\mu} q_i\,s^{\mu-i}\,y(t) = \sum_{i=1}^{\mu} P_i\,s^i\,u(t) \tag{1}$$

where P_i ($i = 1, \ldots, \mu$) are constant $q \times p$ matrices. The obtained representation is clearly of the type (2.3.2, 3). □

The following similar result for discrete-time systems is derived by simply replacing s with z.

Corollary 2.4-1.
System (2.2.17, 18) admits an IO representation of the type (2.3.5,6), in which μ is the degree of the minimal polynomial of A_d.

From (1) the transfer matrix $G(s)$ is immediately derived. Collecting $y(t)$ on the left and dividing by the minimal polynomial yields

$$G(s) = \frac{\sum_{i=0}^{\mu} P_i \, s^i}{s^{\mu} + \sum_{i=1}^{\mu} q_i \, s^{\mu-i}} \tag{2}$$

Note that every element of the matrix on the right of (2) is a strictly proper rational function.

Another procedure to derive the system transfer matrix is the following. Write (2.2.15, 16) in the form

$$s \, x(t) = A \, x(t) + B \, u(t)$$
$$y(t) = C \, x(t) + D \, u(t)$$

Thus
$$x(t) = (sI - A)^{-1} B \, u(t)$$
$$y(t) = \left(C \, (sI - A)^{-1} B + D \right) u(t)$$

hence

$$G(s) = C \, (sI - A)^{-1} B + D$$
$$= \frac{1}{\det(sI - A)} \, C \, \mathrm{adj}(sI - A) \, B + D \tag{3}$$

Rational functions in matrix (3) are strictly proper, since $\det(sI-A)$ is a polynomial with degree n and $C \, \mathrm{adj}(sI-A) \, B$ is a polynomial matrix whose elements have degrees less than or equal to $n-1$. The possible difference between the maximal degree of the polynomial at the denominator of (2) and that of (3) is due to possible cancellations of common factors in numerator and denominator of polynomial fractions, which have not been considered in deriving (3).

2.4.1 The Realization Problem

We now consider the inverse problem, i.e., the passage from IO to ISO representation. In literature this is called the *realization problem*. Given polynomial matrices $P(s)$, $Q(s)$ which appear in (2.3.2) or transfer matrix $G(s) = Q^{-1}(s) \, P(s)$, derive matrices A, B, C, D of a corresponding ISO representation of type (2.2.15, 16). The solution to the realization problem is not unique: even the state-space dimension n may be different in different realizations of the same transfer matrix. A *minimal realization* is one in which n

Sect. 2.4 Relations Between IO and ISO Representations

is a minimum. In this section we present a constructive procedure to derive a convenient realization, but not a minimal one, at least in the multi-input multi-output case.

First refer to a single-input single-output (SISO) system with transfer function $G(s)$, which is assumed to be proper rational of the type

$$G(s) = \frac{P(s)}{Q(s)} = k_0 + \frac{M(s)}{Q(s)} \tag{4}$$

where $M(s)/Q(s)$ is strictly proper. Denote by m the degree of $P(s)$ and by n that of $Q(s)$, which is assumed to be monic without any loss of generality: if not, divide both numerator and denominator by the coefficient of its greater power in s. If $m < n$, set $k_0 := 0$, $M(s) := P(s)$; if $m = n$, divide $P(s)$ by $Q(s)$ and set k_0 equal to the quotient and $M(s)$ to the remainder. Furthermore, denote by λ_i, m_i ($i = 1, \ldots, h$) the roots of the polynomial equation $Q(s) = 0$ and their multiplicities.

It is well known that the strictly proper function $M(s)/Q(s)$ admits the partial fraction expansion

$$\frac{M(s)}{Q(s)} = \sum_{i=1}^{h} \sum_{\ell=1}^{m_i} \frac{k_{i\ell}}{(s - \lambda_i)^\ell}$$

where

$$k_{i\ell} := \frac{1}{(m_i - \ell)!} \frac{d^{m_i - \ell}}{ds^{m_i - \ell}} \left((s - \lambda_i)^{m_i} \frac{M(s)}{Q(s)} \right) \bigg|_{s = \lambda_i}$$
$$(i = 1, \ldots, h; \ell = 1, \ldots, m_i) \tag{5}$$

so that the transfer function can be written as

$$G(s) = k_0 + \sum_{i=1}^{h} \sum_{\ell=1}^{m_i} \frac{k_{i\ell}}{(s - \lambda_i)^\ell} \tag{6}$$

It is convenient to transform (6) into a form with real coefficients: suppose $\lambda_1, \ldots, \lambda_r$ are real and $\lambda_{r+1}, \ldots, \lambda_c$ are complex poles with the imaginary part positive, and $\lambda_{c+1}, \ldots, \lambda_h$ are their conjugates. Denote with σ_i, ω_i the real and imaginary part of λ_i ($i = r+1, \ldots, c$), with $u_{i\ell}$, $v_{i\ell}$ the real and imaginary part of $k_{i\ell}$ ($i = r+1, \ldots, c; \ell = 1, \ldots, m_i$), i.e.,

$$\lambda_i = \begin{cases} \sigma_i + j\omega_i & (i = r+1, \ldots, c) \\ \sigma_i - j\omega_i & (i = c+1, \ldots, h) \end{cases}$$

$$k_{i\ell} = \begin{cases} u_{i\ell} + j\,v_{i\ell} & (i = r+1, \ldots, c; \ell=1, \ldots, m_i) \\ u_{i\ell} - j\,v_{i\ell} & (i = c+1, \ldots, h; \ell=1, \ldots, m_i) \end{cases}$$

The equivalence of (6) to the real coefficients expression

$$G(s) = k_0 + \sum_{i=1}^{r}\sum_{\ell=1}^{m_i} \frac{k_{i\ell}}{(s - \lambda_i)^\ell} + \sum_{i=r+1}^{c}\sum_{\ell=1}^{m_i} \frac{\alpha_{i\ell}\, s + \beta_{i\ell}}{(s^2 - a_i\, s + b_i)^\ell} \tag{7}$$

where

$$\begin{aligned} a_i &:= 2\,\sigma_i \\ b_i &:= \sigma_i^{\,2} + \omega_i^{\,2} \end{aligned} \tag{8}$$

and $\alpha_{i\ell}$, $\beta_{i\ell}$ are functions of σ_i, ω_i, u_{ik}, v_{ik} ($k=1,\ldots,\ell$), can be proved by direct check.

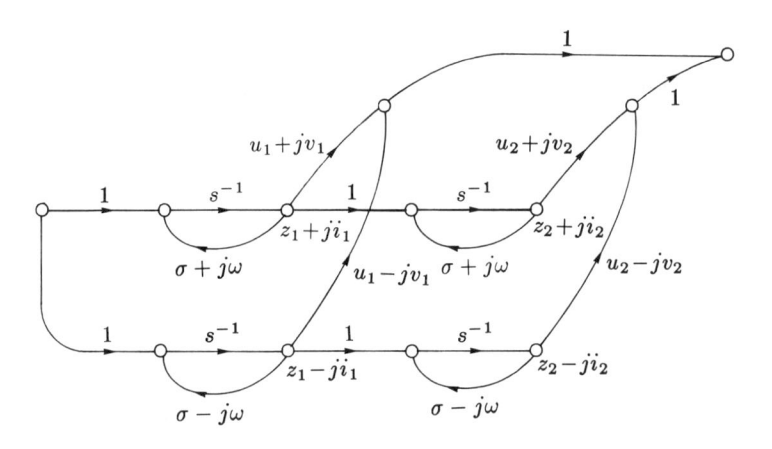

Figure 2-9 The Jordan-type complex realization.

An Example. Consider expansion (6) and, in particular, suppose that a double complex pole is present in it with its conjugate, and refer to the corresponding terms

$$\frac{u_1 + j v_1}{s - \sigma - j\omega} + \frac{u_2 + j v_2}{(s - \sigma - j\omega)^2} + \frac{u_1 - j v_1}{s - \sigma + j\omega} + \frac{u_2 - j v_2}{(s - \sigma + j\omega)^2} \tag{9}$$

A signal-flow graph realization for these is shown in Fig. 2-9. It represents the Jordan realization in the complex field. Note that the involved signals z_1+ji_1, z_2+ji_2, z_1-ji_1, z_2-ji_2, are complex conjugate.

The real and imaginary parts of all signals can be separated as shown in the equivalent flow graph of Fig. 2-10, which refers to the Jordan realization in the real field.

The corresponding *block-companion form* in the present case has the structure shown in Fig. 2-11. Terms (9) are replaced by

$$\frac{\alpha_1 s + \beta_1}{s^2 - a s + b} + \frac{\alpha_2 s + \beta_2}{(s^2 - a s + b)^2} \tag{10}$$

Sect. 2.4 Relations Between IO and ISO Representations

where

$$a = 2\sigma, \quad b = \sigma^2 + \omega^2$$
$$\alpha_1 = 2u_1, \quad \beta_1 = -2u_1\sigma - 2v_1\omega + 2u_2$$
$$\alpha_2 = -4v_2\omega, \quad \beta_2 = -4u_2\omega^2 + 4v_2\sigma\omega$$

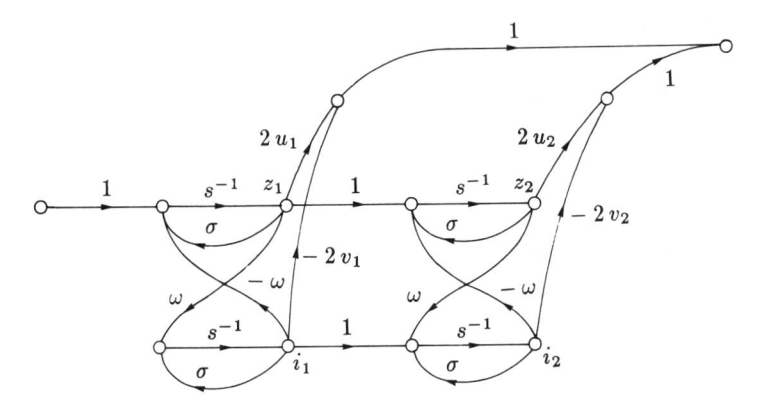

Figure 2-10 The Jordan-type real realization.

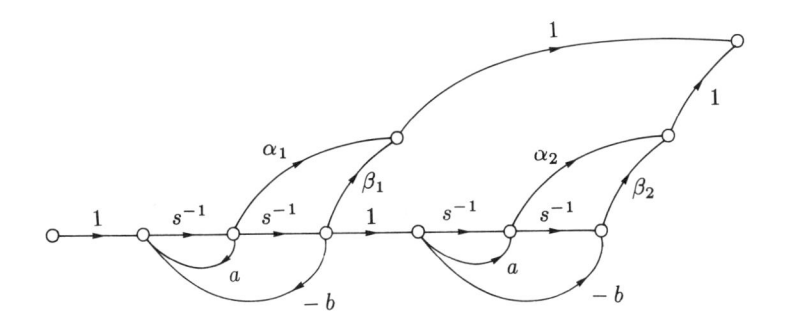

Figure 2-11 The block-companion real realization.

The aforementioned identities can be derived by means of the following general procedure:

1. Express as a single fraction each pair of complex conjugate terms of the original expansion;
2. Reduce the degree of the polynomial to one at the numerator of each term by means of repeated divisions: for instance, when the denominator is squared, use

$$\frac{N(s)}{D^2(s)} = \frac{Q(s)}{D(s)} + \frac{R(s)}{D^2(s)}$$

where $Q(s)$ and $R(s)$ denote the quotient and the remainder of the division of $N(s)$ by $D(s)$;

3. Collect terms with the same denominator and equate numerator coefficients.

We are now ready to derive the realization. We shall first present a corresponding signal-flow graph in which the meaning of the state variables is particularly stressed, then associate to it a state-space mathematical model, i.e., an ISO mathematical description. In the signal-flow graph the dynamic behavior is concentrated in the basic element *integrator*. The counterpart of the integrator in discrete-time systems is the *unit delay*. These elements in signal-flow graphs are denoted with the single branches represented in Fig. 2-12, while the corresponding elementary differential and difference equations are

$$\dot{y}(t) = u(t) \quad \text{and} \quad y(i+1) = u(i)$$

and the transfer functions, i.e., the transmittances of the branches

$$g(s) = \frac{1}{s} \quad \text{and} \quad g_d(z) = \frac{1}{z}$$

```
    s⁻¹                        z⁻¹
u(t) ────────► y(t)       u(i) ────────► y(i)
      (a)                        (b)
```

Figure 2-12 An integrator and a unit delay element.

The continuous-time system with transfer function (7) can be represented by the signal-flow graph whose structure is shown in Fig. 2-13, with a possible input to output direct algebraic link having a constant k_0 as transmittance, and a chain of standard dynamic elements for each real pole and each pair of complex conjugate poles, the length of the chain being equal to pole multiplicity. The elements of the chains are shown in Fig. 2-14(a,b) respectively for a real pole and a pair of complex conjugate poles cases; each element introduces a state variable x_v in the former case, or a pair of state variables x_v, x_{v+1} in the latter.

Refer to the overall signal-flow graph shown in Fig. 2-13 and number all the states subsequently from right to left and from the top downward. Matrices A, B, C, D of the corresponding realization are obtained from the signal-flow graph by expressing the state time derivatives as functions of state vector x and input u. Since these functions are available as nodes, they are directly provided by the graph.

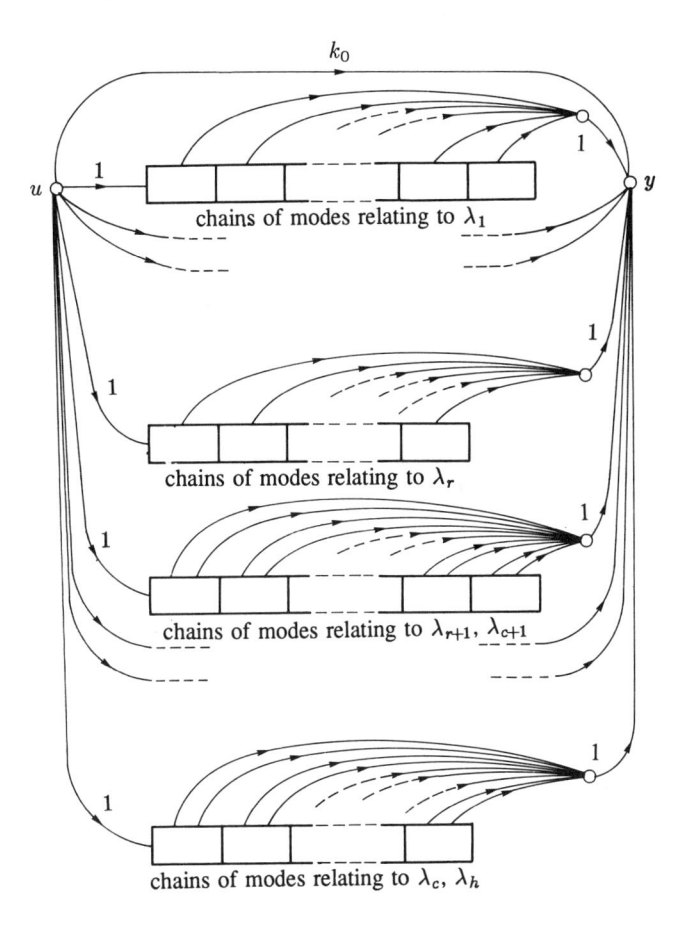

Sect. 2.4 Relations Between IO and ISO Representations

Figure 2-13 The general signal-flow graph for the single-input single-output case.

For instance, in a particular case in which the transfer function has a triple real pole and a pair of double complex conjugate poles, the realization has the following structure:

$$A = \begin{bmatrix} \lambda_1 & 1 & 0 & & & & \\ 0 & \lambda_1 & 1 & & O & & \\ 0 & 0 & \lambda_1 & & & & \\ & & & 0 & 1 & 0 & 0 \\ & O & & b_2 & a_2 & 1 & 0 \\ & & & 0 & 0 & 0 & 1 \\ & & & 0 & 0 & b_2 & a_2 \end{bmatrix}, \quad B = \begin{bmatrix} 0 \\ 0 \\ 1 \\ 0 \\ 0 \\ 0 \\ 1 \end{bmatrix},$$

$$C = [\, k_{13} \quad k_{12} \quad k_{11} \quad \beta_{12} \quad \alpha_{12} \quad \beta_{11} \quad \alpha_{11} \,], \quad D = k_0$$

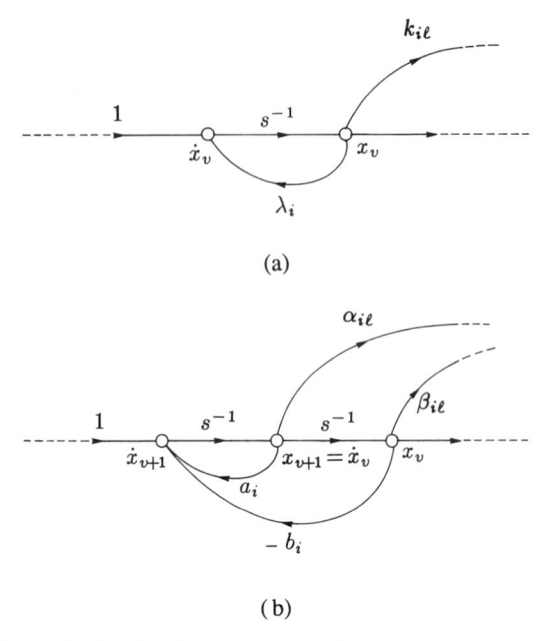

Figure 2-14 The elements of real chains and complex chains.

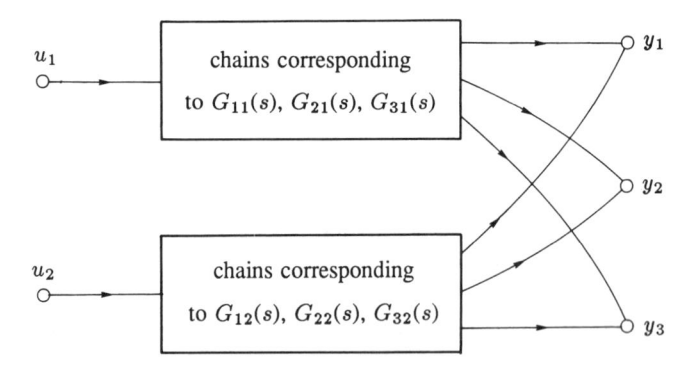

Figure 2-15 A flow-graph realization of a 3×2 transfer matrix.

Exactly the same procedure applies in the discrete-time case, with the only difference being that integrators have to be replaced with unit delays.

It is easily shown that the procedure extends to the multi-input multi-output (MIMO) case. Since this extension is quite straightforward, we will present it through an example: the particular case of a system with two inputs and three outputs, i.e., represented by a 3×2 transfer matrix. For each input, consider all the transfer functions relating it to the outputs, which are the elements of a column of the transfer matrix, and take into account each involved pole, possibly common to several transfer functions, with the maximal order of multiplicity; then implement chains as in Fig. 2-14, with connections to all outputs by means of branches having as transmittances the coefficients of the partial fraction expansion of the corresponding transfer function. The signals of these connections are summed together for each output, i.e., are merged into a single node in the graph, as shown in Fig. 2-15.

Here again it is very straightforward to derive matrices A, B, C, D of a corresponding ISO description, which is a realization of the given transfer matrix but, in general, not a minimal one: however, it can easily be transformed into a minimal realization by means of a simple algorithm, which will be presented in Subsection 3.3.1 after introduction of the Kalman canonical decomposition.

In literature the previously proposed realization is called a *parallel realization*, since it is composed of elementary blocks connected in parallel. Matrix A has a block-companion form. The most significant advantage offered by this realization over most of the other known ones is robustness with respect to parameter variations. In fact, eigenvalues of A, on which the system stability depends, are related to the matrix nonzero coefficients by explicit simple formulae, so that their sensitivity to variations of these coefficients is very direct and in any case can be easily computed and taken into account.

2.5 STABILITY

The term *stability* in the broad sense denotes the capability of a dynamic system to react with bounded variations of its motion or response to bounded initial state, input, or parameter perturbations. The notion of stability implies that the vector spaces of input, state, output, and parameter vectors and functions are metric. This will be true for all dynamic systems considered in the sequel. Stability plays a fundamental role in approaching most linear system analysis and synthesis problems, since the property of being stable is required for all actual control system implementations. Hence, in this section we will review the most important mathematical definitions and properties of stability.

2.5.1 Linear Time-Varying Systems

Refer to the linear time-varying free system

$$\dot{x}(t) = A(t)\, x(t) \tag{1}$$

The concept of stability is introduced through the following definition which, for linear systems, specializes the well-known concept of stability in the sense of Liapunov.

Definition 2.5-1. Stability in the Sense of Liapunov

Linear system (1) is said to be *stable in the sense of Liapunov* if for all t_0 and for all $\epsilon > 0$ there exists an $\eta > 0$ such that

$$\|x(t_0)\| < \eta \quad \Rightarrow \quad \|x(t)\| < \epsilon \quad \forall t \geq t_0 \tag{2}$$

It is said to be *asymptotically stable in the sense of Liapunov* if, in addition to (2),

$$\lim_{t \to \infty} \|x(t)\| = 0 \tag{3}$$

The following theorems express stability of system (1) in terms of transition matrix properties.

Theorem 2.5-1.

Linear system (1) is stable in the sense of Liapunov if and only if for all t_0 there exists a real number M such that

$$\|\Phi(t, t_0)\| \leq M < \infty \quad \forall t \geq t_0 \tag{4}$$

Proof. If. From

$$x(t) = \Phi(t, t_0) \, x(t_0)$$

it follows that

$$\|x(t)\| \leq \|\Phi(t, t_0)\| \, \|x(t_0)\| \leq M \|x(t_0)\|$$

hence, by assuming $\eta := \epsilon/M$, we derive $\|x(t)\| < \epsilon$, $t \geq t_0$, if $\|x(t_0)\| < \eta$.

Only if. If at a time t_1 no value of M exists such that (4) holds, by virtue of the matrix norm inequality

$$\|A\| \leq \sum_{i=1}^{n} \sum_{j=1}^{n} |a_{ij}|$$

there exists at least one element $\varphi_{ij}(t_1, t_0)$ of $\Phi(t_1, t_0)$ whose absolute value is unbounded; by assuming an initial state $x(t_0)$ with the j-th component equal to η and the others equal to zero, an $x(t_1)$ is obtained with the i-th component unbounded, i.e., an $x(t_1)$ unbounded in norm for any value of η; hence, the system is not stable. □

Theorem 2.5-2.

System (1) is asymptotically stable in the sense of Liapunov if and only if (4) holds for all t_0 and

$$\lim_{t \to \infty} \|\Phi(t, t_0)\| = 0 \tag{5}$$

Proof. If. Let $\eta := \|x(t_0)\|$. Since

$$\|x(t)\| \leq \|\Phi(t, t_0)\| \, \|x(t_0)\| = \|\Phi(t, t_0)\| \, \eta \tag{6}$$

if (5) holds, the system is asymptotically stable in the origin.

Sect. 2.5 Stability

Only if. Since for a proper choice of $x(t_0)$ relation (6) holds with the equality sign, if (5) were not satisfied it would not be true that $\lim_{t\to\infty} \|x(t)\|=0$ for all $x(t_0)$ such that $\|x(t_0)\|>0$. □

Let us refer now to equation (1.3.8) and consider system stability at zero state with respect to input function perturbations. Since linear systems stability with respect to input function perturbations does not depend on the particular equilibrium state or on the particular motion referred to, it is possible to define bounded input-bounded state stability as follows.

Definition 2.5-2. BIBS Stability

Linear system (1.3.8) is said to be *stable with respect to input function perturbations* or *bounded input-bounded state (BIBS) stable* if for all t_0 and for all $\epsilon>0$ there exists an $\eta>0$ such that from $\|u(t)\|<\eta$, $t\geq t_0$ the following holds:

$$\|x(t)\| = \left\| \int_{t_0}^{t} \Phi(t,\tau) B(\tau) u(\tau)\, d\tau \right\| < \epsilon \quad \forall t \geq t_0 \tag{7}$$

Theorem 2.5-3.

Linear system (1.3.8) is BIBS stable if and only if

$$\int_{t_0}^{t} \|V(t,\tau)\|\, d\tau := \int_{t_0}^{t} \|\Phi(t,\tau) B(\tau)\|\, d\tau \leq M < \infty \quad \forall t \geq t_0 \tag{8}$$

Proof. If. Norms satisfy

$$\|x(t)\| = \left\| \int_{t_0}^{t} \Phi(t,\tau) B(\tau) u(\tau)\, d\tau \right\| \leq \int_{t_0}^{t} \|V(t,\tau)\| \|u(\tau)\|\, d\tau$$

and, since $\|u(t)\| \leq \eta$, $t \geq t_0$, it follows that $\|x(t)\| \leq M\eta$, $t \geq t_0$. On the assumption $\eta := \epsilon/M$ equation (8) clearly holds.

Only if. If equation (8) is not satisfied, i.e., if there exists a time t_1 such that the integral

$$\int_{t_0}^{t_1} \|V(t_1,\tau)\|\, d\tau$$

is unbounded, the integral

$$\int_{t_0}^{t_1} |v_{ij}(t_1,\tau)|\, d\tau$$

is unbounded for at least one pair of indices i,j. In fact

$$\int_{t_0}^{t_1} \|V(t_1,\tau)\|\, d\tau \leq \int_{t_0}^{t_1} \sum_{r=1}^{n} \sum_{s=1}^{p} |v_{rs}(t_1,\tau)|\, d\tau$$

$$= \sum_{r=1}^{n} \sum_{s=1}^{p} \int_{t_0}^{t_1} |v_{rs}(t_1,\tau)|\, d\tau$$

$$\leq np \sup_{r,s} \int_{t_0}^{t_1} |v_{rs}(t_1,\tau)|\, d\tau$$

Assume an input function $u(t)$ with the j-th component defined as
$$u_j(t) := \eta \operatorname{sign}(v_{ij}(t_1, t))$$
and the other components identically zero. Since
$$x_i(t_1) = \int_{t_0}^{t_1} v_{ij}(t_1, \tau) u_j(\tau) d\tau = \eta \int_{t_0}^{t_1} |v_{ij}(t_1, \tau)| d\tau$$
for such an input function, the i-th component of $x(t_1)$ is unbounded, i.e., $x(t_1)$ is unbounded in norm for any value of η; hence, the system is not BIBS stable in the zero state. $\square$

Bounded input-bounded output stability can be approached in a similar way: the following definition and theorem are derived.

Definition 2.5-3. BIBO Stability
Linear system (1.3.8,9) is said to be *bounded input-bounded output (BIBO) stable* if for all t_0 and all $\epsilon > 0$ there exists an $\eta > 0$ such that from $\|u(t)\| < \eta$, $t \geq t_0$ it follows that
$$\|y(t)\| = \left\| C(t) \int_{t_0}^{t} \Phi(t, \tau) B(\tau) u(\tau) d\tau + D(t) u(t) \right\| < \epsilon \quad \forall t \geq t_0 \qquad (9)$$

Theorem 2.5-4.
Linear system (1.3.8,9) is BIBO stable if and only if
$$\int_{t_0}^{t} \|W(t, \tau)\| d\tau = \int_{t_0}^{t} \|C(t) \Phi(t, \tau) B(\tau)\| d\tau \leq M < \infty \quad \forall t \geq t_0 \qquad (10)$$

Proof. (Hint) An argument similar to that reported above for Theorem 2.5-3 can be used. Furthermore, it is also necessary to take into account the possible direct action of input on output through matrix $D(t)$. This does not cause any problem, since this matrix has been assumed to be bounded in norm for all t. $\square$

The previous definitions and theorems refer to linear time-varying continuous systems of type (1.3.8,9). Their extension to discrete systems of type (1.3.10,11) in the same section is straightforward.

2.5.2 Linear Time-Invariant Systems

The results derived in the previous section for general linear time-varying systems correspond to a more direct computational framework in the particular case of linear time-invariant systems.

Consider the linear time-invariant free system
$$\dot{x}(t) = A x(t) \qquad (11)$$
The main result, which relates stability to the eigenvalues of the system matrix, is stated in the following theorem.

Theorem 2.5-5.
Linear system (11) is stable in the sense of Liapunov if and only if
1. no eigenvalue of A has positive real part;
2. the eigenvalues of A with zero real part are simple zeros of the minimal polynomial.

Proof. *If.* Recall that every element of the matrix exponential is a linear combination of the time functions

$$e^{\lambda_1 t}, t e^{\lambda_1 t}, \ldots, t^{m_1-1} e^{\lambda_1 t}, \ldots, e^{\lambda_h t}, t e^{\lambda_h t}, \ldots, t^{m_h-1} e^{\lambda_h t} \qquad (12)$$

where λ_i, m_i ($i=1,\ldots,h$) denote the distinct eigenvalues of A and their multiplicities as zeros of the minimal polynomial. Hence, if conditions 1 and 2 hold, i.e., if all modes are stable, it follows that

$$\|e^{At}\| \leq M < \infty \quad \forall t \geq 0 \qquad (13)$$

and, by virtue of Theorem 2.5-1, the system is stable in the zero state.

Only if. It is necessary to prove that all functions (12) appear in the elements of the matrix exponential, so that (13) holds only if 1 and 2 are satisfied: if not, the absolute value of at least one element of the matrix exponential would be unbounded at the limit for t approaching infinity. Denote by B the Jordan form of A. Since $B = T^{-1} A T$, from (13) it follows that

$$\|e^{Bt}\| \leq M' < \infty \quad \forall t \geq 0$$

Since the multiplicity of an eigenvalue as a zero of the minimal polynomial is equal to the dimension of the greatest Jordan block of this eigenvalue (see below for proof), from relations (2.1.41, 42) it follows that all functions (12) are elements of e^{Bt}; hence, 1 and 2 are necessary. The minimal polynomial of A can be written, in factored form, as

$$m(\lambda) = (\lambda - \lambda_1)^{m_1} (\lambda - \lambda_2)^{m_2} \ldots (\lambda - \lambda_h)^{m_h} \qquad (14)$$

Since similar matrices have the same minimal polynomial, $m(\lambda)$ is the minimal polynomial of B also. Consider the identity

$$m(B) = (B - \lambda_1 I)^{m_1} (B - \lambda_2 I)^{m_2} \ldots (B - \lambda_h I)^{m_h} = O \qquad (15)$$

Factor $(B - \lambda_1 I)^{m_1}$ has the block-diagonal form

$$\begin{bmatrix} (B_{11} - \lambda_1 I)^{m_1} & \cdots & O & \cdots & O \\ \vdots & \ddots & \vdots & \ddots & \vdots \\ O & \cdots & (B_{1,k_1} - \lambda_1 I)^{m_1} & \cdots & O \\ \vdots & \ddots & \vdots & \ddots & \vdots \\ O & \cdots & O & \cdots & (B_{h,k_h} - \lambda_1 I)^{m_1} \end{bmatrix} \qquad (16)$$

where the first k_1 matrices on the main diagonal, which in the Jordan form correspond to the eigenvalue λ_1, have the structure

$$\begin{bmatrix} 0 & 1 & 0 & \ldots & 0 & 0 \\ 0 & 0 & 1 & \ldots & 0 & 0 \\ 0 & 0 & 0 & \ldots & 0 & 0 \\ \vdots & \vdots & \vdots & \ddots & \vdots & \vdots \\ 0 & 0 & 0 & \ldots & 0 & 1 \\ 0 & 0 & 0 & \ldots & 0 & 0 \end{bmatrix}$$

Denote the dimensions of such matrices with ℓ_{1j} ($j=1,\ldots,k_1$): they satisfy

$$(B_{1j} - \lambda_1 I)^{\ell_{1j}} = O, \quad (B_{1j} - \lambda_1 I)^{\ell_{1j}-1} \neq O$$

or, in other terms, they are nilpotent of orders ℓ_{1j}. It follows that, if $m_1 \geq \ell_{1j}$ ($j=1,\ldots,k_1$) all submatrices in (16) are zero; hence, the first factor in (15) is zero. The other factors are nonzero, being powers of upper triangular matrices with nonzero elements on the main diagonal. Thus, relation $m(B) = O$ implies that the value of each m_i ($i=1,\ldots,h$) is not less than the dimension of the greater Jordan block corresponding to λ_i. Actually it must be exactly equal since, if not, a polynomial nulled by B would exist with lower degree than the minimal polynomial: that obtained by substituting, in (14), m_i with the dimension of the greatest Jordan block corresponding to λ_i. □

Corollary 2.5-1.
Linear system (11) is stable in the sense of Liapunov if and only if
1. no eigenvalue of A has positive real part;
2. all Jordan blocks corresponding to eigenvalues with zero real part have dimensions equal to one.

Proof. The proof is contained in that of the previous theorem. □

Theorem 2.5-6.
Linear system (11) is asymptotically stable in the sense of Liapunov if and only if all eigenvalues of A have negative real part.

Proof. Also in this case similarity transformation into the Jordan form provides the most direct proof argument. Recall Theorem 2.5-2 and definitions of norms in finite-dimensional vector spaces: it is clear that system (11) is zero-state stable if and only if every element of e^{Bt} tends to zero for t approaching infinity.
Let $\lambda = \sigma + j\omega$ be a generic eigenvalue of A. Hence

$$\lim_{t \to \infty} t^k e^{\lambda t} = \lim_{t \to \infty} t^k e^{\sigma t} e^{j\omega t} = \lim_{t \to \infty} t^k e^{\sigma t}(\cos \omega t + j \operatorname{sen} \omega t) = 0$$

if and only if

$$\lim_{t \to \infty} t^k e^{\sigma t} = 0 \tag{17}$$

It is easily seen by using De L'Hospital's rule that (17) holds for all nonnegative integers k if and only if $\sigma = 0$. □

Sect. 2.5 Stability

As far as BIBS and BIBO stability are concerned, two interesting results will be stated in Subsection 3.3.1, after introduction of the Kalman canonical decomposition.

Linear time-invariant discrete systems. The above theorems are easily extended to linear discrete systems. First, recall that in the discrete-time case modes, instead of functions of type (12), there are sequences of the type

$$\lambda_1^k, k\lambda_1^{k-1}, \ldots, k!\binom{m_1}{k}\lambda_1^{k-m_1}, \ldots, \lambda_h^k, k\lambda_h^{k-1}, \ldots, k!\binom{m_h}{k}\lambda_h^{k-m_h} \qquad (18)$$

In connection with stability of the free system

$$x(i+1) = A_d\, x(i) \qquad (19)$$

the following results hold. They can be proved by a procedure similar to that used earlier for continuous systems.

Theorem 2.5-7.
Linear system (19) is stable in the sense of Liapunov if and only if
1. no eigenvalue of A has absolute value greater than one;
2. the eigenvalues of A with absolute value equal to one are simple zeros of the minimal polynomial.

Theorem 2.5-8.
Linear system (19) is asymptotically stable in the sense of Liapunov if and only if all eigenvalues of A have absolute value less than one.

2.5.3 The Liapunov and Sylvester Equations

The concept of Liapunov function is basic in the stability theory of nonlinear systems and will herein briefly be recalled in order to state some interesting results for the very particular case of linear time-invariant systems. Refer to the free system

$$\dot{x}(t) = f(x(t)) \qquad (20)$$

where $x \in \mathbf{R}^n$ and function f is continuous and satisfies $f(0)=0$. A continuous function $V : \mathbf{R}^n \to \mathbf{R}$ is said to be *positive definite* in a domain $\mathcal{D}$ containing the origin if $V(0)=0$ and $V(x)>0$ for all $x \in \mathcal{D}$, $x \neq 0$. It is called a *Liapunov function* if, moreover, $\dot{V}(x) := \langle \operatorname{grad} V, f(x) \rangle \leq 0$ for all $x \in \mathcal{D}$, $x \neq 0$. Let $\mathcal{D}_h := \{x : V(x) < h\}$, $\mathcal{D}_h \subseteq \mathcal{D}$. A system that admits a Liapunov function is simply stable at the origin for every initial state belonging to $\mathcal{D}_h$. Furthermore, it is asymptotically stable at the origin for every initial state belonging to $\mathcal{D}_h$ if $\dot{V}(x)$ is strictly negative for all $x \in \mathcal{D}_h$, $x \neq 0$.

The Liapunov Equation. Refer to free system (11) and consider the quadratic form

$$V(x) = \langle x, Px \rangle \qquad (21)$$

where P denotes a $n \times n$ symmetric, real positive definite matrix. The time derivative along a generic trajectory is

$$\begin{aligned} \dot{V}(x) &= \langle Ax, Px \rangle + \langle x, PAx \rangle \\ &= \langle x, A^T Px \rangle + \langle x, PAx \rangle \\ &= -\langle x, Mx \rangle \end{aligned} \qquad (22)$$

with

$$M := -(A^T P + PA) \qquad (23)$$

Note that, P being symmetric, M is symmetric. If the quadratic form (22) is negative definite, function (21) is a Liapunov function and system (11) is globally (i.e., for all initial states) asymptotically stable in the zero state. The equation

$$A^T X + XA = C \qquad (24)$$

is called a *Liapunov matrix equation*. The following results point out the importance of Liapunov equations in connection with stability of linear time-invariant systems.

Lemma 2.5-1.
Consider the functional

$$\Gamma := \int_0^\infty \langle x(t), M\, x(t) \rangle\, dt \qquad (25)$$

where M denotes any real symmetric matrix, and suppose that matrix A of free system (11) is strictly stable. The value of Γ along the trajectory of (11) starting at x_0 is

$$\Gamma_0 = \langle x_0, P x_0 \rangle \qquad (26)$$

where P is the unique solution of the Liapunov equation

$$A^T P + PA = -M \qquad (27)$$

Proof. Existence and uniqueness of the solution of (27) are a consequence of Theorem 2.5.10 reported herein, concerning the Sylvester equation. In fact, A being nonsingular by assumption, matrices A and $-A^T$ have no common eigenvalues (their eigenvalues are nonzero and have opposite sign). Furthermore, the solution is a symmetric matrix because the linear function on the left transforms a symmetric matrix P into a symmetric matrix M and, by uniqueness, the inverse transformation has the same property. Function

$$s(t) := \langle e^{At} x_0, P e^{At} x_0 \rangle$$

with P satisfying (27), is an indefinite integral of

$$\langle e^{At} x_0, M e^{At} x_0 \rangle$$

In fact,

$$\begin{aligned} \dot{s}(t) &= \langle A e^{At} x_0, P e^{At} x_0 \rangle + \langle e^{At} x_0, P A e^{At} x_0 \rangle \\ &= \langle e^{At} x_0, A^T P e^{At} x_0 \rangle + \langle e^{At} x_0, P A e^{At} x_0 \rangle \\ &= -\langle e^{At} x_0, M e^{At} x_0 \rangle \end{aligned}$$

Hence

$$\Gamma_0 = s(t) \Big|_{t=0}^{t=\infty} = \langle x_0, P x_0 \rangle \qquad \square$$

Theorem 2.5-9.

The Liapunov equation (27) admits a unique, symmetric positive definite solution P for any symmetric positive definite M if and only if matrix A has all eigenvalues with the negative real part.

Proof. If. The statement directly follows from Lemma 2.5-1, since the function under the integral sign in (25) is strictly positive for all $x(t) \neq 0$; hence, (27) is strictly positive for all $x_0 \neq 0$.

Only if. If (27) has a positive definite solution P with M positive definite, (21) is a Liapunov function, and system (11) is asymptotically stable in the zero state. Thus, A has all eigenvalues with negative real part by virtue of Theorem 2.5-6. □

The Sylvester Equation. Consider the matrix equation

$$AX - XB = C \qquad (28)$$

where A is an $m \times m$ matrix, B an $n \times n$ matrix, and X and C are both $m \times n$. Equation (28) is very basic in linear system theory: it is a generalization of the Liapunov equation and expresses the complementability condition of an invariant subspace, which recurs in numerous instances in regulation theory. Equation (28) can also be written as the following set of mn scalar equations:

$$\sum_{k=1}^{m} a_{ik} x_{kj} - \sum_{k=1}^{n} x_{ik} b_{kj} = c_{ij} \quad (i=1,\ldots,m;\ j=1,\ldots,n)$$

or, with a different matrix notation

$$\begin{bmatrix} A - b_{11} I_m & -b_{21} I_m & \cdots & -b_{n1} I_m \\ -b_{12} I_m & A - b_{22} I_m & \cdots & -b_{n2} I_m \\ \vdots & \vdots & \ddots & \vdots \\ -b_{1n} I_m & -b_{2n} I_m & \cdots & A - b_{nn} I_m \end{bmatrix} \begin{bmatrix} X_1 \\ X_2 \\ \vdots \\ X_n \end{bmatrix} = \begin{bmatrix} C_1 \\ C_2 \\ \vdots \\ C_n \end{bmatrix}$$

where X_j and C_j ($j=1,\ldots,n$) denote the j-th columns of X and C.

By properly redefining matrices, previous equations can be written in the form

$$\hat{A} \hat{x} = \hat{b} \qquad (29)$$

where $\hat{A}$ is $(mn) \times (mn)$, while $\hat{x}$ and $\hat{b}$ are both $(mn) \times 1$. A well-known necessary and sufficient condition for equation (29) to have a solution is

$$\hat{b} \in \operatorname{im} \hat{A}$$

A necessary and sufficient condition for the existence and uniqueness of the solution is stated in the following theorem.

Theorem 2.5-10.
Equation (28) admits a unique solution if and only if A and B have no common eigenvalues.

Proof.[1] *Only if.* Let A and B be respectively $m \times m$ and $n \times n$. Equation (28) is equivalent to a set of nm linear equations with nm unknowns. Its solution is unique if and only if the corresponding homogeneous equation

$$AX - XB = O \qquad (30)$$

admits $X = O$ as the unique solution. Let λ be a common eigenvalue, so that there exist nonzero vectors (column matrices) u, v such that $Av = \lambda v$, $B^T u = \lambda u$, or, for transposes, $u^T B = \lambda u^T$. The nonzero matrix $X := v u^T$ satisfies equation (30). In fact,

$$A v u^T - v u^T B = \lambda v u^T - v \lambda u^T = O$$

If. Let $\sigma(B) = \{\mu_1, \ldots, \mu_h\}$ be the spectrum of B and

$$(\lambda - \mu_i)^{m_{ij}} \quad (i = 1, \ldots, h; \; j = 1, \ldots, k_i)$$

the elementary divisors of B: k_i is the number of Jordan blocks corresponding to the eigenvalue μ_i and m_{ij} $(j = 1, \ldots, k_i)$ their dimensions. It is well known that B admits n linearly independent generalized eigenvectors $v_{ij\ell}$ $(i = 1, \ldots, h; \; j = 1, \ldots, k_i; \; \ell = 1, \ldots, m_{ij})$, satisfying

$$B v_{ij1} = \mu_i v_{ij1}$$
$$B v_{ij\ell} = \mu_i v_{ij\ell} + v_{ij,\ell-1}$$
$$(i = 1, \ldots, h; \; j = 1, \ldots, k_i; \; \ell = 2, \ldots, m_{ij})$$

If X is a nonzero solution of (30), there exists at least one generalized eigenvector $v_{ij\ell}$ such that $X v_{ij\ell} \neq 0$: choose ℓ in the corresponding chain in such a way that

$$X v_{ij\ell} \neq 0, \quad X v_{ij,\ell-1} = 0$$

hence

$$0 = (AX - XB) v_{ij\ell} = A X v_{ij\ell} - X \mu_i v_{ij\ell} = (A - \mu_i I) X v_{ij\ell}$$

That is, μ_i is an eigenvalue of A. □

Numerical solution of Sylvester equation (hence of Liapunov equation, which is a particular case) can be obtained through the Schur decomposition in the following terms.[2]

---

[1] The proof herein reported is due to Ostrowski and Schneider [24].

[2] This method and Fortran programs for its implementation were proposed by Bartels and Stewart [2].

Algorithm 2.5-1. Solution of the Sylvester Equation
Consider equation (24) and perform the Schur decomposition of A and B:

$$U M U^* X - X V N V^* = C$$

where U and V are unitary, M and N upper-triangular complex matrices. Premultiply by U^* and postmultiply by V, thus obtaining

$$M D - D N = G \quad \text{with} \quad D := U^* X V, \quad G := U^* C V$$

which can be directly solved by

$$D_1 = (M - n_{11} I_m)^{-1} G_1$$

$$D_i = (M - n_{ii} I_m)^{-1} \left(G_i + \sum_{j=1}^{i-1} n_{ji} D_j \right) \quad (i=2,\ldots,n)$$

where D_i and G_i ($i=1,\ldots,n$) denote the i-th columns of D and G. Then compute $X = U D V^*$.

2.6 CONTROLLABILITY AND OBSERVABILITY

It will be shown that the sets defined in Section 1.4 assume particular structures for linear systems, so that procedures to solve general control and observation problems can easily be derived.

2.6.1 Linear Time-Varying Systems

A basic property of the sets of states reachable from the origin and of states controllable to the origin is as follows.

Property 2.6-1.
In the case of linear time-varying systems the *reachable set from the origin* $\mathcal{R}^+(t_0,t_1,0)$ and the *controllable set to the origin* $\mathcal{R}^-(t_0,t_1,0)$ are subspaces of the state space $\mathcal{X}$.

Proof. The set $\mathcal{R}^+(t_0,t_1,0)$ is a subspace, being the image of the linear transformation $\varphi(t_1,t_0,0,u(\cdot))$ from $\mathcal{U}_f$ to $\mathcal{X}$. $\mathcal{R}^-(t_0,t_1,0)$ can be defined as

$$\left\{ x : (x, u(\cdot)) \in \mathcal{N} \right\} \quad (1)$$

where $\mathcal{N}$ denotes the subset of $\mathcal{X} \times \mathcal{U}_f$ defined by

$$\mathcal{N} := \left\{ (x, u(\cdot)) : 0 = \varphi(t_1, t_0, x, u(\cdot)) \right\}$$

which is a subspace, being the kernel of a linear transformation. Set (1) is a subspace, being the projection of a subspace. □

On the other hand, sets $\mathcal{W}^+(t_0,t_1,0)$ and $\mathcal{W}^-(t_0,t_1,0)$ are not generally subspaces in the case of time-varying systems.[1] It will be shown in the next subsection that, on the contrary, they are subspaces in the case of linear time-invariant systems.

As a consequence of Properties 2.6-1 and 1.3-4 (decomposability of motion), the following statement holds.

Property 2.6-2.
In the case of linear time-varying systems set $\mathcal{R}^+(t_0,t_1,x_0)$ is the linear variety defined as the sum of subspace $\mathcal{R}^+(t_0,t_1,0)$ and any state reachable from x_0 in $[t_0,t_1]$, while set $\mathcal{R}^-(t_0,t_1,x_1)$ is the linear variety defined as the sum of subspace $\mathcal{R}^-(t_0,t_1,0)$ and any state from which x_1 can be reached in $[t_0,t_1]$.

Now we shall examine some computational aspects of control. Refer to linear time-varying system (1.3.8,9) and consider the problem of computing basis matrices for subspaces $\mathcal{R}^+(t_0,t_1,0)$ and $\mathcal{R}^-(t_0,t_1,0)$.[2]

The following lemma will be the basic tool to derive the main results.

Lemma 2.6-1.
Let $F(\cdot)$ be an $n \times m$ matrix whose elements are piecewise continuous functions in $[t_0,t_1]$ with values in $\mathbf{R}^m$. The equality

$$F^T(t)\,x = 0 \quad \forall t \in [t_0,t_1] \tag{2}$$

holds if and only if $x \in \ker G(t_0,t_1)$, where $G(t_0,t_1)$ denotes the *Gramian matrix* defined as

$$G(t_0,t_1) := \int_{t_0}^{t_1} F(t)\,F^T(t)\,dt \tag{3}$$

---

[1] This can be shown with a simple example. Consider a linear discrete system described by equations (1.3.10, 11), with

$$A_d(0) := A_d(1) := \begin{bmatrix} 0 & 0 \\ 1 & 0 \end{bmatrix}, \quad B_d(0) := \begin{bmatrix} 1 \\ 0 \end{bmatrix}, \quad B_d(1) := \begin{bmatrix} 0 \\ 1 \end{bmatrix}$$

Let $A_d := A_d(0) = A_d(1)$; clearly

$$\mathcal{R}^+(0,1,0) = \operatorname{im} B_d(0)$$
$$\mathcal{R}^+(0,2,0) = A_d \operatorname{im} B_d(0) + \operatorname{im} B_d(1)$$

and, since $A_d \operatorname{im} B_d(0) = \operatorname{im} B_d(1)$, it follows that

$$\mathcal{W}^+(0,2,0) := \mathcal{R}^+(0,1,0) \cup \mathcal{R}^+(0,2,0) = \operatorname{im} B_d(0) \cup \operatorname{im} B_d(1)$$

which shows that $\mathcal{W}^+(0,2,0)$ is not a subspace.

[2] A generic subspace $\mathcal{S} \in \mathbf{R}^n$ is numerically determined by means of a matrix S such that $\mathcal{S} = \operatorname{im} S$, which is called a *basis matrix* of $\mathcal{S}$.

Sect. 2.6 Controllability and Observability

Proof. Matrix $G(t_0, t_1)$ is symmetric, being the integral of a symmetric matrix. Furthermore

$$\langle x, G(t_0, t_1) x \rangle = \int_{t_0}^{t_1} \|F^T(t) x\|_2^2 \, dt \quad \forall \; x \in \mathbf{R}^n$$

hence $G(t_0, t_1)$ is positive semidefinite and any $x \in \mathbf{R}^n$ such that $\langle x, G(t_0, t_1) x \rangle = 0$ also satisfies the equality $F^T(t) x = 0$ for all $t \in [t_0, t_1]$ and vice versa. On the other hand, by virtue of Theorem A.5-4, from $G(t_0, t_1)$ being positive semidefinite it follows that $\langle x, G(t_0, t_1) x \rangle = 0$ if and only if $x \in \ker G(t_0, t_1)$. □

If relation (2) holds for $x \neq 0$, the rows of $F(\cdot)$ are linearly dependent in $[t_0, t_1]$ by definition, so that the rows of $F(\cdot)$ are linearly independent in $[t_0, t_1]$ if and only if $G(t_0, t_1)$ is nonsingular.

Theorem 2.6-1.
Refer to system (1.3.8,9). The following equalities hold.

$$\mathcal{R}^+(t_0, t_1, 0) = \text{im} P(t_0, t_1) \tag{4}$$

$$\mathcal{R}^-(t_0, t_1, 0) = \Phi^{-1}(t_1, t_0) \text{im} P(t_0, t_1) = \Phi(t_0, t_1) \text{im} P(t_0, t_1) \tag{5}$$

where $P(t_0, t_1)$ denotes the symmetric positive semidefinite matrix

$$P(t_0, t_1) := \int_{t_0}^{t_1} \Phi(t_1, \tau) B(\tau) B^T(\tau) \Phi^T(t_1, \tau) \, d\tau \tag{6}$$

Proof. Since $\ker P(t_0, t_1) = (\text{im} P(t_0, t_1))^\perp$, to prove (4) it is sufficient to show that the nonzero states belonging to $\ker P(t_0, t_1)$ are not reachable from the origin in $[t_0, t_1]$, while those belonging to $\text{im} P(t_0, t_1)$ are reachable. Suppose that a state x_1 is reachable from the origin in $[t_0, t_1]$; hence there exists an input function $u(\cdot)$ such that

$$x_1 = \int_{t_0}^{t_1} \Phi(t_1, \tau) B(\tau) u(\tau) \, d\tau \tag{7}$$

Let $x_1 \in \ker P(t_0, t_1)$. The left scalar product of both members of the previous relation by x_1 gives

$$\langle x_1, x_1 \rangle = \int_{t_0}^{t_1} \langle B^T(\tau) \Phi^T(t_1, \tau) x_1, u(\tau) \rangle \, d\tau$$

By virtue of Lemma 2.6-1, $B^T(t) \Phi^T(t_1, t) x_1 = 0$ for all $t \in [t_0, t_1]$, so that for any input function $u(\cdot)$ condition $x_1 \in \ker P(t_0, t_1)$ implies $x_1 = 0$. On the other hand, let $x_1 \in \text{im} P(t_0, t_1)$, so that

$$x_1 = P(t_0, t_1) P^+(t_0, t_1) x_1$$

where $P^+(t_0, t_1)$ denotes the pseudoinverse of $P(t_0, t_1)$; by using (6) the previous expression can be written as

$$x_1 = \int_{t_0}^{t_1} \Phi(t_1, \tau) B(\tau) B^T(\tau) \Phi^T(t_1, \tau) P^+(t_0, t_1) x_1 \, d\tau$$

which, by comparison with (7), states that x_1 is reachable from the origin with the particular input

$$u(t) := B^T(t)\, \Phi^T(t_1, t)\, P^+(t_0, t_1)\, x_1 \tag{8}$$

To prove (5), note that relation $x_0 \in \mathcal{R}^-(t_0, t_1, 0)$ implies the existence of an input function $u(\cdot)$ such that

$$0 = \Phi(t_1, t_0)\, x_0 + \int_{t_0}^{t_1} \Phi(t_1, \tau)\, B(\tau)\, u(\tau)\, d\tau$$

i.e.

$$-\Phi(t_1, t_0)\, x_0 \in \mathcal{R}^+(t_0, t_1, 0)$$

hence

$$x_0 \in \Phi^{-1}(t_1, t_0)\, \mathcal{R}^+(t_0, t_1, 0) = \Phi(t_0, t_1)\, \mathcal{R}^+(t_0, t_1, 0)$$

This completes the proof. □

In conclusion, system (1.3.8,9) is completely controllable and completely reachable in $[t_0, t_1]$ (i.e., $\mathcal{R}^+(t_0, t_1, 0) = \mathcal{R}^-(t_0, t_1, 0) = \mathbf{R}^n$, hence $\mathcal{R}^+(t_0, t_1, x) = \mathcal{R}^-(t_0, t_1, x) = \mathbf{R}^n$ for all $x \in \mathbf{R}^n$) if and only if matrix $P(t_0, t_1)$, defined in (6), is nonsingular (strictly positive definite).

Problem 2.6-1. Control Between Two Given States

Refer to system (1.3.8,9). Determine an input function $u(\cdot)$ which produces transition between two arbitrarily given states x_0, x_1 in the time interval $[t_0, t_1]$.

Solution. By linearity, the input function that solves the problem also drives the system from the origin to $x_2 := x_1 - \Phi(t_1, t_0)\, x_0$, so that, for the problem to have a solution, relation $x_2 \in \mathcal{R}^+(t_0, t_1, 0)$ must be satisfied. Recall (8): it is possible to assume

$$u(t) := B^T(t)\, \Phi^T(t_1, t)\, P^+(t_0, t_1)\, x_2, \quad t \in [t_0, t_1] \tag{9}$$

By virtue of a well-known feature of the pseudoinverse matrix, such a control function provides the best approximation (in euclidean norm) to the final state x_1 when this is not reachable from x_0. □

Matrix $P(t_0, t_1)$ can be computed by the following procedure: denote by $\hat{\Phi}(\cdot, \cdot)$ the state transition matrix corresponding to the system matrix

$$\hat{A}(t) := \begin{bmatrix} A(t) & B^T(t) B(t) \\ O & -A^T(t) \end{bmatrix} \tag{10}$$

and by M the submatrix corresponding to the first n rows and the last n columns of $\hat{\Phi}(t_1, t_0)$. Then

$$P(t_0, t_1) = M\, \Phi^T(t_1, t_0) \tag{11}$$

Sect. 2.6 Controllability and Observability

where $\Phi(\cdot,\cdot)$ denotes the transition matrix corresponding to $A(t)$. Computation is particularly simple in the case of time-invariant systems, where

$$\hat{\Phi}(T,0) = e^{\hat{A}T}, \quad \Phi^T(T,0) = e^{A^T T}$$

See Problem 3.5-2 for further details.

Extension of the preceding material on the state controllability to the output controllability is relatively simple, since "pointwise" output controllability is easily reconducted to state controllability. Continuous or "functional" output controllability will be investigated in Section 4.3 by using the geometric approach.

We shall now consider observability of linear time-invariant systems. Sets $Q^-(t_0,t_1,u(\cdot),y(\cdot))$ and $Q^+(t_0,t_1,u(\cdot),y(\cdot))$, defined in Section 1.4, have special features in the linear system case, by virtue of Property 1.3-4 (decomposability of the response function). The following two properties will be proved together.

Property 2.6-3.

Like any other system, a linear system completely observable in $[t_0,t_1]$ is also reconstructable in $[t_0,t_1]$. The converse is also true in the case of linear continuous systems.

Property 2.6-4.

A linear system completely observable by a diagnosis experiment or reconstructable by a homing experiment is simply completely observable or reconstructable. In other words, for linear systems there is no advantage in using a special input function to derive the initial or final state from the response function.

Proof. The response decomposition property implies that

$$Q^-(t_0,t_1,u(\cdot),y(\cdot)) = \\ \{x : y(\tau) = \gamma(\tau,t_0,x,0) + \gamma(\tau,t_0,0,u(\cdot)), \tau \in [t_0,t_1]\}$$

Assume $y_1(t) = \gamma(t,t_0,0,u(\cdot)) + D(t)u(t)$. It follows that

$$Q^-(t_0,t_1,u(\cdot),y(\cdot)) = Q^-(t_0,t_1,0,y(\cdot)-y_1(\cdot)) = Q^-(t_0,t_1,0,y_0(\cdot)) \quad (12)$$

where $y_0(\cdot) := y(\cdot) - y_1(\cdot)$ denotes the free response, which depends only on the state at time t_0. Then, from (1.4.10)

$$Q^+(t_0,t_1,u(\cdot),y(\cdot)) = \Phi(t_1,t_0) Q^-(t_0,t_1,0,y_0(\cdot)) + \{\varphi(t_1,t_0,0,u(\cdot))\} \quad (13)$$

If set (12) reduces to a single element, (13) also does, while the contrary is true only if matrix $\Phi(t_1,t_0)$ is nonsingular, i.e., in particular, for continuous systems. If set (12) or set (13) reduces to a single element, this occurs independently on input $u(\cdot)$. □

Property 2.6-5.
In the case of linear systems sets $Q^-(t_0,t_1,0,0)$, $Q^+(t_0,t_1,0,0)$ are subspaces of $\mathcal{X}$.

Proof. Set $Q^-(t_0,t_1,0,0)$ is a subspace, being the kernel of the linear transformation from $\mathcal{X}$ to $\mathcal{Y}_f$ which associates to any $x \in \mathcal{X}$ function $y(\tau) = \gamma(\tau,t_0,x,0)$, $\tau \in [t_0,t_1]$. Set $Q^+(t_0,t_1,0,0)$ is a subspace, being its image in the linear transformation from $\mathcal{X}$ to $\mathcal{X}$ corresponding to the transition matrix $\Phi(t_1,t_0)$. $\square$

The following result is a consequence of zero-state response linearity.

Property 2.6-6.
In the case of linear systems set $Q^-\bigl(t_0,t_1,0,y_0(\cdot)\bigr)$ is the linear variety defined as the sum of any initial state corresponding to free response $y_0(\cdot)$ in $[t_0,t_1]$ and subspace $Q^-(t_0,t_1,0,0)$, while set $Q^+\bigl(t_0,t_1,0,y_0(\cdot)\bigr)$ is the linear variety defined as the sum of any initial state corresponding to free response $y_0(\cdot)$ in $[t_0,t_1]$ and subspace $Q^+(t_0,t_1,0,0)$.

Computational aspects of observation problems for linear systems are similar to those of control problems. First of all, we shall consider determination of subspaces $Q^-(t_0,t_1,0,0)$ and $Q^+(t_0,t_1,0,0)$.

Theorem 2.6-2.
Refer to system (1.3.8,9). The following equalities hold.

$$Q^-(t_0,t_1,0,0) = \ker Q(t_0,t_1) \tag{14}$$
$$Q^+(t_0,t_1,0,0) = \Phi(t_1,t_0)\ker Q(t_0,t_1) \tag{15}$$

where $Q(t_0,t_1)$ denotes the symmetric positive semidefinite matrix

$$Q(t_0,t_1) := \int_{t_0}^{t_1} \Phi^T(\tau,t_0) C^T(\tau) C(\tau) \Phi(\tau,t_0) \, d\tau \tag{16}$$

Proof. By virtue of Lemma 2.6-1, the relation

$$y_0(t) = C(t)\Phi(t,t_0) x_0 = 0 \quad \forall t \in [t_0,t_1]$$

is satisfied if and only if $x_0 \in \ker Q(t_0,t_1)$. On the other hand, if the initial state x_0 belongs to $\mathrm{im} Q(t_0,t_1)$, it can be uniquely determined from the free response $y_0(t)$. In fact, by using (16) in the equality $x_0 = Q^+(t_0,t_1) Q(t_0,t_1) x_0$, it follows that

$$x_0 = Q^+(t_0,t_1) \int_{t_0}^{t_1} \Phi^T(\tau,t_0) C^T(\tau) C(\tau) \Phi(\tau,t_0) x_0 \, d\tau$$

i.e.

$$x_0 = Q^+(t_0,t_1) \int_{t_0}^{t_1} \Phi^T(\tau,t_0) C^T(\tau) y_0(\tau) \, d\tau \tag{17}$$

Relation (15) directly follows from (13). $\square$

Sect. 2.6 Controllability and Observability

As a consequence of the previous theorem, it can be stated that system (1.3.8,9) is completely observable and reconstructable in $[t_0, t_1]$ (i.e., $Q^-(t_0,t_1,0,0) = Q^+(t_0,t_1,0,0) = \{0\}$) if and only if matrix $Q(t_0,t_1)$ defined in (16) is nonsingular (strictly positive definite).

Problem 2.6-2. Observing the Initial State
Refer to system (1.3.8,9). Given functions $u(\cdot)$, $y(\cdot)$ in the time interval $[t_0, t_1]$, determine the initial state x_0.

Solution. Derive the free response

$$y_0(t) = y(t) - C(t) \int_{t_0}^{t} \Phi(t,\tau) B(\tau) u(\tau) \, d\tau - D(t) u(t)$$

and use (17). By virtue of a property of the pseudoinverse matrix, the right side of (17) directly provides the orthogonal projection of the initial state on the orthogonal complement of $Q^-(t_0,t_1,0,0)$; this coincides with x_0 if the system is completely observable. □

The problem of reconstructing final state x_1, or its orthogonal projection x_2 on the orthogonal complement of $Q^+(t_0,t_1,0,0)$ when the system is not completely reconstructable, can be solved in a similar way. If the system is completely observable, i.e., if the problem of determining the initial state has a unique solution, from (2.2.2) with $t = t_1$ it is possible to derive x_1 if x_0 is known.

To solve Problem 2.6-2 from a computational standpoint it is still convenient to use the adjoint system. In fact, consider the system

$$\dot{p}(t) = -A^T(t) p(t) + C^T(t) y_0(t) \tag{18}$$

with initial condition $p(t_0) = 0$. Solving in $[t_0, t_1]$ provides

$$p(t_1) = \int_{t_0}^{t_1} \Psi(t_1, \tau) C^T(\tau) y_0(\tau) \, d\tau$$

$$= \Phi^T(t_0, t_1) \int_{t_0}^{t_1} \Phi^T(\tau, t_0) C^T(\tau) y_0(\tau) \, d\tau$$

By comparison with (17), if the system is completely observable (hence matrix $Q(t_0, t_1)$ is invertible), the initial state can be derived as

$$x_0 = Q^{-1}(t_0, t_1) \Psi(t_0, t_1) p(t_1) \tag{19}$$

The observation or reconstruction of the state can be realized "on line" as follows: by means of a model of system (1.3.8,9) with zero initial state determine the forced response $y_1(\cdot)$ and subtract it from the output function to obtain the free response $y_0(\cdot)$, which, in turn, is the input function to the adjoint system (18), whose solution with zero initial condition provides the value $p(t_1)$ to use in (19). The final state by virtue of (2.2.2) is expressed as

$$x_1 = \Phi(t_1, t_0) Q^{-1}(t_0, t_1) \Psi(t_0, t_1) p(t_1) + \varphi_1(t_1)$$

where $\varphi_1(t_1)$ is the final value of the forced motion.

Extension to Discrete Systems. The extension of the above to discrete systems is straightforward, so that we shall report the main results without proof.

Refer to system (1.3.10, 11). The following equalities hold.

$$\mathcal{R}^+(j, i, 0) = \text{im} P(j, i) \tag{20}$$
$$\mathcal{R}^-(j, i, 0) = \Phi^{-1}(i, j) \text{im} P(j, i) \tag{21}$$

where $P(j, i)$ denotes the symmetric positive semidefinite matrix

$$P(j, i) := \sum_{k=j}^{i-1} \Phi(i, k+1) B_d(k) B_d^T(k) \Phi^T(i, k+1)$$

The problem of controlling the state trajectory between two given states x_0, x_1 in the time interval $[j, i]$ is solved by

$$u(k) = B_d^T(k) \Phi^T(i, k+1) P^+(j, i) x_2, \quad k \in [j, i-1] \tag{22}$$

with $x_2 := x_1 - \Phi(i, j) x_0$.

The subspaces of the initial and final states corresponding to zero response in the time interval $[j, i]$ for system (1.3.10, 11) are, respectively

$$\mathcal{Q}^-(j, i, 0, 0) = \ker Q(j, i) \tag{23}$$
$$\mathcal{Q}^+(j, i, 0, 0) = \Phi(i, j) \ker Q(j, i) \tag{24}$$

where $Q(j, i)$ denotes the symmetric positive semidefinite matrix

$$Q(j, i) := \sum_{k=j}^{i} \Phi^T(k, j) C_d^T(k) C_d(k) \Phi(k, j)$$

The problem of determining the initial state x_0 (or its orthogonal projection on the orthogonal complement of $\mathcal{Q}^-(j, i, 0, 0)$ when the system is not completely observable) from functions $u(\cdot)$, $y(\cdot)$ given in the time interval $[j, i]$, can be solved by means of

$$y_0(k) = y(k) - C_d(k) \sum_{h=j}^{k-1} \Phi(k, h+1) B_d(h) u(h) - D_d(k) u(k), \quad k \in [j, i] \tag{25}$$

$$x_0 = Q^+(j, i) \sum_{k=j}^{i} \Phi^T(k, j) C_d^T(k) y_0(k) \tag{26}$$

2.6.2 Linear Time-Invariant Systems

Consistent with notation introduced in Section 1.4, denote by $\mathcal{R}_{t_1}^+(x)$ the reachable set from x in $[0, t_1]$ and by $\mathcal{R}_{t_1}^-(x)$ the controllable set to x in $[0, t_1]$. By Property 2.6-1 $\mathcal{R}_{t_1}^+(0)$ and $\mathcal{R}_{t_1}^-(0)$ are subspaces of $\mathcal{X}$, while by Property 2.6-2 $\mathcal{R}_{t_1}^+(x)$ and $\mathcal{R}_{t_1}^-(x)$ are linear varieties contained in $\mathcal{X}$.

Property 2.6-7.
In the case of linear time-invariant systems the following equalities hold:

$$\mathcal{R}_{t_1}^+(0) \subseteq \mathcal{R}_{t_2}^+(0) \quad \text{for } t_1 \leq t_2 \tag{27}$$

$$\mathcal{R}_{t_1}^-(0) \subseteq \mathcal{R}_{t_2}^-(0) \quad \text{for } t_1 \leq t_2 \tag{28}$$

Proof. Let $x_1 \in \mathcal{R}_{t_1}^+(0)$, so that a control function $u_1(t)$, $t \in [0, t_1]$ exists, which drives the state from zero to x_1 at time t_1; the control function defined as $u_2(t) = 0$ for $t \in [0, t_2 - t_1]$ and $u_2(t) = u_1(t - t_2 + t_1)$ for $t \in [t_2 - t_1, t_2]$ clearly drives the state from zero to x_1 at time t_2, hence (27) holds. A similar argument proves (28). □

The following corollary is an immediate consequence of Property 2.6-7.

Corollary 2.6-1.
In the case of linear time-invariant systems the following equalities hold:

$$\mathcal{W}_{t_1}^+(0) = \mathcal{R}_{t_1}^+(0) \tag{29}$$

$$\mathcal{W}_{t_1}^-(0) = \mathcal{R}_{t_1}^-(0) \tag{30}$$

Let us consider, in particular, linear time-invariant continuous systems. The basic result for controllability is stated in the following theorem.

Theorem 2.6-3.
Refer to system (2.2.15, 16). The reachable set from the origin in $[0, t_1]$ is

$$\mathcal{R}_{t_1}^+(0) = \operatorname{im} P \quad \text{with} \quad P := [B|AB| \ldots |A^{n-1}B] \tag{31}$$

Proof. From Lemma 2.6-1 and Theorem 2.6-1 it follows that any x_1 such that

$$x_1 \in \mathcal{R}_{t_1}^+(0) \tag{32}$$

satisfies

$$B^T \Phi^T(t_1, \tau) x_1 = 0 \quad \forall \tau \in [0, t_1] \tag{33}$$

and, conversely, (32) holds for all x_1 satisfying (33). Let $t := t_1 - \tau$, so that (33) is written as

$$B^T e^{A^T t} x_1 = 0 \quad \forall t \in [0, t_1] \tag{34}$$

Since the function on the left of (34) is analytic and identically zero in the interval $[0, t_1]$, all its derivatives are also identically zero. Differentiating at $t = 0$ yields

$$B^T (A^T)^i x_1 = 0 \quad (i = 0, \ldots, n-1) \tag{35}$$

In (35) it is not necessary to consider powers higher than $n-1$. In fact, the Cayley-Hamilton theorem implies

$$A^n = -(a_1 A^{n-1} + a_2 A^{n-2} + \ldots + a_n I)$$

hence

$$A^n B = -(a_1 A^{n-1} B + a_2 A^{n-2} B + \ldots + a_n B) \tag{36}$$

Transpose and multiply on the right by x_1. Then

$$B^T (A^T)^n x_1 = -(a_1 B^T (A^T)^{n-1} x_1 + \ldots + a_n B^T x_1)$$

so that (35) are satisfied also for $i \geq n$. Recall the matrix exponential power series expansion: clearly relations (35) are not only necessary, but also sufficient for (34) to hold. On the other hand, (35) are equivalent to $x_1 \in \ker P^T$. Since (34) is satisfied if and only if (32) holds, (31) is proved. □

The following property expresses Theorem 2.6-3 in coordinate-free form and can be considered as a first step towards a geometric settling of the controllability and observability concepts.

Property 2.6-8.
Refer to system (2.2.15, 16). The reachable set from the origin in $[0, t_1]$ can be expressed as

$$\mathcal{R}^+_{t_1}(0) = \min \mathcal{J}(A, \mathcal{B}) \tag{37}$$

where $\min \mathcal{J}(A, \mathcal{B})$ denotes the minimal A-invariant containing $\mathcal{B} := \mathrm{im} B$.[3]

Proof. It has already been proved that $\mathrm{im} P$ is an A-invariant. Since all the columns of matrix P clearly belong to all other A-invariants containing $\mathcal{B}$, $\mathrm{im} P$ coincides with $\min \mathcal{J}(A, \mathcal{B})$. □

Property 2.6-9.
Refer to system (2.2.15, 16). The controllable set to the origin is equal to the reachable set from the origin, i.e.

$$\mathcal{R}^-_{t_1}(0) = \mathcal{R}^+_{t_1}(0) \tag{38}$$

Proof. Consider (5), which in this case can be written as

$$\mathcal{R}^-_{t_1}(0) = e^{-At_1} \mathcal{R}^+_{t_1}(0)$$

and note that any A-invariant is also an invariant with respect to e^{At} and e^{-At}, as immediately follows from the power series expansion of the matrix exponential; equality (38) is implied by e^{-At_1} being nonsingular. □

---

[3] The set of all A-invariants containing a given subspace $\mathcal{B}$ is a non-distributive lattice with respect to $\subseteq, +, \cap$, which admits a supremum, the whole space $\mathcal{X}$, and an infimum, $\min \mathcal{J}(A, \mathcal{B})$.

Sect. 2.6 Controllability and Observability

It is remarkable that the expressions for $\mathcal{R}_{t_1}^+(0)$ and $\mathcal{R}_{t_1}^-(0)$ derived earlier are independent of t_1, i.e., in the case of linear time-invariant continuous systems the reachable subspace and the controllable subspace do not depend on the length of time for control, provided it is nonzero.

From now on, the simple symbol $\mathcal{R}$ will be used for many sets referring to controllability of linear time-invariant continuous systems:

$$\mathcal{R} := \min \mathcal{J}(A, B) = \mathcal{R}_{t_1}^+(0) = \mathcal{R}_{t_1}^-(0) = \mathcal{W}_{t_1}^+(0) = \mathcal{W}_{t_1}^-(0) \tag{39}$$

Subspace $\mathcal{B} := \mathrm{im}\, B$ will be called the *forcing actions subspace* and $\mathcal{R}$ the *controllability set* or *controllability subspace*. System (2.2.15, 16) will be said to be *completely controllable* if $\mathcal{R} = \mathcal{X}$. In this case it is also customary to say that *the pair* (A, B) is controllable.

Observability of linear time-invariant systems is now approached in a similar way. Denote by $\mathcal{Q}_{t_1}^-(u(\cdot), y(\cdot))$ and $\mathcal{Q}_{t_1}^+(u(\cdot), y(\cdot))$ the sets of all initial and final states compatible with input function $u(\cdot)$ and output function $y(\cdot)$ in $[0, t_1]$, also called the unobservable set and the unreconstructable set in $[0, t_1]$ with respect to the given input and output functions. By Property 2.6-5 $\mathcal{Q}_{t_1}^-(0, 0)$ and $\mathcal{Q}_{t_1}^+(0, 0)$ are subspaces of $\mathcal{X}$, while by Property 2.6-6 $\mathcal{Q}_{t_1}^-(u(\cdot), y(\cdot))$ and $\mathcal{Q}_{t_1}^+(u(\cdot), y(\cdot))$ are linear varieties contained in $\mathcal{X}$.

Property 2.6-10.
In the case of linear time-invariant systems the following equalities hold:

$$\mathcal{Q}_{t_1}^-(0,0) \supseteq \mathcal{Q}_{t_2}^-(0,0) \quad \text{for } t_1 \leq t_2 \tag{40}$$

$$\mathcal{Q}_{t_1}^+(0,0) \supseteq \mathcal{Q}_{t_2}^+(0,0) \quad \text{for } t_1 \leq t_2 \tag{41}$$

Proof. Initial states belonging to $\mathcal{Q}_{t_2}^-(0,0)$ cause zero free response in the time interval $[0, t_2]$, which contains $[0, t_1]$, so that they also belong to $\mathcal{Q}_{t_1}^-(0,0)$ and (40) is proved. A similar argument proves (41). □

Let us consider, in particular, linear time-invariant continuous systems. The most important result of observability is stated in the following theorem, dual of Theorem 2.6-3.

Theorem 2.6-4.
Refer to system (2.2.15, 16). The zero-input unobservable set in $[0, t_1]$ is

$$\mathcal{Q}_{t_1}^-(0,0) = \ker Q \quad \text{with} \quad Q^T := [C^T | A^T C^T | \ldots | (A^T)^{n-1} C^T] \tag{42}$$

Proof. From Lemma 2.6-1 and Theorem 2.6-2 it follows that any x_0 such that

$$x_0 \in \mathcal{Q}_{t_1}^-(0,0) \tag{43}$$

satisfies
$$C\Phi(\tau,0)x_0 = 0 \quad \forall \tau \in [0,t_1] \tag{44}$$
and, conversely, (43) holds for all x_0 satisfying (44). Relation (44) can also be written as
$$Ce^{A\tau}x_0 = 0 \quad \forall \tau \in [0,t_1] \tag{45}$$
and by the argument considered in the proof of Theorem 2.6-3 condition $x_0 \in \ker Q$ is proved to be necessary and sufficient for (43) to hold. □

Theorem 2.6-4 can be stated in coordinate-free form as follows.

Property 2.6-11.
Refer to system (2.2.15,16). The zero-input unobservable set in $[0,t_1]$ can be expressed as
$$\mathcal{Q}_{t_1}^-(0) = \max \mathcal{J}(A,\mathcal{C}) \tag{46}$$
where $\max \mathcal{J}(A,\mathcal{C})$ denotes the maximal A-invariant contained in $\mathcal{C} := \ker C$.[4]

Proof. From the proof of Property 2.6-8 it follows that $\mathrm{im} Q^T$ is the minimal A^T-invariant containing $\mathrm{im} C^T$, so that its orthogonal complement $\ker Q$ is the maximal A-invariant contained in $\ker C$. □

Property 2.6-12.
Refer to system (2.2.15,16). The zero-input unreconstructable set in $[0,t_1]$ is equal to the zero-input unobservable set, i.e.
$$\mathcal{Q}_{t_1}^+(0,0) = \mathcal{Q}_{t_1}^-(0,0) \tag{47}$$

Proof. Consider (15), which in this case can be written as
$$\mathcal{Q}_{t_1}^+(0,0) = e^{At_1} \mathcal{Q}_{t_1}^-(0,0)$$
Equality (47) follows from $\mathcal{Q}_{t_1}^-(0,0)$ being an A-invariant, hence an invariant also with respect to e^{At_1} and from the matrix exponential being nonsingular. □

The above expressions for $\mathcal{Q}_{t_1}^-(0,0)$ and $\mathcal{Q}_{t_1}^+(0,0)$ are independent of t_1, so that in the case of linear time-invariant continuous systems the zero-input unobservable subspace and the zero-input unreconstructable subspace do not depend on the length of time for observation, provided it is nonzero.

In the following, the simple symbol $\mathcal{Q}$ will be used for both:
$$\mathcal{Q} := \max \mathcal{J}(A,\mathcal{C}) = \mathcal{Q}_{t_1}^-(0,0) = \mathcal{Q}_{t_1}^+(0,0) \tag{48}$$

The subspace $\mathcal{C} := \ker C$ will be called the *inaccessible states subspace* and $\mathcal{Q}$ the *unobservability set* or *unobservability subspace*. System (2.2.15,16) will

---
[4] The set of all A-invariants contained in a given subspace $\mathcal{C}$ is a nondistributive lattice with respect to $\subseteq, +, \cap$, which admits a supremum, $\max \mathcal{J}(A,\mathcal{C})$, and an infimum, the origin $\{0\}$.

Sect. 2.6 Controllability and Observability

be said to be *completely observable* if $Q=\{0\}$. In this case it is also customary to say that *the pair (A, C) is observable*.

Extension to Discrete Systems. We shall now extend the results on controllability and observability to linear time-invariant discrete systems. Refer to system (2.2.17,18): let $\mathcal{R}_i^+(0)$ be the reachable set from the origin in i steps. It is easy to see that

$$\mathcal{R}_i^+(0) = \text{im} P_i \quad \text{with} \quad P_i := [B_d | A_d B_d | \ldots | A_d^{i-1} B_d] \quad (i=1,2,\ldots) \quad (49)$$

In fact, let $x_1 \in \mathcal{R}_i^+(0)^\perp$. From (20) and Theorem A.5-4 it follows that $B_d^T (A_d^T)^i x_1 = 0$, hence $x_1 \in \ker P_i^T$. The Cayley-Hamilton theorem implies that the maximal reachable subspace is attained in a number of steps not greater than n. It coincides with the minimal A_d-invariant containing the forcing action subspace $\mathcal{B}_d := \text{im} B_d$, i.e., $\mathcal{R}^+(0) := \lim_{i \to \infty} \mathcal{R}_i^+(0) = \mathcal{R}_n^+(0) = \min \mathcal{J}(A_d, \mathcal{B}_d)$.

The controllable set to the origin in i steps is determined from (21) as $\mathcal{R}_i^-(0) = A_d^{-i} \mathcal{R}_i^+(0)$. Moreover, $\mathcal{R}^-(0) := \lim_{i \to \infty} \mathcal{R}_i^-(0) = \mathcal{R}_n^-(0) = A_d^{-n} \min \mathcal{J}(A_d, \mathcal{B}_d)$.

The zero-input unobservable subspace in i steps $Q_i^-(0,0)$ $(i=0,1,\ldots)$ is expressed by

$$Q_i^-(0,0) = \ker Q_i \quad \text{with} \quad Q_i^T := [C_d^T | A_d^T C_d^T | \ldots | (A_d^T)^i C_d^T] \quad (i=0,1,\ldots) \quad (50)$$

In fact, let $x_0 \in Q_i^-(0,0)^\perp$. From (23) and Theorem A.5-4 it follows that $C_d A_d^i x_0 = 0$, hence $x_0 \in \ker Q_i^T$. The Cayley-Hamilton theorem implies that the minimal unobservable subspace corresponds to a number of steps not greater than $n-1$. It coincides with the maximal A_d-invariant contained in the inaccessible states subspace $\mathcal{C}_d := \ker C_b$, i.e., $Q^-(0,0) := \lim_{i \to \infty} Q_i^-(0,0) = Q_{n-1}^-(0,0) = \max \mathcal{J}(A_d, \mathcal{C}_d)$.

The zero-input unobservable subspace in i steps is determined from (24) as $Q_i^+(0,0) = A_d^i Q_i^-(0,0)$. Moreover, $Q^+(0,0) := \lim_{i \to \infty} Q_i^+(0,0) = Q_n^+(0,0) = A_d^n \max \mathcal{J}(A_d, \mathcal{C}_d)$.

As for the continuous system, we define $\mathcal{R} := \mathcal{R}^+(0) = \min \mathcal{J}(A_d, \mathcal{B}_d)$ and $\mathcal{Q} := \mathcal{Q}^-(0,0) = \max \mathcal{J}(A_d, \mathcal{C}_d)$. However, for discrete systems the equalities $\mathcal{R}^-(0) = \mathcal{R}^+(0)$ and $\mathcal{Q}^-(0,0) = \mathcal{Q}^+(0,0)$ are not true in general. They are replaced by the inclusions $\mathcal{R}^-(0) \supseteq \mathcal{R}^+(0)$ and $\mathcal{Q}^-(0,0) \supseteq \mathcal{Q}^+(0,0)$, which derive from the general property that if $\mathcal{J}$ is an A-invariant, both $A\mathcal{J}$ and $A^{-1}\mathcal{J}$ are A-invariants, hence, by a recursion argument, $A^i \mathcal{J}$ and $A^{-i} \mathcal{J}$ $(i=2,3,\ldots)$ are A-invariants.

REFERENCES

1. AIZERMAN, M.A., and GANTMACHER, F.R., *Absolute Stability of Regulator Systems*, Holden-Day, San Francisco, 1964.
2. BARTELS, R.H., and STEWART, G.W., "Solution of the matrix equation $AX+XB=C$," *Communications of the ACM*, vol. 15, no. 9, pp. 820–826, 1972.
3. BELLMAN, R., *Stability Theory of Differential Equations*, McGraw-Hill, New York, 1953.
4. BROCKETT, R.W., *Finite Dimensional Linear Systems*, Wiley, New York, 1970.
5. BYERS, R., "A LINPACK-style condition estimator for the equation $AX - XB^T = C$," *IEEE Trans. on Aut. Control*, vol. AC-29, pp. 926–928, 1984.
6. CALLIER, F.M., and DESOER, C.A., *Multivariable Feedback Systems*, Springer-Verlag, New York, 1982.
7. CHEN, C.T., *Introduction to Linear System Theory*, Holt, Rinehart & Winston, New York, 1970.
8. —, *Linear System Theory and Design*, Holt, Rinehart & Winston, New York, 1984.
9. DE RUSSO, P.M., ROY, R.I., and CLOSE, C.M., *State Variables for Engineers*, Wiley, New York, 1967.
10. HAHN, W., *Theory and Application of Liapunov's Direct Method*, Prentice Hall, Englewood Cliffs, N.J., 1963.
11. HALE, J.K., *Oscillation in Nonlinear Systems*, McGraw-Hill, New York, 1963.
12. KALMAN, R.E., "On structural properties of linear, constant, multivariable systems," *Proceedings of the 3rd IFAC Congress*, paper 6a, 1966.
13. KALMAN, R.E., and BERTRAM, J.E., "Control system analysis and design via the second method of Liapunov - I Continuous-time systems," *Trans. of the ASME, J. of Basic Engineering*, vol. 82, pp. 371–393, 1960.
14. —, "Control system analysis and design via the second method of Liapunov - II Discrete-time systems," *Trans. of the ASME, J. of Basic Engineering*, vol. 82, pp. 394–400, 1960.
15. LA SALLE, J., and LEFSCHETZ, S., *Stability by Liapunov's Direct Method with Applications*, Academic, New York, 1961.
16. LEFSCHETZ, S., *Stability of Nonlinear Control Systems*, Academic, New York, 1965.
17. LIAPUNOV, A.M., "Problème général de la stabilité du mouvement," *Ann. Fac. Sci. Toulouse*, vol. 9, pp. 203–474, 1907.
18. LUENBERGER, D.G., "Invertible solutions to the operator equation TA-BT=C," *Proc. Am. Math. Soc.*, vol. 16, pp. 1226–1229, 1965.

19. MACFARLANE, A.G.J., "System matrices," *Proc. IEE*, vol. 115, no. 5, pp. 749–754, 1968.
20. —, "The development of frequency-response methods in automatic control," *IEEE Trans. Autom. Control*, vol. AC-24, no. 2, pp. 250–265, 1979.
21. MINORSKY, N., *Introduction to Nonlinear Mechanics*, J. W. Edwards, Ann Arbor, 1947.
22. MOLINARI, B.E., "Algebraic solution of matrix linear equations in control theory," *Proc. IEE*, vol. 116, pp. 1748–1754.
23. OGATA, K., *State Space Analysis of Control Systems*, Prentice Hall, Englewood Cliffs, N.J., 1967.
24. OSTROWSKI, A., and SCHNEIDER, H., "Some theorems on the inertia of general matrices," *J. of Math. Analysis and Applications*, vol. 4, pp. 72–84, 1962.
25. ROSENBROCK, H.H., "Transformation of linear constant system equations," *Proc. IEE*, vol. 114, no. 4, pp. 541–544, 1967.
26. —, "On linear system theory," *Proc. IEE*, vol. 114, no. 9, pp. 1353–1359, 1967.
27. —, "Computation of minimal representations of a rational transfer-function matrix," *Proc. IEE*, vol. 115, no. 2, pp. 325–327, 1968.
28. —, *State-space and Multivariable Theory*, Nelson, London, 1970.
29. ROSENBROCK, H.H., and STOREY, C., *Mathematics of Dynamical Systems*, Nelson, London, 1970.
30. SANSONE, G., and CONTI, R., *Equazioni Differenziali Non Lineari*, Edizioni Cremonese, Rome, 1956.
31. SCHULTZ, D.G., and MELSA, J.L., *State Function and Linear Control Systems*, McGraw-Hill, New York, 1967.
32. TOU, J.T., *Modern Control Theory*, McGraw-Hill, New York, 1964.
33. VARAH, J.M., "On the separation of two matrices," *SIAM J. Numer. Anal.*, vol. 16, pp. 216–222, 1979.
34. VIDYASAGAR, M., *Nonlinear System Analysis*, Prentice Hall, Englewood Cliffs, N.J., 1978.
35. VIDYASAGAR, M., *Control System Synthesis: A Factorization Approach*, The MIT Press, Cambridge, Mass., 1985.
36. WILLEMS, J.L., *Stability Theory of Dynamical Systems*, Wiley, New York, 1970.

3
THE GEOMETRIC APPROACH: CLASSIC FOUNDATIONS

3.1 INTRODUCTION

The essence of the geometric approach consists of developing most of the mathematical support in coordinate-free form, to take advantage of simpler and more elegant results, which facilitate insight into the actual meaning of statements and procedures; the computational aspects are considered independently of the theory and handled by means of the standard methods of matrix algebra, once a suitable coordinate system is defined. The cornerstone of the approach is the concept of invariance of a subspace with respect to a linear transformation. In this chapter the properties and geometric meaning of invariants are presented and investigated, and their connection with the most important classical system theory problems like controllability, observability, and pole assignability is pointed out.

3.1.1 Some Subspace Algebra

The coordinate-free approach used from now on in this book requires a computational background in terms of operations and transformations involving subspaces, which are reflected, of course, in numerical procedures referring to their basis matrices. The most important of these operations and some of their properties are briefly reviewed in this subsection.

Consider subspaces $\mathcal{X}, \mathcal{Y}, \mathcal{Z}$ of finite-dimensional inner product vector spaces $\mathbf{R}^n, \mathbf{R}^m$ ($\mathbf{C}^n, \mathbf{C}^m$) and denote with A both an $m \times n$ real (complex) matrix and its corresponding linear transformation from $\mathbf{R}^n$ to $\mathbf{R}^m$ (from $\mathbf{C}^n$ to $\mathbf{C}^m$).

The basic operations on subspaces are:

1. *Sum:*
$$\mathcal{Z} = \mathcal{X} + \mathcal{Y} := \{z : z = x + y,\ x \in \mathcal{X},\ y \in \mathcal{Y}\} \tag{1}$$

2. *Linear transformation:*
$$\mathcal{Y} = A\mathcal{X} := \{y : y = Ax,\ x \in \mathcal{X}\} \tag{2}$$

3. *Orthogonal complementation:*
$$\mathcal{Y} = \mathcal{X}^\perp := \{y : \langle y, x \rangle = 0,\ x \in \mathcal{X}\} \tag{3}$$

4. *Intersection:*
$$\mathcal{Z} = \mathcal{X} \cap \mathcal{Y} := \{z : z \in \mathcal{X},\ z \in \mathcal{Y}\} \tag{4}$$

5. *Inverse linear transformation:*
$$\mathcal{X} = A^{-1}\mathcal{Y} := \{x : y = Ax,\ y \in \mathcal{Y}\} \tag{5}$$

The set of all the subspaces of a given vector space $\mathcal{W}$ is a nondistributive lattice with $\subseteq$ as the partial ordering relation and $+, \cap$ as the binary operations. Its universal bounds are $\mathcal{W}$ and $\{0\}$.

The following relations are often considered in algebraic manipulations regarding subspaces. Their proofs, all quite simple, are here omitted for the sake of brevity.

$$\mathcal{X} \cap (\mathcal{Y} + \mathcal{Z}) \supseteq (\mathcal{X} \cap \mathcal{Y}) + (\mathcal{X} \cap \mathcal{Z}) \tag{6}$$
$$\mathcal{X} + (\mathcal{Y} \cap \mathcal{Z}) \subseteq (\mathcal{X} + \mathcal{Y}) \cap (\mathcal{X} + \mathcal{Z}) \tag{7}$$
$$(\mathcal{X}^\perp)^\perp = \mathcal{X} \tag{8}$$
$$(\mathcal{X} + \mathcal{Y})^\perp = \mathcal{X}^\perp \cap \mathcal{Y}^\perp \tag{9}$$
$$(\mathcal{X} \cap \mathcal{Y})^\perp = \mathcal{X}^\perp + \mathcal{Y}^\perp \tag{10}$$
$$A(\mathcal{X} \cap \mathcal{Y}) \subseteq A\mathcal{X} \cap A\mathcal{Y} \tag{11}$$
$$A(\mathcal{X} + \mathcal{Y}) = A\mathcal{X} + A\mathcal{Y} \tag{12}$$
$$A^{-1}(\mathcal{X} \cap \mathcal{Y}) = A^{-1}\mathcal{X} \cap A^{-1}\mathcal{Y} \tag{13}$$
$$A^{-1}(\mathcal{X} + \mathcal{Y}) \supseteq A^{-1}\mathcal{X} + A^{-1}\mathcal{Y} \tag{14}$$

Relations (6, 7) show that the lattice of all subspaces of a given vector space is nondistributive. A particular case in which they hold with the equality sign is considered in the following property.

Property 3.1-1.
Relations (6, 7) hold with the equality sign if any one of the involved subspaces $\mathcal{X}, \mathcal{Y}, \mathcal{Z}$ is contained in any of the others.

Proof. First, consider (6). Let $\mathcal{Y} \subseteq \mathcal{X}$ and x be any vector belonging to the subspace on the left, so that $x \in \mathcal{X}$ and there exist two vectors $y \in \mathcal{Y}$ and $z \in \mathcal{Z}$ such that $x = y + z$ but, since $y \in \mathcal{X}$, also $z \in \mathcal{X}$, then $x \in (\mathcal{X} \cap \mathcal{Y}) + (\mathcal{X} \cap \mathcal{Z})$. If $\mathcal{X} \subseteq \mathcal{Y}$, both members reduce to $\mathcal{X}$, while, if $\mathcal{Y} \subseteq \mathcal{Z}$, both members reduce to $\mathcal{X} \cap \mathcal{Z}$.
Now consider (7). Let $\mathcal{X} \subseteq \mathcal{Y}$ and y be any vector belonging to the subspace on the right, so that $y \in \mathcal{Y}$ and there exist two vectors $x \in \mathcal{X}$ and $z \in \mathcal{Z}$ such that $y = x + z$ but, since $x \in \mathcal{Y}$, also $z \in \mathcal{Y}$, then $y \in \mathcal{X} + (\mathcal{Y} \cap \mathcal{Z})$. If $\mathcal{Y} \subseteq \mathcal{X}$, both members reduce to $\mathcal{X}$, while, if $\mathcal{Z} \subseteq \mathcal{Y}$, both members reduce to $\mathcal{X} + \mathcal{Z}$. □

Two other properties, which relate a generic subspace to its orthogonal complement and allow interesting dualities to be set, are presented. One states, in particular, that the orthogonal complement of an A-invariant is an A^T-invariant in the real field and an A^*-invariant in the complex field, and the other suggests a procedure to compute the inverse transform of a subspace.

Property 3.1-2.
Consider a linear map $A : \mathcal{F}^n \to \mathcal{F}^m$ ($\mathcal{F} = \mathbf{R}$ or $\mathcal{F} = \mathbf{C}$) and any two subspaces $\mathcal{X} \subseteq \mathcal{F}^n$ and $\mathcal{Y} \subseteq \mathcal{F}^m$. In the real and complex fields the following relations hold:

$$A\mathcal{X} \subseteq \mathcal{Y} \Leftrightarrow A^T \mathcal{Y}^\perp \subseteq \mathcal{X}^\perp \qquad (15)$$

$$A\mathcal{X} \subseteq \mathcal{Y} \Leftrightarrow A^* \mathcal{Y}^\perp \subseteq \mathcal{X}^\perp \qquad (16)$$

Proof. Refer to (14). Inclusion on the left implies $\langle Ax, y \rangle = 0$ for all $x \in \mathcal{X}$ and for all $y \in \mathcal{Y}^\perp$ or, equivalently, $\langle x, A^T y \rangle$ for all $x \in \mathcal{X}$ and for all $y \in \mathcal{Y}^\perp$ which, in turn, implies and is implied by $A^T \mathcal{Y}^\perp \subseteq \mathcal{X}^\perp$. □

Property 3.1-3.
Consider a linear map $A : \mathcal{F}^n \to \mathcal{F}^m$ ($\mathcal{F} = \mathbf{R}$ or $\mathcal{F} = \mathbf{C}$) and a subspace $\mathcal{Y} \subseteq \mathcal{F}^m$. In the real and complex fields the following relations hold:

$$(A^{-1} \mathcal{Y})^\perp = A^T \mathcal{Y}^\perp \qquad (17)$$

$$(A^{-1} \mathcal{Y})^\perp = A^* \mathcal{Y}^\perp \qquad (18)$$

Proof. Refer to (17). Let Y be a basis matrix of $\mathcal{Y}^\perp$, so that $\operatorname{im} Y = \mathcal{Y}^\perp$, $\ker Y^T = \mathcal{Y}$. It follows that

$$A^T \mathcal{Y}^\perp = A^T \operatorname{im} Y = \operatorname{im}(A^T Y) = \left(\ker(Y^T A)\right)^\perp$$
$$= (A^{-1} \ker Y^T)^\perp = (A^{-1} \mathcal{Y})^\perp \quad □$$

To implement computational procedures for operations on subspaces, the Gauss-Jordan elimination method or the Gram-Schmidt orthonormalization process - both provided with a suitable linear dependence test - can be used.

For instance, computations with the Gram-Schmidt process are realized as follows.

1. *Sum of two subspaces.* Let X, Y be basis matrices of subspaces $\mathcal{X}, \mathcal{Y} \subseteq \mathcal{F}^n$, with $\mathcal{F} = \mathbf{R}$ or $\mathcal{F} = \mathbf{C}$. A basis matrix Z of $\mathcal{Z} := \mathcal{X} + \mathcal{Y}$ is obtained by orthonormalizing the columns of matrix $[X\ Y]$.

2. *Linear transform of a subspace.* Let X be a basis matrix of subspace $\mathcal{X} \subseteq \mathcal{F}^n$. A basis matrix Y of $\mathcal{Y} := A\mathcal{X}$ is obtained by orthonormalizing the columns of matrix AX.

3. *Orthogonal complement of a subspace.* Let X, $n \times h$, be a basis matrix of subspace $\mathcal{X} \subseteq \mathcal{F}^n$. A basis matrix Y of $\mathcal{Y} := \mathcal{X}^\perp$ is obtained by orthonormalizing the columns of matrix $[X\ I_n]$ and selecting the last $n - h$ of the n obtained vectors.

4. *Intersection of two subspaces.* It is reduced to sum and orthogonal complementation by virtue of (8, 10).

5. *Inverse linear transform of a subspace.* It is reduced to direct transform and orthogonal complementation by virtue of (8, 17, 18).

Some useful geometric-approach-oriented matlab routines based on the above subspace computations are reported in Section B.4.

3.2 INVARIANTS

Consider a linear transformation $A : \mathcal{X} \to \mathcal{X}$ with $\mathcal{X} := \mathcal{F}^n$. Recall that an A-invariant is a subspace $\mathcal{J} \subseteq \mathcal{X}$ such that

$$A\mathcal{J} \subseteq \mathcal{J} \tag{1}$$

Property 3.2-1.
A subspace $\mathcal{J}$ with basis matrix V is an A-invariant if and only if there exists a matrix X such that
$$AV = VX \tag{2}$$

Proof. Let v_i ($i = 1, \ldots, r$) be the columns of V: $\mathcal{J}$ is an A-invariant if and only if each transformed column is a linear combination of all columns, i.e., if and only if there exist vectors x_i such that $Av_i = Vx_i$ ($i = 1, \ldots, r$); relation (2) expresses these equalities in compact form. □

3.2.1 Invariants and Changes of Basis

Refer to a linear function $A : \mathcal{F}^n \to \mathcal{F}^m$, with $\mathcal{F} = \mathbf{R}$ or $\mathcal{F} = \mathbf{C}$, represented by an $m \times n$ real or complex matrix A with respect to the main bases of $\mathcal{F}^n$ and $\mathcal{F}^m$. Recall that a change of basis or similarity transformation [1] is defined by

---

[1] Changes of coordinates are also treated in Section A.2. They are briefly recalled here for the sake of completeness and in simpler form (the first reference bases are main bases).

two nonsingular real or complex matrices P, Q whose columns are the vectors of the new bases expressed with respect to the main ones. If x, y are the old coordinates and ξ, η the new ones, so that $x = P\xi$, $y = Q\eta$, it follows that

$$\eta = Q^{-1} A P \xi = A' \xi, \quad \text{with} \quad A' := Q^{-1} A P$$

If A maps a vector space $\mathcal{F}^n$ into itself, so that it is possible to assume a unique change of basis represented by transformation $T := Q = P$, we obtain, as a special case

$$\xi = T^{-1} A T \xi = A' \xi \quad \text{with} \quad A' := T^{-1} A T$$

A suitable choice of matrix T is often used to point out structural features of the involved linear transformation. Typical examples are the Jordan canonical form and the Schur decomposition presented in Appendix A.

By using (2) it can be immediately shown that invariance is a coordinate-free concept. Let V be a basis matrix of $\mathcal{J}$ and W the transformed basis matrix in the new coordinates defined by T. Clearly $W = T^{-1} V$. The identity

$$T^{-1} A T (T^{-1} V) = (T^{-1} V) X$$

which is easily derived from (2), is equivalent to

$$A' W = W X \tag{3}$$

which proves the assertion.

Theorem 3.2-1.

Let $A : \mathcal{X} \to \mathcal{X}$, $\mathcal{X} := \mathcal{F}^n$, be a linear map and $\mathcal{J} \subseteq \mathcal{X}$ be an A-invariant subspace. There exists a similarity transformation T such that

$$A' := T^{-1} A T = \begin{bmatrix} A'_{11} & A'_{12} \\ O & A'_{22} \end{bmatrix} \tag{4}$$

where A'_{11} is an $h \times h$ matrix with $h := \dim \mathcal{J}$.

Proof. Assume $T := [T_1 \; T_2]$, with $\operatorname{im} T_1 = \mathcal{J}$. Clearly

$$W := T^{-1} T_1 = \begin{bmatrix} I_h \\ O \end{bmatrix}$$

which, together with (3), implies structure (4). $\square$

3.2.2 Lattices of Invariants and Related Algorithms

We shall now investigate some specific properties of invariant subspaces. Let $A : \mathcal{X} \to \mathcal{X}$ be a linear transformation and $\mathcal{B}$ any subspace of $\mathcal{X}$: the set of all A-invariants containing $\mathcal{B}$ is a nondistributive lattice with respect to $\subseteq, +, \cap$. The supremum of the lattice is clearly $\mathcal{X}$, while the infimum is the intersection of all the A-invariants containing $\mathcal{B}$. It will be called the *minimal A-invariant containing* $\mathcal{B}$ and denoted by the symbol $\min \mathcal{J}(A, \mathcal{B})$. It can be determined by means of the following algorithm.

Algorithm 3.2-1. The Minimal A-Invariant Containing imB
Subspace $\min \mathcal{J}(A, \mathcal{B})$ coincides with the last term of the sequence

$$\mathcal{Z}_0 := \mathcal{B} \qquad (5)$$
$$\mathcal{Z}_i := \mathcal{B} + A\mathcal{Z}_{i-1} \quad (i=1,\ldots,k) \qquad (6)$$

where the value of $k \leq n-1$ is determined by condition $\mathcal{Z}_{k+1} = \mathcal{Z}_k$.

Proof. First, note that $\mathcal{Z}_i \supseteq \mathcal{Z}_{i-1}$ ($i=1,\ldots,k$). In fact, instead of (6), consider the recursion expression

$$\mathcal{Z}'_i := \mathcal{Z}'_{i-1} + A\mathcal{Z}'_{i-1} \quad (i=1,\ldots,k)$$

with $\mathcal{Z}'_0 := \mathcal{B}$, which defines a sequence such that $\mathcal{Z}'_i \supseteq \mathcal{Z}'_{i-1}$ ($i=1,\ldots,k$); hence, $A\mathcal{Z}'_i \supseteq A\mathcal{Z}'_{i-1}$ ($i=1,\ldots,k$). This sequence is equal to (6): by induction, note that if $\mathcal{Z}'_j = \mathcal{Z}_j$ ($j=1,\ldots,i-1$), also $\mathcal{Z}'_i = \mathcal{B} + A\mathcal{Z}_{i-2} + A\mathcal{Z}_{i-1} = \mathcal{Z}_i$ (since $A\mathcal{Z}_{i-2} \subseteq A\mathcal{Z}_{i-1}$). If $\mathcal{Z}_{k+1} = \mathcal{Z}_k$, also $\mathcal{Z}_j = \mathcal{Z}_k$ for all $j > k+1$ and $\mathcal{Z}_k$ is an A-invariant containing $\mathcal{B}$. In fact, in such a case $\mathcal{Z}_k = \mathcal{B} + A\mathcal{Z}_k$; hence, $\mathcal{B} \subseteq \mathcal{Z}_k$, $A\mathcal{Z}_k \subseteq \mathcal{Z}_k$. Since two subsequent subspaces are equal if and only if they have equal dimensions and the dimension of the first subspace is at least one, an A-invariant is obtained in at most $n-1$ steps. The last subspace of the sequence is the minimal A-invariant containing $\mathcal{B}$, as can be proved again by induction. Let $\mathcal{J}$ be another A-invariant containing $\mathcal{B}$: if $\mathcal{J} \supseteq \mathcal{Z}_{i-1}$, it follows that $\mathcal{J} \supseteq \mathcal{Z}_i$. In fact, $\mathcal{J} \supseteq \mathcal{B} + A\mathcal{J} \supseteq \mathcal{B} + A\mathcal{Z}_{i-1} = \mathcal{Z}_i$. □

These results are easily dualized: let $A : \mathcal{X} \to \mathcal{X}$ be a linear transformation and $\mathcal{C}$ any subspace of $\mathcal{X}$: the set of all A-invariants contained in $\mathcal{C}$ is a nondistributive lattice with respect to $\subseteq, +, \cap$. The infimum of the lattice is clearly $\{0\}$, while the supremum is the sum of all the A-invariants contained in $\mathcal{C}$. It will be called the *maximal A-invariant contained in* $\mathcal{C}$ and denoted by the symbol $\max \mathcal{J}(A, \mathcal{C})$. It can be determined as follows (in the real case for the sake of simplicity): from

$$A\mathcal{J} \subseteq \mathcal{J} \Leftrightarrow A^T \mathcal{J}^\perp \subseteq \mathcal{J}^\perp$$
$$\mathcal{C} \supseteq \mathcal{J} \Leftrightarrow \mathcal{C}^\perp \subseteq \mathcal{J}^\perp$$

it follows that

$$\max \mathcal{J}(A,\mathcal{C}) = \left(\min \mathcal{J}(A^T, \mathcal{C}^\perp)\right)^\perp \qquad (7)$$

This reduces computation of $\max \mathcal{J}(A,\mathcal{C})$ to that of $\min \mathcal{J}(A,\mathcal{B})$.

Relation (7) can be used to prove the following algorithm, dual of Algorithm 3.2-1.

Algorithm 3.2-2. The Maximal A-Invariant Contained in kerC
Subspace $\max \mathcal{J}(A,\mathcal{C})$ coincides with the last term of the sequence

$$\mathcal{Z}_0 := \mathcal{C} \tag{8}$$
$$\mathcal{Z}_i := \mathcal{C} \cap A^{-1}\mathcal{Z}_{i-1} \quad (i=1,\ldots,k) \tag{9}$$

where the value of $k \leq n-1$ is determined by condition $\mathcal{Z}_{k+1} = \mathcal{Z}_k$.

Proof. Relations (8,9) are equivalent to

$$\mathcal{Z}_0^\perp := \mathcal{C}^\perp$$
$$\mathcal{Z}_i^\perp := (\mathcal{C} \cap A^{-1}\mathcal{Z}_{i-1})^\perp = \mathcal{C}^\perp + A^T \mathcal{Z}_{i-1}^\perp$$

which, by virtue of Algorithm 3.2-1, converge to the orthogonal complement of $\min \mathcal{J}(A^T,\mathcal{C}^\perp)$, which is $\max \mathcal{J}(A,\mathcal{C})$ by (7). □

3.2.3 Invariants and System Structure

Invariant subspaces define the structure of linear transformations, thus playing an important role in linear dynamic system analysis.

Definition 3.2-1. Restriction of a Linear Map
Consider a linear map $A: \mathcal{X} \to \mathcal{X}$, $\mathcal{X} := \mathcal{F}^n$, and an A-invariant subspace $\mathcal{J} \subseteq \mathcal{X}$. The *restriction of A to $\mathcal{J}$* is the linear map $\rho: \mathcal{J} \to \mathcal{J}$ defined by

$$\rho(x) = A(x) \quad \forall x \in \mathcal{J}$$

The restriction of A to $\mathcal{J}$ is usually denoted by $A|_{\mathcal{J}}$.

Let $h := \dim \mathcal{J}$: by virtue of Theorem 3.2-1, $A|_{\mathcal{J}}$ is represented in a suitable basis by an $h \times h$ matrix.

Definition 3.2-2. Induced Map on a Quotient Space
Consider a linear map $A: \mathcal{X} \to \mathcal{X}$, $\mathcal{X} := \mathcal{F}^n$, and an A-invariant subspace $\mathcal{J} \subseteq \mathcal{X}$. The *map induced by A on the quotient space* $\mathcal{X}/\mathcal{J}$ is the map $\sigma: \mathcal{X}/\mathcal{J} \to \mathcal{X}/\mathcal{J}$ defined by

$$\sigma(\{x\}+\mathcal{J}) = \{A(x)\}+\mathcal{J} \quad \forall \ \{x\}+\mathcal{J} \in \mathcal{X}/\mathcal{J}$$

The function induced by A on the quotient space $\mathcal{X}/\mathcal{J}$ is usually denoted by $A|_{\mathcal{X}/\mathcal{J}}$.

Let $n := \dim \mathcal{X}$, $h := \dim \mathcal{J}$: by virtue of Theorem 3.2-1, $A|_{\mathcal{X}/\mathcal{J}}$ is represented in a suitable basis by an $(n-h) \times (n-h)$ matrix.

The following corollary is an immediate consequence of Theorem 3.2-1.

Sect. 3.2 Invariants   133

Corollary 3.2-1.
Let $A: \mathcal{X} \to \mathcal{X}$, $\mathcal{X} := \mathcal{F}^n$, be a linear map and $\mathcal{J}, \mathcal{K} \subseteq \mathcal{X}$ be a pair of A-invariant subspaces such that $\mathcal{J} \oplus \mathcal{K} = \mathcal{X}$. There exists a similarity transformation T such that

$$A' := T^{-1}AT = \begin{bmatrix} A'_{11} & O \\ O & A'_{22} \end{bmatrix} \quad (10)$$

where A'_{11} is an $h \times h$ matrix with $h := \dim \mathcal{J}$ and A'_{22} an $(n-h) \times (n-h)$ matrix.

Any pair of A-invariants $\mathcal{J}, \mathcal{K}$ such that $\mathcal{J} \oplus \mathcal{K} = \mathcal{X}$ is said to *decompose* linear map $A: \mathcal{X} \to \mathcal{X}$ into two restrictions $\rho_1: \mathcal{J} \to \mathcal{J}$ and $\rho_2: \mathcal{K} \to \mathcal{K}$, defined by

$$\rho_1(x) = A(x) \quad \forall x \in \mathcal{J}, \quad \rho_2(x) = A(x) \quad \forall x \in \mathcal{K}$$

Let $h := \dim \mathcal{J}$, $k := \dim \mathcal{K}$: by virtue of Corollary 3.2-1 these restrictions are represented in a suitable basis by an $h \times h$ matrix and a $k \times k$ matrix respectively. According to the previously considered notation, they can be denoted with symbols $A|_\mathcal{J}$ and $A|_\mathcal{K}$. Consider the projections P and Q introduced in Definition A.3-8: the relation

$$\rho_1\big(P(x)\big) + \rho_2\big(Q(x)\big) = A(x) \quad \forall x \in \mathcal{X}$$

clarifies the origin of the expression "to decompose."

Definition 3.2-3. Complementability of an Invariant
Let $A: \mathcal{X} \to \mathcal{X}$, $\mathcal{X} := \mathcal{F}^n$, be a linear map: an A-invariant $\mathcal{J} \subseteq \mathcal{X}$ is said to be *complementable* if there exists an A-invariant $\mathcal{K}$ such that $\mathcal{J} \oplus \mathcal{K} = \mathcal{X}$. If so, $\mathcal{K}$ is called a *complement* of $\mathcal{J}$.

Complementability of invariant subspaces is a very basic concept for system theory applications of linear algebra, since it may be necessary for particular subspaces of the state space to be complementable in order that some control problems have a solution. Necessary and sufficient conditions for complementability are stated in the next two theorems.

Theorem 3.2-2.
Let $A: \mathcal{X} \to \mathcal{X}$, $\mathcal{X} := \mathcal{F}^n$, be a linear map and V be a basis matrix of an A-invariant subspace $\mathcal{J} \subseteq \mathcal{X}$, so that (4) holds with $T := [V \ T_2]$. $\mathcal{J}$ is complementable if and only if the *Sylvester equation* in X

$$A'_{11}X - XA'_{22} = -A'_{12} \quad (11)$$

admits a solution.

Proof. Let $\mathcal{J}$ be complementable and denote by $\mathcal{J}_c$ a complement of $\mathcal{J}$. In the new reference system considered in the proof of Theorem 3.2-1

$$V' = \begin{bmatrix} I_h \\ O \end{bmatrix} \quad \text{and} \quad V'_c = \begin{bmatrix} X \\ I_{n-h} \end{bmatrix}$$

are basis matrices for $\mathcal{J}$ and $\mathcal{J}_c$ respectively. Assuming an identity submatrix in the last $n-h$ rows of the second basis matrix does not affect generality. In fact, this submatrix must be nonsingular (since $[V'\ V_c']$ is nonsingular); postmultiplying by its inverse the basis matrix again provides a basis matrix with the structure of V_c'. Equation (11) immediately derives from the well-known condition (Property 3.2-1) that there exists an F such that

$$A' V' = V_c' F \quad \square$$

In the old reference system a basis matrix for a complement $\mathcal{J}_c$ of $\mathcal{J}$ is given by $T = T_1 X + T_2$, where X is a solution of (11).

Theorem 3.2-3.
Let $A: \mathcal{X} \to \mathcal{X}$, $\mathcal{X} := \mathcal{F}^n$, be a linear map and T_1 a basis matrix of an A-invariant subspace $\mathcal{J} \subseteq \mathcal{X}$, so that (9) holds with $T := [T_1\ T_2]$. $\mathcal{J}$ is complementable if and only if the elementary divisors[2] of A', hence of A, are the union of those of A_{11}' and A_{22}'.

Proof. Only if. Suppose $A_{12}' = O$ and apply a block diagonal similarity transformation which takes both A_{11}' and A_{22}' into the Jordan canonical form. The complete matrix is also in Jordan canonical form and its elementary divisors are the union of those of the submatrices.
If. Apply a block upper-triangular similarity transformation that takes A' into the Jordan canonical form: if the elementary divisors are separated, the off-diagonal matrices are zero, hence $\mathcal{J}$ is complementable. $\square$

3.2.4 Invariants and State Trajectories

Only linear time-invariant systems will be considered in what follows. To be more concise in notation, a purely dynamic system like

$$\dot{x}(t) = A\, x(t) + B\, u(t) \tag{12}$$

$$y(t) = C\, x(t) \tag{13}$$

will be simply called the *three-map system* or the *triple* (A, B, C), while a nonpurely dynamic system like

$$\dot{x}(t) = A\, x(t) + B\, u(t) \tag{14}$$

$$y(t) = C\, x(t) + D\, u(t) \tag{15}$$

will be called the *four-map system* or the *quadruple* (A, B, C, D).
For the sake of simplicity, most of the analysis that follows will be referred to triples: its extension to quadruples is often straightforward.

---

[2] Elementary divisors are defined in Subsection A.4.5 in connection with the Jordan canonical form.

Sect. 3.2 Invariants

In (12,13) and (14,15) $u \in \mathbf{R}^p$, $y \in \mathbf{R}^q$, $x \in \mathcal{X} = \mathbf{R}^n$ and A, B, C, D denote properly dimensioned real matrices. We assume that input function $u(\cdot)$ belongs to the class of piecewise continuous functions. A triple can be represented with the block diagram shown in Fig. 3-1, where the algebraic operators B and C, which provide respectively the *forcing action* $f \in \mathcal{X} = \mathbf{R}^n$ as a function of u and the output y as a function of x are shown as separate from the strictly dynamic part of the system. Matrices A, B, C are called respectively the *system matrix*, the *input distribution matrix*, and the *output distribution matrix*. Usually $p < n$ and $q < n$, so that matrices B and C are nonsquare, hence noninvertible: if B were square and invertible, forcing action f and, consequently, state velocity $\dot{x}$ could be arbitrarily assigned at any instant of time by means of input u, thus it could be possible to follow any arbitrary continuous and continuously differentiable state trajectory. Similarly, if C were square and invertible, state x would be completely known at any instant of time by simply observing output y.

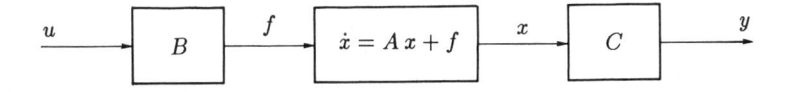

$$u \longrightarrow \boxed{B} \xrightarrow{f} \boxed{\dot{x} = Ax + f} \xrightarrow{x} \boxed{C} \longrightarrow y$$

Figure 3-1 A block diagram representation of a triple (A, B, C).

In the geometric approach it is very important to state whether or not state trajectories exist that completely belong to given subsets of the state space, especially to subspaces. For this purpose we present a useful lemma, which will be referred to very often thereafter.

Lemma 3.2-1. The Fundamental Lemma of the Geometric Approach
Any state trajectory $x|_{[t_0, t_1]}$ of (12) or (14) belongs to a subspace $\mathcal{L} \subseteq \mathcal{X}$ if and only if $\dot{x}(t) \in \mathcal{L}$ almost everywhere in $[t_0, t_1]$.

Proof. Recall that a Lebesgue measurable and integrable function is zero almost everywhere in $[t_0, t_1]$ if and only if its integral in any subinterval of $[t_0, t_1]$ is zero. Apply this property to function $Y^T \dot{x}(t)$, where Y denotes a basis matrix of $\mathcal{L}^\perp$: the function is zero in $[t_0, t_1]$ if and only if

$$Y^T \int_{t_0}^{t} \dot{x}(\tau) \, d\tau = Y^T \big(x(t) - x(t_0)\big) = 0 \quad \forall t \in [t_0, t_1]$$

This clearly implies $x(t) \in \mathcal{L}$ for all $t \in [t_0, t_1]$. □

Refer now to the free system

$$\dot{x}(t) = A x(t) \tag{16}$$

Theorem 3.2-4.
A subspace $\mathcal{L} \subseteq \mathcal{X}$ is a locus of trajectories of system (16) if and only if it is an A-invariant.

Proof. *If.* Let $\mathcal{L}$ be an A-invariant: at every $x \in \mathcal{L}$ the corresponding state velocity Ax belongs to $\mathcal{L}$, so that, by virtue of the fundamental lemma, every trajectory of system (16) which originates at a point of $\mathcal{L}$ completely belongs to $\mathcal{L}$.

Only if. Consider a trajectory $x(\cdot)$ of system (16), denote by $\mathcal{L}$ the subspace of minimal dimension in which it is contained, and let $k := \dim \mathcal{L}$: there exist k instants of time $t_1, \ldots, t_k$ such that $\{x(t_1), \ldots, x(t_k)\}$ is a basis of $\mathcal{L}$. By virtue of the fundamental lemma it is necessary that

$$\dot{x}(t_i) = A\, x(t_i) \in \mathcal{L} \quad (i = 1, \ldots, k)$$

so that $\mathcal{L}$ is an A-invariant. □

3.2.5 Stability and Complementability

Still referring to the free system (16), we shall now introduce the concept of *stability of an invariant*. We recall that system (16) is (asymptotically)[3] stable if and only if all the eigenvalues of matrix A have negative real part. By extension, in this case A is said to be a *stable matrix*.

Since an A-invariant $\mathcal{J} \subseteq \mathcal{X}$ is a locus of trajectories, stability can be "split" with respect to $\mathcal{J}$. To clarify this concept, recall change of basis (9) and let $x = Tz$: in the new coordinate we obtain the system

$$\begin{bmatrix} \dot{z}_1(t) \\ \dot{z}_2(t) \end{bmatrix} = \begin{bmatrix} A'_{11} & A'_{12} \\ O & A'_{22} \end{bmatrix} \begin{bmatrix} z_1(t) \\ z_2(t) \end{bmatrix} \quad (17)$$

which is equivalent to (16). Consider an initial state $x'_0 \in \mathcal{J}$: the corresponding transformed state $z'_0 = T^{-1} x'_0$ decomposes into $(z'_{01}, 0)$. The motion on $\mathcal{J}$ is described by

$$\dot{z}_1(t) = A'_{11}\, z_1(t), \quad z_1(0) = z'_{01}$$

while $z_2(t)$ remains identically zero. Therefore, the motion on $\mathcal{J}$ is stable if and only if submatrix A'_{11} is stable. This situation is represented by trajectory 1 in Fig. 3-2.

On the other hand, consider an initial state $x''_0 \notin \mathcal{J}$, so that $z''_{02} \neq 0$; the time evolution of the second component of the transformed state is described by

$$\dot{z}_2(t) = A'_{22}\, z_2(t), \quad z_2(0) = z''_{02}$$

This means that the projection of the state along $\mathcal{J}$ over any complement of $\mathcal{J}$ has a stable behavior if and only if A'_{22} is a stable matrix. In other words, in

---

[3] From now on stability is always tacitly assumed to be strict or asymptotic.

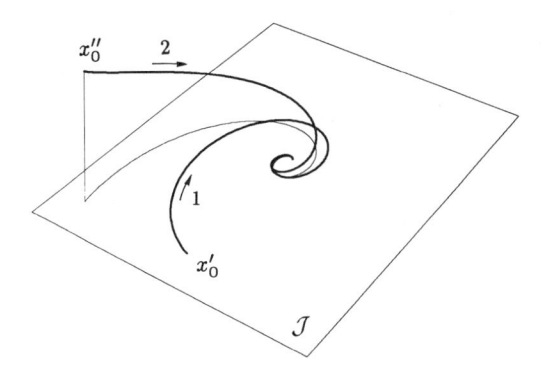

Figure 3-2 Internal and external stability of an invariant.

this case the canonical projection of the state on the quotient space $\mathcal{X}/\mathcal{J}$ tends to the origin as t approaches infinity: this means that the linear variety parallel to $\mathcal{J}$, which contains the state, tends to coincide with $\mathcal{J}$ for t approaching infinity. This situation is represented by trajectory 2 in Fig. 3-2.

Invariant $\mathcal{J}$ is said to be *internally stable* if submatrix A'_{11} in (4) is stable and *externally stable* if A'_{22} is stable. A more formal, coordinate-free definition is stated as follows.

Definition 3.2-4. Internally (Externally) Stable Invariant
Consider a linear map $A : \mathcal{X} \to \mathcal{X}$: an A-invariant $\mathcal{J} \subseteq \mathcal{X}$ is said to be *internally stable* if $A|_{\mathcal{J}}$ is stable, *externally stable* if $A|_{\mathcal{X}/\mathcal{J}}$ is stable.

By virtue of the Laplace expansion of determinants, from (4) it follows that
$$\det A = \det A' = \det A'_{11} \cdot \det A'_{22}$$
hence a partition of the eigenvalues of A is associated to every A-invariant $\mathcal{J}$: the *eigenvalues internal with respect to* $\mathcal{J}$ (those of A'_{11} or of $A|_{\mathcal{J}}$) and the *eigenvalues external with respect to* $\mathcal{J}$ (those of A'_{22} or of $A|_{\mathcal{X}/\mathcal{J}}$).

Relative Stability. Internal and/or external stability of invariants can be referred to other invariants: let $\mathcal{J}$ and $\mathcal{J}_c$ be A-invariants and define
$$\mathcal{J}_1 := \mathcal{J} \cap \mathcal{J}_c \tag{18}$$
$$\mathcal{J}_2 := \mathcal{J} + \mathcal{J}_c \tag{19}$$
$\mathcal{J}_1$ and $\mathcal{J}_2$ are A-invariants too, as the intersection and the sum of A-invariants respectively. Perform the change of basis defined by $T := [T_1\ T_2\ T_3\ T_4]$, with $\operatorname{im} T_1 = \mathcal{J}_1$, $\operatorname{im}[T_1\ T_2] = \mathcal{J}$, $\operatorname{im}[T_1\ T_3] = \mathcal{J}_c$. From Theorem 3.2-1 it follows that
$$A' := T^{-1}AT = \begin{bmatrix} A'_{11} & A'_{12} & A'_{13} & A'_{14} \\ O & A'_{22} & O & A'_{24} \\ O & O & A'_{33} & A'_{34} \\ O & O & O & A'_{44} \end{bmatrix} \tag{20}$$

Clearly $\mathcal{J}$ is internally stable (or, in coordinate-free notation, $A|_{\mathcal{J}}$ is stable) if and only if matrices A'_{11} and A'_{22} are stable; it is externally stable (i.e., $A|_{\mathcal{X}/\mathcal{J}}$ is stable) if and only if matrices A'_{33} and A'_{44} are stable. Similarly, $\mathcal{J}_c$ is internally stable if and only if matrices A'_{11} and A'_{33} are stable; it is externally stable if and only if matrices A'_{22} and A'_{44} are stable. Structure (20) implies the following properties:

1. the sum of two internally stable invariants is an internally stable invariant;
2. the intersection of two externally stable invariants is an externally stable invariant;
3. the intersection of an internally stable invariant and any other invariant is internally stable;
4. the sum of an externally stable invariant and any other invariant is externally stable.

External stability of an invariant with respect to another invariant containing it can be easily defined. For instance, $\mathcal{J}_1$ is externally stable with respect to $\mathcal{J}$ if matrix A'_{22} is stable (i.e., if $A|_{\mathcal{J}/\mathcal{J}_1}$ is stable), externally stable with respect to $\mathcal{J}_c$ if A'_{33} is stable (i.e., if $A|_{\mathcal{J}_c/\mathcal{J}_1}$ is stable), externally stable with respect to $\mathcal{J}_2$ if both A'_{22} and A'_{33} are stable (i.e., if $A|_{\mathcal{J}_2/\mathcal{J}_1}$ is stable).

The set of all internally stable invariants and that of all externally stable (with respect to the whole space) invariants, possibly subject to the constraint of containing a given subspace $\mathcal{B} \subseteq \mathcal{X}$ and/or of being contained in a given subspace $\mathcal{C} \subseteq \mathcal{X}$, are lattices with respect to $\subseteq, +, \cap$.

Relative Complementability. Let $\mathcal{J}_1$, $\mathcal{J}$ and $\mathcal{J}_2$ be A-invariants satisfying

$$\mathcal{J}_1 \subseteq \mathcal{J} \subseteq \mathcal{J}_2 \tag{21}$$

$\mathcal{J}$ is said to be *complementable with respect to* $(\mathcal{J}_1, \mathcal{J}_2)$ if there exists at least one invariant $\mathcal{J}_c$ that satisfies (18, 19). To search for such an invariant, perform the change of basis defined by $T := [T_1 \, T_2 \, T_3 \, T_4]$, with $\text{im} T_1 = \mathcal{J}_1$, $\text{im}[T_1 \, T_2] = \mathcal{J}$, $\text{im}[T_1 \, T_2 \, T_3] = \mathcal{J}_2$. It follows that

$$A' := T^{-1} A T = \begin{bmatrix} A'_{11} & A'_{12} & A'_{13} & A'_{14} \\ O & A'_{22} & A'_{23} & A'_{24} \\ O & O & A'_{33} & A'_{34} \\ O & O & O & A'_{44} \end{bmatrix} \tag{22}$$

Note that the structure of (22) is different from (20) only because submatrix A'_{23} is in general nonzero; however it may happen that a suitable choice of T_3, performed within the above specified constraint, implies that $A'_{23} = O$. If such a matrix T'_3 exists, $\mathcal{J}$ is complementable with respect to $(\mathcal{J}_1, \mathcal{J}_2)$ and a *complement* $\mathcal{J}_c$ of $\mathcal{J}$ is provided by

$$\mathcal{J}_c := \text{im}[T_1 \, T'_3] \tag{23}$$

By virtue of Theorem 3.2-2, $\mathcal{J}$ is complementable if and only if the Sylvester equation

$$A'_{22} X - X A'_{33} = -A'_{23} \tag{24}$$

admits at least one solution X. In such a case it is possible to assume in (24) $T'_3 := T_2 X + T_3$. In this way a complement $\mathcal{J}_c$ of $\mathcal{J}$ with respect to $(\mathcal{J}_1, \mathcal{J}_2)$ is determined.

Sect. 3.2 Invariants

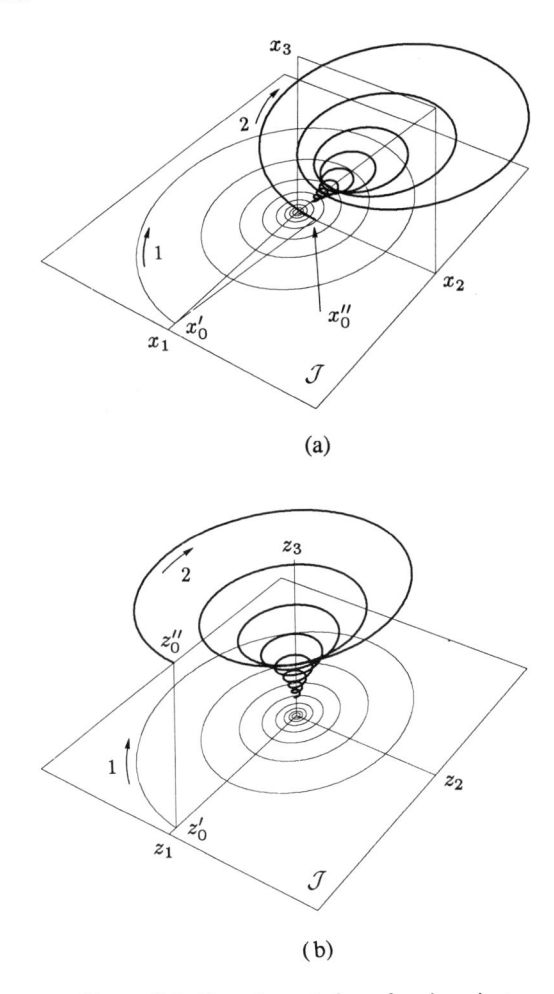

Figure 3-3 Complementation of an invariant.

Example. Consider in $\mathcal{X} := \mathbf{R}^3$ the system

$$\begin{bmatrix} \dot{x}_1(t) \\ \dot{x}_2(t) \\ \dot{x}_3(t) \end{bmatrix} = \begin{bmatrix} \sigma & \omega & a_1 \\ -\sigma & \omega & a_2 \\ 0 & 0 & \rho \end{bmatrix} \begin{bmatrix} x_1(t) \\ x_2(t) \\ x_3(t) \end{bmatrix} \tag{25}$$

Clearly the first coordinate plane is an invariant $\mathcal{J}$: we want to check its complementability with respect to $(\{0\}, \mathcal{X})$: the Sylvester equation in $[t_1 \; t_2]^T$

$$\begin{bmatrix} \sigma & \omega \\ -\omega & \sigma \end{bmatrix} \begin{bmatrix} t_1 \\ t_2 \end{bmatrix} - \begin{bmatrix} t_1 \\ t_2 \end{bmatrix} \rho = - \begin{bmatrix} a_1 \\ a_2 \end{bmatrix}, \tag{26}$$

provides

$$\begin{bmatrix} t_1 \\ t_2 \end{bmatrix} = \frac{1}{(\sigma - \rho)^2 + \omega^2} \begin{bmatrix} a_1(\rho - \sigma) + a_2 \omega \\ a_1(-\omega) + a_2(\rho - \sigma) \end{bmatrix} \tag{27}$$

By means of the similarity transformation $x=Tz$, with

$$T := \begin{bmatrix} 1 & 0 & t_1 \\ 0 & 1 & t_2 \\ 0 & 0 & 1 \end{bmatrix}$$

we obtain the equivalent system

$$\begin{bmatrix} \dot{z}_1(t) \\ \dot{z}_2(t) \\ \dot{z}_3(t) \end{bmatrix} = \begin{bmatrix} \sigma & \omega & 0 \\ -\sigma & \omega & 0 \\ 0 & 0 & \rho \end{bmatrix} \begin{bmatrix} z_1(t) \\ z_2(t) \\ z_3(t) \end{bmatrix} \quad (28)$$

where $\mathcal{J}$ is complemented. In Figs. 3-3(a) and (b), trajectories 1 and 2 refer to free motions, the former of which evolves on $\mathcal{J}$, represented respectively in coordinate systems (x_1, x_2, x_3) and (z_1, z_2, z_3). The values of the parameters used for the plots are $\omega = 2\pi$, $\sigma = -0,5$, $\rho = -0,25$, $a_1 = -\omega$, $a_2 = \rho - \sigma$. The complement $\mathcal{J}_c$ of $\mathcal{J}$ in coordinate system (x_1, x_2, x_3) is

$$\mathcal{J}_c = \mathrm{im} \begin{bmatrix} t_1 \\ t_2 \\ 1 \end{bmatrix} = \mathrm{im} \begin{bmatrix} 0 \\ 1 \\ 1 \end{bmatrix}$$

while it coincides with the third axis in coordinate system (z_1, z_2, z_3).

3.3 CONTROLLABILITY AND OBSERVABILITY

Refer to a triple (A, B, C). We shall again derive Property 2.6-8 geometrically. The argument here presented will also be used in Section 4.1 to introduce the concept of controlled invariance.

Theorem 3.3-1.
For triple (A, B, C) and any finite t_1 the following holds:

$$\mathcal{R} := \mathcal{R}_{t_1}^+ = \min \mathcal{J}(A, \mathcal{B}) \quad \text{with} \quad \mathcal{B} := \mathrm{im}\, B \quad (1)$$

Proof. It will be proved that not only the final state $x(t_1)$, but also all the intermediate states $x(t)$, $t \in [0, t_1]$ of all the admissible trajectories starting at the origin, belong to $\mathcal{R}$: in fact, consider a generic point $x_1(t_a)$ of trajectory $x_1(\cdot)$ corresponding to input function $u_1(\cdot)$. The input function

$$u(t) := \begin{cases} 0 & \text{for } 0 \leq t < t_0 - t_a \\ u_1(t - t_1 + t_a) & \text{for } t_1 - t_a \leq t \leq t_1 \end{cases}$$

corresponds to a trajectory that terminates in $x_1(t_a)$, so that, by definition, $x_1(t_a)$ belongs to $\mathcal{R}$. Let $h := \dim \mathcal{R}$ and $x_i(\cdot)$ $(i = 1, \ldots, h)$ be trajectories such that vectors $x_i(t_1)$ $(i = 1, \ldots, h)$ are a basis of $\mathcal{R}$: since motions are continuous functions, there exists an $\epsilon > 0$ such that states $x_i(t_1 - \epsilon)$ $(i = 1, \ldots, h)$ are still a basis for $\mathcal{R}$. In

Sect. 3.3 Controllability and Observability 141

these states all the admissible velocities must belong to $\mathcal{R}$ because, if not, it would be possible to maintain the velocity out of $\mathcal{R}$ for a finite time and reach points not belonging to $\mathcal{R}$. This implies the inclusion $A\mathcal{R}+\mathcal{B}\subseteq\mathcal{R}$, which means that $\mathcal{R}$ is an A-invariant containing $\mathcal{B}$. Furthermore, $\mathcal{R}$ is the minimal A-invariant containing $\mathcal{B}$ because at no point of it is it possible to impose velocities not belonging to it, hence to drive out the state. □

The dual result concerning observability, already stated as Property 2.6-11, is geometrically approached as follows.

Corollary 3.3-1.
For triple (A, B, C) and any finite t_1 the following holds:

$$\mathcal{Q} := \mathcal{Q}_{t_1}^- = \max \mathcal{J}(A, \mathcal{C}) \quad \text{with} \quad \mathcal{C} := \ker C \tag{2}$$

Proof. The statement is an immediate consequence of Theorem 3.2-3. □

Example. Refer again to the example presented in the previous section, to point out the importance of the concept of invariance in connection with controllability and observability. First, consider controllability and assume

$$A := \begin{bmatrix} \sigma & \omega & a_1 \\ -\omega & \sigma & a_2 \\ 0 & 0 & \rho \end{bmatrix} \quad B := \begin{bmatrix} 1 & 0 \\ 0 & 0 \\ 0 & 1 \end{bmatrix} \tag{3}$$

Denote, as before, with $\mathcal{J}$ the invariant consisting of the first coordinate plane and $\mathcal{J}_c$ its complement, previously determined by the Sylvester equation.

Figure 3-4(a) shows the trajectories corresponding to three free motions: 1 and 2 initiate at $(.9, 0, 0)$ and $(0, .9, 0)$, hence they completely belong to $\mathcal{J}$, while 3, which initiates at $(0, 0, .9)$, tends to lie on $\mathcal{J}$ by winding itself round $\mathcal{J}_c$: a thinner line denotes its projection on $\mathcal{J}$. The assumed values of the parameters are $\omega = 2\pi$, $\sigma = -2$, $\rho = -4$, $a_1 = a_2 = 3$. The system is not completely controllable with only input u_1 since the image of the first column of B is not contained in $\mathcal{J}$, while it is completely controllable with only u_2 since the minimal A-invariant containing the second column of B is the whole state space $\mathcal{X}$.

Figure 3-4(b) shows three controlled trajectories leaving the origin, corresponding to the time interval $t \in [0, T]$ with $T = 1.5$: trajectory 1 corresponds to a piecewise constant control function $u_1(t)$ ($u_1(t)=3$ for $t \in [0, T/6)$, $u_1(t)=-3$ for $t \in [T/6, T/2)$, $u_1(t)=3$ for $t \in [T/2, 5T/6)$, $u_1(t)=-3$ for $t \in [5T/6, T]$), with $u_2(t)$ identically zero; trajectory 2 also corresponds to a piecewise constant control function $u_1(t)$ ($u_1(t)=-3$ for $t \in [0, T/3)$, $u_1(t)=3$ for $t \in [T/3, 2T/3)$, $u_1(t)=-3$ for $t \in [2T/3, T]$), with $u_2(t)$ identically zero; trajectory 3 to u_1 identically zero and u_2 constant (equal to 3). Note that the three final states are linearly independent as a proof of the system being completely controllable. Any other state, being expressable as a linear combination of them, is reachable by means of the control action obtained as the linear combination with the same coefficients of the three above specified control actions. This suggests a simple procedure to solve the problem of controlling the system from the origin to any given state.

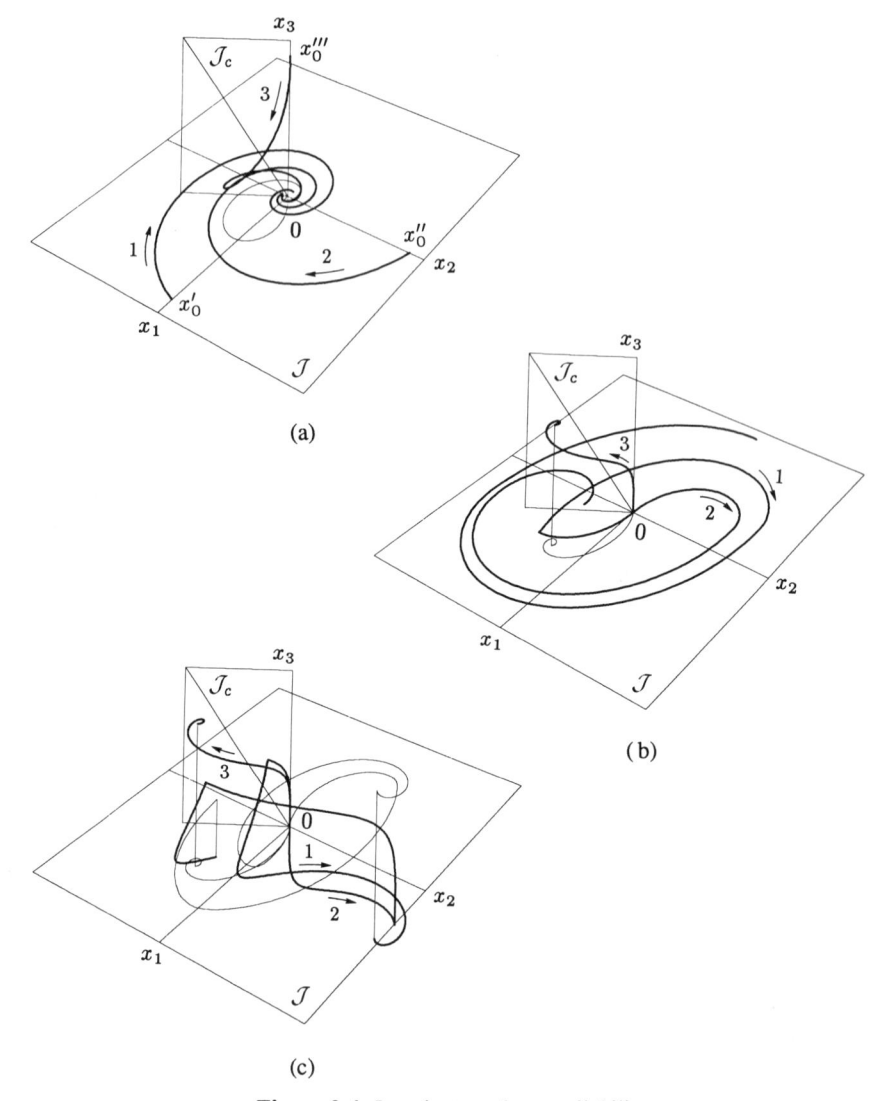

Figure 3-4 Invariants and controllability.

As proof that the system is completely controllable with only input u_2, Fig. 3-4(c) shows three trajectories in time interval $t \in [0, T]$ with $T = 1.5$, all corresponding to $u_1(t)$ identically zero: trajectory 1 corresponds to $u_2(t) = 4$ for $t \in [0, T/6)$, $u_2(t) = -4$ for $t \in [T/6, T]$; trajectory 2 to $u_2(t) = -4$ for $t \in [0, T/3)$, $u_2(t) = 4$ for $t \in [T/3, 5T/6)$, $u_2(t) = -4$ for $t \in [5T/6, T]$; trajectory 3 to $u_2(t) = 3.5$ in the whole time interval. Thinner lines denote the orthogonal projections of the trajectories on $\mathcal{J}$. The final states are linearly independent.

As far as observability is concerned, note that the system is completely observable provided neither $\mathcal{J}$ nor $\mathcal{J}_c$ is contained in ker$C$. Consider $\mathcal{J}$: observability is not complete if the output distribution matrix is

$$c = [0 \ 0 \ c_3] \quad \text{with } c_3 \text{ arbitrary} \tag{4}$$

Consider $\mathcal{J}_c$: in basis (z_1, z_2, z_3) observability is not complete if

$$c' := cT = [c_1 \ c_2 \ 0] \quad \text{with } c_1, c_2 \text{ arbitrary}$$

so that

$$c = c'T^{-1} = [c_1 \ | \ c_2 \ | \ -c_1 t_1 - c_2 t_2] \tag{5}$$

The system is not completely observable if all rows of matrix C are like (4) or are all like (5).

3.3.1 The Kalman Canonical Decomposition

Invariance in connection with controllability and observability properties plays a key role in deriving the *Kalman canonical decomposition*, which provides a relevant insight into linear time-invariant system structure.

Property 3.3-1.
A generic quadruple (A, B, C, D) is equivalent to quadruple (A', B', C', D), where matrices A', B', and C' have the structures

$$A' = \begin{bmatrix} A'_{11} & A'_{12} & A'_{13} & A'_{14} \\ O & A'_{22} & O & A'_{24} \\ O & O & A'_{33} & A'_{34} \\ O & O & O & A'_{44} \end{bmatrix} \quad B' = \begin{bmatrix} B'_1 \\ B'_2 \\ O \\ O \end{bmatrix}$$

$$C' = [O \ C'_2 \ O \ C'_4] \tag{6}$$

Proof. Perform the change of basis $x = Tz$, $z = T^{-1}x$, where submatrices of $T := [T_1 \ T_2 \ T_3 \ T_4]$ satisfy $\text{im} T_1 = \mathcal{R} \cap \mathcal{Q}$, $\text{im}[T_1 \ T_2] = \mathcal{R}$, $\text{im}[T_1 \ T_3] = \mathcal{Q}$. The structure of $A' := T^{-1}AT$ is due to $\mathcal{R}$ and $\mathcal{Q}$ being A-invariants, that of $B' := T^{-1}B$ to inclusion $\mathcal{B} \subseteq \mathcal{R}$ and that of $C' := CT$ to $\mathcal{Q} \subseteq \mathcal{C}$. □

Consider the system expressed in the new basis, i.e.

$$\dot{z}(t) = A' z(t) + B' u(t) \tag{7}$$
$$y(t) = C' z(t) + D u(t) \tag{8}$$

Because of the particular structure of matrices A', B', C' the system can be decomposed into one purely algebraic and four dynamic subsystems, interconnected as shown in Fig. 3-5. Signal paths in the figure show that subsystems 1 and 2 are controllable by input u, while subsystems 1 and 3 are unobservable

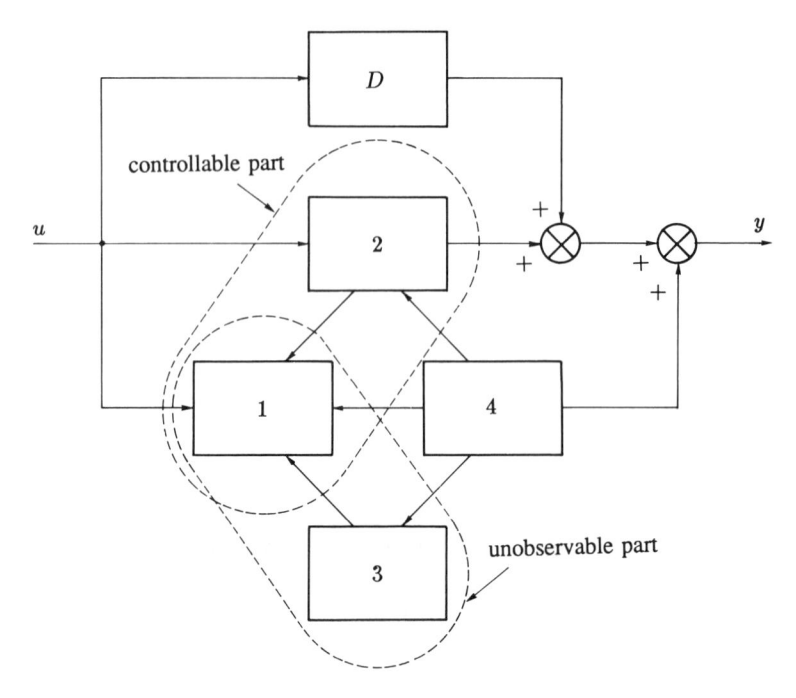

Figure 3-5 The Kalman canonical decomposition.

from output y. Subsystems 2, 4 and D are all together a minimal form of the given system.

Subsystem 2 is the sole completely controllable and observable, and the only one which, with memoryless system D, determines the input-output correspondence, i.e., the zero-state response of the overall system. In fact

$$W(t) = C\, e^{At}\, B = C'\, e^{A't}\, B' = C'_2\, e^{A'_{22}t}\, B'_2 \tag{9}$$

The Kalman decomposition is an ISO representation of linear time-invariant systems that provides complete information about controllability and observability: in particular, if the system is completely controllable and observable, parts 1, 3, and 4 are not present (the corresponding matrices in (6) have zero dimensions).[1]

An interesting application of the Kalman canonical decomposition is to derive a *minimal realization* of an impulse response function $W(t)$ or a transfer matrix $G(s)$. This problem was introduced in Subsection 2.4.1 and can be solved by means of the previously considered change of basis: in fact, subsystem 2 is a minimal realization as a consequence of the following property.

---

[1] Furthermore, the system is stabilizable and detectable (see Section 3.4) respectively if and only if A'_{33}, A'_{44} are stable and if and only if A'_{11}, A'_{33} are stable.

Sect. 3.3 Controllability and Observability 145

Property 3.3-2.
A triple (A, B, C) is a minimal realization if and only if it is completely controllable and observable.

Proof. *Only if.* A system that is not completely controllable and observable cannot be minimal since, clearly, only subsystem 2 of the Kalman canonical decomposition influences its input-output behavior.

If. It will be proved that, given a system of order n completely controllable and observable, no system of order $n_1 < n$ exists with the same transfer function $G(s)$, hence with the same impulse response $W(t)$. In fact, suppose the considered system is controlled by a suitable input function segment $u|_{[0,t_1)}$ to an arbitrary state $x_1 \in \mathbf{R}^n$ at t_1 and the output function, with zero input, is observed in a subsequent finite time interval $[t_1, t_2]$; since the system is completely observable, state $x(t_1)$ and response $y_{[t_1, t_2]}$ are related by an isomorphism: this means that with respect to a suitable basis of the output functions space (whose n elements are each a q-th of functions consisting of modes and linear combinations of modes) the components of $y_{[t_1, t_2]}$ are equal to those of state $x(t_1)$ with respect to the main basis of $\mathbf{R}^n$. In other terms, the zero-input output functions in $[t_1, t_2]$ belong to an n-dimensional vector space, which cannot be related to $\mathbf{R}^{n_1}$ by a similar isomorphism. □

A similar argument applies to prove the following property.

Property 3.3-3.
Any two different minimal realizations (A, B, C) and (A', B', C') of the same impulse response $W(t)$ or of the same transfer matrix $G(s)$ are equivalent, i.e., there exists a nonsingular matrix T such that $A' = T^{-1}AT$, $B' = T^{-1}B$, $C' = CT$.

How to Derive a Minimal Realization. In Subsection 2.4.1 a procedure based on partial fraction expansion was introduced to derive the so-called parallel realization of a transfer matrix. The derived realization consists of several subsystems in parallel for each input; each subsystem corresponds to a single pole, with multiplicity equal to the maximum multiplicity in all transfer functions concerning the considered input (a column of the transfer matrix). The parallel realization is completely controllable by construction, but may not be completely observable, hence not minimal.

Matrices A, B of the parallel realization have numerous elements equal to zero. In particular, A has a structure similar to the real Jordan form, and therefore particularly suitable to emphasize the structural features of the considered system. We shall now describe a simple procedure to obtain a minimal realization from it which preserves such a simple structure.

By means of Algorithm 3.2-2, derive a basis matrix Q of $\mathcal{Q} = \min \mathcal{J}(A, C)$. Let n_q be the number of columns of Q and Π the set of indices of those vectors among e_i ($i = 1, \ldots, n$) (the columns of the identity matrix I_n), which are linearly independent of the columns of Q: these indices can be determined by applying the Gram-Schmidt algorithm to the columns of $[Q\ I_n]$. The number of elements of Π is clearly $n - n_q$. Let P be the permutation matrix such that PI_n has vectors e_i, $i \in \Pi$ as first columns. In matrix

$$PQ = \begin{bmatrix} Q_1 \\ Q_2 \end{bmatrix}$$

submatrix Q_2 ($n_q \times n_q$) is nonsingular. In fact the first n columns of $P[Q\ I_n]$ are linearly independent so that matrix $[PQ|I'_n]$ (where I'_n is defined as the matrix formed by the first $n-n_q$ columns of I_n) is nonsingular, as well as $[I'_n|PQ]$. PQ is a basis matrix of Q with respect to a new basis obtained by applying the above permutation to the vectors of the previous one, as is PQQ_2^{-1}, having the same image. It follows that by the transformation

$$T := P \begin{bmatrix} I_{n-n_q} & Q_1 Q_2^{-1} \\ O & I_{n_q} \end{bmatrix}, \quad \text{whence} \quad T^{-1} = \begin{bmatrix} I_{n-n_q} & -Q_1 Q_2^{-1} \\ O & I_{n_q} \end{bmatrix} P^T$$

an equivalent system is obtained with the structure

$$A' = T^{-1}AT = \begin{bmatrix} A'_{11} & O \\ A'_{21} & A'_{22} \end{bmatrix} \quad B' = T^{-1}B = \begin{bmatrix} B'_1 \\ B'_2 \end{bmatrix}$$

$$C' = CT = [\, C'_1 \quad O \,]$$

Subsystem (B'_1, A'_{11}, C'_1) is a minimal realization. Since it has been obtained from the parallel realization through transformation matrices with numerous zero elements, it generally maintains a simple structure.

The following theorems on BIBS and BIBO stability, which are stated referring to the Kalman canonical form, complete the results of Subsection 2.5.2 on the stability of linear time-invariant systems.

Theorem 3.3-2.
A quadruple (A, B, C, D) is BIBS stable if and only if the eigenvalues of its controllable part have negative real part or, in other terms, if and only if $\mathcal{R}$ is an internally stable A-invariant.

Proof. If. Refer to the Kalman canonical decomposition and consider the controllable part of the system, i.e., pair (A, B) with

$$A := \begin{bmatrix} A'_{11} & A'_{12} \\ O & A'_{22} \end{bmatrix} \quad B := \begin{bmatrix} B'_1 \\ B'_2 \end{bmatrix} \tag{10}$$

which clearly is the only part to influence BIBS stability. We shall prove that the necessary and sufficient condition

$$\int_0^t \|e^{A(t-\tau)} B\|\, d\tau = \int_0^t \|e^{A\tau} B\|\, d\tau \leq M < \infty \quad \forall t \geq 0 \tag{11}$$

(stated by Theorem 2.5-3) holds if the eigenvalues of A have the real part negative. Recall that a matrix norm is less than or equal to the sum of the absolute values of the matrix elements (Property A.6-2) which, in this case, are linear combinations of modes, i.e., of functions of the types $t^r e^{\sigma t}$ and $t^r e^{\sigma t} \cos(\omega t + \varphi)$. But, for $\sigma < 0$

$$\int_0^\infty |t^r e^{\sigma t} \cos(\omega t + \varphi)|\, dt \leq \int_0^\infty t^r e^{\sigma t}\, dt = \frac{n!}{(-\sigma)^{n+1}} \tag{12}$$

Sect. 3.3 Controllability and Observability

so that the preceding linear combinations are all less than or equal to sums of positive finite terms and condition (11) holds. To prove the identity on the right of (12), consider the family of integrals

$$I_r(t) = \int_0^t e^{\sigma \tau} \tau^r \, d\tau \quad \text{with } \sigma < 0$$

Denote the function under the integral sign as $f(\tau)\dot{g}(\tau)$, with $f(\tau) := \tau^r$ (so that $\dot{f} = r\tau^{r-1}$), and $\dot{g}\, d\tau := e^{\sigma\tau}\, d\tau$ (so that $g = (1/\sigma) e^{\sigma\tau}$). Integration by parts yields the recursion formula

$$I_r(t) = \left.\frac{\tau^r e^{\sigma\tau}}{\sigma}\right|_{\tau=0}^{\tau=t} - \frac{r}{\sigma} I_{r-1}(t) \quad \text{with } I_0(t) = \left.\frac{e^{\sigma\tau}}{\sigma}\right|_{\tau=0}^{\tau=t}$$

from which (12) is derived as t approaches infinity.

Only if. We prove that for a particular bounded input, relation (11) does not hold if at least one eigenvalue of A has nonnegative real part. Refer to the real Jordan form and suppose that by a suitable bounded input function segment $u|_{[t_0,t_1)}$, only one state component has been made different from zero at t_1 (this is possible because of the complete controllability assumption) and that input is zero from t_1 on, so that integral (11) is equal to

$$Q_1 + k \int_{t_1}^t \left|\left(\sum_{k=0}^h \frac{\tau^k}{k!}\right) e^{\sigma\tau} \cos(\omega\tau + \varphi)\right| d\tau \tag{13}$$

where Q_1 denotes integral (11) restricted to $[t_0, t_1)$ (a positive finite real number), k a positive finite real number depending on the bound on input and h an integer less or equal to $m-1$, where m is the multiplicity of the considered eigenvalue in the minimal polynomial of A. Denote by t_2 any value of time greater than t_1 such that

$$\sum_{k=0}^h \frac{\tau^k}{k!} \geq 1$$

Since this sum is positive and monotonically increasing in time, the integral on the right of (13) is equal to

$$k Q_2 + k \int_{t_2}^t \left|e^{\sigma\tau} \cos(\omega\tau + \varphi)\right| d\tau$$

where Q_2 denotes the integral in (13) restricted to $[t_1, t_2)$, again a positive real number. The integral in the previous formula is clearly unbounded as t approaches infinity if $\sigma \geq 0$. □

Theorem 3.3-3.
A quadruple (A, B, C, D) is BIBO stable if and only if the eigenvalues of its controllable and observable part have negative real part or, in other terms, if and only if $\mathcal{R} \cap \mathcal{Q}$ is an A-invariant externally stable with respect to $\mathcal{R}$ (this means that the induced map $A|_{\mathcal{R}/\mathcal{Q} \cap \mathcal{R}}$ is stable).

Proof. (Hint) Refer to the Kalman canonical decomposition and consider the controllable and observable part of the system, i.e., triple (A, B, C) with $A := A'_{22}$, $B := B'_{22}$, $C := C'_{22}$, which is clearly the only part to influence BIBO stability. A very slight modification of the argument used to prove Theorem 3.3-2 can be used to prove that the necessary and sufficient condition

$$\int_0^t \|C\,e^{A(t-\tau)}\,B\|\,d\tau = \int_0^t \|C\,e^{A\tau}\,B\|\,d\tau \leq M < \infty \quad \forall t \geq 0 \tag{14}$$

(stated by Theorem 2.5-4) holds if and only if the eigenvalues of A have negative real part. □

3.3.2 Referring to the Jordan Form

Since the Jordan form provides good information about the structural features of linear dynamic systems, it may be convenient to consider complete controllability and observability with respect to this form.

Theorem 3.3-4.
Given a triple (A, B, C) derive, by a suitable transformation in the complex field, the equivalent system [2]

$$\dot{z}(t) = J\,z(t) + B'\,u(t) \tag{15}$$
$$y(t) = C'\,z(t) \tag{16}$$

where J denotes an $n \times n$ matrix in Jordan form. Pair (A, B) is controllable if and only if:
1. the rows of B' corresponding to the last row of every Jordan block are nonzero;
2. the above rows of B' which, furthermore, correspond to Jordan blocks related to the same eigenvalue, are linearly independent.

Pair (A, C) is observable if and only if:
1. the columns of C' corresponding to the first column of every Jordan block are nonzero;
2. the above columns of C' which, furthermore, correspond to Jordan blocks related to the same eigenvalue, are linearly independent.

Proof. By virtue of Lemma 2.6-1 and Theorem 2.6-1, system (15, 16) is completely controllable if and only if the rows of matrix

$$e^{Jt}\,B' \tag{17}$$

---

[2] The results stated in Theorem 3.3-4 can easily be extended to the real Jordan form, which may be more convenient in many cases.

Sect. 3.3 Controllability and Observability

which are vectors of C^p functions of time, are linearly independent in any finite time interval and, by virtue of Theorem 2.6-2, it is completely observable if and only if the columns of matrix

$$C' e^{Jt} \tag{18}$$

which are vectors of C^q functions of time, are linearly independent in any finite time interval.

Conditions 1 are clearly necessary. To show that conditions 2 are necessary and sufficient, note that, since functions

$$e^{\lambda_1 t}, t e^{\lambda_1 t}, \ldots, t^{m_1-1} e^{\lambda_1 t}, \ldots, e^{\lambda_h t}, t e^{\lambda_h t}, \ldots, t^{m_h-1} e^{\lambda_h t}$$

are linearly independent in any finite time interval, it is possible to have linearly independent rows in matrix (17) or linearly independent columns in matrix (18) only if two or more Jordan blocks correspond to the same eigenvalue. Nevertheless, this possibility is clearly excluded if and only if conditions 2 hold. □

Controllability and Observability After Sampling. Theorem 3.3-4 can be extended, in practice without any change, to the discrete triple (A_d, B_d, C_d). If this is a model of a sampled continuous triple, i.e., the corresponding matrices are related to each other as specified by (2.2.23-25), the question arises whether controllability and observability are preserved after sampling. A very basic sufficient condition is stated in the following theorem.[3]

Theorem 3.3-5.

Suppose that triple (A, B, C) is completely controllable and/or observable and denote by $\lambda_i = \sigma_i + j\omega_i$ $(i=1,\ldots,h)$ the distinct eigenvalues of A. The corresponding sampled triple (A_d, B_d, C_d) is completely controllable and/or observable if

$$\omega_i - \omega_j \neq \frac{2\nu\pi}{T} \quad \text{whenever} \quad \sigma_i = \sigma_j \tag{19}$$

where ν stands for any integer, positive or negative.

Proof. We refer to the Jordan canonical form (15, 16). Recall that $A'_d := e^{JT}$ has the structure displayed in (2.1.41, 42). Note that the structure of J is preserved in A'_d and distinct eigenvalues of A'_d correspond to distinct eigenvalues of J if (19) holds. To be precise, A'_d is not in Jordan form, but only in block-diagonal form, with all blocks upper-triangular. However, every block can be transformed into Jordan form by using an upper-triangular transformation matrix with ones in the main diagonal, which does not affect the corresponding last row in B' and the corresponding first column in C'. Hence (A'_d, B') or (A_d, B) is controllable and/or (A'_d, C') or $(A_d, C) = (A_d, C_d)$ observable. Furthermore, again from (2.1.41, 42), by taking the matrix integral it follows that

$$\det f(J, T) = \det f(A, T) = \prod_{i=1}^{m_i} \rho_i \quad \text{with} \quad \rho_i = \begin{cases} \dfrac{e^{\lambda_i T} - 1}{\lambda_i} & \text{if } \lambda_i \neq 0 \\ T & \text{if } \lambda_i = 0 \end{cases}$$

---

[3] This result is due to Kalman, Ho, and Narendra [18].

where m_i denotes the multiplicity of λ_i in the characteristic polynomial of A. Thus, $f(A,T)$ is nonsingular. Since it commutes with A_d [see expansions (2.2.36,37)], it follows that

$$[B_d \mid A_d B_b \mid \ldots \mid A_d^{n-1} B_d] = f(A,T) [B \mid A_d B \mid \ldots \mid A_d^{n-1} B]$$

The controllability matrix on the left has maximal rank since that on the right has. □

Note that loss of controllability and observability after sampling can be avoided by choosing the sampling frequency $1/T$ sufficiently high.

3.3.3 SISO Canonical Forms and Realizations

First, we consider two *canonical forms relative to input*. Consider a controllable pair (A,b), where A and b are respectively an $n \times n$ and an $n \times 1$ real matrix. We shall derive for A,b a canonical structure, called the *controllability canonical form*. Assume the coordinate transformation matrix $T_1 := [p_1 \ p_2 \ \ldots \ p_n]$ with

$$\begin{aligned} p_1 &= b \\ p_2 &= Ab \\ &\ldots\ldots \\ p_n &= A^{n-1}b \end{aligned} \quad (20)$$

Vectors p_i ($i=1,\ldots,n$) are linearly independent because of the controllability assumption. In other words, controllability in this case implies that the linear map A is *cyclic* in $\mathcal{X}$ and b is a *generating vector* of $\mathcal{X}$ with respect to A. Denote by $(-\alpha_0, -\alpha_1, \ldots, -\alpha_{n-1})$ the components of $A^n b$ with respect to this basis, i.e.

$$A^n b = -\sum_{i=1}^{n} \alpha_{i-1} p_i = -\sum_{i=0}^{n-1} \alpha_i A^i b$$

Matrix AT_1, partitioned columnwise, can be written as

$$\begin{aligned} AT_1 &= [\ Ap_1 \mid Ap_2 \mid \ldots \mid Ap_{n-1} \mid Ap_n\ ] \\ &= [\ p_2 \mid p_3 \mid \ldots \mid p_n \mid -\sum_{i=1}^{n} \alpha_{i-1} p_i\ ] \end{aligned}$$

Since the columns of the transformed matrix A_1 coincide with those of AT_1 expressed in the new basis, and b is the first vector of the new basis, $A_1 := T_1^{-1} A T_2$ and $b_1 := T_1^{-1} b$ have the structures

$$A_1 = \begin{bmatrix} 0 & 0 & 0 & \ldots & 0 & -\alpha_0 \\ 1 & 0 & 0 & \ldots & 0 & -\alpha_1 \\ 0 & 1 & 0 & \ldots & 0 & -\alpha_2 \\ \vdots & \vdots & \vdots & \ddots & \vdots & \vdots \\ 0 & 0 & 0 & \ldots & 0 & -\alpha_{n-2} \\ 0 & 0 & 0 & \ldots & 1 & -\alpha_{n-1} \end{bmatrix} \quad b_1 = \begin{bmatrix} 1 \\ 0 \\ 0 \\ \vdots \\ 0 \\ 0 \end{bmatrix} \quad (21)$$

Sect. 3.3 Controllability and Observability

Let us now derive for A, b another structure related to the controllability assumption, called the *controller canonical form*. Assume the coordinate transformation matrix $T_2 := [q_1 \ q_2 \ \ldots \ q_n]$ with

$$
\begin{aligned}
q_n &= b \\
q_{n-1} &= Aq_n + \alpha_{n-1}q_n = Ab + \alpha_{n-1}b \\
&\ldots\ldots \\
q_2 &= Aq_3 + \alpha_2 q_n = A^{n-2}b + \alpha_{n-1}A^{n-3}b + \ldots + \alpha_2 b \\
q_1 &= Aq_2 + \alpha_1 q_n = A^{n-1}b + \alpha_{n-1}A^{n-2}b + \ldots + \alpha_1 b
\end{aligned}
\tag{22}
$$

This can be obtained from the previous one by means of the transformation $T_2 = T_1 Q$, with

$$
Q := \begin{bmatrix}
\alpha_1 & \ldots & \alpha_{n-2} & \alpha_{n-1} & 1 \\
\alpha_2 & \ldots & \alpha_{n-1} & 1 & 0 \\
\alpha_3 & \ldots & 1 & 0 & 0 \\
\vdots & \ddots & \vdots & \vdots & \vdots \\
1 & \ldots & 0 & 0 & 0
\end{bmatrix}
\tag{23}
$$

which is clearly nonsingular (the absolute value of its determinant is equal to one). Columns of Q express the components of basis T_2 with respect to basis T_1. Since

$$
Aq_1 = A^n b + \sum_{i=1}^{n-1} \alpha_i A^i b = -\alpha_0 b = -\alpha_0 q_n
$$

$$
Aq_i = q_{i-1} - \alpha_{i-1}q_n \quad (i = 2, \ldots, n)
$$

matrix AT_2, partitioned columnwise, is

$$
AT_2 = \begin{bmatrix} -\alpha_0 q_n \mid q_1 - \alpha_1 q_n \mid \ldots \mid q_{n-2} - \alpha_{n-2}q_n \mid q_{n-1} - \alpha_{n-1}q_n \end{bmatrix}
$$

so that, in the new basis, $A_2 := T_2^{-1} A T_2$ and $b_2 := T_2^{-1} b$ have the structures

$$
A_2 = \begin{bmatrix}
0 & 1 & 0 & \ldots & 0 & 0 \\
0 & 0 & 1 & \ldots & 0 & 0 \\
0 & 0 & 0 & \ldots & 0 & 0 \\
\vdots & \vdots & \vdots & \ddots & \vdots & \vdots \\
0 & 0 & 0 & \ldots & 0 & 1 \\
-\alpha_0 & -\alpha_1 & -\alpha_2 & \ldots & -\alpha_{n-2} & -\alpha_{n-1}
\end{bmatrix}
\quad b_2 = \begin{bmatrix} 0 \\ 0 \\ 0 \\ \vdots \\ 0 \\ 1 \end{bmatrix}
\tag{24}
$$

Matrices A_1 and A_2 in (21) and (24) are said to be in *companion form*. The controllability and the controller canonical form can easily be dualized, if pair (A, c) is observable, to obtain the *canonical forms relative to output*, which are the *observability canonical form* and the *observer canonical form*.

Four SISO Canonical Realizations. All the preceding canonical forms can be used to derive SISO canonical realizations whose coefficients are directly related to those of the corresponding transfer function. Refer to a SISO system described by the transfer function

$$G(s) = \frac{\beta_n s^n + \beta_{n-1} s^{n-1} + \ldots + \beta_0}{s^n + \alpha_{n-1} s^{n-1} + \ldots + \alpha_0} \quad (25)$$

and consider the problem of deriving a realization (A, b, c, d) with (A, b) controllable. To simplify notation, assume $n = 4$. According to the above derived controllability canonical form, the realization can be expressed as

$$\dot{z}(t) = A_1 z(t) + b_1 u(t)$$
$$y(t) = c_1 z(t) + d u(t)$$

with

$$A_1 = \begin{bmatrix} 0 & 0 & 0 & -\alpha_0 \\ 1 & 0 & 0 & -\alpha_1 \\ 0 & 1 & 0 & -\alpha_2 \\ 0 & 0 & 1 & -\alpha_3 \end{bmatrix} \quad b_1 = \begin{bmatrix} 1 \\ 0 \\ 0 \\ 0 \end{bmatrix} \quad (26)$$

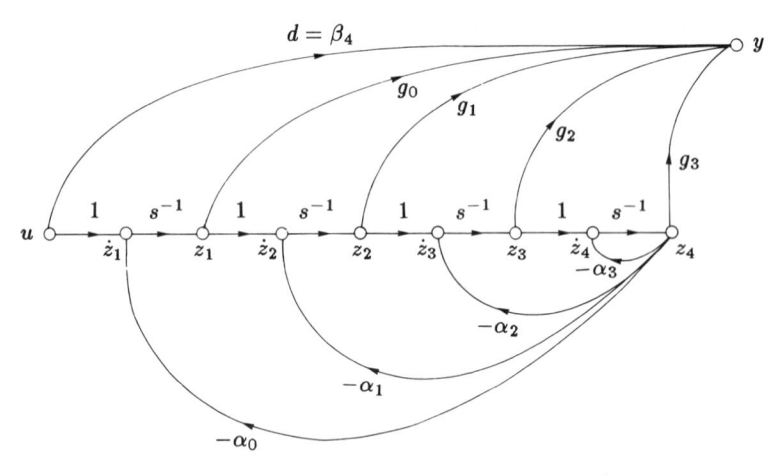

$$c_1 = [\, g_0 \quad g_1 \quad g_2 \quad g_3 \,] \quad d = \beta_4$$

Figure 3-6 The controllability canonical realization.

The realization is called *controllability canonical realization*. The corresponding signal-flow graph is represented in Fig. 3-6. Coefficients g_i are related to α_i, β_i ($i = 0, \ldots, 3$) by simple linear relations, as will be shown. By applying the similarity transformation $A_2 := Q^{-1} A_1 Q$, $b_2 := Q^{-1} b_1$, $c_2 := c_1 Q$, with

$$Q := \begin{bmatrix} \alpha_1 & \alpha_2 & \alpha_3 & 1 \\ \alpha_2 & \alpha_3 & 1 & 0 \\ \alpha_3 & 1 & 0 & 0 \\ 1 & 0 & 0 & 0 \end{bmatrix} \quad (27)$$

Sect. 3.3 Controllability and Observability

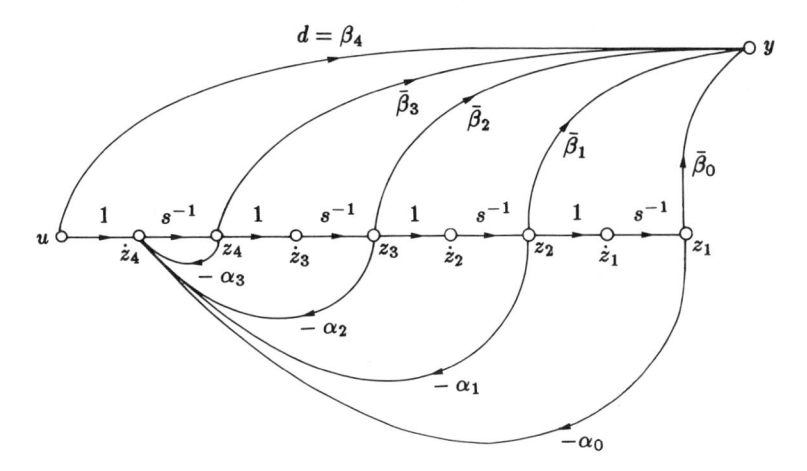

Figure 3-7 The controller canonical realization.

we obtain the *controller canonical realization*, expressed by

$$\dot{z}(t) = A_2 z(t) + b_2 u(t)$$
$$y(t) = c_2 z(t) + d u(t)$$

with

$$A_2 = \begin{bmatrix} 0 & 1 & 0 & 0 \\ 0 & 0 & 1 & 0 \\ 0 & 0 & 0 & 1 \\ -\alpha_0 & -\alpha_1 & -\alpha_2 & -\alpha_3 \end{bmatrix} \quad b_2 = \begin{bmatrix} 0 \\ 0 \\ 0 \\ 1 \end{bmatrix} \tag{28}$$

$$c_2 = [\bar{\beta}_0 \quad \bar{\beta}_1 \quad \bar{\beta}_2 \quad \bar{\beta}_3] \quad d = \beta_4$$

The corresponding signal-flow graph is represented in Fig. 3-7. By using Mason's formula it is shown that the components of c_2 are related to the coefficients on the right of (25) by $\bar{\beta}_i = \beta_i - \beta_4 \alpha_i$ ($i=0,\ldots,3$). Thus, coefficients g_i ($i=0,\ldots,3$) of the controllability realization can be derived from $c_1 = c_2 Q^{-1}$.

The *observability canonical realization* and the *observer canonical realization*, with (A, c) observable, are easily derived by duality (simply use A^T, c^T instead of A, b in the first transformation and transpose the obtained matrices). The former, whose signal-flow graph is represented in Fig. 3-8, is described by

$$\dot{z}(t) = A_3 z(t) + b_3 u(t)$$
$$y(t) = c_3 z(t) + d u(t)$$

with

$$A_3 = \begin{bmatrix} 0 & 1 & 0 & 0 \\ 0 & 0 & 1 & 0 \\ 0 & 0 & 0 & 1 \\ -\alpha_0 & -\alpha_1 & -\alpha_2 & -\alpha_3 \end{bmatrix} \quad b_3 = \begin{bmatrix} f_0 \\ f_1 \\ f_2 \\ f_3 \end{bmatrix} \tag{29}$$

$$c_3 = [1 \quad 0 \quad 0 \quad 0] \quad d = \beta_4$$

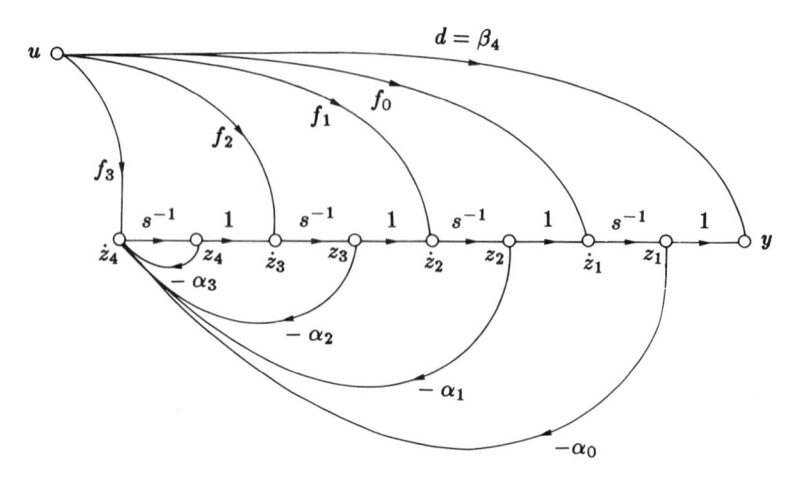

Figure 3-8 The observability canonical realization.

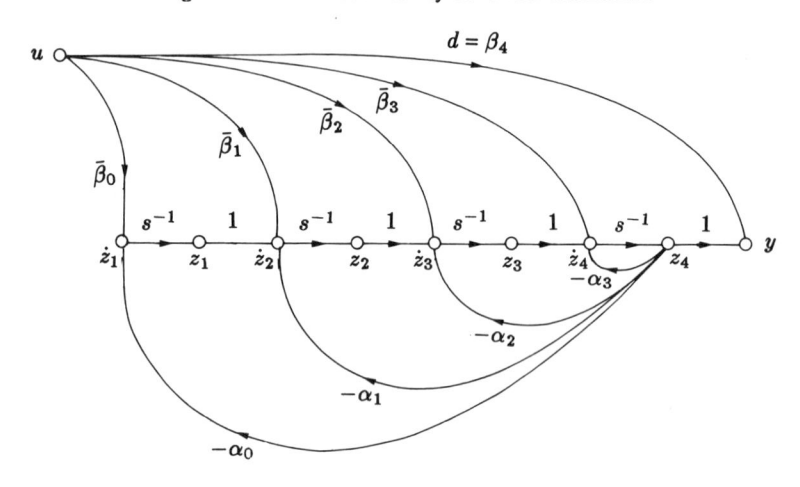

Figure 3-9 The observer canonical realization.

and the latter, whose signal-flow graph is represented in Fig. 3-9, is described by

$$\dot{z}(t) = A_4 \, z(t) + b_4 \, u(t)$$
$$y(t) = c_4 \, z(t) + d \, u(t)$$

with $A_4 := Q \, A_3 \, Q^{-1}$, $b_4 := Q \, b_3$, and $c_4 := c_3 \, Q^{-1}$ having the structures

$$A_4 = \begin{bmatrix} 0 & 0 & 0 & -\alpha_0 \\ 1 & 0 & 0 & -\alpha_1 \\ 0 & 1 & 0 & -\alpha_2 \\ 0 & 0 & 1 & -\alpha_3 \end{bmatrix} \quad b_4 = \begin{bmatrix} \bar{\beta}_0 \\ \bar{\beta}_1 \\ \bar{\beta}_2 \\ \bar{\beta}_3 \end{bmatrix} \quad (30)$$

$$c_4 = [0 \ \ 0 \ \ 0 \ \ 1] \quad d = \beta_4$$

where $\bar{\beta}_i = \beta_i - \beta_4 \alpha_i$ $(i = 0, \ldots, 3)$ are the same as in the controller canonical relization.

The identity of β_i ($i=0,\ldots,3$) (the components of b_4) with the corresponding β_i's on the left of (25) is proved again by using Mason's formula. Coefficients f_i ($i=0,\ldots,3$) of the observer realization are consequently derived from $b_3 = Q^{-1} b_4$.

3.3.4 Structural Indices and MIMO Canonical Forms

The concepts of controllability and controller canonical forms will now be extended to multi-input systems. Consider a controllable pair (A, B), where A and B are assumed to be respectively $n \times n$ and $n \times p$. We shall denote by $b_1, \ldots, b_p$ the columns of B and by $\mu \leq p$ its rank. Vectors $b_1, \ldots, b_\mu$ are assumed to be linearly independent. This does not imply any loss of generality, since, if not, inputs can be suitably renumbered. Consider the table

$$\begin{array}{cccc} b_1 & b_2 & \ldots & b_\mu \\ A b_1 & A b_2 & \ldots & A b_\mu \\ A^2 b_1 & A^2 b_2 & \ldots & A^2 b_\mu \\ \ldots & & & \end{array} \qquad (31)$$

which is assumed to be constructed by rows: each column ends when a vector is obtained that can be expressed as a linear combination of all the previous ones. This vector is not included in the table and the corresponding column is not continued because, as will be shown herein, also all subsequent vectors would be linear combinations of the previous ones. By the controllability assumption, a table with exactly n linearly independent elements is obtained. Denote by r_i ($i = 1, \ldots, \mu$) the numbers of the elements of the i-th column: the above criterion to end columns implies that vector $A^{r_i} b_i$ is a linear combination of all previous ones, i.e.,

$$A^{r_i} b_i = -\sum_{j=1}^{\mu} \sum_{h=0}^{r_i - 1} \alpha_{ijh} A^h b_j - \sum_{j<i} \alpha_{ijr_i} A^{r_i} b_j \quad (i=1,\ldots,\mu) \qquad (32)$$

where the generic coefficient α_{ijh} is zero for $h \geq r_j$. The following property holds.

Property 3.3-4.

If $A^{r_i} b_i$ is a linear combination of the previous vectors in table (31), also $A^{r_i + k} b_i$ for all positive integers k, is a linear combination of the previous vectors.

Proof. For $k = 1$ the property is proved by multiplying on the right by A both members of the i-th of (32) and by eliminating, in the sums at the right side member, all vectors that, according to (32), can be expressed as linear combinations of the previous ones, in particular $A^{r_i} b_i$. Proof is extended by induction to the case $k > 1$. □

Another interesting property of table (31) is the following.

Property 3.3-5.

The set $\{r_i \ (i=1,\ldots,\mu)\}$ does not depend on the ordering assumed for columns of B in table (31).

Proof. The number of columns of table (31) with generic length i is equal to the integer $\Delta\rho_{i-1} - \Delta\rho_i$, with

$$\Delta\rho_i := \rho([B|AB|\ldots|A^i B]) - \rho([B|AB|\ldots|A^{i-1} B])$$

which is clearly invariant with respect to permutations of columns of B. $\square$

It is worth noting that, although the number of columns of table (31) with generic length i is invariant under a permutation of columns of B, the value of the r_i's corresponding to column b_i may commute with each other.

Constants r_i ($i = 1,\ldots,\mu$) are called the *input structural indices* and represent an important characterization of dynamic systems. The value of the greatest input structural index is called the *controllability index*.

It is now possible to extend to multivariable systems the SISO canonical forms and realizations described in Subsection 3.2.1. We shall consider only the controllability and the controller form, since the others, related to observability, can easily be derived by duality.

To maintain expounding at a reasonable level of simplicity we refer to a particular case, corresponding to $n=9$, $p=5$, $\mu=3$, in which table (31) is assumed to be

$$\begin{array}{ccc} b_1 & b_2 & b_3 \\ Ab_1 & Ab_2 & Ab_3 \\ & A^2 b_2 & A^2 b_3 \\ & A^3 b_2 & \end{array} \tag{33}$$

The input structural indices, presented in decreasing order, are in this case $4,3,2$, and the controllability index is 4. Relations (32) can be written as

$$\begin{aligned}
A^2 b_1 &= -\alpha_{110} b_1 - \alpha_{111} Ab_1 - \alpha_{120} b_2 - \alpha_{121} Ab_2 - \alpha_{130} b_3 - \alpha_{131} Ab_3 \\
A^4 b_2 &= -\alpha_{210} b_1 - \alpha_{211} Ab_1 - \alpha_{220} b_2 - \alpha_{221} Ab_2 - \alpha_{222} A^2 b_2 - \alpha_{223} A^3 b_2 - \\
& \quad \alpha_{230} b_3 - \alpha_{231} Ab_3 \\
A^3 b_3 &= -\alpha_{310} b_1 - \alpha_{311} Ab_1 - \alpha_{320} b_2 - \alpha_{321} Ab_2 - \alpha_{322} A^2 b_2 - \alpha_{323} A^3 b_2 - \\
& \quad \alpha_{330} b_3 - \alpha_{331} Ab_3
\end{aligned} \tag{34}$$

The controllability canonical form is obtained through the coordinate transformation defined by

$$T_1 := [b_1 \,|\, Ab_1 \,|\, A^2 b_1 \,|\, b_2 \,|\, Ab_2 \,|\, A^2 b_2 \,|\, A^3 b_2 \,|\, b_3 \,|\, Ab_3 \,|\, A^2 b_3]$$

Matrices $A_1 := T_1^{-1} A T_1$ and $B_1 := T_1^{-1} B$ have the structures shown in (35, 36), as can easily be proved with arguments similar to the single-input case. Note that the submatrices on the main diagonal of A_1 are in companion form. In matrix B_1, $\epsilon_{11}, \epsilon_{21}, \epsilon_{31}$ and $\epsilon_{12}, \epsilon_{22}, \epsilon_{32}$ denote, respectively, the components of b_4 and b_5 with respect to b_1, b_2, b_3.

Sect. 3.3 Controllability and Observability

$$A_1 = \begin{bmatrix} 0 & -\alpha_{110} & 0 & 0 & 0 & -\alpha_{210} & 0 & 0 & -\alpha_{310} \\ 1 & -\alpha_{111} & 0 & 0 & 0 & -\alpha_{211} & 0 & 0 & -\alpha_{311} \\ \hline 0 & -\alpha_{120} & 0 & 0 & 0 & -\alpha_{220} & 0 & 0 & -\alpha_{320} \\ 0 & -\alpha_{121} & 1 & 0 & 0 & -\alpha_{221} & 0 & 0 & -\alpha_{321} \\ 0 & 0 & 0 & 1 & 0 & -\alpha_{222} & 0 & 0 & -\alpha_{322} \\ 0 & 0 & 0 & 0 & 1 & -\alpha_{223} & 0 & 0 & -\alpha_{323} \\ \hline 0 & -\alpha_{130} & 0 & 0 & 0 & -\alpha_{230} & 0 & 0 & -\alpha_{330} \\ 0 & -\alpha_{131} & 0 & 0 & 0 & -\alpha_{231} & 1 & 0 & -\alpha_{331} \\ 0 & 0 & 0 & 0 & 0 & -\alpha_{232} & 0 & 1 & -\alpha_{332} \end{bmatrix} \quad (35)$$

$$B_1 = \begin{bmatrix} 1 & 0 & 0 & \epsilon_{11} & \epsilon_{12} \\ 0 & 0 & 0 & 0 & 0 \\ \hline 0 & 1 & 0 & \epsilon_{21} & \epsilon_{22} \\ 0 & 0 & 0 & 0 & 0 \\ 0 & 0 & 0 & 0 & 0 \\ 0 & 0 & 0 & 0 & 0 \\ \hline 0 & 0 & 1 & \epsilon_{31} & \epsilon_{32} \\ 0 & 0 & 0 & 0 & 0 \\ 0 & 0 & 0 & 0 & 0 \end{bmatrix} \quad (36)$$

We shall now extend the controller form to multi-input systems. First, write (32) as

$$A^{r_i}\left(b_i + \sum_{j<i} \alpha_{ijr_i} b_j\right) = -\sum_{j=1}^{\mu} \sum_{h=0}^{r_i-1} \alpha_{ijh} A^h b_j \quad (i=1,\ldots,\mu) \quad (37)$$

By means of the assumption

$$b_i' := b_i + \sum_{j<i} \alpha_{ijr_i} b_j \quad (i=1,\ldots,\mu)$$

relations (37) can be put in the form

$$A^{r_i} b_i' = -\sum_{j=1}^{\mu} \sum_{h=0}^{r_i-1} \beta_{ijh} A^h b_j' \quad (i=1,\ldots,\mu) \quad (38)$$

If vectors (31) are linearly independent, the vectors of the similar table obtained from b_i' ($i=1,\ldots,\mu$) are also so, since each one of them is expressed by the sum of a vector of (31) with a linear combination of the previous ones. We assume vectors of the new table as a new basis. In the particular case of (33) these are

$$b_1, \; Ab_1, \; A^2b_1, \; b_2, \; Ab_2, \; A^2b_2, \; A^3b_2, \; b'_3, \; Ab'_3, \; A^2b'_3$$

with $b'_3 := b_3 + \alpha_{323} b_2$. Matrices A'_1 and B'_1 that express A and B with respect to the new basis are obtained as $A'_1 = F_1^{-1} A_1 F_1$, $B'_1 = F_1^{-1} B_1$, with

$$F_1 := \left[\begin{array}{cc|cccc|ccc} 1 & 0 & 0 & 0 & 0 & 0 & 0 & 0 & 0 \\ 0 & 1 & 0 & 0 & 0 & 0 & 0 & 0 & 0 \\ \hline 0 & 0 & 1 & 0 & 0 & 0 & \alpha_{323} & 0 & 0 \\ 0 & 0 & 0 & 1 & 0 & 0 & 0 & \alpha_{323} & 0 \\ 0 & 0 & 0 & 0 & 1 & 0 & 0 & 0 & \alpha_{323} \\ 0 & 0 & 0 & 0 & 0 & 1 & 0 & 0 & 0 \\ \hline 0 & 0 & 0 & 0 & 0 & 0 & 1 & 0 & 0 \\ 0 & 0 & 0 & 0 & 0 & 0 & 0 & 1 & 0 \\ 0 & 0 & 0 & 0 & 0 & 0 & 0 & 0 & 1 \end{array}\right] \qquad (39)$$

We obtain matrices having the structures

$$A'_1 = \left[\begin{array}{cc|cccc|ccc} 0 & -\beta_{110} & 0 & 0 & 0 & -\beta_{210} & 0 & 0 & -\beta_{310} \\ 1 & -\beta_{111} & 0 & 0 & 0 & -\beta_{211} & 0 & 0 & -\beta_{311} \\ \hline 0 & -\beta_{120} & 0 & 0 & 0 & -\beta_{220} & 0 & 0 & -\beta_{320} \\ 0 & -\beta_{121} & 1 & 0 & 0 & -\beta_{221} & 0 & 0 & -\beta_{321} \\ 0 & 0 & 0 & 1 & 0 & -\beta_{222} & 0 & 0 & -\beta_{322} \\ 0 & 0 & 0 & 0 & 1 & -\beta_{223} & 0 & 0 & 0 \\ \hline 0 & -\beta_{130} & 0 & 0 & 0 & -\beta_{230} & 0 & 0 & -\beta_{330} \\ 0 & -\beta_{131} & 0 & 0 & 0 & -\beta_{231} & 1 & 0 & -\beta_{331} \\ 0 & 0 & 0 & 0 & 0 & -\beta_{232} & 0 & 1 & -\beta_{332} \end{array}\right] \qquad (40)$$

$$B'_1 = \left[\begin{array}{ccccc} 1 & 0 & 0 & \epsilon'_{11} & \epsilon'_{12} \\ 0 & 0 & 0 & 0 & 0 \\ \hline 0 & 1 & -\alpha_{323} & \epsilon'_{21} & \epsilon'_{22} \\ 0 & 0 & 0 & 0 & 0 \\ 0 & 0 & 0 & 0 & 0 \\ 0 & 0 & 0 & 0 & 0 \\ \hline 0 & 0 & 1 & \epsilon'_{31} & \epsilon'_{32} \\ 0 & 0 & 0 & 0 & 0 \\ 0 & 0 & 0 & 0 & 0 \end{array}\right] \qquad (41)$$

Sect. 3.3 Controllability and Observability

In (41) $\epsilon'_{11}, \epsilon'_{21}, \epsilon'_{31}$ and $\epsilon'_{12}, \epsilon'_{22}, \epsilon'_{32}$ are the components of b_4 and b_5 with respect to b'_1, b'_2, b'_3. As the last step, express A and B with respect to the basis

$$
\begin{aligned}
q_1 &= A\, q_2 + \beta_{111}\, q_2 + \beta_{211}\, q_6 + \beta_{311}\, q_9 \\
q_2 &= b'_1 = b_1 \\
q_3 &= A\, q_4 + \beta_{121}\, q_2 + \beta_{221}\, q_6 + \beta_{321}\, q_9 \\
q_4 &= A\, q_5 \phantom{ + \beta_{121}\, q_2} + \beta_{222}\, q_6 + \beta_{322}\, q_9 \\
q_5 &= A\, q_6 \phantom{ + \beta_{121}\, q_2} + \beta_{223}\, q_6 + \beta_{323}\, q_9 \quad\quad (42) \\
q_6 &= b'_2 = b_3 \\
q_7 &= A\, q_8 + \beta_{131}\, q_2 + \beta_{231}\, q_6 + \beta_{331}\, q_9 \\
q_8 &= A\, q_9 \phantom{ + \beta_{131}\, q_2} + \beta_{232}\, q_6 + \beta_{332}\, q_9 \\
q_9 &= b'_3
\end{aligned}
$$

The corresponding transformation T_2 and the new matrices $A_2 = T_2^{-1} A'_1 T_2$, $B_2 = T_2^{-1} B'_1$ are

$$
T_2 := \left[\begin{array}{cc|cccc|ccc}
\beta_{111} & 1 & \beta_{211} & 0 & 0 & 0 & \beta_{311} & 0 & 0 \\
1 & 0 & 0 & 0 & 0 & 0 & 0 & 0 & 0 \\
\hline
\beta_{121} & 0 & \beta_{221} & \beta_{222} & \beta_{223} & 1 & \beta_{321} & \beta_{322} & 0 \\
0 & 0 & \beta_{222} & \beta_{223} & 1 & 0 & \beta_{322} & 0 & 0 \\
0 & 0 & \beta_{223} & 1 & 0 & 0 & 0 & 0 & 0 \\
0 & 0 & 1 & 0 & 0 & 0 & 0 & 0 & 0 \\
\hline
\beta_{131} & 0 & \beta_{231} & \beta_{232} & 0 & 0 & \beta_{331} & \beta_{332} & 1 \\
0 & 0 & \beta_{232} & 0 & 0 & 0 & \beta_{332} & 0 & 0 \\
0 & 0 & 0 & 0 & 0 & 0 & 1 & 0 & 0
\end{array}\right] \quad (43)
$$

$$
A_2 = \left[\begin{array}{cc|cccc|ccc}
0 & 1 & 0 & 0 & 0 & 0 & 0 & 0 & 0 \\
-\beta_{110} & -\beta_{111} & -\beta_{120} & -\beta_{121} & 0 & 0 & -\beta_{130} & -\beta_{131} & 0 \\
\hline
0 & 0 & 0 & 1 & 0 & 0 & 0 & 0 & 0 \\
0 & 0 & 0 & 0 & 1 & 0 & 0 & 0 & 0 \\
0 & 0 & 0 & 0 & 0 & 1 & 0 & 0 & 0 \\
-\beta_{210} & -\beta_{211} & -\beta_{220} & -\beta_{221} & -\beta_{222} & -\beta_{223} & -\beta_{230} & -\beta_{231} & -\beta_{232} \\
\hline
0 & 0 & 0 & 0 & 0 & 0 & 0 & 1 & 0 \\
0 & 0 & 0 & 0 & 0 & 0 & 0 & 0 & 1 \\
-\beta_{310} & -\beta_{311} & -\beta_{320} & -\beta_{321} & -\beta_{322} & -\beta_{323} & -\beta_{330} & -\beta_{331} & -\beta_{332}
\end{array}\right]
$$

$$(44)$$

$$B_2 = \begin{bmatrix} 0 & 0 & 0 & 0 & 0 \\ 1 & (-\alpha_{214}) & (-\alpha_{313}) & \epsilon'_{11} & \epsilon'_{12} \\ \hline 0 & 0 & 0 & 0 & 0 \\ 0 & 0 & 0 & 0 & 0 \\ 0 & 0 & 0 & 0 & 0 \\ 0 & 1 & -\alpha_{323} & \epsilon'_{21} & \epsilon'_{22} \\ \hline 0 & 0 & 0 & 0 & 0 \\ 0 & 0 & 0 & 0 & 0 \\ 0 & 0 & 1 & \epsilon'_{31} & \epsilon'_{32} \end{bmatrix} \qquad (45)$$

Vectors (42) are linearly independent since, considered in proper order (i.e., in the reverse order in each chain), they can be obtained as the sum of a vector of table (33) with a linear combination of the previous ones. The elements in round brackets in (45) vanish in this particular case, but could be nonzero in the most general case.

The controllability canonical form and the controller canonical form are *canonical forms relative to input*. By duality from pair (A, C) it is possible to derive the observability canonical form and the observer canonical form, also called the *canonical forms relative to output*. The *output structural indices* can be derived likewise. The value of the greatest output structural index is called the *observability index* of the system referred to.

3.4 STATE FEEDBACK AND OUTPUT INJECTION

The term *feedback* denotes an external connection through which effects are brought back to influence corresponding causes. It is the main tool at the disposal of designers to adapt system features to a given task, hence it is basic for all synthesis procedures. In this section an important theorem concerning eigenvalues assignability will be stated and discussed. It directly relates controllability and observability with the possibility of arbitrarily assigning the system eigenvalues through a suitable feedback connection.

Refer to triple (A, B, C) whose structure is represented in Fig. 3-1. In Fig. 3-10 two basic feedback connections are shown: the *state-to-input feedback*, often called simply *state feedback*, and the *output-to-forcing action feedback*, also called *output injection*. In the former, the state is brought to act on the input u through a purely algebraic linear connection, represented by the $p \times n$ real matrix F, while in the latter the output is brought to act on forcing action f again through a purely algebraic linear connection, represented by the $n \times q$ real matrix G. The part shown in the dashed box in figures is the original

Sect. 3.4 State Feedback and Output Injection

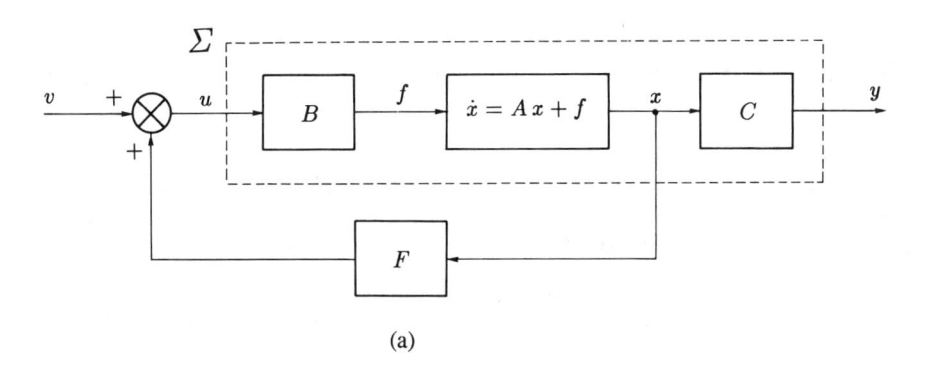

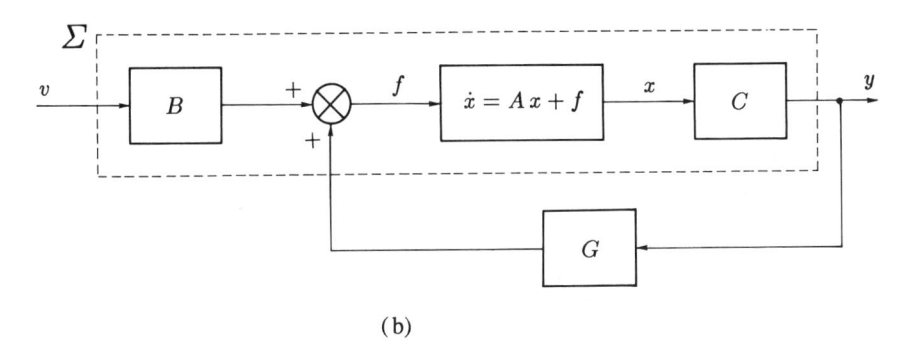

Figure 3-10 State feedback and output injection.

three-map system Σ.

In actual physical systems neither connection is implementable, since neither is the state accessible for direct measurement nor is the forcing action accessible for direct intervention. Nevertheless, these two schemes are useful to state some basic properties that will be referred to in the synthesis of more complex, but physically implementable, feedback connections, such as the output-to-input dynamic feedback, which will be examined later in this section.

We shall first refer to the standard state feedback connection, which is considered the basic one, since its properties are generally easily extendible to the output injection by duality. The system represented in Fig. 3-10(a) has a new input $v \in \mathbf{R}^p$ and is described by the equations

$$\dot{x}(t) = (A + BF)\, x(t) + B\, v(t) \tag{1}$$
$$y(t) = C\, x(t) \tag{2}$$

in which matrix A has been replaced by $A+BF$. By a suitable choice of F a new system (1,2) can be obtained with features significantly different from those of the original one. One of these features is the spectrum of matrix $A+BF$, which can be completely assigned when pair (A, B) is controllable.

The eigenvalue assignability theorem will be presented in two steps: first in restricted form for SISO systems, then in the most general case of MIMO systems.[1]

Theorem 3.4-1. Pole Assignment: SISO Systems

Refer to a SISO system (A, b, c). Let $\sigma = \{\lambda_1, \ldots, \lambda_n\}$ be an arbitrary set of n complex numbers such that $\rho \in \sigma$ implies $\rho^* \in \sigma$. There exists at least one row matrix f such that the spectrum of $A+bf$ coincides with σ if and only if (A, b) is controllable.

Proof. If. Let (A, b) be controllable. There exists a similarity transformation T such that $A' := T^{-1}AT$ and $b' := T^{-1}b$ have the same structures as A_2, b_2 in (3.3.28). Parameters α_i ($i = 0, \ldots, n-1$) are the coefficients of the characteristic polynomial of A, which, in monic form, can in fact be written as

$$\sum_{i=0}^{n-1} \alpha_i \lambda^i + \lambda^n = 0$$

Let β_i ($i = 0, \ldots, n-1$) be the corresponding coefficients of the monic polynomial having the assigned eigenvalues as zeros; they are defined through the identity

$$\sum_{i=0}^{n-1} \beta_i \lambda^i + \lambda^n = \prod_{i=1}^{n} (\lambda - \lambda_i)$$

Clearly the row matrix

$$f' := [\alpha_0 - \beta_0 \mid \alpha_1 - \beta_1 \mid \ldots \mid \alpha_{n-1} - \beta_{n-1}]$$

is such that in the new basis $A' + b'f'$ has the elements of σ as eigenvalues. The corresponding matrix in the main basis

$$A + bf \quad \text{with} \quad f := f' T^{-1}$$

has the same eigenvalues, being similar to it.

Only if. See the only if part of the MIMO case (Theorem 3.4-2). □

Theorem 3.4-2. Pole Assignment: MIMO Systems

Refer to a MIMO system (A, B, C). Let $\sigma = \{\lambda_1, \ldots, \lambda_n\}$ be an arbitrary set of n complex numbers such that $\rho \in \sigma$ implies $\rho^* \in \sigma$. There exists at least one $p \times n$ matrix F such that the spectrum of $A + BF$ coincides with σ if and only if (A, B) is controllable.

Proof. If. For the sake of simplicity, a procedure for deriving a feedback matrix F that solves the pole assignment problem will be presented in the particular case where pair (A, B) is transformed by $T := T_2 F_1 T_1$ into pair (A_2, B_2) whose structure

---

[1] The theorem on eigenvalue assignability is due to Langenhop [25] in the SISO case and Wonham [38] in the MIMO case.

Sect. 3.4 State Feedback and Output Injection

is shown in (3.3.44, 45). The involved successive similarity transformations were derived in Subsection 3.3.4. We shall show that by a suitable choice of feedback matrix F it is possible to obtain for matrix $B_2 FT$, which represents BF in the new basis, the structure

$$B_2 FT = \begin{bmatrix} 0 & 0 & 0 & 0 & 0 & 0 & 0 & 0 & 0 \\ \times & \times & \times & \times & \times & \times & \times & \times & \times \\ 0 & 0 & 0 & 0 & 0 & 0 & 0 & 0 & 0 \\ 0 & 0 & 0 & 0 & 0 & 0 & 0 & 0 & 0 \\ 0 & 0 & 0 & 0 & 0 & 0 & 0 & 0 & 0 \\ \times & \times & \times & \times & \times & \times & \times & \times & \times \\ 0 & 0 & 0 & 0 & 0 & 0 & 0 & 0 & 0 \\ 0 & 0 & 0 & 0 & 0 & 0 & 0 & 0 & 0 \\ \times & \times & \times & \times & \times & \times & \times & \times & \times \end{bmatrix} \qquad (3)$$

where the elements denoted by $\times$ are arbitrary. Suppose, for a moment, that the last two columns of (3.3.45) are not present, i.e., that the system has only three inputs, corresponding to linearly independent forcing actions. It is possible to decompose B_2 as

$$B_2 = MN \qquad (4)$$

with

$$M := \begin{bmatrix} 0 & 0 & 0 \\ 1 & 0 & 0 \\ 0 & 0 & 0 \\ 0 & 0 & 0 \\ 0 & 0 & 0 \\ 0 & 1 & 0 \\ 0 & 0 & 0 \\ 0 & 0 & 0 \\ 0 & 0 & 1 \end{bmatrix} \qquad N := \begin{bmatrix} 1 & (-\alpha_{214}) & (-\alpha_{313}) \\ 0 & 1 & -\alpha_{323} \\ 0 & 0 & 1 \end{bmatrix}$$

where the elements in round brackets could be nonzero in general, but vanish in this particular case. Let W be the three-row matrix formed by the significant elements in (3) (those denoted by $\times$). From

$$B_2 FT = MW$$

and taking (4) into account, it follows that

$$F = N^{-1} W T^{-1} \qquad (5)$$

This relation can be used to derive a feedback matrix F for any choice of the significant elements in (3). In the general case in which B_2 has some columns linearly dependent on the others like the last two in (3.3.45), apply the preceding procedure to the submatrix of B_2 obtained by deleting these columns, then insert in the obtained F zero rows in the same places as the deleted columns.

The eigenvalues coincide with those of matrix $A_2 + B_2 FT$, which has a structure equal to that of A_2, but with all significant elements (i.e., those of the second, third, and ninth row) arbitrarily assignable. It is now easily shown that the eigenvalues are also arbitrarily assignable. In fact, it is possible to set equal to zero all the elements external to the 2×2, 4×4, and 3×3 submatrices on the main diagonal, so that the union of their eigenvalues clearly coincides with the spectrum of the overall matrix. On the other hand, the eigenvalues of these submatrices are easily assignable by a suitable choice of the elements in the last rows (which are the coefficients, with sign changed, of the corresponding characteristic polynomials in monic form).

Only if. Suppose that (A, B) is not controllable, so that the dimension of $\mathcal{R} = \min \mathcal{J}(A, B)$ is less than n. The coordinate transformation $T = [T_1 \, T_2]$ with $\text{im} T_1 = \mathcal{R}$ yields

$$A' := T^{-1} A T = \begin{bmatrix} A'_{11} & A'_{12} \\ O & A'_{22} \end{bmatrix} \quad B' := T^{-1} B = \begin{bmatrix} B'_1 \\ O \end{bmatrix} \tag{6}$$

where, in particular, the structure of B' depends on $\mathcal{B}$ being contained in $\mathcal{R}$. State feedback matrix F corresponds, in the new basis, to

$$F' := F T^{-1} = [F_1 \, F_2] \tag{7}$$

This influences only submatrices on the first row of A', so that the eigenvalues of A'_{22} cannot be varied. □

If there exists a state feedback matrix F such that $A+BF$ is stable, the pair (A, B) is said to be *stabilizable*. By virtue of Theorem 3.4-2 a completely controllable pair is always stabilizable. Nevertheless, the converse is not true: complete controllability is not necessary for (A, B) to be stabilizable, as the following corollary states.

Corollary 3.4-1.
Pair (A, B) is stabilizable if and only if $\mathcal{R} := \min \mathcal{J}(A, B)$ is externally stable.

Proof. Refer to relations (6,7) in the only if part of the proof of Theorem 3.4-2: since (A'_{11}, B'_1) is controllable, by a suitable choice of F'_1 it is possible to obtain $A'_{11} + B'_1 F'_1$ having arbitrary eigenvalues, but it is impossible to influence the second row of A'. Therefore, the stability of A'_{22} is necessary and sufficient to make $A'+B'F'$ (hence $A+BF$) stable. □

Similar results concerning output injection are easily derived by duality. Refer to the system represented in Fig. 3-10(b), described by

$$\dot{x}(t) = (A + GC) x(t) + B v(t) \tag{8}$$
$$y(t) = C x(t) \tag{9}$$

The more general result on pole assignment by state feedback (Theorem 3.4-2) is dualized as follows.

Sect. 3.4 State Feedback and Output Injection

Theorem 3.4-3.
Refer to a MIMO system (A, B, C). Let $\sigma = \{\lambda_1, \ldots, \lambda_n\}$ be an arbitrary set of n complex numbers such that $\rho \in \sigma$ implies $\rho^* \in \sigma$. There exists at least one $n \times q$ matrix G such that the spectrum of $A+GC$ coincides with σ if and only if (A, C) is observable.

Proof. Since

$$\max \mathcal{J}(A, \ker C) = \{0\} \Leftrightarrow \min \mathcal{J}(A^T, \operatorname{im} C^T) = \mathcal{X}$$

pair (A, C) is observable if and only if (A^T, C^T) is controllable. In such a case, by virtue of Theorem 3.4-2 there exists at least one $q \times n$ matrix G^T such that the spectrum of $A^T + C^T G^T$ coincides with the elements of σ. The statement follows from the eigenvalues of any square matrix being equal to those of the transpose matrix, which in this case is $A+GC$. □

If there exists an output injection matrix G such that $A+GC$ is stable, pair (A, C) is said to be *detectable*. By virtue of Theorem 3.4-3 an observable pair is always detectable. Nevertheless, the converse is not true: complete observability is not necessary for (A, C) to be detectable, as stated by the following corollary, which can easily be derived by duality from Corollary 3.4-1.

Corollary 3.4-2.
Pair (A, C) is detectable if and only if $\mathcal{Q} := \max \mathcal{J}(A, C)$ is internally stable.

State feedback and output injection through eigenvalue variation influence stability. It is quite natural at this point to investigate whether they influence other properties, as for instance controllability and observability themselves. We note that:
1. state feedback does not influence controllability, i.e., $\min \mathcal{J}(A, \mathcal{B}) = \min \mathcal{J}(A+BF, \mathcal{B})$;
2. output injection does not influence observability, i.e., $\max \mathcal{J}(A, \mathcal{C}) = \max \mathcal{J}(A+GC, \mathcal{C})$.

These properties can be proved in several ways. Referring to the former (the latter follows by duality), note that the feedback connection in Fig. 3-10(a) does not influence the class of the possible input functions $u(\cdot)$ (which, with or without feedback, is the class of all piecewise continuous functions with values in $\mathbf{R}^p$), hence the reachable set. Otherwise refer to Algorithm 3.2-1 and note that it provides a sequence of subspaces that does not change if A is replaced by $A+BF$. In fact, if this is the case, term $BF\mathcal{Z}_{i-1}$ is added on the right of the definition formula of generic $\mathcal{Z}_i$: this term is contained in $\mathcal{B}$, hence is already a part of $\mathcal{Z}_i$.

On the other hand, state feedback can influence observability and output injection can influence controllability. For instance, by state feedback the greatest $(A, \mathcal{B})$-controlled invariant (see next section) contained in $\mathcal{C}$ can be transformed into an $(A+BF)$-invariant. Since it is, in general, larger than $\mathcal{Q}$, the unobservability subspace is extended.

Furthermore, feedback can make the structures of matrices $A+BF$ and $A+GC$ different from that of A, in the sense that the number and dimensions of the Jordan blocks can be different. To investigate this point we shall refer again to the canonical forms, in particular to the proof of Theorem 3.4-2. First, consider the following lemma.

Lemma 3.4-1.

A companion matrix has no linearly independent eigenvectors corresponding to the same eigenvalue.

Proof. Consider the single-input controller form (A_2, b_2) defined in (3.3.23). Let λ_1 be an eigenvalue of A_2 and $x=(x_1,\ldots,x_n)$ a corresponding eigenvector, so that

$$(A_2 - \lambda_1 I)x = 0$$

or, in detail

$$-\lambda_1 x_1 + x_2 = 0$$
$$-\lambda_1 x_2 + x_3 = 0$$
$$\ldots$$
$$-\lambda_1 x_{n-1} + x_n = 0$$
$$-\alpha_0 x_1 - \alpha_1 x_2 - \ldots - (\alpha_{n-1}+\lambda_1) x_n = 0$$

These relations, considered as equations with $x_1,\ldots,x_n$ as unknowns, admit a unique solution, which can be worked out, for instance, by setting $x_1=1$ and deriving $x_2,\ldots,x_n$ from the first $n-1$ equations. The last equation has no meaning because, by substitution of the previous ones, it becomes

$$(\alpha_0 + \alpha_1 \lambda_1 + \ldots + \alpha_{n-1} \lambda_1^n) x_1 = 0$$

which is an identity, since λ_1 is a zero of the characteristic polynomial. □

This lemma is used to prove the following result, which points out the connection between input structural indices and properties of the system matrix in the presence of state feedback.

Theorem 3.4-4.

Let (A, B) be controllable. A suitable choice of F allows, besides the eigenvalues to be arbitrarily assigned, the degree of the minimal polynomial of $A+BF$ to be made equal, at least, to the controllability index of (A, B).[2]

---

[2] Recall that the controllability index is the minimal value of i such that

$$\rho([B|AB|\ldots|A^i B]) = n$$

Sect. 3.4 State Feedback and Output Injection

Proof. The proof of Theorem 3.4-2 has shown that by a suitable choice of the feedback matrix F a system matrix can be obtained in the *block-companion form*, i.e., with companion matrices on the main diagonal and the remaining elements equal to zero: the dimension of each matrix can be made equal to but not less than the value of the corresponding input structural index. The eigenvalues of these matrices can be arbitrarily assigned: if they are all made equal to one another, by virtue of Lemma 3.4-1 a Jordan block of equal dimension corresponds to every companion matrix. Since multiplicity of an eigenvalue as a zero of the minimal polynomial coincides with the dimension of the corresponding greatest Jordan block (see the proof of Theorem 2.5-5), the multiplicity of the unique zero of the minimal polynomial is equal to the greatest input structural index, i.e., to the controllability index. On the other hand, if the assigned eigenvalues were not equal to each other, the degree of the minimal polynomial, which has all the eigenvalues as zeros, could not be less, since the eigenvalues of any companion matrix (hence of that with greatest dimension) have a multiplicity at least equal to the dimension of the corresponding Jordan block in this matrix, which is unique by virtue of Lemma 3.4-1. In other words, in any case the degree of the minimal polynomial of a companion matrix is equal to its dimension. □

The theorem just presented is very useful for synthesis as a complement on "structure assignment" of Theorem 3.4-2 on pole assignment: in fact, it states a lower bound on the eigenvalue multiplicity in the minimal polynomial. For instance, in the case of discrete-time systems, by a suitable state feedback (which sets all the eigenvalues to zero) the free motion can be made to converge to zero in a finite time. The minimal achievable transient time is specified in the following corollary.

Corollary 3.4-3.
Let (A, B) be controllable. By a suitable choice of F, matrix $A+BF$ can be made nilpotent of order equal, at least, to the controllability index of (A, B).

Proof. Apply the procedure described in the proof of Theorem 3.4-4 to obtain an $A+BF$ similar to a block companion matrix with all eigenvalues zero and with blocks having dimensions equal to the values of the input structural indices. This matrix coincides with the Jordan form: note that a Jordan block corresponding to a zero eigenvalue is nilpotent of order equal to its dimension. □

Theorem 3.4-4 and Corollary 3.4-3 are dualized as follows.

Theorem 3.4-5.
Let (A, C) be observable. A suitable choice of G allows, besides the eigenvalues to be arbitrarily assigned, the degree of the minimal polynomial of $A+GC$ to be made equal, at least, to the observability index of (A, C).

---

while the observability index is the minimal value of j such that

$$\rho([C^T | A^T C^T | \ldots | (A^T)^j C^T]) = n$$

Corollary 3.4-4.
Let (A,C) be controllable. By a suitable choice of G, matrix $A+GC$ can be made nilpotent of order equal, at least, to the observability index of (A,C).

3.4.1 Asymptotic State Observers

Special dynamic devices, called *asymptotic observers*, are used to solve numerous synthesis problems.[3] These are auxiliary linear time-invariant dynamic systems that are connected to the input and output of the observed system and provide an asymptotic estimate of its state, i.e., provide an output z that asymptotically approaches the observed system state. In practice, after a certain settling time from the initial time (at which the observer is connected to the system), $z(t)$ will reproduce the time evolution of the system state $x(t)$. The asymptotic state observer theory is strictly connected to that of the eigenvalue assignment presented earlier. In fact, for an asymptotic observer to be realized, a matrix G must exist such that $A+GC$ is stable, i.e., pair (A,C) must be observable or, at least, detectable.

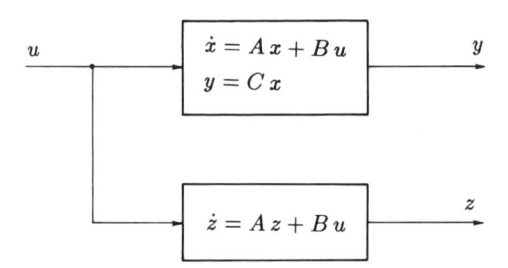

Figure 3-11 State estimate obtained through a model.

Consider a triple (A, B, C). If A is asymptotically stable (has all the eigenvalues with negative real part), a state asymptotic estimate $z(t)$ (state reconstruction in real time) can be achieved by applying the same input signal to a *model* of the system, i.e., another system, built "ad hoc," with a state $z(t)$ whose time evolution is described by the same matrix differential equation, i.e.

$$\dot{z}(t) = A\,z(t) + B\,u(t) \qquad (10)$$

The corresponding connection is shown in Fig. 3-11. This solution has two main drawbacks:

---

[3] Although the word "observability" usually refers to the ability to derive the initial state, following the literature trend we shall indifferently call "observer" or "estimator" a special dynamic device that provides an asymptotic estimate of the current state of a system to whose input and output it is permanently connected.

Sect. 3.4 State Feedback and Output Injection

1. It is not feasible if the observed system is unstable;
2. It does not allow settling time to be influenced.
 In fact, let e be the *estimate error*, defined by

$$e(t) := z(t) - x(t) \tag{11}$$

Subtracting the system equation

$$\dot{x}(t) = A\,x(t) + B\,u(t) \tag{12}$$

from (10) yields

$$\dot{e}(t) = A\,e(t)$$

from which it follows that the estimate error has a time evolution depending only on matrix A and converges to zero whatever its initial value is if and only if all the eigenvalues of A have negative real part.

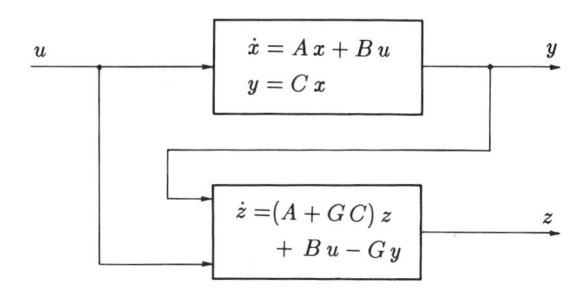

Figure 3-12 State estimate obtained through an asymptotic observer.

A more general asymptotic observer, where both the above drawbacks can be eliminated, is shown in Fig. 3-12. It is named *identity observer* and is different from that of Fig. 3-11 because it also derives information from the system output. It is described by

$$\dot{z}(t) = (A + G\,C)\,z(t) + B\,u(t) - G\,y(t) \tag{13}$$

Matrix G in (13) is arbitrary. The model of Fig. 3-11 can be derived as a particular case by setting $G = O$. Subtracting (12) from (13) and using (11) yields the differential equation

$$\dot{e}(t) = (A + G\,C)\,e(t) \tag{14}$$

from which, by virtue of Corollary 3.4-2, it follows that if (A, C) is observable the convergence of the estimate to the actual state can be made arbitrarily fast. These considerations lead to the following statement.

Property 3.4-1.

For any triple (A, B, C) with (A, C) observable there exists a state observer whose estimate error evolves in time as the solution of a linear, homogeneous, constant-coefficient differential equation of order n with arbitrarily assignable eigenvalues.

If (A, C) is not observable, a suitable choice of G can modify only the eigenvalues of A that are not internal to the unobservability subspace $Q := \max \mathcal{J}(A, C)$: an asymptotic estimation of the state is possible if and only if the system is detectable.

Asymptotic Observers and Complementability. We shall now show that, under very general conditions, any stable dynamic system connected to the output of a free system behaves as an asymptotic observer, because it provides an asymptotic estimate of a linear function of the system state.[4]

Consider the free system

$$\dot{x}(t) = A\, x(t) \tag{15}$$
$$y(t) = C\, x(t) \tag{16}$$

and suppose a generic linear time-invariant dynamic system is connected to its output. The time evolution of state z of this system is assumed to be described by

$$\dot{z}(t) = N\, z(t) + M\, y(t) \tag{17}$$

with N stable. The problem is to state conditions under which there exists a matrix T such that

$$z(t) = T\, x(t) \quad \forall\, t > 0 \tag{18}$$

if the initial conditions satisfy

$$z(0) = T\, x(0) \tag{19}$$

From

$$\dot{z}(t) - T\, \dot{x}(t) = N\, z(t) + M\, C\, x(t) - T\, A\, x(t)$$
$$= (N\, T + M\, C - T\, A)\, x(t)$$

it follows that, because $x(0)$ is generic, (18) holds if and only if $NT + MC - TA = 0$, i.e., if and only if T satisfies the Sylvester equation

$$N\, T - T\, A = -M\, C \tag{20}$$

We recall that this equation has a unique solution T if and only if A and N have no common eigenvalues (Theorem 2.5-10). If (20) holds it follows that

$$\dot{z}(t) - T\, \dot{x}(t) = N\, \bigl(z(t) - T\, x(t)\bigr)$$

---

[4] The general theory of asymptotic observers, including these results, are due to Luenberger [28, 29, 31].

Sect. 3.4 State Feedback and Output Injection

Hence

$$z(t) = T\,x(t) + e^{Nt}\left(z(0) - T\,x(0)\right)$$

which means that, when initial condition (19) is not satisfied, (18) does not hold identically in time, but tends to be satisfied as t approaches infinity.

The obtained result is susceptible to geometric interpretation: by introducing an extended state $\hat{x}$, equations (15–17) can be written together as

$$\dot{\hat{x}}(t) = \hat{A}\,\hat{x}(t)$$

where

$$\hat{x} := \begin{bmatrix} x \\ z \end{bmatrix} \quad \hat{A} := \begin{bmatrix} A & O \\ MC & N \end{bmatrix}$$

In the extended state space $\hat{\mathcal{X}}$, the subspace

$$\hat{\mathcal{Z}} := \{\hat{x} : x = 0\}$$

(the z coordinate hyperplane) is clearly an $\hat{A}$-invariant: it corresponds to an asymptotic observer if and only if it is complementable.

If, instead of the free system (15, 16), we consider a system with forcing action $B\,u(t)$, the asymptotic estimation of the same linear function of state can be obtained by applying a suitable linear function of input also to the observer. In this case (17) is replaced by

$$\dot{z}(t) = N\,z(t) + M\,y(t) + T\,B\,u(t) \tag{21}$$

It may appear at this point that the identity observer, where

$$N := A + G\,C \qquad M = -G$$

with arbitrary G, is a very particular case. This is not true because any observer of order n providing a complete estimate of the state is equivalent to it (i.e., has a state isomorphic to its state). In fact, let T be the corresponding matrix in (20) which, in this case, is nonsingular: from (20) it follows that

$$N = T\,A\,T^{-1} - M\,C\,T^{-1} = T\,(A - T^{-1}M\,C)\,T^{-1} \tag{22}$$

In the above arguments no assumption has been considered on system observability which, actually, is needed only as far as pole assignability of the observer is concerned. Nevertheless, to obtain an asymptotic state estimate it is necessary that the observed system be detectable, since $\mathcal{Q} \subseteq \ker T$ for all T satisfying (20).

```
        v  +              u      ẋ = A x + B u           y
        ───▶⊗──────────┬────────▶                    ┬──────▶
            +▲         │        y = C x             │
             │         │                            │
             │         │                            │
             │         │   ż = (A + G C) z           │     z
             │         └──▶                          ├──────▶
             │             + B u − G y              │
             │                                      │
             │              ┌───────┐               │
             └──────────────┤   F   ├───────────────┘
                            └───────┘
```

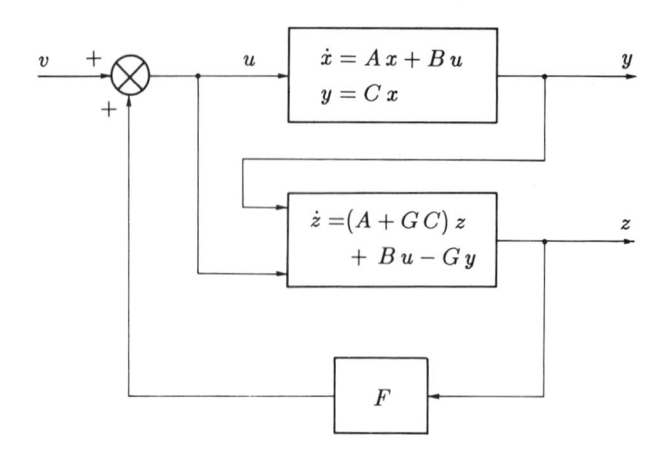

Figure 3-13 Using an asymptotic observer to realize state feedback.

3.4.2 The Separation Property

At the beginning of this section state feedback has been presented as a means to influence some linear system features, in particular eigenvalues, hence stability; on the other hand it has been remarked that such feedback is often practically unfeasible since usually state is not directly accessible for measurement.

It is quite natural to investigate whether it is possible to overcome this drawback by using the state estimate provided by an identity observer instead of the state itself. This corresponds to the feedback connection shown in Fig. 3-13, which no longer refers to an *algebraic* state-to-input feedback, but to a *dynamic* output-to-input one. This connection produces an overall system of order $2n$ described by the equations

$$\dot{x}(t) = A\,x(t) + B\,F\,z(t) + B\,v(t) \tag{23}$$
$$\dot{z}(t) = (A + B\,F + G\,C)\,z(t) - G\,C\,x(t) + B\,v(t) \tag{24}$$
$$y(t) = C\,x(t) \tag{25}$$

The following very basic result relates the eigenvalues of the overall system to those of the system with the purely algebraic state feedback and those of the observer.

Theorem 3.4-6. The Separation Property
The eigenvalues of the overall system corresponding to a state feedback connection through an observer are the union with repetition of those of the system with the simple algebraic state feedback and those of the observer.

Proof. Let $e(t) := x(t) - z(t)$. By the transformation

$$\begin{bmatrix} x \\ e \end{bmatrix} = T \begin{bmatrix} x \\ z \end{bmatrix} \quad \text{with } T = T^{-1} = \begin{bmatrix} I_n & O \\ I_n & -I_n \end{bmatrix}$$

Sect. 3.4 State Feedback and Output Injection

from (23, 24) we derive

$$\begin{bmatrix} \dot{x}(t) \\ \dot{e}(t) \end{bmatrix} = \begin{bmatrix} A+BF & -BF \\ O & A+GC \end{bmatrix} \begin{bmatrix} x(t) \\ e(t) \end{bmatrix} + \begin{bmatrix} B \\ O \end{bmatrix} v(t) \qquad (26)$$

The spectrum of the system matrix in (26) is clearly $\sigma(A+BF) \uplus \sigma(A+GC)$: it is equal to that of the original system (23, 24), from which (26) has been obtained by means of a similarity transformation. □

As a consequence of Theorems 3.4-2, 3.4-3, and 3.4-6, it follows that, if triple (A, B, C) is completely controllable and completely observable, the eigenvalues of the overall system represented in Fig. 3-13 are all arbitrarily assignable. In other words, any completely controllable and observable dynamic system of order n is stabilizable with an output-to-input feedback (or, simply, *output feedback*) through a suitable dynamic system, also of order n.

The duality between control and observation, a characteristic feature of linear time-invariant systems, leads to the introduction of the so-called *dual observers* or *dynamic precompensators*, which are also very important to characterize numerous control system synthesis procedures.

To introduce dynamic precompensators, it is convenient to refer to the block diagram represented in Fig. 3-14(a), which, like that of Fig. 3-11, represents the connection of the observed system with a model; here in the model the purely algebraic operators have been represented as separated from the dynamic part, pointing out the three-map structure. The identity observer represented in Fig. 3-12 is obtained through the connections shown in Fig. 3-14(b), in which signals obtained by applying the same linear transformation G to the outputs of both the model and the system are added to and subtracted from the forcing action. These signals, of course, have no effect if the observer is tracking the system, but influence time behavior and convergence to zero of a possible estimate error.

The identity dynamic precompensator is, on the contrary, obtained by executing the connections shown in Fig. 3-14(c), from the model state to both the model and system inputs. Also in this case, since contributions to inputs are identical, if the system and model states are equal at the initial time, their subsequent evolutions in time will also be equal. The overall system, represented in Fig. 3-14(c), is described by the equations

$$\dot{x}(t) = A\,x(t) + B\,F\,z(t) + B\,v(t) \qquad (27)$$

$$\dot{z}(t) = (A + B\,F)\,z(t) + B\,v(t) \qquad (28)$$

from which, by difference, it follows that

$$\dot{e}(t) = A\,e(t) \qquad (29)$$

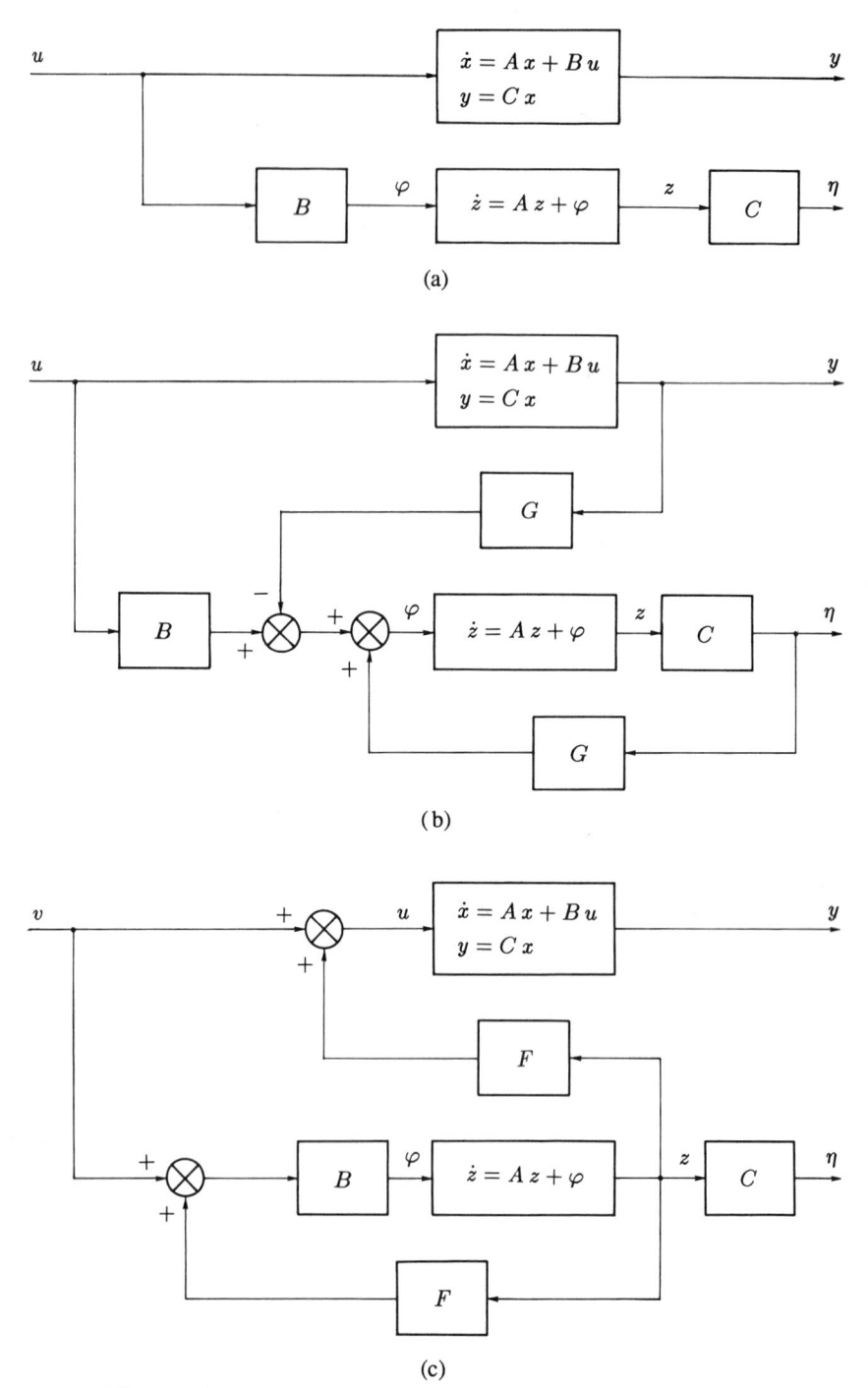

Figure 3-14 Model, asymptotic observer and dynamic precompensator (dual observer).

If triple (A, B, C) is asymptotically stable (this assumption is not very restrictive because, as previously shown, under complete controllability and observability assumption eigenvalues are arbitrarily assignable by means of a dynamic feedback), once the transient due to the possible difference in the initial states is finished, system and precompensator states will be identical at every instant of time. If the considered system is completely controllable, the dynamic behavior of the precompensator can be influenced through a suitable choice of matrix F. For instance, an arbitrarily fast response can be obtained by aptly assigning the eigenvalues of $A+BF$.

Note that the dynamic feedback shown in Fig. 3-13 can be obtained by performing both connections of Fig. 3-14(b) (G blocks) and of Fig. 3-14(c) (F blocks): in the dynamic precompensator case it is equivalent to a purely algebraic feedback G from the output difference $\eta - y$ to forcing action φ.

Extension to Discrete Systems. All the previous results on pole assignment, asymptotic state estimation, and the separation property can easily be extended to discrete systems. A specific feature of these systems is the possibility to extinguish the free motion in a finite number of transitions by assigning the value zero to all eigenvalues. This feature suggests alternative solutions to some typical control problems. Two significant examples are reported in the following.

Problem 3.4-1. Control to the Origin from a Known Initial State

Refer to a discrete pair (A_d, B_d), which is assumed to be controllable. Find a control sequence $u|_{[0,k]}$ that causes the transition from an arbitrary initial state $x(0)$ to the origin in the minimal number of steps compatible with any initial state.[5]

Solution. Find a state feedback F such that A_d+B_dF is nilpotent of order equal to the controllability index of (A_d, B_d). This is possible by virtue of Corollary 3.4-3. Solution of the problem reduces to determination of the free motion of a discrete free system, i.e., to a simple iterative computation. □

Problem 3.4-2. Control to the Origin from an Unknown Initial State

Refer to a discrete triple (A_d, B_d, C_d), which is assumed to be controllable and observable. Determine a control sequence $u|_{[0,k]}$ whose elements can be functions of the observed output, which causes the transition from an unknown initial state to the origin.

Solution. Realize an output-to-input dynamic feedback (through a state observer) of the type shown in Fig. 3-13. From Theorem 3.4-6 and Corollaries 3.4-3 and 3.4-4 it follows that by a suitable choice of F and G the overall system matrix can be made nilpotent of order equal to the sum of controllability and observability indices, so that the number of steps necessary to reach the origin is not greater than this sum. □

---

[5] This problem was also solved in Subsection 2.6.2, but in the more general case of linear time-varying discrete systems and with assigned control time.

3.5 SOME GEOMETRIC ASPECTS OF OPTIMAL CONTROL

In this section we shall present a geometric framework for dynamic optimization problems. Consider the linear time-varying system

$$\dot{x}(t) = A(t)\, x(t) + B(t)\, u(t) \tag{1}$$

where $A(\cdot)$ and $B(\cdot)$ are known, piecewise continuous real matrices, functions of time. The state space is $\mathbf{R}^n$ and the input space $\mathbf{R}^m$. We shall denote by $[t_0, t_1]$ the *optimal control time interval*, i.e., the time interval to which the optimal control problem is referred, by $x_0 := x(t_0)$ the *initial state* and by $x_1 := x(t_1)$ the *final state*, or the extreme point of the *optimal trajectory*.

In many optimal control problems the control function is assumed to be bounded: for instance, it may be *bounded in magnitude* component by component, through the constraints

$$|u_j(t)| \le H \quad (j=1,\ldots,m), \quad t \in [t_0, t_1] \tag{2}$$

or, more generally, by

$$u(t) \in \Omega, \quad t \in [t_0, t_1] \tag{3}$$

where Ω denotes a convex, closed, and bounded subset of $\mathbf{R}^p$ containing the origin. Note that (2) expresses a constraint on the ∞-norm of control function segment $u_{[t_0,t_1]}$. Of course, different individual bounds on every control component and unsymmetric bounds can be handled through suitable manipulations of reference coordinates, such as translation of the origin and scaling.

In order to state an optimization problem a measure of control "goodness" is needed: this is usually expressed by a functional to be minimized, called the *performance index*. We shall here consider only two types of performance indices: the *linear function of the final state*

$$\Gamma = \langle \gamma, x(t_1) \rangle \tag{4}$$

where $\gamma \in \mathbf{R}^n$ is a given vector, and the *integral performance index*

$$\Gamma = \int_{t_0}^{t_1} f\big(x(\tau), u(\tau), \tau\big)\, d\tau \tag{5}$$

where f is a continuous function, in most cases convex. In (4) and (5) symbol Γ stands for *cost*: optimization problems are usually formulated in terms of achieving a minimum cost; of course, maximization problems are reduced to minimization ones by simply changing the sign of the functional.

Sect. 3.5 Some Geometric Aspects of Optimal Control

In addition to the performance index, optimal control problems require definition of an *initial state set* $\mathcal{X}_0$, and a *final state set* $\mathcal{X}_1$ (which may reduce to a single point or extend to the whole space). A typical dynamic optimization problem consists of searching for an initial state $x_0 \in \mathcal{X}_0$ and an admissible control function $u(\cdot)$ such that the corresponding terminal state satisfies $x_1 \in \mathcal{X}_1$ and the performance index is minimal (with respect to all the other admissible choices of x_0 and $u(\cdot)$). The initial time t_0 and/or the final time t_1 are given a priori or must also be optimally derived.

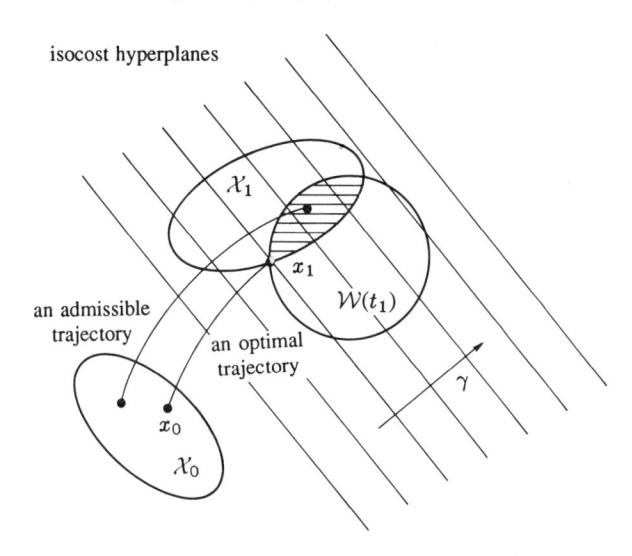

Figure 3-15 The geometric meaning of an optimization problem with performance index defined as a linear function of the final state.

Figure 3-15 shows a geometric interpretation of a dynamic optimization problem with the performance index of the former type and fixed control time interval. $\mathcal{W}(t_1)$ is the set of all states reachable at time t_1 from event (x_0, t_0). Due to the superposition property, it can be expressed as

$$\mathcal{W}(t_1) := \Phi(t_1, t_0) \mathcal{X}_0 + \mathcal{R}^+(t_0, t_1, 0) \tag{6}$$

where

$$\mathcal{R}^+(t_0, t_1, 0) := \left\{ x_1 \: : \: x_1 = \int_{t_0}^{t_1} \Phi(t_1, \tau) B(\tau) u(\tau) \, d\tau, \quad u(\tau) \in \Omega \right\} \tag{7}$$

is the reachable set from the origin with bounded control. Note the layout of the isocost hyperplanes: the final state x_1 of an optimal trajectory belongs to the intersection of $\mathcal{X}_1$ and $\mathcal{W}(t_1)$ and corresponds to a minimal cost.

3.5.1 Convex Sets and Convex Functions

In dealing with optimization in the presence of constraints, we need the concepts and some properties of convex sets and convex functions.[1]

Refer to the vector space $\mathbf{R}^n$ with the standard euclidean norm induced by the inner product. The concepts of subspace and linear variety are considered in Appendix A; however, since in dealing with convexity special emphasis on orthogonality and topological properties is called for, it is convenient to use a slightly different notation, more suitable for the particular topic at hand.

Given an $n \times p$ matrix B having rank p, the set

$$\mathcal{M} := \{ x \,:\, x = B\mu,\ \mu \in \mathbf{R}^p \} \tag{8}$$

is a subspace or a linear variety through the origin, for which B is a basis matrix. In an inner product space an alternative way of defining a subspace is

$$\mathcal{M} := \{ x \,:\, Ax = 0 \} \tag{9}$$

where A is an $(n-p) \times n$ matrix such that $B^T A = O$, i.e., a basis matrix for $\ker B^T = (\operatorname{im} B)^\perp$.

Similarly, the "shifted" linear variety $\mathcal{M}_s := \{x_0\} + \mathcal{M}$, parallel to $\mathcal{M}$ and passing through x_0, is defined as

$$\mathcal{M}_s := \{ x \,:\, x = x_0 + B\mu,\ \mu \in \mathbf{R}^p \} \tag{10}$$

or

$$\mathcal{M}_s := \{ x \,:\, A(x - x_0) = 0 \} \tag{11}$$

Relations (8, 10) are said to define linear varieties in *parametric form*, (9, 11) in *implicit form*.

A particularization of (10) is the *straight line* through two points x_0, x_1:

$$\mathcal{L} := \{ x \,:\, x = x_0 + \mu(x_1 - x_0),\ \mu \in \mathbf{R} \} \tag{12}$$

while a particularization of (11) is the *hyperplane* with normal a passing through x_0:

$$\mathcal{P} := \{ x \,:\, \langle a, (x - x_0) \rangle = 0 \} \tag{13}$$

The hyperplane through n points $x_0, \ldots, x_{n-1} \in \mathbf{R}^n$, such that $\mathcal{B} := \{x_1 - x_0, \ldots, x_{n-1} - x_0\}$ is a linearly independent set, is defined as

$$\mathcal{P} := \{ x \,:\, x = x_0 + B\mu,\ \mu \in \mathbf{R}^{n-1} \}$$

---

[1] Extended treatment of convexity is beyond the scope of this book, so we report only the basic definitions and properties. Good references are the books by Eggleston [6] and Berge [2].

where B is the $n \times (n-1)$ matrix having the elements of $\mathcal{B}$ as columns.

The *line segment* joining any two points $x_1, x_2 \in \mathbf{R}^n$ is defined by

$$\mathcal{R}(x_1, x_2) := \{ x \,:\, x = x_1 + \mu(x_2 - x_1),\ 0 \leq \mu \leq 1 \} \tag{14}$$

The sets

$$\bar{\mathcal{H}}_+(\mathcal{P}) := \{ x \,:\, \langle a, (x-x_0) \rangle \geq 0 \} \tag{15}$$
$$\bar{\mathcal{H}}_-(\mathcal{P}) := \{ x \,:\, \langle a, (x-x_0) \rangle \leq 0 \} \tag{16}$$

are called *closed half-spaces* bounded by $\mathcal{P}$, the hyperplane defined in (13), while the corresponding sets without the equality sign in the definition are called *open half-spaces* bounded by $\mathcal{P}$ and denoted by $\mathcal{H}_+(\mathcal{P})$, $\mathcal{H}_-(\mathcal{P})$.

Definition 3.5-1. Convex Set

A set $\mathcal{X} \subseteq \mathbf{R}^n$ is said to be *convex* if for any two points $x_1, x_2 \in \mathcal{X}$ the straight line segment joining x_1 and x_2 is contained in $\mathcal{X}$. In formula, $\alpha x_1 + (1-\alpha) x_2 \in \mathcal{X}$ for all $x_1, x_2 \in \mathcal{X}$ and all $\alpha \in [0, 1]$.

The following properties of convex sets are easily derived:
1. The intersection of two convex sets is a convex set;
2. The sum or, more generally, any linear combination of two convex sets is a convex set;[2]
3. The cartesian product of two convex sets is a convex set;
4. The image of a convex set in a linear map is a convex set.

Since $\mathbf{R}^n$ is metric with the norm induced by the inner product, it is possible to divide the points of any given set $\mathcal{X} \subseteq \mathbf{R}^n$ into interior, limit, and isolated points according to Definitions A.6-5, A.6-6, and A.6-7. Clearly, a convex set cannot have any isolated point.

Definition 3.5-2. Dimension of a Convex Set

The *dimension* of a convex set $\mathcal{X}$ is the largest integer m for which there exist $m+1$ points $x_i \in \mathcal{X}$ ($i = 0, \ldots, m$) such that the m vectors $x_1 - x_0, \ldots, x_m - x_0$ are linearly independent.

A convex set of dimension m is contained in the linear variety $\mathcal{M} := \{x_0\} + B\mu$, $\mu \in \mathbf{R}^n$, whose basis matrix B has the above m vectors as columns. A convex set whose interior is not empty has dimension n.

Given a convex set $\mathcal{X}$ of dimension m, with $m < n$, it is possible to define the *relative interior* of $\mathcal{X}$ (denoted by rint$\mathcal{X}$) as the set of all interior points of $\mathcal{X}$ considered in the linear variety $\mathcal{M}$, i.e., referring to an m-dimensional vector space.

---

[2] Given any two sets $\mathcal{X}, \mathcal{Y} \subseteq \mathbf{R}^n$, their linear combination with coefficients α, β is defined as

$$\alpha \mathcal{X} + \beta \mathcal{Y} := \{ z \,:\, z = \alpha x + \beta y,\ x \in \mathcal{X},\ y \in \mathcal{Y} \}$$

Definition 3.5-3. Support Hyperplane of a Convex Set

A hyperplane $\mathcal{P}$ that intersects the closure of a convex set $\mathcal{X}$ and such that there are no points of $\mathcal{X}$ in one of the open half-spaces bounded by $\mathcal{P}$ is called a *support hyperplane* of $\mathcal{X}$. In other words, let x_0 be a frontier point of $\mathcal{X}$: $\mathcal{P}$ defined in (13) is a support hyperplane of $\mathcal{X}$ at x_0 if $\mathcal{X} \subseteq \mathcal{H}_+(\mathcal{P})$ or $\mathcal{X} \subseteq \mathcal{H}_-(\mathcal{P})$, i.e.

$$\langle a,(x-x_0)\rangle \geq 0 \quad \text{or} \quad \langle a,(x-x_0)\rangle \leq 0 \quad \forall x \in \mathcal{X} \tag{17}$$

When considering a particular support hyperplane, it is customary to refer to the outer normal, i.e., to take the sign of a in such a way that the latter of (17) holds.

Property 3.5-1.

Any convex set $\mathcal{X}$ admits at least one support hyperplane $\mathcal{P}$ through every point x_0 of its boundary. Conversely, if through every boundary point of $\mathcal{X}$ there exists a support hyperplane of $\mathcal{X}$, $\mathcal{X}$ is convex.

Hence, any convex set is the envelope of all its support hyperplanes.

Definition 3.5-4. Cone

A *cone* with *vertex* in the origin is a set C such that for all $x \in C$ the half-line or *ray* αx, $\alpha \geq 0$, is contained in C. A cone with vertex in x_0 is a set C such that for all $x \in C$ the ray $x_0 + \alpha(x-x_0)$, $\alpha \geq 0$, is contained in C.

Definition 3.5-5. Polar Cone of a Convex Set

Let x_0 be any point of the convex set $\mathcal{X}$. The *polar cone* of $\mathcal{X}$ at x_0 (which will be denoted by $C_p(\mathcal{X}-x_0)$) is defined as

$$C_p(\mathcal{X}-x_0) := \{ p : \langle p,(x-x_0)\rangle \leq 0 \quad \forall x \in \mathcal{X} \} \tag{18}$$

If x_0 is a boundary point of $\mathcal{X}$, $C_p(\mathcal{X}-x_0)$ is the locus of the outer normals of all the support hyperplanes of $\mathcal{X}$ at x_0. If $\dim \mathcal{X} = n$ and x_0 is an interior point of $\mathcal{X}$, $C_p(\mathcal{X}-x_0)$ clearly reduces to the origin. It is easy to prove that any polar cone of a convex set is convex.

Definition 3.5-6. Convex Function

A function $f : \mathcal{D} \to \mathbf{R}$, where $\mathcal{D}$ denotes a convex subset of $\mathbf{R}^n$, is said to be a *convex function* if for any two points $x_1, x_2 \in \mathcal{D}$ and any $\alpha \in [0,1]$

$$f\bigl(\alpha x_1 + (1-\alpha)x_2\bigr) \leq \alpha f(x_1) + (1-\alpha)f(x_2) \tag{19}$$

If the preceding relation holds with the strict inequality sign, function f is said to be *strictly convex*; if f is a (strictly) convex function, $-f$ is said to be *(strictly) concave*.

For example, in Fig. 3-16 the graph of a possible convex function with domain in $\mathbf{R}$ is represented and the meaning of condition (19) is pointed out: note that the function cannot be constant on any finite segment of its domain if the value at some other point is less.

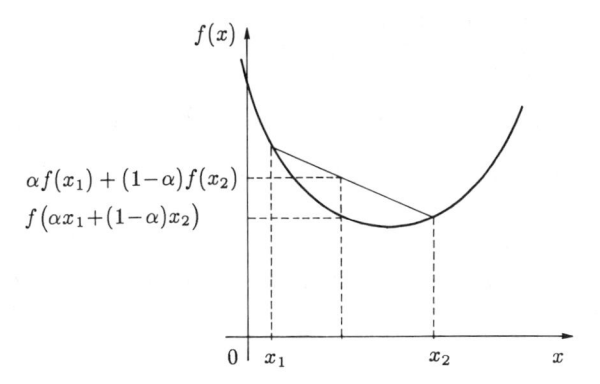

Sect. 3.5 Some Geometric Aspects of Optimal Control

Figure 3-16 A convex function with $\mathcal{D} \subseteq \mathbf{R}$.

Property 3.5-2.
The sum or, more generally, any linear combination with nonnegative coefficients of two convex functions is a convex function.

Proof. Let f, g be two convex functions with the same domain $\mathcal{D}$ and $\varphi := \beta f + \gamma g$ with $\beta, \gamma \geq 0$ a linear combination of them. It follows that

$$\varphi\big(\alpha x_1 + (1-\alpha)x_2\big) = \beta f\big(\alpha x_1 + (1-\alpha)x_2\big) + \gamma g\big(\alpha x_1 + (1-\alpha)x_2\big)$$
$$\leq \alpha\beta f(x_1) + (1-\alpha)\beta f(x_2) + \alpha\gamma g(x_1) + (1-\alpha)\gamma g(x_2)$$
$$= \alpha \varphi(x_1) + (1-\alpha) \varphi(x_2) \quad \square$$

It is easy to prove that a linear combination with positive coefficients of two convex functions is strictly convex if at least one of them is so.

Property 3.5-3.
Let f be a convex function with a sufficiently large domain and k any real number. The set
$$\mathcal{X}_1 := \big\{ x : f(x) \leq k \big\}$$
is convex or empty.

Proof. Let $x_1, x_2 \in \mathcal{X}_1$ and $k_1 := f(x_1)$, $k_2 := f(x_2)$. Then $f(\alpha x_1 + (1-\alpha)x_2) \leq \alpha f(x_1) + (1-\alpha) f(x_2) = \alpha k_1 + (1-\alpha) k_2 \leq k$. $\square$

Property 3.5-4.
A positive semidefinite (positive definite) quadratic form is a convex (strictly convex) function.

Proof. Let $f(x) := \langle x, Ax \rangle$: by assumption $f(x) > 0$ ($f(x) \geq 0$) for all $x \neq 0$. Hence

$$\alpha f(x_1) + (1-\alpha) f(x_2) = \alpha \langle x_1, Ax_1 \rangle + (1-\alpha)\langle x_2, Ax_2 \rangle$$
$$f\big(\alpha x_1 + (1-\alpha)x_2\big) = \alpha^2 \langle x_1, Ax_1 \rangle + 2\alpha(1-\alpha)\langle x_1, Ax_2 \rangle + (1-\alpha)^2 \langle x_2, Ax_2 \rangle$$

By subtraction on the right one obtains

$$\alpha(1-\alpha)\langle x_1, Ax_1\rangle - 2\alpha(1-\alpha)\langle x_1, Ax_2\rangle + \alpha(1-\alpha)\langle x_2, Ax_2\rangle$$
$$= \alpha(1-\alpha)\langle (x_1-x_2), A(x_1-x_2)\rangle > 0 \ (\geq 0) \ \forall x_1, x_2, \ x_1 \neq x_2$$

Since the same inequality must hold for the difference of the left, the property is proved. □

Property 3.5-5.

Let f be a continuously differentiable function defined on a convex domain and x_0 any point of this domain. Denote by

$$g(x_0) := \text{grad} f|_{x_0}$$

the gradient of f at x_0. Then f is convex if and only if

$$f(x) \geq f(x_0) + \langle g(x_0), (x-x_0)\rangle \qquad (20)$$

In other words, a function is convex if and only if it is greater than or equal to all its local linear approximations.

Proof. Only if. From

$$f\big(x_0 + \alpha(x-x_0)\big) = f\big(\alpha x + (1-\alpha)x_0\big) \leq \alpha f(x) + (1-\alpha)f(x_0) \qquad (21)$$

it follows that

$$f(x) \geq f(x_0) + \frac{f\big(x_0+\alpha(x-x_0)\big) - f(x_0)}{\alpha}$$

which converges to (20) as α approaches zero from the right.

If. Consider any two points x_1, x_2 in the domain of f and for any α, $0 \leq \alpha \leq 1$, define $x_0 := \alpha x_1 + (1-\alpha)x_2$. Multiply relation (20) with $x:=x_1$ by α and with $x:=x_2$ by $1-\alpha$ and sum: it follows that

$$\alpha f(x_1) + (1-\alpha)f(x_2) \geq f(x_0) + \langle g(x_0), \big(\alpha x_1+(1-\alpha)x_2 - x_0\big)\rangle$$
$$= f\big(\alpha x_1 + (1-\alpha)x_2\big) \quad \square$$

3.5.2 The Pontryagin Maximum Principle

The maximum principle is a contribution to the calculus of variations developed by the Soviet mathematician L.S. Pontryagin to solve variational problems in the presence of constraints on the control effort.[3] It can be simply and clearly interpreted geometrically, especially in particular cases of linear systems with performance index (4) or (5). We shall present it here referring only to these cases.

---

[3] See the basic book by Pontryagin, Boltyanskii, Gamkrelidze, and Mishchenko [33].

Sect. 3.5 Some Geometric Aspects of Optimal Control

Property 3.5-6.
The reachable set $\mathcal{R}^+(t_0, t_1, 0)$ of system (1) with control function $u(\cdot)$ subject to constraint (3) is convex.

Proof. Let x_1, x_2 be any two terminal states belonging to $\mathcal{R}^+(t_0, t_1, 0)$, corresponding to the admissible control functions $u_1(\cdot), u_2(\cdot)$. Since $u(t)$ is constrained to belong to a convex set for all $t \in [t_0, t_1]$, also $\alpha u_1(\cdot) + (1-\alpha) u_2(\cdot)$ is admissible, so that the corresponding terminal state $\alpha x_1 + (1-\alpha) x_2$ belongs to $\mathcal{R}^+(t_0, t_1, 0)$. □

Remark. Refer to Fig. 3-15: if the initial and final state sets $\mathcal{X}_0, \mathcal{X}_1$ are convex, the set of all admissible $x(t_1)$ is still convex, it being obtained through linear transformations, sums, and intersections of convex sets.

It is easily shown that for any finite control interval $[t_0, t_1]$ the reachable set $\mathcal{R}^+(t_0, t_1, 0)$ is also closed, bounded, and symmetric with respect to the origin if Ω is so.

Theorem 3.5-1. The Maximum Principle. Part I
Consider system (1) in a given control interval $[t_0, t_1]$ with initial state x_0, constraint (3) on the control effort, and performance index (4). A state trajectory $\bar{x}(\cdot)$ with initial state $\bar{x}(t_0) = x_0$, corresponding to an admissible control function $\bar{u}(\cdot)$, is optimal if and only if the solution $p(t)$ of the *adjoint system*

$$\dot{p}(t) = -A^T(t) p(t) \qquad (22)$$

with final condition $p(t_1) := -\gamma$, satisfies the *maximum condition*[4]

$$\langle p(t), B(t)(u - \bar{u}(t)) \rangle \leq 0 \quad \forall u \in \Omega \quad \text{a.e. in } [t_0, t_1] \qquad (23)$$

Variable p is usually called the *adjoint variable*.

Proof. If. Let $u(\cdot)$ be another admissible control function and $x(\cdot)$ the corresponding state trajectory. By difference we derive

$$\langle p(t_1), (x(t_1) - \bar{x}(t_1)) \rangle = \langle p(t_1), \int_{t_0}^{t_1} \Phi(t_1, \tau) B(\tau)(u(\tau) - \bar{u}(\tau)) d\tau \rangle$$

$$= \int_{t_0}^{t_1} \langle \Phi^T(t_1, \tau) p(t_1), B(\tau)(u(\tau) - \bar{u}(\tau)) \rangle d\tau$$

$$= \int_{t_0}^{t_1} \langle p(\tau), B(\tau)(u(\tau) - \bar{u}(\tau)) \rangle d\tau$$

---

[4] Some particular terms, which derive from the classical calculus of variations, are often used in optimal control theory. Function $H(p, x, u, t) := \langle p, (A(t) x + B(t) u) \rangle$ is called the *Hamiltonian function* and the overall system (1, 22), consisting of the controlled system and adjoint system equations, is called the *Hamiltonian system*. It can be derived in terms of the Hamiltonian function as

$$\dot{x}(t) = \frac{\partial H}{\partial p}, \quad \dot{p}(t) = -\frac{\partial H}{\partial x}$$

The maximum condition requires the Hamiltonian function to be maximal at the optimal control $\bar{u}(t)$ with respect to any other admissible control action $u \in \Omega$ at every instant of time.

Condition (23) implies that the inner product under the integral sign on the right is nonpositive for all $t \in [t_0, t_1]$, so that the inner product on the left is also nonpositive. Since $p(t_1) = -\gamma$, any admissible variation of the trajectory corresponds to a nonnegative variation of the cost, hence $\bar{x}(\cdot)$ is optimal.

Only if. Suppose there exists a subset $T \subseteq [t_0, t_1]$ with nonzero measure such that (23) does not hold for all $t \in T$: then it is possible to choose an admissible control function $u(\cdot)$ (possibly different from $\bar{u}(\cdot)$ only in T) such that the corresponding state trajectory $x(\cdot)$ satisfies $\langle p(t_1), (x(t_1) - \bar{x}(t_1)) \rangle > 0$ or $\langle \gamma, (x(t_1) - \bar{x}(t_1)) \rangle < 0$, so that $\bar{x}(\cdot)$ is nonoptimal. □

A Geometric Interpretation. The maximum principle can be interpreted in strict geometric terms as a necessary and sufficient condition for a given vector φ to belong to the polar cone of the reachable set at some boundary point $\bar{x}(t_1)$. It can be used to derive the reachable set as an envelope of hyperplanes (see next subsection).

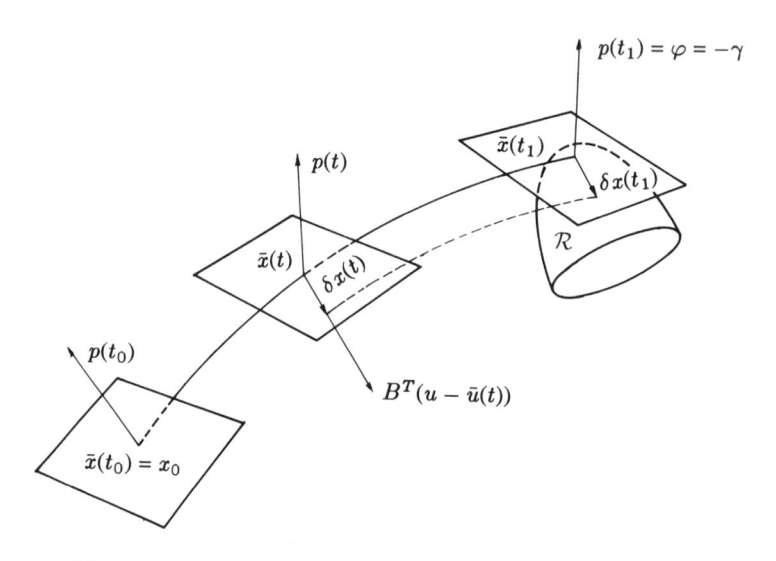

Figure 3-17 A geometric interpretation of the maximum principle: case in which cost Γ is defined as a linear function of the final state.

Refer to Fig. 3-17, where $\mathcal{R}$ denotes the (convex) reachable set of system (1) from the initial state x_0, in the time interval $[t_0, t_1]$, with the control effort constrained by (3). Recall that the inner product of a solution of the free system

$$\dot{x}(t) = A(t) x(t) \qquad (24)$$

and a solution of the adjoint system (22) is a constant (Property 2.1-3): since any variation $\delta x(t)$ of trajectory $\bar{x}(\cdot)$ at time t (due to an admissible pulse variation of the control function) is translated at the final time t_1 as $\delta x(t_1) = \Phi(t_1, t)\, \delta x(t)$, the inner product $\langle p(t_1), \delta x(t_1) \rangle$ is nonpositive (so that $p(t_1)$ is the outer normal of a support hyperplane of $\mathcal{R}$ at $\bar{x}(t_1)$) if and only if $p(t)$ belongs to the polar cone of $B(t)(\Omega - \bar{u}(t))$ almost everywhere in $[t_0, t_1]$. These remarks lead to the following statement.

Corollary 3.5-1.

Let $\mathcal{R}$ be the reachable set of system (1) under the constraints stated in Theorem 3.5-1. Denote by $\bar{x}(\cdot), \bar{u}(\cdot)$ an admissible state trajectory and the corresponding control function. For any given vector $\varphi \in \mathbf{R}^n$ relation

$$\varphi \in C_p\big(\mathcal{R} - \bar{x}(t_1)\big) \tag{25}$$

holds if and only if

$$p(t) \in C_p\big(B(t)(\Omega - \bar{u}(t))\big) \quad \text{a.e. in } [t_0, t_1] \tag{26}$$

with

$$\dot{p}(t) = -A^T(t)\, p(t), \quad p(t_1) = \varphi \tag{27}$$

which is equivalent to the maximum condition expressed by (23).

Theorem 3.5-1 allows immediate solution of a class of optimization problems: refer to system (1) with control function subject to constraints (2), the initial state x_0, and the control interval $[t_0, t_1]$ given, the final state free and performance index (4). Assume that the system is completely controllable.

Algorithm 3.5-1.

An extremal trajectory of system (1) from initial state x_0 at a given time t_0 with control action subject to saturation constraints (2), corresponding to a minimum of cost (4) (where time t_1 is given and the final state is completely free), is determined as follows:

1. Solve the adjoint system (22) with $p(t_1) := -\gamma$, i.e., compute $p(t) = \Phi^T(t_1, t) p(t_1)$, where $\Phi(\cdot, \cdot)$ is the state transition matrix of homogeneous system (24);
2. Determine the optimal control function by means of the maximum condition as

$$u_j(t) = H\, \text{sign}\big(B^T p(t)\big) \quad (j = 1, \ldots, m), \quad t \in [t_0, t_1] \tag{28}$$

Note that the control function is of the so-called *bang-bang* type: every component switches from one to the other of its extremal values. If the argument of function *sign* is zero for a finite time interval, the corresponding value of u_j is immaterial: the reachable set has more than one point in common with its support hyperplane, so that the optimal trajectory is not unique.

Also note that both the adjoint system and the maximum condition are homogeneous in $p(\cdot)$, so that scaling γ and $p(\cdot)$ by an arbitrary positive factor does not change the solution of the optimal control problem.

We shall now derive the maximum principle for the case of integral performance index (5). Let us extend the state space by adding to (1) the nonlinear differential equation

$$\dot{c}(t) = f\big(x(t), u(t), t\big), \quad c(t_0) = 0 \tag{29}$$

where f, the function appearing in (5), is assumed to be convex. Denote by $\hat{x}=(c,x)$ the extended state: by using this artifice we still have the performance index expressed as a linear function of the (extended) terminal state, since clearly

$$\Gamma = c(t_1) = \langle \hat{e}_0, \hat{x}(t_1) \rangle \tag{30}$$

where $\hat{e}_0$ denotes the unit vector in the direction of c axis. Let $\hat{\mathcal{R}} \subseteq \mathbf{R}^{n+1}$ be the reachable set in the extended state space. The standard reachable set $\mathcal{R} \subseteq \mathbf{R}^n$ is related to it by

$$\mathcal{R} = \{x : (c,x) \in \hat{\mathcal{R}}\} \tag{31}$$

In order to extend the above stated maximum principle to the case at hand, the following definition is needed.

Definition 3.5-7. Directionally Convex Set

Let z be any vector of $\mathbf{R}^n$. A set $\mathcal{X} \subseteq \mathbf{R}^n$ is said to be *z-directionally convex*[5] if for any $x_1, x_2 \in \mathcal{X}$ and any $\alpha \in [0,1]$ there exists a $\beta \geq 0$ such that $\alpha x_1 + (1-\alpha)x_2 + \beta z \in \mathcal{X}$. A point x_0 is a *z-directional boundary point* of $\mathcal{X}$ if it is a boundary point of $\mathcal{X}$ and all $x_0 + \beta z$, $\beta \geq 0$, do not belong to $\mathcal{X}$. If set $\mathcal{X}$ is z-directionally convex, the set

$$\mathcal{X}_s := \{y : y = x - \beta z, \ x \in \mathcal{X}, \ \beta \geq 0\} \tag{32}$$

which is called *z-shadow* of $\mathcal{X}$, is convex.

Property 3.5-7.

$\hat{\mathcal{R}}^+(t_0, t_1, 0)$, the reachable set of extended system (29, 1) with control function $u(\cdot)$ subject to constraint (3), is $(-\hat{e}_0)$-directionally convex.

Proof. Let $u_1(\cdot), u_2(\cdot)$ be any two admissible control functions and $\hat{x}_1(\cdot), \hat{x}_2(\cdot)$ the corresponding extended state trajectories. Apply the control function $u(\cdot) := \alpha u_1(\cdot) + (1-\alpha)u_2(\cdot)$: the corresponding trajectory $\hat{x}(\cdot)$ is such that

$$c(t_1) = \int_{t_0}^{t_1} f\big(\alpha x_1(\tau) + (1-\alpha)x_2(\tau), \alpha u_1(\tau) + (1-\alpha)u_2(\tau), \tau\big) d\tau$$

$$\leq \alpha \int_{t_0}^{t_1} f\big(x_1(\tau), u_1(\tau), \tau\big) d\tau + (1-\alpha) \int_{t_0}^{t_1} f\big(x_2(\tau), u_2(\tau), \tau\big) d\tau$$

$$= \alpha\, c_1(t_1) + (1-\alpha)\, c_2(t_1)$$

Hence

$$c(t_1) = \alpha\, c_1(t_1) + (1-\alpha)\, c_2(t_1) - \beta, \quad \beta \geq 0 \quad \square$$

---

[5] Directional convexity was introduced by Holtzmann and Halkin [12].

Theorem 3.5-2. The Maximum Principle. Part II

Consider system (1) in a given control interval $[t_0, t_1]$ with initial state x_0, constraint (3) on the control effort, and performance index (5), where function f is assumed to be convex. A state trajectory $\bar{x}(\cdot)$ with initial state $\bar{x}(t_0) = x_0$ and a given final state $\bar{x}(t_1)$ strictly internal to the reachable set $\mathcal{R}$ is optimal if and only if for any real constant $\psi < 0$ there exists a solution $p(\cdot)$ of the adjoint system

$$\dot{p}(t) = -A^T(t) p(t) - \psi \operatorname{grad}_x f \Big|_{\substack{\bar{x}(t) \\ \bar{u}(t)}} \qquad (33)$$

which satisfies the maximum condition

$$\langle p(t), B(t)(u - \bar{u}(t)) \rangle + \psi \left(f(\bar{x}(t), u, t) - f(\bar{x}(t), \bar{u}(t), t) \right) \leq 0$$

$$\forall u \in \Omega \quad \text{a.e. in } [t_0, t_1] \qquad (34)$$

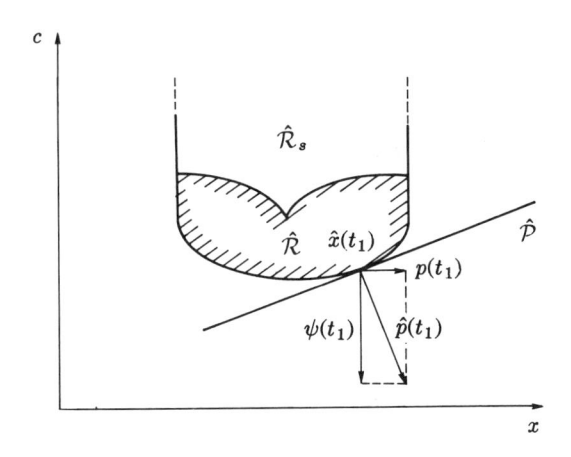

Figure 3-18 The reachable set in the extended state space: case in which cost Γ is defined as the integral of a convex functional of the state and control trajectories.

Proof. *Only if.* Apply Theorem 3.5-1 locally (in a small neighborhood of trajectory $\bar{c}(\cdot), \bar{x}(\cdot)$ in the extended state space). For a trajectory to be optimal the necessary conditions for the linear case must be satisfied for the first variation: in fact, if not, there would exist a small, admissible control function variation causing decrease in cost. The extended adjoint system corresponding to local linearization of (24, 1) defines the extended adjoint variable $\hat{p} = (\psi, p)$, where $\psi(\cdot)$ satisfies the differential equation

$$\dot{\psi}(t) = 0 \qquad (35)$$

and $p(\cdot)$ satisfies (33). In (35) the right side member is zero because variable c does not appear in the Jacobian matrix of (24, 1). Equation (35) implies that $\psi(\cdot)$ is constant over $[t_0, t_1]$. At an optimal terminal point $\bar{\bar{x}}(t_1)$, which is a $(-\hat{e}_0)$-directional boundary point of the extended reachable set $\hat{\mathcal{R}}$ (see Fig. 3-18), a support hyperplane

$\hat{\mathcal{P}}$ of $\hat{\mathcal{R}}_s$ has the outer normal with negative component in the direction of axis c, so that constant ψ is negative: furthermore, it is arbitrary because all conditions are homogeneous in $\hat{p}(\cdot)$. The "local" maximum condition

$$\langle p(t), B(t)(u-\bar{u}(t))\rangle + \psi \left\langle \text{grad}_u f \Big|_{\substack{\bar{x}(t)\\ \bar{u}(t)}}, (u-\bar{u}(t))\right\rangle \le 0$$
$$\forall u \in \Omega \quad \text{a.e. in } [t_0, t_1] \qquad (36)$$

is equivalent to (34) by virtue of Property 3.5-5.

If. Due to convexity of f, any finite variation of the control function and trajectory with respect to $\bar{u}(\cdot)$ and $\bar{x}(\cdot)$ (with $\delta x(t_1) = 0$ since the terminal state is given) will cause a nonnegative variation $\delta c(t_1)$ of the performance index. Thus, if the stated conditions are satisfied, $\hat{\mathcal{P}}$ is a support hyperplane of $\hat{\mathcal{R}}_s$. □

We consider now some computational aspects of dynamic optimization problems. Cases in which solution is achievable by direct computation, without any trial-and-error search procedure, are relatively rare. Algorithm 3-5.1 refers to one of these cases: direct solution is possible because the final state has been assumed completely free. However, when the final state is given, as in the case of Theorem 3.5-2, we have a typical *two-point boundary value problem*: it is solved by assuming, for instance, $\psi := -1$ and adjusting $p(t_0)$, both in direction and magnitude, until the state trajectory, obtained by solving together (1) and (33) with initial conditions $x(t_0) := x_0, p(t_0)$ and with the control function provided at every instant of time by the maximum condition (34) or (36), reaches x_1 at time t_1. Two particular optimal control problems that can be solved with this procedure are the *minimum-time control* and the *minimum-energy control*, which are briefly discussed hereafter. In both cases system (1) is assumed to be completely controllable in a sufficiently large time interval starting at t_0.

Problem 3.5-1. Minimum-Time Control

Consider system (1) with initial state zero, initial time t_0, final state x_1 and constraint (3) on the control action. Derive a control function $u(\cdot)$ which produces transition from the origin to x_1 in minimum time.

Solution. Denote, as before, by $\mathcal{R}^+(t_0, t_1, 0)$ the reachable set of (1) under constraint (3). Since Ω contains the origin

$$\mathcal{R}^+(t_0, t_1, 0) \subseteq \mathcal{R}^+(t_0, t_2, 0) \quad \text{for } t_1 < t_2 \qquad (37)$$

In fact a generic point x' of $\mathcal{R}^+(t_0, t_1, 0)$, reachable at t_1 by applying some control function $u'(\cdot)$, can also be reached at t_2 by

$$u''(t) = \begin{cases} 0 & \text{for } 0 \le t < t_2 - t_1 \\ u'(t + t_2 - t_1) & \text{for } t_2 - t_1 \le t \le t_2 \end{cases} \qquad (38)$$

In most cases (37) is a strict inclusion and, since $\mathcal{R}^+(t_0, t_1, 0)$ is closed for any finite control interval, all its boundary points are reachable at minimal time t_1. The boundary is called an *isochronous surface* (corresponding to time t_1). The problem is solved by searching (by a trial-and-error or steepest descent procedure) for a value of $p(t_0)$ (only direction has to be varied, since magnitude has no influence) such that the simultaneous solution of (1) and (22) with control action $u(t)$ chosen at every instant of time to maximize $\langle p(t), B(t)u(t) \rangle$ over Ω, provides a state trajectory passing through x_1: the corresponding time t_1 is the minimal time. This means that x_1 belongs to the boundary of $\mathcal{R}^+(t_0, t_1, 0)$ and the corresponding $p(t_1)$ is the outer normal of a support hyperplane of $\mathcal{R}^+(t_0, t_1, 0)$ at x_1. The same procedure, based on trial-and-error search for $p(t_0)$, can also be used for any given initial state x_0 (not necessarily coinciding with the origin). If the control action is constrained by (2) instead of (3), $u(\cdot)$ is of the bang-bang type and is given by (28) at every instant of time as a function of $p(t)$. □

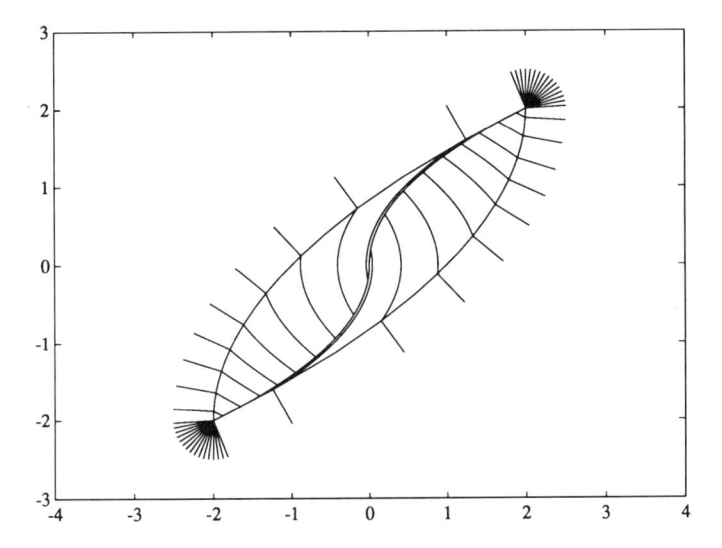

Figure 3-19 The reachable set $\mathcal{R}^+(0, 2, 0)$ of system (39).

Example 3.5-1.
Consider the linear time-invariant system corresponding to

$$A := \begin{bmatrix} 0 & 1 \\ 0 & 0 \end{bmatrix} \quad B := \begin{bmatrix} 0 \\ 1 \end{bmatrix} \tag{39}$$

in the control interval $[0, 2]$ and with the control effort constrained by $-1 \leq u \leq 1$. The boundary of $\mathcal{R}^+(0, 2, 1)$, i.e., the isochronous curve for $t_1 = 2$, is represented in Fig. 3-19: it has been obtained by connecting 50 terminal points, each of which has been computed by applying Algorithm 3.5-1 to one of 50 equally angularly spaced unit vectors $p_i(t_1) := \varphi_i$ ($i = 1, \ldots, 50$). The corresponding control actions, of the bang-bang type, are

$$u_i(t) = \text{sign}\left(B^T p_i(t)\right) \quad \text{with} \quad p_i(t) = e^{A^T t} \varphi_i \quad (i = 1, \ldots, 50) \tag{40}$$

Figure 3-20 shows four different isochronous curves obtained with this procedure.

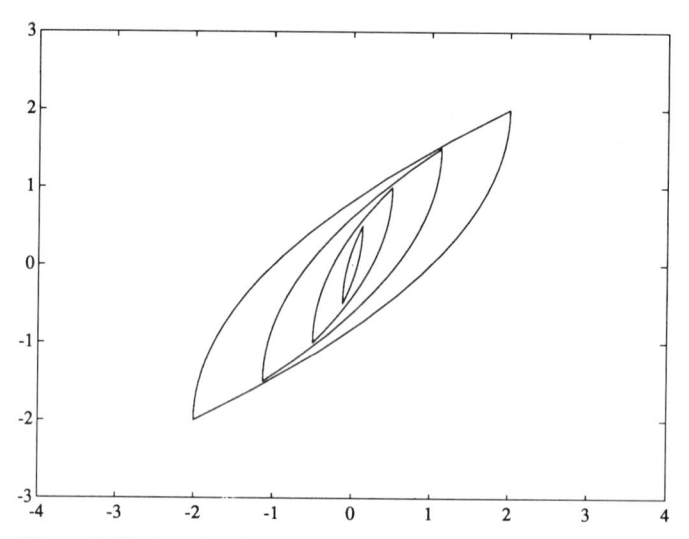

Figure 3-20 Four isochronous curves of system (39), corresponding to the final times $.5, 1, 1.5, 2$.

Problem 3.5-2. Minimum-Energy Control

Consider system (1) with initial state x_0, initial time t_0, final state x_1, and control interval $[t_0, t_1]$. Derive a control function $u(\cdot)$ that produces transition from x_0 to x_1 with minimum energy. Energy is defined as

$$e = \left(\int_{t_0}^{t_1} \|u(\tau)\|_2^2 \, d\tau \right)^{\frac{1}{2}} \tag{41}$$

i.e., as the euclidean norm of control function segment $u|_{[t_0, t_1]}$.

Solution. By virtue of Property 3.5-4 functional (41) is convex and the problem can be solved by applying Theorem 3.5-2 with $\Gamma := e^2/2$. First, note that adjoint system (33) coincides with (22) since in this case function f does not depend on x and the maximum condition (36) requires that function

$$\langle p(t), B(t) u(t) \rangle - \frac{1}{2} \langle u(t), u(t) \rangle \tag{42}$$

is maximized with respect to $u(t)$. This leads to

$$u(t) = B^T(t) p(t) \tag{43}$$

The initial condition $p(t_0)$ has to be chosen (both in direction and magnitude) in such a way that the corresponding state trajectory (starting from x_0 at time t_0) with the control function provided by (43) reaches x_1 at time t_1. This is still a two-point boundary value problem, but easily solvable because the overall system is linear. In fact it is described by the homogeneous matrix differential equation

$$\begin{bmatrix} \dot{x}(t) \\ \dot{p}(t) \end{bmatrix} = \begin{bmatrix} A(t) & B(t) B^T(t) \\ O & -A^T(t) \end{bmatrix} \begin{bmatrix} x(t) \\ p(t) \end{bmatrix} \tag{44}$$

Sect. 3.5 Some Geometric Aspects of Optimal Control

and in terms of the overall state transition matrix, accordingly partitioned, we have

$$\begin{bmatrix} x(t_1) \\ p(t_1) \end{bmatrix} = \begin{bmatrix} \Phi(t_1, t_0) & M \\ O & \Phi^{-T}(t_1, t_0) \end{bmatrix} \begin{bmatrix} x(t_0) \\ p(t_0) \end{bmatrix} \quad (45)$$

with $\Phi^{-T} := (\Phi^T)^{-1}$ and

$$M := \int_{t_0}^{t_1} \Phi(t_1, \tau) B(\tau) B^T(\tau) \Phi^{-T}(\tau, t_0) \, d\tau \quad (46)$$

which is nonsingular because of the controllability assumption: in fact matrix $P(t_0, t_1)$ defined by (2.6.6) is related to it by

$$P(t_0, t_1) = M \Phi^{-T}(t_0, t_1) = M \Phi^T(t_1, t_0) \quad (47)$$

From (45) we obtain

$$p(t_0) = M^{-1} x_2 \quad \text{with} \quad x_2 := x_1 - \Phi(t_1, t_0) x_0 \quad (48)$$

and substitution into (43) yields the solution in the form

$$u(t) = B^T(t) \Phi^{-T}(t, t_0) M^{-1} x_2 = B^T(t) \Phi^T(t_0, t) M^{-1} x_2, \quad t \in [t_0, t_1] \quad \square \quad (49)$$

Control law (49) coincides with (2.6.9).[6] Hence (2.6.9) solves the problem of controlling the state from x_0 to x_1 with the minimum amount of energy.

Remark. If in Problem 3.5-2 the control action is constrained by (2), the maximum principle gives

$$u_j(t) = \begin{cases} H & \text{for } u_j^\circ(t) \geq H \\ u_j^\circ(t) & \text{for } |u_j^\circ(t)| < H \\ -H & \text{for } u_j^\circ(t) \leq -H \end{cases} \quad (j=1,\ldots,p), \quad u^\circ(t) := B^T(t) p(t) \quad (50)$$

However in this case, (50) being nonlinear, $p(t_0)$ is not directly obtainable as a linear function of x_0, x_1.

---

[6] This is proved by:

$$\begin{aligned} u(t) &= B^T(t) \Phi^{-T}(t, t_0) M^{-1} x_2 \\ &= B^T(t) \Phi^{-T}(t, t_0) \Phi^{-T}(t_0, t_1) P^{-1}(t_0, t_1) x_2 \\ &= B^T(t) \Phi^T(t_1, t) P^{-1}(t_0, t_1) x_2 \end{aligned}$$

Problem 3.5-3. The Reachable Set with Bounded Energy[7]

Determine the reachable set of system (1) from the origin in time interval $[t_0, t_1]$ under the control energy constraint

$$\left(\int_{t_0}^{t_1} \|u(\tau)\|_2^2 \, d\tau \right)^{\frac{1}{2}} \leq H \tag{51}$$

Solution. Denote by $\mathcal{R}^+(t_0, t_1, 0, H)$ the reachable set of system (1) with constraint (51). The functional on the left of (51) is the euclidean norm of function segment $u|_{[t_0,t_1]}$ and, like any other norm, satisfies the triangle inequality

$$\|\alpha u_1(\cdot) + (1-\alpha) u_2(\cdot)\| \leq \alpha \|u_1(\cdot)\| + (1-\alpha) \|u_2(\cdot)\|$$

This means that (51) defines a convex set in the functional space of all piecewise continuous control functions $u|_{[t_0,t_1]}$. Hence $\mathcal{R}^+(t_0, t_1, 0, H)$ is convex as the image of a convex set in a linear map. Furthermore, in the extended state space with cost defined by

$$\dot{c}(t) = \frac{1}{2} \langle u(t), u(t) \rangle, \quad c(t_0) = 0$$

from $\hat{\mathcal{R}}^+(t_0, t_1, 0)$ being $(-\hat{e}_0)$-directionally convex it follows that

$$\mathcal{R}^+(t_0, t_1, 0, H_1) \subseteq \mathcal{R}^+(t_0, t_1, 0, H_2) \quad \text{for } H_1 < H_2 \tag{52}$$

(note that a generic $\mathcal{R}^+(t_0, t_1, 0, H)$ is obtained by intersecting $\hat{\mathcal{R}}^+(t_0, t_1, 0)$ with a hyperplane orthogonal to the cost axis, called an *isocost hyperplane*). It is also clear that relation (51) holds with the equality sign at every boundary point of $\mathcal{R}^+(t_0, t_1, 0, H)$. By virtue of the maximum principle, a given vector φ is the outer normal of a support hyperplane of $\mathcal{R}^+(t_0, t_1, 0, H)$ if and only if

$$u(t) = k \, B^T(t) \, \Phi^T(t_1, t) \, \varphi \tag{53}$$

where constant k has to be chosen to satisfy (51). This requirement leads to

$$u(t) = \frac{H \, B^T(t) \, \Phi^T(t_1, t) \, \varphi}{\sqrt{\langle \varphi, P(t_0, t_1) \, \varphi \rangle}} \tag{54}$$

The boundary of $\mathcal{R}^+(t_0, t_1, 0, H)$ is the hyperellypsoid defined by

$$\langle x_1, P^{-1}(t_0, t_1) \, x_1 \rangle = H^2 \tag{55}$$

This can easily be checked by direct substitution of

$$x_1 = \int_{t_0}^{t_1} \Phi(t_1, \tau) \, B(\tau) \, u(\tau) \, d\tau = \frac{H \, P(t_0, t_1) \, \varphi}{\sqrt{\langle \varphi, P(t_0, t_1) \, \varphi \rangle}} \quad \square$$

---

[7] The geometry of the reachable set with a bound on the generic p-norm of the control function was investigated in the early 1960s by Kreindler [23] and Kranc and Sarachik [22].

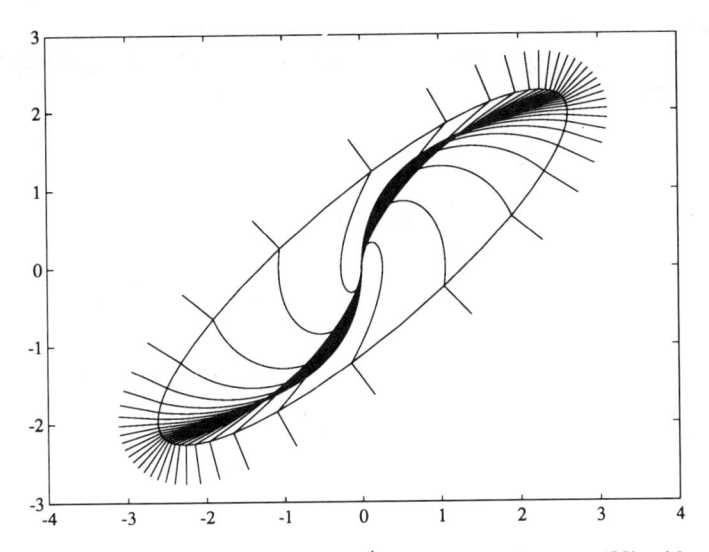

Figure 3-21 The reachable set $\mathcal{R}^+(0,2,0,1.6)$ of system (39) with energy constraint (51).

Example 3.5-2.

Consider again the linear time-invariant system corresponding to matrices (39) in the control interval $[0,2]$ and with the energy bound $H := 1.6$. The reachable set $\mathcal{R}^+(t_0, t_1, 0, E)$ is shown in Fig. 3-21: also in this case it has been obtained by connecting 50 terminal points, each of which has been computed by considering one of 50 equally angularly spaced unit vectors $p_i(t_1) := \varphi_i$ ($i=1,\ldots,50$). The corresponding control actions have been computed by means of (54) with $\varphi := \varphi_i$.

3.5.3 The Linear-Quadratic Regulator

The linear-quadratic regulator problem, also called the *LQR problem* or the *Kalman regulator*, can be considered as an extension of the minimum-energy control problem considered in the previous subsection.[8]

Problem 3.5-4. The LQR Problem

Consider system (1) with initial state x_0, initial time t_0, final state x_1 and control interval $[t_0, t_1]$. Derive a control function $u(\cdot)$ which produces transition from x_0 to x_1 while minimizing the performance index

$$\Gamma = \frac{1}{2} \int_{t_0}^{t_1} \left(\langle x(\tau), Q(\tau) x(\tau) \rangle + \langle u(\tau), R(\tau) u(\tau) \rangle \right) d\tau \tag{56}$$

where matrices $Q(\tau)$ and $R(\tau)$ are respectively symmetric positive semidefinite and symmetric positive definite for all $\tau \in [t_0, t_1]$.

---

[8] Most of the results on the linear-quadratic regulator are due to Kalman [13, 14].

Solution. By virtue of Property 3.5-4 functional (56) is convex and the problem can be solved by applying Theorem 3.5-2. Assume $\psi := -1$: the adjoint system (33) in this case can be written as

$$\dot{p}(t) = -A^T(t)\, p(t) + Q(t)\, x(t) \tag{57}$$

and the maximum condition (36) requires that

$$\langle p(t), B(t)\, u(t) \rangle - \frac{1}{2}\big(\langle x(t), Q(t)\, x(t)\rangle + \langle u(t), R(t)\, u(t)\rangle\big) \tag{58}$$

is maximized with respect to $u(t)$. This yields

$$u(t) = R^{-1}(t)\, B^T(t)\, p(t) \tag{59}$$

The Hamiltonian system is

$$\begin{bmatrix} \dot{x}(t) \\ \dot{p}(t) \end{bmatrix} = \begin{bmatrix} A(t) & B(t)\, R^{-1}(t)\, B^T(t) \\ Q(t) & -A^T(t) \end{bmatrix} \begin{bmatrix} x(t) \\ p(t) \end{bmatrix} \tag{60}$$

Denote by $\hat{A}(t)$ the overall system matrix in (60): in terms of the corresponding state transition matrix $\hat{\Phi}(t_1, t_0)$, accordingly partitioned, we have

$$\begin{bmatrix} x(t_1) \\ p(t_1) \end{bmatrix} = \begin{bmatrix} \Phi_1(t_1, t_0) & \Phi_2(t_1, t_0) \\ \Phi_3(t_1, t_0) & \Phi_4(t_1, t_0) \end{bmatrix} \begin{bmatrix} x(t_0) \\ p(t_0) \end{bmatrix} \tag{61}$$

It can be proved that under the assumption that system (1) is completely controllable in time interval $[t_0, t_1]$, matrix $\Phi_2(t_1, t_0)$ is nonsingular, so that the problem can be solved by deriving

$$p(t_0) = \Phi_2^{-1}(t_1, t_0)\big(x_1 - \Phi_1(t_1, t_0)\, x_0\big) \tag{62}$$

then using (60) with initial condition $x_0, p(t_0)$. □

Remark. In Problem 3.5-4 both the initial and the final state are given. It is possible, however, to formulate the LQR problem also with the final state completely free. Since $\hat{\mathcal{R}}$, the extended reachable set from $(0, x_0)$ in $[t_0, t_1]$, is $(-e_0)$-directionally convex, it presents a $(-e_0)$-directional boundary point corresponding to a globally minimum cost. At this point $\hat{\mathcal{R}}_*$ has a support hyperplane $\hat{\mathcal{P}}$ (see Fig. 3-18) orthogonal to the c axis. The corresponding globally optimal final state is detected by condition $p(t_1) = 0$ on the final value of the adjoint variable, which leads to

$$p(t_0) = -\Phi_4^{-1}(t_1, t_0)\, \Phi_3(t_1, t_0)\, x_0 \tag{63}$$

to be used instead of (62) to solve the problem.

3.5.4 The Time-Invariant LQR Problem

The time-invariant LQR problem is a particular case of the previous LQR problem. It refers to the time-invariant system

$$\dot{x}(t) = A\, x(t) + B\, u(t) \tag{64}$$

with initial state x_0 given, the final state completely free, the infinite optimal control interval $[0, \infty]$ and performance index

$$\Gamma = \frac{1}{2} \int_0^\infty \left(\langle x(\tau), Q\, x(\tau) \rangle + \langle u(\tau), R\, u(\tau) \rangle \right) d\tau \tag{65}$$

where matrices Q and R are assumed to be respectively (symmetric) positive semidefinite and positive definite. It is remarkable that in this case control $u(t)$ is a linear function of state $x(t)$ for all $t \geq 0$.

Theorem 3.5-3. (Kalman)

Consider system (64) in control interval $[0, \infty]$, with initial state x_0, final state free, and performance index (65). Assume that (A, B) is controllable and that Q can be expressed as $C^T C$ with (A, C) observable. The optimal control is given by

$$u(t) = F\, x(t) \quad \text{with} \quad F := -R^{-1} B^T P \tag{66}$$

where P is the unique (symmetric) positive definite solution of the *algebraic Riccati equation*

$$PA + A^T P - PBR^{-1}B^T P + Q = O \tag{67}$$

Proof. First of all, note that if a matrix P satisfies (67), so does P^T, so that we can assume that P is symmetric without any loss of generality. Consider the Hamiltonian system

$$\begin{bmatrix} \dot{x}(t) \\ \dot{p}(t) \end{bmatrix} = \begin{bmatrix} A & BR^{-1}B^T \\ Q & -A^T \end{bmatrix} \begin{bmatrix} x(t) \\ p(t) \end{bmatrix} \tag{68}$$

and denote by A_H the corresponding system matrix (called the *Hamiltonian matrix*). We shall prove that if λ is an eigenvalue of A_H, $-\lambda$ also is. Define $S := BR^{-1}B^T$ and consider the equalities

$$\det\left(\begin{bmatrix} A - \lambda I_n & S \\ Q & -A^T - \lambda I_n \end{bmatrix} \right) = \det\left(\begin{bmatrix} A^T - \lambda I_n & Q \\ S & -A - \lambda I_n \end{bmatrix} \right)$$

$$= \det\left(\begin{bmatrix} -Q & -A^T + \lambda I_n \\ -A - \lambda I_n & S \end{bmatrix} \right) = \det\left(\begin{bmatrix} A + \lambda I_n & S \\ Q & -A^T + \lambda I_n \end{bmatrix} \right)$$

which prove the above assertion. The first follows from M and M^T having the same eigenvalues for any square real M, while the interchanges of rows and columns and related changes of signs in the other equalities have been obtained by multiplying on the right and on the left by $L_1 L_2$, with

$$L_1 := \begin{bmatrix} O & I_n \\ I_n & O \end{bmatrix} \quad \text{and} \quad L_2 := \begin{bmatrix} -I_n & O \\ O & I_n \end{bmatrix}$$

whose determinants are both 1 if n is even and both -1 if it is odd. Due to the controllability assumption, the state can be controlled to the origin in finite time and, due to the observability assumption, equilibrium at the origin is the only motion corresponding to zero differential cost. Hence the solution of the optimal control problem converges to the origin: this implies that n eigenvalues of the Hamiltonian matrix have strictly negative real parts while the remaining n have strictly positive real parts. Relation (63) shows that in any LQR problem with the final state free $p(t_0)$ depends on both the initial state x_0 and the control interval $[t_0, t_1]$ (in this case on the control time T). Since T is infinity, p_0 is related only to x_0: in other words there exists a matrix P such that

$$p(t_0) = -P x_0 \qquad (69)$$

and, provided solution of the control problem cannot depend on the initial time because the system is time-invariant, it follows that the same equality holds *at any time*, i.e.

$$p(t) = -P x(t) \quad \forall t \geq 0 \qquad (70)$$

This means that the subspace

$$\mathcal{J}_H := \text{im}\left(\begin{bmatrix} I_n \\ -P \end{bmatrix}\right) \qquad (71)$$

is an A_H-invariant, so that in the product

$$\begin{bmatrix} I_n & O \\ P & I_n \end{bmatrix} \begin{bmatrix} A & BR^{-1}R \\ Q & -A^T \end{bmatrix} \begin{bmatrix} I_n & O \\ -P & I_n \end{bmatrix}$$
$$= \begin{bmatrix} A - BR^{-1}B^T P & BR^{-1}B^T \\ PA - PBR^{-1}B^T P + Q + A^T P & PBR^{-1}B^T - A^T \end{bmatrix} \qquad (72)$$

the first submatrix in the second row must be zero. This is expressed by equation (67). Since the final state is zero, $\mathcal{J}_H$ must coincide with the subspace of the stable modes of A_H which, due to the previously shown property of the eigenvalues, has dimension n and, since the problem admits a solution for any initial state by the controllability assumption, $\mathcal{J}_H$ projects into the whole controlled system state space. The internal eigenvalues of $\mathcal{J}_H$ are those of the first submatrix in the first row of (72), so that

$$A + BF = A - BR^{-1}B^T P \qquad (73)$$

is a stable matrix. Define

$$M := Q + F^T R F \quad \text{with} \quad F := -R^{-1} B^T P \qquad (74)$$

Sect. 3.5 Some Geometric Aspects of Optimal Control

and consider Lemma 2.5-1: it is easily seen that Liapunov equation (2.5.27) with $A+BF$ instead of A coincides with Riccati equation (67), so that P is, at least, positive semidefinite, M being positive semidefinite. Actually, it is positive definite, since the differential cost is maintained at zero only at the origin. Any other solution P of the Riccati equation is related to an A_H-invariant of type (71) which is internally unstable (matrix $A+BF$ with F defined as in (74) is unstable) and cannot be positive semidefinite or positive definite. In fact, due to instability, any state x_1 such that $V_1 := \langle x_1, Px_1 \rangle$ is positive and arbitrarily large would be reached from states $x(t)$ such that $V(t) := \langle x(t), Px(t) \rangle$ is less than V_1 by an overall system trajectory on J_H with $\dot{V}(t) = -\langle x(t), M x(t)\rangle$ nonpositive at every instant of time, which is clearly a contradiction. □

Corollary 3.5-2.

If in the above Theorem 3.5-3 assumptions are relaxed to (A, B) being stabilizable and (A, C) detectable, the statement remains valid, but with P being possibly positive semidefinite instead of positive definite.

Proof. Only minor changes in the above proof of Theorem 3.5-3 are necessary. Since the uncontrollable modes are stable and the arcs of trajectory corresponding to zero differential cost (hence to zero control function) belong to the unobservability subspace (hence converge to the origin), the optimal control problem still admits a solution converging to the origin, the Hamiltonian matrix has n eigenvalues with strictly negative real parts and $A+BF$ is strictly stable. However in this case there are nonzero initial states (all the unobservable ones) which correspond to optimal trajectories with zero differential cost, so that in this case matrix P is positive semidefinite. □

Corollary 3.5-3.

For any given initial state x_0 the optimal value of performance index (65) is

$$\Gamma_0 = \frac{1}{2}\langle x_0, P x_0 \rangle$$

where P denotes, as before, the positive semidefinite or positive definite solution of Riccati equation (67).

Proof. Consider Lemma 2.5-1 with $A+BF$ instead of A and M defined as in (74). □

Remark. Solution $P=O$ is not excluded. For instance, if A is stable and Q is zero (minimum-energy control to the origin) this is the only positive semidefinite solution of the Riccati equation. In fact, in this case the most convenient control policy to reach the origin in infinite time is clearly not to apply any control, i.e., to choose $u(\cdot)=0$, which corresponds to zero energy.

We shall now briefly consider computational aspects of the Riccati equation. The most direct method derives from the above proof of Theorem 3.5-3: assume that A_H has distinct eigenvalues (hence linearly independent eigenvectors) and let

$$\begin{bmatrix} T_1 \\ T_2 \end{bmatrix}$$

be the $2n \times n$ matrix having as columns the eigenvectors corresponding to stable eigenvalues. From (71) $P = -T_2 T_1^{-1}$ directly follows. This procedure has two (minor) drawbacks: it requires computations in the complex field and is not applicable when the eigenvalues of A_H are not distinct, since standard computational routines for eigenvalues and eigenvectors in general do not provide generalized eigenvectors. On the other hand, note that, due to genericity, this case is relatively rare.

An alternative, completely different computational procedure, based on iterative solution of a Liapunov equation, is set in the following algorithm, which, due to good convergence, is quite interesting in computational practice.

Algorithm 3.5-2. (Kleinman) [9]

The positive semidefinite or positive definite solution of Riccati equation (67) can be computed through the following steps:
1. choose any F_0 such that $A_0 := A + B F_0$ is stable;
2. perform the recursive computations:

$$A_i^T P_i + P_i A_i + Q + F_i^T R F_i = O \quad (i = 0, 1, \ldots)$$

where

$$F_{i+1} := -R^{-1} B^T P_i, \quad A_{i+1} := A + B F_{i+1} \tag{75}$$

and stop when the difference in norm between two consecutive P_i is less than a small real number (for instance $100\,\epsilon$, where ϵ denotes the machine zero).

Proof. By virtue of Lemma 2.5-1

$$P_k = \int_0^\infty e^{A_k^T \tau} (Q + F_k^T R F_k) e^{A_k \tau} d\tau \tag{76}$$

is positive semidefinite or positive definite if A_k is stable. Let X_1, X_2 be any two positive semidefinite or positive definite symmetric matrices. We understand that the inequality $X_1 \geq X_2$ means that $\langle x, X_1 x \rangle \geq \langle x, X_2 x \rangle$ for all x. Let S be any positive definite symmetric matrix, so that

$$(X_1 - X_2) S (X_1 - X_2) \geq O \tag{77}$$

The same inequality can also be written

$$X_1 S X_1 \geq X_1 S X_2 + X_2 S X_1 - X_2 S X_2 \tag{78}$$

or

$$X_1 S X_1 = X_1 S X_2 + X_2 S X_1 - X_2 S X_2 + M \tag{79}$$

---

[9] A very complete and formal treatment of this algorithm was presented by Kleinman [20, 21] and Vit [35].

with $M \geq O$. The equality holds in (78) or $M = O$ in (79) if $X_1 = X_2$. The recursion formula (75) can be put in the form

$$A_i^T P_i + P_i A_i + Q + P_{i-1} S P_{i-1} = O \quad \text{with} \quad S := BR^{-1}B^T \tag{80}$$

or, by using (79) with $X_1 := P_{i-1}$ and $X_2 := P_i$,

$$A_i^T P_i + P_i A_i + Q + P_{i-1} S P_i + P_i S P_{i-1} - P_i S P_i + M = O$$

and, being $A_{i+1} := A - S P_i = A_i + S P_{i-1} - S P_i$,

$$A_{i+1}^T P_i + P_i A_{i+1} + Q + P_i S P_i + M = O$$

The use of this and the subsequent recursion formula

$$A_{i+1}^T P_{i+1} + P_{i+1} A_{i+1} + Q + P_i S P_i = O$$

in integral (76) yields the desired result. Let $P_i = P$ with P satisfying Riccati equation (67): it is easily shown that in this case $P_{i+1} = P_i$ and the recursion Liapunov equation (80) coincides with Riccati equation (67). Furthermore, $P_i > P$ for any P_i that does not satisfy the Riccati equation. This is proved by using the above argument with P_i, P instead of P_i, P_{i+1}. Hence, the limit of sequence $\{P_i\}$ is P. □

Problem 3.5-5. The Infinite-Time Reachable Set with Bounded Quadratic Cost

Consider system (64) with matrix A strictly stable. Determine the reachable set from the origin under the constraint

$$\int_0^\infty \left(\langle x(\tau), Q\, x(\tau) \rangle + \langle u(\tau), R\, u(\tau) \rangle \right) d\tau \leq H^2 \tag{81}$$

where matrices Q and R are assumed to be respectively (symmetric) positive semidefinite and positive definite.

Solution. We shall show that the reachable set is bounded by the hyperellypsoid

$$\langle x_1, (-P) x_1 \rangle = H^2 \tag{82}$$

where P is the unique (symmetric) negative definite solution of Riccati equation (67). The proof closely follows that of Theorem 3.5-3: in fact the problem can be reconducted to the same optimization problem, but with modified extremal conditions. Let x_1 be a boundary point of the reachable set, which is convex: then, there exists an infinite-time state trajectory $x(\cdot)$ from the origin to x_1 such that cost (65) is minimal (with respect to the other trajectories from the origin to x_1) and its value is precisely $H^2/2$. Refer to Hamiltonian system (68): along this trajectory relation (70) is still valid since the control interval is infinite and at every instant of time the control effort depends only on the current state. This implies that the trajectory belongs to an A_H-invariant, which in this case coincides with the subspace of the unstable modes of A_H (any trajectory, considered backward in time, tends to the origin). It is proved to be, at least, negative semidefinite by an argument similar to that presented in the proof of Theorem 3.5-3 (which uses Lemma 2.5-1), considering matrix $-A - BF$ (which is strictly stable) instead of $A + BF$. As for Corollary 3.5-1, the same argument proves that the quadratic form on the left of (82) is the related cost: since the differential cost corresponding to a finite arc of trajectory from the origin cannot be zero due to strict stability, P cannot be negative semidefinite, but strictly negative definite. Relation (82) follows from the reachable set being bounded by an isocost surface. □

Problem 3.5-6. The Infinite-Time Reachable Set with Bounded Energy

Consider system (64) with matrix A strictly stable. Determine the reachable set from the origin under the constraint

$$\int_0^\infty \|u(\tau)\|_2^2 \, d\tau \leq H^2 \tag{83}$$

Solution. This problem is a particular case of the previous one. However in this case it is not necessary to solve an algebraic Riccati equation to derive P, but only a Liapunov equation whose solution is unique. Consider (67) and assume $R := I_m$, $Q := O$; by multiplying on the left and right by P^{-1} we obtain

$$A P^{-1} + P^{-1} A^T - B B^T = O \tag{84}$$

The boundary of the reachable set in this case is still provided by (82), but with P (negative definite) obtainable through equation (84). A remark is in order: recall Problem 3.5-3, which refers to the finite-time case and, in particular, relation (55). By comparison, it follows that the infinite-time Gramian

$$P(0,\infty) := \int_0^\infty e^{At} B B^T e^{A^T t} \, dt \tag{85}$$

satisfies

$$A P(0,\infty) + P(0,\infty) A^T + B B^T = O \quad \square \tag{86}$$

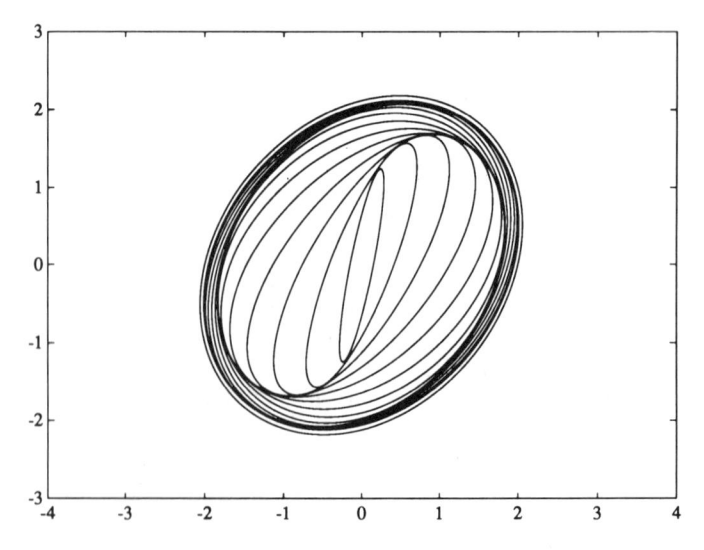

Figure 3-22 The reachable set $\mathcal{R}^+(0, T, 0, 3)$ of system (87) with energy constraint (88) ($T = .2, .4, \ldots, 3, T = \infty$).

Example 3.5-3.
Consider the asymptotically stable linear time-invariant system with matrices

$$A := \begin{bmatrix} -.5 & 2 \\ -2 & -.5 \end{bmatrix} \quad B := \begin{bmatrix} 0 \\ 1 \end{bmatrix} \tag{87}$$

in the control interval $[0, T]$, under the control energy constraint

$$\int_0^T \|u(\tau)\|_2^2 \, d\tau \leq H^2 \quad \text{with} \quad H = 3 \tag{88}$$

The boundary of the reachable set $\mathcal{R}^+(0, T, 0, H)$ is

$$\langle x_1, P^{-1}(0, T) x_1 \rangle = H^2$$

To compute $P(0, T)$, if T is finite, consider the Hamiltonian matrix

$$A_H := \begin{bmatrix} A & B B^T \\ O & -A^T \end{bmatrix}$$

and denote by M the submatrix of $e^{A_H T}$ corresponding to the first two rows and the last two columns: then $P(0, T) = M e^{A^T T}$. If T is infinite, use (86). Some reachable sets referring to finite values of T and that corresponding to $T = \infty$ are shown in Fig. 3-22.

REFERENCES

1. ATHANS, M., and FALB, P.L., *Optimal Control*, McGraw-Hill, New York, 1966.
2. BERGE, C., *Topological Spaces*, Oliver & Boyd, London, 1963.
3. BRUNOVSKÝ, P., "A classification of linear controllable systems," *Kybernetika* (Prague), vol. 3, no. 6, pp. 173–187, 1970.
4. CHEN, C.T., and DESOER, C.A., "Controllability and observability of composite systems," *IEEE Trans. Autom. Contr.*, vol. AC-12, pp. 402–409, 1967.
5. DELLON, F., and SARACHICK, P.E., "Optimal control of unstable linear plants with unaccessible states," *IEEE Trans. on Autom. Contr.*, vol. AC-13, no. 5, pp. 491–495, 1968.
6. EGGLESTON, H.G., *Convexity*, Cambridge University Press, London, 1958.
7. GILBERT, E.G., "Controllability and observability in multivariable control systems," *SIAM J. Control*, vol. 2, no. 1, pp. 128–161, 1963.
8. GOPINATH, B., "On the control of linear multiple input-output systems," *Bell System Techn. J.*, vol. 50, no. 3, pp. 1063–1081, 1971.
9. HAUTUS, M.L.J., "A simple proof of Heymann's lemma," *IEEE Trans. Autom. Contr.*, vol. AC-22, no. 5, pp. 825–826, 1977.

10. HEYMANN, M., "Comments on 'Pole assignment in multi-input controllable linear systems'," *IEEE Trans. Autom. Contr.*, vol. AC-13, pp. 748–749, 1968.
11. HIJAB, O., *Stabilization of Control Systems*, Springer-Verlag, New York, 1987.
12. HOLTZMAN, J.M., and HALKIN, H., "Directional convexity and the maximum principle for discrete systems," *J. SIAM Control*, vol. 4, no. 2, pp. 263–275, 1966.
13. KALMAN, R.E., "Contributions to the theory of optimal control," *Bulletin de la Sociedad Matematica Mexicana*, vol. 5, pp. 102–119, 1960.
14. —, "On the general theory of control systems," *Proceedings of the 1st IFAC Congress*, vol. 1, pp. 481–492, Butterworths, London, 1961.
15. —, "Canonical structure of linear dynamical systems," *Proc. Natl. Acad. Sciences*, vol. 48, no. 4, pp. 596–600, 1962.
16. —, "Mathematical description of linear dynamical systems," *SIAM J. Control*, vol. 1, no. 2, pp. 152–192, 1963.
17. —, "Kronecker invariants and feedback," in *Ordinary Differential Equations*, edited by L. Weiss, Academic, New York, 1972.
18. KALMAN, R.E., HO, Y.C., and NARENDRA, K.S., "Controllability of linear dynamical systems," *Contributions to Differential Equations*, vol. 1, no. 2, pp. 189–213, 1962.
19. KIPINIAK, W., *Dynamic Optimization and Control*, the MIT Press and Wiley, New York, 1961.
20. KLEINMAN, D.L., "On the linear regulator problem and the matrix Riccati equation," M.I.T. Electronic System Lab., Cambridge, Mass., Rept. 271, 1966.
21. —, "On an iterative technique for Riccati equation computations," *IEEE Trans. on Autom. Contr.*, vol. AC-13, no. 1, pp. 114–115, 1968.
22. KRANC, G.M., and SARACHIK, P.E., "An application of functional analysis to the optimal control problem," *Trans. ASME, J. Basic Engrg.*, vol. 85, no. 6, pp. 143–150, 1963.
23. KREINDLER, E., "Contributions to the theory of time-optimal control," *J. of the Franklin Institute*, no. 4, pp. 314–344, 1963.
24. KREINDLER, A., and SARACHIK, P.E., "On the concepts of controllability and observability of linear systems," *IEEE Trans. Autom. Contr.*, vol. AC-9, no. 2, pp. 129–136, 1964.
25. LANGENHOP, C.E., "On the stabilization of linear systems," *Proc. Am. Math. Society*, vol. 15, pp. 735–742, 1964.
26. LEE, E.B., and MARKUS, L., *Foundations of Optimal Control Theory*, Wiley, New York, 1967.
27. LEITMANN, G., *Topics in Optimization*, Academic, New York, 1967.
28. LUENBERGER, D.G., "Observing the state of a linear system," *IEEE Trans. Mil. Electron.*, vol. MIL-8, pp. 74–80, 1964.
29. —, "Observers for multivariable systems," *IEEE Trans. Autom. Contr.*, vol. AC-11, pp. 190–197, 1966.

30. —, "Canonical forms for linear multivariable systems," *IEEE Trans. Autom. Contr.*, vol. AC-13, pp. 290–293, 1967.
31. —, "An introduction to observers," *IEEE Trans. Autom. Contr.*, vol. AC-16, no. 5, pp. 596–692, 1971.
32. O' REILLY, J., *Observers for Linear Systems*, Academic, New York, 1983.
33. PONTRYAGIN, L.S., BOLTYANSKII, V.G., GAMKRELIDZE, R.V., and MISHCHENKO, E.F., *The Mathematical Theory of Optimal Processes*, Interscience (Wiley), New York, 1962.
34. POTTER, J.E., "Matrix quadratic solutions," *SIAM J. Appl. Math.*, vol. 14, no. 3, pp. 496–501, 1966.
35. VIT, K., "Iterative solution of the Riccati equation," *IEEE Trans. Autom. Contr.*, vol. AC-17, no. 2, pp. 258-259, 1972.
36. WEISS, L., and KALMAN, R.E., "Contributions to linear system theory," *Int. J. Engin. Sci.*, vol. 3, pp. 161–176, 1975.
37. WOLOVICH, W.A., and FALB, P.L., "On the structure of multivariable systems," *SIAM J. Control*, vol. 7, no. 3, pp. 437–451, 1969.
38. WONHAM, W.M., "On pole assignment in multi-input controllable linear systems," *IEEE Trans. Autom. Contr.*, vol. AC-12, no. 6, pp. 660–665, 1967.

4

THE GEOMETRIC APPROACH: ANALYSIS

4.1 CONTROLLED AND CONDITIONED INVARIANTS

The extensions of invariance, namely controlled and conditioned invariance, provide means for further developments of linear system analysis: in this chapter properties like constrained controllability and observability, unknown-input observability, system left and right invertibility, and the concept of transmission zero, are easily handled with these new mathematical tools.

Consider a three-map system (A, B, C). It has been proved that in the absence of control action (i.e., when function $u(\cdot)$ is identically zero) a subspace of the state space $\mathcal{X}$ is a locus of trajectories if and only if it is an A-invariant (Theorem 3.2-4). The extension of this property to the case in which the control is present and suitably used to steer the state along a convenient trajectory leads to the concept of $(A, \mathcal{B})$-controlled invariant and to the following formal definition.

Definition 4.1-1. Controlled Invariant

Consider a pair (A, B). A subspace $\mathcal{V} \subseteq \mathcal{X}$ is said to be an (A, B)-*controlled invariant*[1] if

$$A\mathcal{V} \subseteq \mathcal{V} + \mathcal{B} \quad \text{with} \quad \mathcal{B} := \text{im } B \tag{1}$$

---

[1] The concepts of controlled invariance and related computational algorithms were introduced by Basile and Marro [4], Wonham and Morse [44]. The concept of conditioned invariance was contemporarily introduced in [4].

Sect. 4.1 Controlled and Conditioned Invariants

The dual of the controlled invariant is the conditioned invariant, which is defined as follows.

Definition 4.1-2. Conditioned Invariant

Consider a pair (A, C). A subspace $\mathcal{S} \subseteq \mathcal{X}$ is said to be an (A, C)-*conditioned invariant* if

$$A(\mathcal{S} \cap \mathcal{C}) \subseteq \mathcal{S} \quad \text{with} \quad \mathcal{C} := \ker C \tag{2}$$

Note that any A-invariant is also an $(A, \mathcal{B})$-controlled invariant for any $\mathcal{B}$ and an $(A, \mathcal{C})$-conditioned invariant for any $\mathcal{C}$: in particular, the origin $\{0\}$ and the whole space $\mathcal{X}$ are so. Furthermore, $(A, \{0\})$-controlled invariants and $(A, \mathcal{X})$-conditioned invariants are, in particular, A-invariants.

The geometric meaning of controlled and conditioned invariants is illustrated by the examples shown in Fig. 4-1. In Fig. 4-1(a) subspace $\mathcal{V}$ is an $(A, \mathcal{B})$-controlled invariant since, if im$A$ and $\mathcal{B}$ are disposed as shown, clearly $A\mathcal{V} \subseteq \mathcal{V} + \mathcal{B}$; however, it is not a controlled invariant with respect to $(A, \mathcal{B}')$. In Fig. 4-1(b), subspace $\mathcal{S}$ is an $(A, \mathcal{C})$-conditioned invariant since, if $A(\mathcal{S} \cap \mathcal{C})$ is disposed as shown, it follows that $A(\mathcal{S} \cap \mathcal{C}) \subseteq \mathcal{S}$; however, it is not an $(A, \mathcal{C}')$-conditioned invariant, because of the different layout of $A(\mathcal{S} \cap \mathcal{C}')$ with respect to $\mathcal{S}$.

The following properties are easily proved by direct check.

Property 4.1-1.

The sum of any two $(A, \mathcal{B})$-controlled invariants is an $(A, \mathcal{B})$-controlled invariant.

Property 4.1-2.

The intersection of any two $(A, \mathcal{C})$-conditioned invariants is an $(A, \mathcal{C})$-conditioned invariant.

In general, however, the intersection of two controlled invariants is not a controlled invariant and the sum of two conditioned invariants is not a conditioned invariant.

As a consequence of Property 4.1-1 the set of all $(A, \mathcal{B})$-controlled invariants contained in a given subspace $\mathcal{E} \subseteq \mathcal{X}$ is an upper semilattice with respect to $\subseteq$, $+$, hence it admits a supremum, the *maximal $(A, \mathcal{B})$-controlled invariant contained in $\mathcal{E}$*, which will be denoted by $\max \mathcal{V}(A, \mathcal{B}, \mathcal{E})$. Similarly, Property 4.1-2 implies that the set of all $(A, \mathcal{C})$-conditioned invariants containing a given subspace $\mathcal{D} \subseteq \mathcal{X}$ is a lower semilattice with respect to $\subseteq$, $\cap$, hence it admits an infimum, the *minimal $(A, \mathcal{C})$-conditioned invariant containing $\mathcal{D}$*, which will be denoted by $\min \mathcal{S}(A, \mathcal{C}, \mathcal{D})$. Algorithms to compute $\max \mathcal{V}(A, \mathcal{B}, \mathcal{E})$ and $\min \mathcal{S}(A, \mathcal{C}, \mathcal{D})$ will be presented in Subsection 4.1.1. Duality between controlled and conditioned invariants is stated in precise terms by the following property.

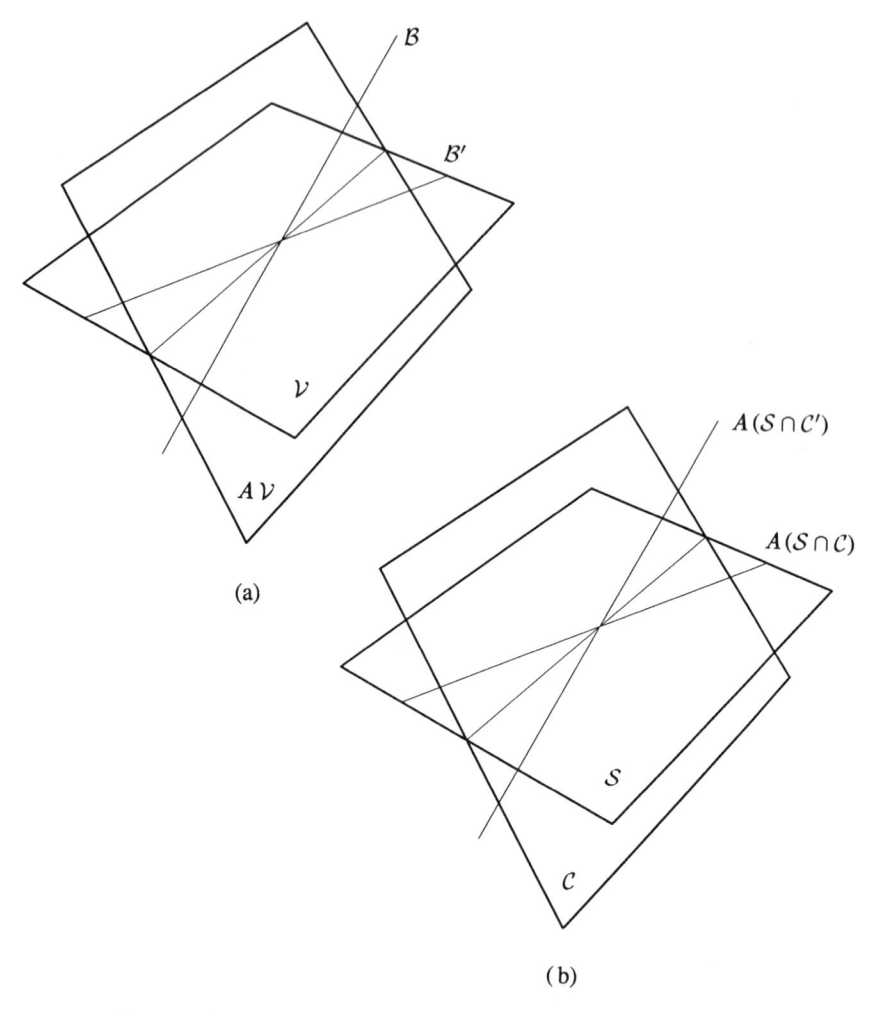

Figure 4-1 The geometric meaning of controlled and conditioned invariants.

Property 4.1-3.

The orthogonal complement of an $(A,\mathcal{L})$-controlled (conditioned) invariant is an $(A^T,\mathcal{L}^\perp)$-conditioned (controlled) invariant.

Proof. By Property 3.1-2 it follows that

$$A\mathcal{V} \subseteq \mathcal{V}+\mathcal{L} \Leftrightarrow A^T(\mathcal{V}+\mathcal{L})^\perp \subseteq \mathcal{V}^\perp$$
$$A(\mathcal{S}\cap\mathcal{L}) \subseteq \mathcal{S} \Leftrightarrow A^T\mathcal{S}^\perp \subseteq (\mathcal{S}\cap\mathcal{L})^\perp$$

and, by (3.1.9, 10)

$$A\mathcal{V} \subseteq \mathcal{V}+\mathcal{L} \Leftrightarrow A^T(\mathcal{V}^\perp \cap \mathcal{L}^\perp) \subseteq \mathcal{V}^\perp$$
$$A(\mathcal{S}\cap\mathcal{L}) \subseteq \mathcal{S} \Leftrightarrow A^T\mathcal{S}^\perp \subseteq \mathcal{S}^\perp + \mathcal{L}^\perp \quad \square$$

Sect. 4.1 Controlled and Conditioned Invariants

The following theorem is basic: it establishes the connection between controlled invariants and dynamic systems.

Theorem 4.1-1.
Consider a pair (A, B). A subspace $\mathcal{V} \subseteq \mathcal{X}$ is a locus of controlled trajectories of (A, B) if and only if it is an $(A, \mathcal{B})$-controlled invariant.

Proof. *If.* Let $\mathcal{V}$ be an $(A, \mathcal{B})$-controlled invariant: by virtue of (1), for any $x \in \mathcal{V}$ there exists at least one value of control u such that $Ax + Bu \in \mathcal{V}$: this means that at any point of $\mathcal{V}$ the state velocity can be maintained on $\mathcal{V}$ by a suitable control action, hence, by virtue of the fundamental lemma (Lemma 3.2-1), for any initial state x_0 in $\mathcal{V}$ there exists an admissible state trajectory starting at x_0 and completely belonging to $\mathcal{V}$.

Only if. Consider a state trajectory $x(\cdot)$ of (A, B) and denote by $\mathcal{V}$ the subspace of minimal dimension in which it is contained. Let $h := \dim \mathcal{V}$: there exist h values of time $t_1, \ldots, t_h$ such that $\{x(t_1), \ldots, x(t_h)\}$ is a basis of $\mathcal{V}$. The fundamental lemma implies

$$\dot{x}(t_i) = A x(t_i) + B u(t_i) \in \mathcal{V} \quad (i = 1, \ldots, h)$$

hence $A\mathcal{V} \subseteq \mathcal{V} + \mathcal{B}$. □

A matrix characterization that extends Property 3.2-1 of simple invariants is the following.

Property 4.1-4.
A subspace $\mathcal{V}$ with basis matrix V is an $(A, \mathcal{B})$-controlled invariant if and only if there exist matrices X, U such that

$$AV = VX + BU \tag{3}$$

Proof. Let v_i $(i = 1, \ldots, r)$ be the columns of V: $\mathcal{V}$ is an $(A, \mathcal{B})$-controlled invariant if and only if each transformed column is a linear combination of columns of V and B, i.e., if and only if there exist vectors x_i, u_i such that $A v_i = V x_i + B u_i$ $(i = 1, \ldots, r)$: relation (3) is the same in compact form. □

To show that controlled invariance is a coordinate-free concept, consider the change of basis corresponding to the nonsingular transformation T. Matrices $A' := T^{-1}AT$, $B' := T^{-1}B$ and $W := T^{-1}V$ correspond to matrices A, B, V in the new basis. Relation (3) can be written as

$$T^{-1}AT(T^{-1}V) = (T^{-1}V) X + T^{-1} B U \quad \text{or} \quad A'W = WX + B'U$$

Controlled and conditioned invariants are very important in connection with synthesis problems because of their feedback properties: in fact a controlled invariant can be transformed into a simple invariant by means of a suitable state feedback, just as a conditioned invariant can be transformed into a simple invariant by means of a suitable output injection.

Theorem 4.1-2.
A subspace $\mathcal{V} \subseteq \mathcal{X}$ is an (A,B)-controlled invariant if and only if there exists at least one matrix F such that $(A+BF)\mathcal{V} \subseteq \mathcal{V}$.

Proof. Only if. Consider Property 4.1-4, in particular relation (3), and assume

$$F := -U(V^T V)^{-1} V^T \tag{4}$$

Simple manipulations yield

$$(A+BF)V = VX$$

hence, by Property 3.2-1, $\mathcal{V}$ is an $(A+BF)$-invariant.

If. Suppose that (1) does not hold: then, there exists at least one vector $x_0 \in \mathcal{V}$ such that Ax_0 cannot be expressed as the sum of two vectors $x_0' \in \mathcal{V}$ and $Bu_0 \in \mathcal{B}$, hence no F exists such that $(A+BF)x_0 \in \mathcal{V}$. □

Theorem 4.1-3.
A subspace $\mathcal{S} \subseteq \mathcal{X}$ is an (A,C)-conditioned invariant if and only if there exists at least one matrix G such that $(A+GC)\mathcal{S} \subseteq \mathcal{S}$.

Proof. The statement is derived, by duality, from Theorem 4.1-2. In fact, by virtue of Property 4.1-1, the defining relation (2) is equivalent to

$$A^T \mathcal{S}^\perp \subseteq \mathcal{S}^\perp + \mathcal{C}^\perp = \mathcal{S}^\perp + \text{im} C^T$$

By Theorem 4.1-2 this is a necessary and sufficient condition for the existence of a matrix G such that $(A^T + C^T G^T)\mathcal{S}^\perp \subseteq \mathcal{S}^\perp$ or, by Property 3.1-2, $(A+GC)\mathcal{S} \subseteq \mathcal{S}$. □

The question now arises whether two or more controlled invariants can be transformed into simple invariants by the same state feedback. An answer is contained in the following property.

Property 4.1-5.
Let $\mathcal{V}_1, \mathcal{V}_2$ be (A,B)-controlled invariants. There exists a matrix F such that $(A+BF)\mathcal{V}_i \subseteq \mathcal{V}_i$ $(i=1,2)$ if and only if $\mathcal{V} := \mathcal{V}_1 \cap \mathcal{V}_2$ is an (A,B)-controlled invariant.

Proof. If. This part can be proved in the same way as the only if part of Theorem 4.1-2. Let V_1 be a basis matrix of $\mathcal{V}$, $[V_1\ V_2]$ a basis matrix of $\mathcal{V}_1$, $[V_1\ V_3]$ a basis matrix of $\mathcal{V}_2$, so that $[V_1\ V_2\ V_3]$ is a basis matrix of $\mathcal{V}_1 + \mathcal{V}_2$. Denote by U_1, U_2, U_3 the corresponding matrices in relation (3). It is easy to check that matrix F defined as in (4) with $U := [U_1\ U_2\ U_3]$ and $V := [V_1\ V_2\ V_3]$, is such that $(A+BF)\mathcal{V}_i \subseteq \mathcal{V}_i$ $(i=1,2)$ and $(A+BF)\mathcal{V} \subseteq \mathcal{V}$.

Only if. If $\mathcal{V}$ is not an (A,B)-controlled invariant, by virtue of Theorem 4.1-2 no F exists such that $(A+BF)\mathcal{V} \subseteq \mathcal{V}$, hence $(A+BF)\mathcal{V}_i \subseteq \mathcal{V}_i$ $(i=1,2)$, provided the intersection of two invariants is an invariant. □

This result is dualized as follows.

Property 4.1-6.
Let $\mathcal{S}_1, \mathcal{S}_2$ be two (A,C)-conditioned invariants. There exists a matrix G such that $(A+GC)\mathcal{S}_i \subseteq \mathcal{S}_i$ $(i=1,2)$ if and only if $\mathcal{S} := \mathcal{S}_1 + \mathcal{S}_2$ is an (A,C)-conditioned invariant.

4.1.1 Some Specific Computational Algorithms

Subspaces $\min \mathcal{S}(A,\mathcal{C},\mathcal{D})$ and $\max \mathcal{V}(A,\mathcal{B},\mathcal{E})$, which are respectively the infimum of the semilattice of all $(A,\mathcal{C})$-conditioned invariants containing a given subspace $\mathcal{D}$ and the supremum of the semilattice of all $(A,\mathcal{B})$-controlled invariants contained in a given subspace $\mathcal{E}$, can be determined with algorithms that extend those presented for simple invariants in Subsection 3.2.2. The basic algorithm is the following.

Algorithm 4.1-1. The Minimal $(A, \ker C)$-Conditioned Inv. Containing imD
Subspace $\min \mathcal{S}(A,\mathcal{C},\mathcal{D})$ coincides with the last term of the sequence

$$\mathcal{Z}_0 := \mathcal{D} \tag{5}$$

$$\mathcal{Z}_i := \mathcal{D} + A(\mathcal{Z}_{i-1} \cap \mathcal{C}) \quad (i=1,\ldots,k) \tag{6}$$

where the value of $k \le n-1$ is determined by condition $\mathcal{Z}_{k+1} = \mathcal{Z}_k$.

Proof. First, note that $\mathcal{Z}_i \supseteq \mathcal{Z}_{i-1}$ $(i=1,\ldots,k)$. In fact, instead of (6), consider the recursion expression

$$\mathcal{Z}'_i := \mathcal{Z}'_{i-1} + A(\mathcal{Z}'_{i-1} \cap \mathcal{C}) \quad (i=1,\ldots,k)$$

with $\mathcal{Z}'_0 := \mathcal{D}$, which defines a sequence such that $\mathcal{Z}'_i \supseteq \mathcal{Z}'_{i-1}$ $(i=1,\ldots,k)$, hence $A(\mathcal{Z}'_i \cap \mathcal{C}) \supseteq A(\mathcal{Z}'_{i-1} \cap \mathcal{C})$ $(i=1,\ldots,k)$. This sequence is equal to (6): by induction, note that if $\mathcal{Z}'_j = \mathcal{Z}_j$ $(j=1,\ldots,i-1)$, also $\mathcal{Z}'_i = \mathcal{D} + A(\mathcal{Z}_{i-2} \cap \mathcal{C}) + A(\mathcal{Z}_{i-1} \cap \mathcal{C}) = \mathcal{Z}_i$ (since $A(\mathcal{Z}_{i-2} \cap \mathcal{C}) \subseteq A(\mathcal{Z}_{i-1} \cap \mathcal{C})$).

If $\mathcal{Z}_{k+1} = \mathcal{Z}_k$, also $\mathcal{Z}_j = \mathcal{Z}_k$ for all $j > k+1$ and $\mathcal{Z}_k$ is an $(A,\mathcal{C})$-conditioned invariant containing $\mathcal{D}$. In fact, in such a case $\mathcal{Z}_k = \mathcal{D} + A(\mathcal{Z}_k \cap \mathcal{C})$ hence $\mathcal{D} \subseteq \mathcal{Z}_k$, $A(\mathcal{Z}_k \cap \mathcal{C}) \subseteq \mathcal{Z}_k$. Since two subsequent subspaces are equal if and only if they have equal dimensions and the dimension of the first subspace is at least one, an $(A,\mathcal{C})$-conditioned invariant is obtained in at most $n-1$ steps.

The last subspace of the sequence is the minimal $(A,\mathcal{C})$-conditioned invariant containing $\mathcal{C}$, as can be again proved by induction. Let $\mathcal{S}$ be another $(A,\mathcal{C})$-conditioned invariant containing $\mathcal{D}$: if $\mathcal{S} \supseteq \mathcal{Z}_{i-1}$, it follows that $\mathcal{S} \supseteq \mathcal{Z}_i$. In fact, $\mathcal{S} \supseteq \mathcal{D} + A(\mathcal{S} \cap \mathcal{C}) \supseteq \mathcal{D} + A(\mathcal{Z}_{i-1} \cap \mathcal{C}) = \mathcal{Z}_i$. □

From Property 4.1-1 and from

$$\mathcal{E} \supseteq \mathcal{V} \iff \mathcal{E}^\perp \subseteq \mathcal{V}^\perp \tag{7}$$

one can derive

$$\max \mathcal{V}(A,\mathcal{B},\mathcal{E}) = \left(\min \mathcal{S}(A^T, \mathcal{B}^\perp, \mathcal{E}^\perp)\right)^\perp \tag{8}$$

which brings determination of $\max \mathcal{V}(A,\mathcal{B},\mathcal{E})$ back to that of $\min \mathcal{S}(A,\mathcal{C},\mathcal{D})$.

From relation (8) it is possible to derive also the following algorithm, dual of Algorithm 4.1-1.

Algorithm 4.1-2. The Maximal (A,imB)-Controlled Inv. Contained in kerE
Subspace $\max \mathcal{V}(A, \mathcal{B}, \mathcal{E})$ coincides with the last term of the sequence

$$\mathcal{Z}_0 := \mathcal{E} \tag{9}$$

$$\mathcal{Z}_i := \mathcal{E} \cap A^{-1}(\mathcal{Z}_{i-1} + \mathcal{B}) \quad (i=1, \ldots, k) \tag{10}$$

where the value of $k \leq n-1$ is determined by condition $\mathcal{Z}_{k+1} = \mathcal{Z}_k$.

Proof. Sequence (9, 10) is equivalent to

$$\mathcal{Z}_0^\perp := \mathcal{E}^\perp$$

$$\mathcal{Z}_i^\perp := \left(\mathcal{E} \cap A^{-1}(\mathcal{Z}_{i-1} + \mathcal{B})\right)^\perp = \mathcal{E}^\perp + A(\mathcal{Z}_{i-1}^\perp \cap \mathcal{B}^\perp)$$

which, by virtue of Algorithm 4.1-1, converges to the orthogonal complement of $\min \mathcal{S}(A^T, \mathcal{B}^\perp, \mathcal{E}^\perp)$, which is $\max \mathcal{V}(A, \mathcal{B}, \mathcal{E})$ by (8). □

Algorithm 4.1-3. Computation of State Feedback Matrix F

Let $\mathcal{V}$ be an $(A, \mathcal{B})$-controlled invariant. We search for a matrix F such that $(A+BF)\mathcal{V} \subseteq \mathcal{V}$. The rank of B is assumed to be maximal: if not, delete linearly dependent columns to obtain matrix B_1, then derive F by adding to the corresponding F_1 an equal number of zero rows in the same places as the deleted columns of B. Let X_1, X_2, X_3, X_4 be basis matrices of subspaces $\mathcal{B} \cap \mathcal{V}$, $\mathcal{V}$, $\mathcal{B}$, $\mathcal{X}$; in particular, we can assume $X_4 := I_n$. Orthonormalize matrix $[X_1\ X_2\ X_3\ X_4]$ (by the Gram-Schmidt process provided with a linear dependency test) and denote by $[M_1\ M_2\ M_3\ M_4]$ the orthonormal matrix obtained, in which the submatrices shown are not necessarily all present. The coordinate transformation $T := [B\ M_2\ M_4]$ yields

$$A' := T^{-1}AT = \begin{bmatrix} A'_{11} & A'_{12} & A'_{13} \\ A'_{21} & A'_{22} & A'_{23} \\ O & O & A'_{33} \end{bmatrix} \qquad B' := T^{-1}B = \begin{bmatrix} I_p \\ O \\ O \end{bmatrix} \tag{11}$$

The state feedback matrix

$$F' := [\,-A'_{11} \quad -A'_{12} \quad O\,] \tag{12}$$

is such that $A' + B'F'$ transforms, in the new basis, vectors of $\mathcal{B} + \mathcal{V}$ into vectors of $\mathcal{V}$, hence fits our needs. The corresponding matrix in the old reference is $F := F'T^{-1}$. Note that $\ker F = (\mathcal{B} + \mathcal{V})^\perp$.

4.1.2 Self-Bounded Controlled Invariants and their Duals

Self-bounded controlled invariants are a particular class of controlled invariants that has interesting properties, the most important of which is to be a lattice instead of a semilattice, hence to admit both a supremum and an infimum. They are introduced through the following argument: given any subspace $\mathcal{E} \subseteq \mathcal{X}$, define

$$\mathcal{V}^* := \max \mathcal{V}(A, \mathcal{B}, \mathcal{E}) \tag{13}$$

Sect. 4.1 Controlled and Conditioned Invariants

(the maximal $(A, \mathcal{B})$-controlled invariant contained in $\mathcal{E}$): it is well known (Theorem 4.1-1) that a trajectory of pair (A, B) can be controlled on $\mathcal{E}$ if and only if its initial state belongs to a controlled invariant contained in $\mathcal{E}$, hence in $\mathcal{V}^*$. In general, for any initial state belonging to a controlled invariant $\mathcal{V}$, it is possible not only to continuously maintain the state on $\mathcal{V}$ by means of a suitable control action, but also to leave $\mathcal{V}$ with a trajectory on $\mathcal{E}$ (hence on $\mathcal{V}^*$) and to pass to some other controlled invariant contained in $\mathcal{E}$ (hence in $\mathcal{V}^*$). On the other hand there exist controlled invariants that are closed with respect to the control, i.e., that cannot be exited by means of any trajectory on $\mathcal{E}$: these will be called self-bounded with respect to $\mathcal{E}$.

The following lemma will be used to introduce a characterizing property of self-bounded controlled invariants.

Lemma 4.1-1.
Consider three subspaces $\mathcal{X}, \mathcal{Y}, \mathcal{Z}$ such that $\mathcal{X} \subseteq \mathcal{Y} + \mathcal{Z}$. For any vector $x_0 \in \mathcal{X}$ all possible decompositions $x_0 = y + z$, $y \in \mathcal{Y}$, $z \in \mathcal{Z}$, are obtainable from any one of them, say $x_0 = y_0 + z_0$, by summing to y_0 and subtracting from z_0 all vectors of $\mathcal{Y} \cap \mathcal{Z}$.

Proof. Let $x_0 = y_1 + z_1$ be another decomposition of x_0: by difference, $0 = (y_0 - y_1) + (z_0 - z_1)$, i.e. $(y_0 - y_1) = -(z_0 - z_1)$: since in this equality the first vector belongs to $\mathcal{Y}$ and the second to $\mathcal{Z}$, both must belong to $\mathcal{Y} \cap \mathcal{Z}$. On the other hand, if a vector belonging to $\mathcal{Y} \cap \mathcal{Z}$ is summed to y_0 and subtracted from z_0, two vectors belonging respectively to $\mathcal{Y}$ and $\mathcal{Z}$ are obtained whose sum is x_0: a decomposition of x_0 is thus derived. □

Definition 4.1-3. Self-Bounded Controlled Invariant
Let $\mathcal{V}$ be an $(A, \mathcal{B})$-controlled invariant contained in a subspace $\mathcal{E} \subseteq \mathcal{X}$: $\mathcal{V}$ is said to be *self-bounded with respect to* $\mathcal{E}$ if

$$\mathcal{V}^* \cap \mathcal{B} \subseteq \mathcal{V} \tag{14}$$

where $\mathcal{V}^*$ is the subspace defined by (13).

It is easily shown that the above definition implies that $\mathcal{V}$ is closed with respect to trajectories lying on $\mathcal{E}$. Let F be a matrix such that $(A + BF)\mathcal{V} \subseteq \mathcal{V}$: given any state $x_0 \in \mathcal{V}$, by virtue of Lemma 4.1-1 the set of all admissible velocities on $\mathcal{V}^*$ is the linear variety

$$T(x) = (A + BF)x_0 + \mathcal{V}^* \cap \mathcal{B}$$

On the other hand the fundamental lemma implies that trajectories belonging to $\mathcal{E}$, hence to $\mathcal{V}^*$, cannot leave $\mathcal{V}$ if and only if

$$T(x) \subseteq \mathcal{V} \quad \forall x \in \mathcal{V}$$

hence if and only if $\mathcal{V}^* \cap \mathcal{B} \subseteq \mathcal{V}$.

To show that the set of all controlled invariants self-bounded with respect to $\mathcal{E}$ is a lattice, let us first introduce the following characterizing properties.

Property 4.1-7.
Let F be a matrix such that $(A+BF)\mathcal{V}^* \subseteq \mathcal{V}^*$. Any controlled invariant $\mathcal{V}$ self-bounded with respect to $\mathcal{E}$ satisfies $(A+BF)\mathcal{V} \subseteq \mathcal{V}$.

Proof. By definition $\mathcal{V}$ satisfies the inclusions

$$A\mathcal{V} \subseteq \mathcal{V}+\mathcal{B} \quad \mathcal{V} \subseteq \mathcal{V}^* \quad \mathcal{V} \supseteq \mathcal{V}^* \cap \mathcal{B} \tag{15}$$

and F is such that

$$(A+BF)\mathcal{V}^* \subseteq \mathcal{V}^* \tag{16}$$

The second of (15) and (16) lead to

$$(A+BF)\mathcal{V} \subseteq \mathcal{V}^* \tag{17}$$

while the trivial inclusion $BF\mathcal{V} \subseteq \mathcal{B}$ and the first of (15) imply

$$(A+BF)\mathcal{V} \subseteq \mathcal{V}+\mathcal{B} \tag{18}$$

Intersecting both members of (17, 18) finally yields

$$(A+BF)\mathcal{V} \subseteq (\mathcal{V}+\mathcal{B}) \cap \mathcal{V}^* = \mathcal{V}+\mathcal{B} \cap \mathcal{V}^* = \mathcal{V} \quad \square$$

Property 4.1-8.
The intersection of any two $(A,\mathcal{B})$-controlled invariants self-bounded with respect to $\mathcal{E}$ is an $(A,\mathcal{B})$-controlled invariant self-bounded with respect to $\mathcal{E}$.

Proof. Let $\mathcal{V}_1, \mathcal{V}_2$ be the controlled invariants considered in the statement. By virtue of Property 4.1-5 there exists a matrix F such that $(A+BF)\mathcal{V}_i \subseteq \mathcal{V}_i$ ($i=1,2$). Define $\mathcal{V} := \mathcal{V}_1 \cap \mathcal{V}_2$: then $(A+BF)\mathcal{V} \subseteq \mathcal{V}$ (since the intersection of two invariants is an invariant). Therefore by Theorem 4.1-2, $\mathcal{V}$ is an $(A,\mathcal{B})$-controlled invariant. $\mathcal{V}$ is self-bounded with respect to $\mathcal{E}$ since from $\mathcal{V}_i \subseteq \mathcal{E}$, $\mathcal{V}_i \supseteq \mathcal{V}^* \cap \mathcal{B}$ ($i=1,2$) it follows that $\mathcal{V} \subseteq \mathcal{E}$, $\mathcal{V} \supseteq \mathcal{V}^* \cap \mathcal{B}$. $\square$

By virtue of this property, the set of all $(A,\mathcal{B})$-controlled invariants self-bounded with respect to $\mathcal{E}$ is closed with respect to the intersection. Being closed also with respect to the sum by Property 4.1-1, it is a lattice (nondistributive) with respect to $\subseteq, +, \cap$; it will be denoted by $\Phi(\mathcal{B}, \mathcal{E})$. Its definition formula is

$$\Phi(\mathcal{B}, \mathcal{E}) := \{\mathcal{V} : A\mathcal{V} \subseteq \mathcal{V}+\mathcal{B}, \mathcal{V} \subseteq \mathcal{E}, \mathcal{V} \supseteq \mathcal{V}^* \cap \mathcal{B}\} \tag{19}$$

The supremum of $\Phi(\mathcal{B}, \mathcal{E})$ is $\mathcal{V}^*$, the maximal controlled invariant contained in $\mathcal{E}$, which is clearly self-bounded (it contains $\mathcal{V}^* \cap \mathcal{B}$), while its infimum will be determined below.

The following theorem defines the infimum of $\Phi(\mathcal{B}, \mathcal{E})$. It is remarkable that it is expressed as the intersection of the supremum, which can be determined by means of Algorithm 4.1-2, with the infimum of a particular semilattice of conditioned invariants, which can be determined by means of Algorithm 4.1-1. Since these algorithms can be reconducted to each other by duality, in practice just one computational procedure is sufficient to determine both limits of $\Phi(\mathcal{B}, \mathcal{E})$.

Theorem 4.1-4.
The infimum of $\Phi(B,\mathcal{E})$ is [2]

$$\mathcal{V}^* \cap \mathcal{S}_2^* \quad \text{with} \quad \mathcal{S}_2^* := \min \mathcal{S}(A,\mathcal{E},B) \tag{20}$$

Proof. Let
$$\bar{\mathcal{S}}_2^* := \min \mathcal{S}(A,\mathcal{V}^*,B) \tag{21}$$

The proof will be developed in three steps:
1. Any element of $\Phi(B,\mathcal{E})$ contains $\mathcal{V}^* \cap \bar{\mathcal{S}}_2^*$;
2. $\mathcal{V}^* \cap \bar{\mathcal{S}}_2^*$ is an element of $\Phi(B,\mathcal{E})$;
3. $\mathcal{V}^* \cap \bar{\mathcal{S}}_2^*$ is equal to $\mathcal{V}^* \cap \mathcal{S}_2^*$.

Step 1. Consider the sequence that defines $\bar{\mathcal{S}}_2^*$:

$$\mathcal{Z}_0' := B$$
$$\mathcal{Z}_i' := B + A(\mathcal{Z}_{i-1}' \cap \mathcal{V}^*) \quad (i=1,\ldots)$$

Let $\mathcal{V}$ be a generic element of $\Phi(B,\mathcal{E})$, so that

$$A\mathcal{V} \subseteq \mathcal{V} + B, \quad \mathcal{V} \supseteq \mathcal{V}^* \cap B$$

We proceed by induction: clearly

$$\mathcal{Z}_0' \cap \mathcal{V}^* \subseteq \mathcal{V}$$

and from
$$\mathcal{Z}_{i-1}' \cap \mathcal{V}^* \subseteq \mathcal{V}$$

it follows that
$$A(\mathcal{Z}_{i-1}' \cap \mathcal{V}^*) \subseteq \mathcal{V} + B \tag{22}$$

since $\mathcal{V}$ is an (A,B)-controlled invariant. Summing B to both members of (22) yields

$$B + A(\mathcal{Z}_{i-1}' \cap \mathcal{V}^*) \subseteq \mathcal{V} + B$$

and, by intersection with $\mathcal{V}^*$,

$$\mathcal{Z}_i' \cap \mathcal{V}^* \subseteq \mathcal{V}$$

which completes the induction argument and the proof of step 1.
Step 2. From

$$A\mathcal{V}^* \subseteq \mathcal{V}^* + B$$
$$A(\bar{\mathcal{S}}_2^* \cap \mathcal{V}^*) \subseteq \bar{\mathcal{S}}_2^*$$

---
[2] Note the symmetry in (20), which defines the reachable set on $\mathcal{E}$ as the intersection of the maximal (A,B)-controlled invariant contained in $\mathcal{E}$ with the minimal $(A,\mathcal{E})$-conditioned invariant containing B. Relation (20) was first derived by Morse [33].

which simply express V^* to be an (A, B)-controlled invariant and $\bar{S}_2^*$ to be an (A, V^*)-conditioned invariant, by intersection it follows that

$$A(\bar{S}_2^* \cap V^*) \subseteq A\bar{S}_2^* \cap AV^* \subseteq \bar{S}_2^* \cap (V^* + B) = \bar{S}_2^* \cap V^* + B$$

thus $\bar{S}_2^* \cap V^*$ is an (A, B)-controlled invariant. It is self-bounded, since $\bar{S}_2^* \supseteq B$, hence $V^* \cap \bar{S}_2^* \supseteq V^* \cap B$.

Step 3. It will be proved that the following holds:

$$\bar{S}_2^* = S_2^* \cap V^* + B \tag{23}$$

from which our thesis $V^* \cap \bar{S}_2^* = V^* \cap S_2^*$ follows.

Equality (23) can be proved by considering the sequences

$$Z_0' := B \qquad\qquad Z_0 := B$$
$$Z_i' := B + A(Z_{i-1}' \cap V^*) \qquad Z_i := B + A(Z_{i-1}' \cap \mathcal{E}) \quad (i=1,\ldots)$$

which converge respectively to $\bar{S}_2^*$ and to S_2^*. It can be shown by induction that

$$Z_i' = Z_i \cap (V^* + B) = Z_i \cap V^* + B \quad (\text{since } B \subseteq Z_i)$$

if

$$Z_{i-1}' = Z_{i-1} \cap (V^* + B)$$

In fact

$$Z_i' = B + A\big(Z_{i-1} \cap (V^* + B) \cap V^*\big) = B + A(Z_{i-1} \cap V^*)$$
$$= B + A\big(Z_i \cap \big(\mathcal{E} \cap A^{-1}(V^* + B)\big)\big) = B + A(Z_{i-1} \cap \mathcal{E}) \cap (V^* + B)$$
$$= Z_i \cap (V^* + B)$$

In previous manipulations relation $V^* = \mathcal{E} \cap A^{-1}(V^* + B)$ (which expresses the limit of the sequence of Algorithm 4.1-2) and the identity $A(\mathcal{X} \cap A^{-1}\mathcal{Y}) = A\mathcal{X} \cap \mathcal{Y}$ have been used. Since

$$Z_0' = Z_0 \cap (V^* + B)$$

the proof by induction of (23) is complete. □

The following corollary, whose proof is contained in the argument just presented for Theorem 4.1-4, provides an alternative expression for the infimum of $\Phi(\mathcal{B}, \mathcal{E})$.

Corollary 4.1-1.
The infimum of $\Phi(\mathcal{B}, \mathcal{E})$ is $V^* \cap \bar{S}_2^*$, with V^* and $\bar{S}_2^*$ defined by (13, 21).

The preceding results will be extended to conditioned invariants by duality. The duals of the self-bounded controlled invariants are the self-hidden conditioned invariants: their characterizing property is the possibility to become all unobservable by means of an output injection of the type shown in Fig. 3-10(b).

In the following, subspace

$$S^* := \min S(A, \mathcal{C}, \mathcal{D}) \tag{24}$$

will be referred to frequently. According to our standard notation, it represents the minimal $(A, \mathcal{C})$-conditioned invariant containing $\mathcal{D}$.

Sect. 4.1 Controlled and Conditioned Invariants

Definition 4.1-4. Self-Hidden Conditioned Invariant

Let S be an (A,C)-conditioned invariant containing a subspace $\mathcal{D} \subseteq \mathcal{X}$: S is said to be *self-hidden with respect to* $\mathcal{D}$ if

$$S \subseteq S^* + C \qquad (25)$$

where S^* is the subspace defined by (24).

Property 4.1-9.

Let G be a matrix such that $(A+GC)S^* \subseteq S^*$. Any conditioned invariant S self-hidden with respect to $\mathcal{D}$ satisfies $(A+GC)S \subseteq S$.

Property 4.1-10.

The sum of any two (A,C)-conditioned invariants self-hidden with respect to $\mathcal{D}$ is an (A,C)-conditioned invariant self-hidden with respect to $\mathcal{D}$.

Refer now to the set

$$\Psi(\mathcal{C},\mathcal{D}) := \{ S \,:\, A(S \cap \mathcal{C}) \subseteq S,\, S \supseteq \mathcal{D},\, S \subseteq S^* + \mathcal{C} \} \qquad (26)$$

which is the lattice of all (A,C)-conditioned invariants self-hidden with respect to $\mathcal{D}$. $\Psi(\mathcal{C},\mathcal{D})$ is a nondistributive lattice with respect to $\subseteq, +, \cap$ whose infimum is S^*. Its supremum is defined by the following theorem, dual of Theorem 4.1-4.

Theorem 4.1-5.

The supremum of $\Psi(\mathcal{C},\mathcal{D})$ is

$$S^* + V_2^* \quad \text{with} \quad V_2^* := \max V(A,\mathcal{D},\mathcal{C}) \qquad (27)$$

As for the infimum of the lattice of self-bounded controlled invariants, it is also possible to give an alternative expression for the supremum of that of self-hidden conditioned invariants. It is stated by the following corollary, dual of Corollary 4.1-1.

Corollary 4.1-2.

The supremum of $\Psi(\mathcal{C},\mathcal{D})$ is $S^* + V_2^*$, with

$$V_2^* := \max V(A,S^*,\mathcal{C}) \qquad (28)$$

4.1.3 Constrained Controllability and Observability

Controlled invariants are subspaces such that, from any initial state belonging to them, at least one state trajectory can be maintained on them by means of a suitable control action. In general, however, it is not possible to reach any point of a controlled invariant from any other point (in particular, from the origin) by a trajectory completely belonging to it. In other words, given a subspace $\mathcal{E} \subseteq \mathcal{X}$, by leaving the origin with trajectories belonging to $\mathcal{E}$, hence to $\mathcal{V}^*$ (the maximal (A,B)-controlled invariant contained in $\mathcal{E}$), it is not possible to reach any point of $\mathcal{V}^*$, but only a subspace of $\mathcal{V}^*$, which is called the *reachable set on* $\mathcal{E}$ (or on $\mathcal{V}^*$) and denoted by $\mathcal{R}_\mathcal{E}$ (or $\mathcal{R}_{\mathcal{V}^*}$). The following theorem holds.

Theorem 4.1-6.
$\mathcal{R}_\mathcal{E}$, the reachable set on $\mathcal{E}$, coincides with the minimal (A,B)-controlled invariant self-bounded with respect to $\mathcal{E}$.

Proof. Consider a state feedback F such that $(A+BF)\mathcal{V}^* \subseteq \mathcal{V}^*$. The set of all admissible state velocities at a generic state $x \in \mathcal{V}^*$ is

$$T(x) = (A+BF)x + \mathcal{V}^* \cap \mathcal{B}$$

and does not depend on F. In fact, for a different choice of F, denoted here by F_1, it becomes

$$T_1(x) = (A+BF_1)x + \mathcal{V}^* \cap \mathcal{B}$$

with both $(A+BF)x$ and $(A+BF_1)x$ belonging to $\mathcal{V}^*$; by difference, $B(F-F_1)x \in \mathcal{V}^*$. But, because of the premultiplication by B, $B(F-F_1)x \in \mathcal{V}^* \cap \mathcal{B}$, so that $T(x) = T_1(x)$. By virtue of Theorem 3.3-1 it follows that

$$\mathcal{R}_\mathcal{E} = \mathcal{R}_{\mathcal{V}^*} = \min \mathcal{J}(A+BF, \mathcal{V}^* \cap \mathcal{B}) \tag{29}$$

which together with Theorem 4.1-2 and Definition 4.1-3 prove the statement. □

A more elegant expression for $\mathcal{R}_\mathcal{E}$, not depending on matrix F, which is not unique, is

$$\mathcal{R}_\mathcal{E} = \mathcal{R}_{\mathcal{V}^*} = \mathcal{V}^* \cap \mathcal{S}_2^* \quad \text{with} \quad \mathcal{S}_2^* := \min \mathcal{S}(A, \mathcal{E}, \mathcal{B}) \tag{30}$$

which directly derives from Theorem 4.1-4.

By duality, given a subspace $\mathcal{D} \subseteq \mathcal{X}$, it is possible to define the *unobservable set containing* $\mathcal{D}$, in symbols $\mathcal{Q}_\mathcal{D}$, as the maximum unobservability subspace with a dynamic pole-assignable observer in the presence of an unknown forcing action belonging to $\mathcal{D}$ (see Section 4.2 for details on this type of observer). The following is the dual of Theorem 4.1-6.

Theorem 4.1-7.
$\mathcal{Q}_\mathcal{D}$, the unobservable set containing $\mathcal{D}$, coincides with the maximal (A,C)-conditioned invariant self-hidden with respect to $\mathcal{D}$. Let $\mathcal{S}^*$ be the minimal (A,C)-conditioned invariant containing $\mathcal{D}$. The following two expressions for $\mathcal{Q}_\mathcal{D}$ are the duals of (29, 30):

$$\mathcal{Q}_\mathcal{D} = \mathcal{Q}_{\mathcal{S}^*} = \max \mathcal{I}(A+GC, \mathcal{S}^* + \mathcal{C}) \tag{31}$$

where G denotes any matrix such that $(A+GC)\mathcal{S}^* \subseteq \mathcal{S}^*$, and

$$\mathcal{Q}_\mathcal{D} = \mathcal{Q}_{\mathcal{S}^*} = \mathcal{S}^* + \mathcal{V}_2^* \quad \text{with} \quad \mathcal{V}_2^* := \max \mathcal{V}(A, \mathcal{D}, \mathcal{C}) \tag{32}$$

4.1.4 Stabilizability and Complementability

The concepts of internal and external stability and complementability of A-invariants, introduced and discussed in Subsection 3.2.5 referring to the asymptotic behavior of trajectories of linear free dynamic systems, will now be extended to controlled and conditioned invariants. In the particular case of self-bounded controlled and self-hidden conditioned invariants the extension is immediate; in fact, it will be shown that in this case a proper similarity transformation reduces controlled and conditioned invariants to simple invariants.

Definition 4.1-5. Internally Stabilizable Controlled Invariant

An (A,B)-controlled invariant $\mathcal{V}$ is said to be *internally stabilizable* if for any $x(0) \in \mathcal{V}$ there exists at least one admissible trajectory of the pair (A,B) belonging to $\mathcal{V}$ and converging to the origin.

Because of linearity, the sum of any two internally stabilizable controlled invariants is clearly internally stabilizable. Therefore, the set of all internally stabilizable controlled invariants, possibly constrained to be contained in a given subspace $\mathcal{E}$ and to contain a given subspace $\mathcal{D} \subseteq \mathcal{V}^*$, is an upper semilattice with respect to $\subseteq, +$.

As in the case of simple invariants, the internal stabilizability of a controlled invariant $\mathcal{V}$ will be checked by means of a simple change of basis. Consider $\mathcal{R}_\mathcal{V}$, the reachable set on $\mathcal{V}$, which can be expressed as $\mathcal{R}_\mathcal{V} = \mathcal{V} \cap \mathcal{S}'$, with

$$\mathcal{S}' := \min \mathcal{S}(A, \mathcal{V}, \mathcal{B}) \tag{33}$$

and perform suitable changes of basis in the state and input spaces: to this end, define the similarity transformations $T := [T_1\ T_2\ T_3\ T_4]$, with $\mathrm{im}\,T_1 = \mathcal{R}_\mathcal{V}$, $\mathrm{im}[T_1\ T_2] = \mathcal{V}$, $\mathrm{im}[T_1\ T_3] = \mathcal{S}'$, and $U := [U_1\ U_2]$, with $\mathrm{im}(BU_1) = \mathcal{V} \cap \mathcal{B}$, $\mathrm{im}(BU) = \mathcal{B}$.

Matrices $A' := T^{-1}AT$ and $B' := T^{-1}BU$, corresponding to A and B in the new bases and accordingly partitioned, have the structures

$$A' = \begin{bmatrix} A'_{11} & A'_{12} & A'_{13} & A'_{14} \\ O & A'_{22} & A'_{23} & A'_{24} \\ A'_{31} & A'_{32} & A'_{33} & A'_{34} \\ O & O & A'_{43} & A'_{44} \end{bmatrix} \qquad B' = \begin{bmatrix} B'_{11} & O \\ O & O \\ O & B'_{32} \\ O & O \end{bmatrix} \tag{34}$$

The structure of B' depends on $\mathcal{B}$ being contained in $\mathcal{S}'$. The first submatrix in the second row of A' is zero because of the particular structure of B' and because $\mathcal{R}_\mathcal{V}$ is an $(A+BF)$-invariant for all F such that $(A+BF)\mathcal{V} \subseteq \mathcal{V}$. Also the zero submatrices in the fourth row are due to the invariance of $\mathcal{V}$ with respect to $A+BF$.

Let $r := \dim \mathcal{R}_\mathcal{V}$, $k := \dim \mathcal{V}$. Denote by $z := T^{-1}x$ and $\alpha := U^{-1}u$ the state and the control in the new bases, accordingly partitioned. For all initial states on $\mathcal{V}$ (so that $z_3(0) = z_4(0) = 0$), at every instant of time it is possible to maintain $\dot{z}_3(t) = \dot{z}_4(t) = 0$ by means of a suitable control action $\alpha_2(t)$. Different choices of $\alpha_2(t)$ clearly do not influence the set of all admissible velocities on $\mathcal{V}$, which can be influenced only by $\alpha_1(t)$. Since (A', B') is controllable, it is possible to obtain a trajectory on $\mathcal{V}$ converging to the origin if and only if A'_{22} is stable. The $k-r$ eigenvalues of this matrix do not depend on the particular basis since both stability and controllability are coordinate-free properties. They will be called the *unassignable internal eigenvalues* of $\mathcal{V}$ and clearly coincide with the elements of $\sigma((A+BF)|_{\mathcal{V}/\mathcal{R}_\mathcal{V}})$, where F is any matrix such that $(A+BF)\mathcal{V} \subseteq \mathcal{V}$. This leads to the following property.

Property 4.1-11.
A controlled invariant $\mathcal{V}$ is internally stabilizable if and only if all its unassignable internal eigenvalues are stable.

In the literature, internal stability of controlled invariants is often defined referring to state feedback. The following property makes the two definitions equivalent.

Property 4.1-12.
A controlled invariant $\mathcal{V}$ is internally stabilizable if and only if there exists at least one real matrix F such that $(A+BF)\mathcal{V} \subseteq \mathcal{V}$ with $(A+BF)|_\mathcal{V}$ stable.

Proof. Matrix F, expressed in the same basis as (34) and accordingly partitioned, is

$$F' := U^{-1}FT = \begin{bmatrix} F'_{11} & F'_{12} & F'_{13} & F'_{14} \\ F'_{21} & F'_{22} & F'_{23} & F'_{24} \end{bmatrix} \quad (35)$$

The elements of the first and third row of $A' + B'F'$ are $A'_{1i} + B'_{11}F'_{1i}$ and $A'_{3i} + B'_{32}F'_{2i}$ ($i = 1, \ldots, 4$), while those of the second and fourth row are equal to the corresponding elements of A'. Since (A'_{11}, B'_{11}) is controllable, F'_{11} can be chosen such that $A'_{11} + B'_{11}F'_{11}$ is stable. Furthermore, $\mathcal{V}$ being a controlled invariant, F'_{21} and F'_{22} can be chosen such that the equalities $B'_{31}F'_{21} = -A'_{31}$ and $B'_{32}F'_{22} = -A'_{32}$ hold. The F' obtained is such that $A' + B'F'$ is stable if and only if A'_{22} is stable. $\square$

A similar approach is used for external stabilizability of controlled invariants.

Definition 4.1-6. Externally Stabilizable Controlled Invariant
An $(A, \mathcal{B})$-controlled invariant $\mathcal{V}$ is said to be *externally stabilizable* if for any $x(0) \in \mathcal{X}$ there exists at least one admissible trajectory of the pair (A, B) converging to $\mathcal{V}$.

Property 4.1-13.
Denote with $\mathcal{R}$ the reachable set of pair (A, B). A controlled invariant $\mathcal{V}$ is externally stabilizable if and only if subspace $\mathcal{V} + \mathcal{R}$, which is an A-invariant, is externally stable, i.e., if and only if $A|_{\mathcal{X}/(\mathcal{V}+\mathcal{R})}$ is stable.

Proof. Only if. Perform the changes of basis in the state and input spaces corresponding to the similarity transformations $T := [T_1 \; T_2 \; T_3]$, with $\text{im} T_1 = \mathcal{V}$,

Sect. 4.1 Controlled and Conditioned Invariants

$\text{im}[T_1 \ T_2] = \mathcal{V} + \mathcal{R}$, and $U := [U_1 \ U_2]$, with $\text{im}(BU_1) = \mathcal{V} \cap \mathcal{B}$, $\text{im}(BU) = \mathcal{B}$.
Matrices $A' := T^{-1}AT$ and $B' := T^{-1}BU$ can be accordingly partitioned as

$$A' = \begin{bmatrix} A'_{11} & A'_{12} & A'_{13} \\ A'_{21} & A'_{22} & A'_{23} \\ O & O & A'_{33} \end{bmatrix} \quad B' = \begin{bmatrix} B'_{11} & O \\ O & B'_{22} \\ O & O \end{bmatrix} \quad (36)$$

The structure of B' depends on $\mathcal{B}$ being contained in $\mathcal{R}$, while the first two submatrices in the third row of A' are zero because $\mathcal{R}$ is an A-invariant. If A'_{33} were not stable, there would be noncontrollable trajectories external to $\mathcal{V}$ and not converging to $\mathcal{V}$.

If. Consider a state feedback matrix F, which in the new basis can be partitioned as

$$F' := U^{-1}FT = \begin{bmatrix} F'_{11} & F'_{12} & F'_{13} \\ F'_{21} & F'_{22} & F'_{23} \end{bmatrix} \quad (37)$$

It is possible to choose F'_{21} such that $B'_{22}F'_{21} = -A'_{21}$: since this particular assumption, like any state feedback, does not influence controllability, pair (A'_{22}, B'_{22}) must be controllable, so that $A'_{22} + B'_{22}F'_{22}$ has the eigenvalues completely assignable by a suitable choice of F'_{22}. □

It follows that the sum of two controlled invariants is externally stabilizable if any one of them is. The preceding argument also proves the following property.

Property 4.1-14.
A controlled invariant $\mathcal{V}$ is externally stabilizable if and only if there exists at least one real matrix F such that $(A+BF)\mathcal{V} \subseteq \mathcal{V}$ with $(A+BF)|_{\mathcal{X}/\mathcal{V}}$ stable.

Since the changes of basis introduced in the proofs of Properties 4.1-12 and 4.1-13 are congruent (in the sense that they could coexist in a finer partition of the basis vectors) and correspond to the same partition of the forcing action, it can easily be checked that internal and external stabilization of a controlled invariant are independent of each other. Thus, the following statement holds.

Property 4.1-15.
A controlled invariant $\mathcal{V}$ is both internally and externally stabilizable if and only if there exists at least one real matrix F such that $(A+BF)\mathcal{V} \subseteq \mathcal{V}$ with $A+BF$ stable.

External stabilizability of controlled invariants is often tacitly assumed, it being assured under general conditions on the controlled system referred to. Regarding this, consider the following property.

Property 4.1-16.
If pair (A, B) is stabilizable, all $(A, \mathcal{B})$-controlled invariants are externally stabilizable.

Proof. We recall that (A, B) is stabilizable if $\mathcal{R}$, which is an A-invariant, is externally stable. Let $\mathcal{V}$ be any $(A, \mathcal{B})$-controlled invariant: all the more reason for $\mathcal{V} + \mathcal{R}$, which is an A-invariant containing $\mathcal{R}$, being externally stable. □

All the previous definitions and properties can be extended to conditioned invariants by duality. For the sake of simplicity, instead of Definitions 4.1-5 and 4.1-6, which refer to state trajectories and are not directly dualizable, we shall assume as definitions the duals of Properties 4.1-12 and 4.1-13.

Definition 4.1-7. Externally Stabilizable Conditioned Invariant
A conditioned invariant S is said to be *externally stabilizable* if there exists at least one real matrix G such that $(A+GC)S \subseteq S$ with $(A+GC)|_{\mathcal{X}/S}$ stable.

The intersection of any two externally stabilizable conditioned invariants is an externally stabilizable conditioned invariant. Therefore, the set of all externally stabilizable conditioned invariants, possibly constrained to be contained in a given subspace $\mathcal{E} \supseteq S^*$ and to contain a given subspace $\mathcal{D}$, is a lower semilattice with respect to $\subseteq, \cap$.

Any (A, C)-conditioned invariant S is externally stabilizable if and only if $S^\perp$ is internally stabilizable as an $(A^T, C^\perp)$-controlled invariant.

The *unassignable external eigenvalues* of S can be defined by referring to a change of basis for matrices (A, C) dual to (34). They are the elements of $\sigma((A+GC)|_{\mathcal{Q}_S/S})$, with G being any matrix such that $(A+GC)S \subseteq S$ and coincide with the unassignable internal eigenvalues of $S^\perp$ as an $(A^T, C^\perp)$-controlled invariant. A conditioned invariant is externally stabilizable if and only if all its unassignable external eigenvalues are stable.

Definition 4.1-8. Internally Stabilizable Conditioned Invariant
A conditioned invariant S is said to be *internally stabilizable* if there exists at least one real matrix G such that $(A+GC)S \subseteq S$ with $(A+BF)|_S$ stable.

Property 4.1-17.
Denote with $\mathcal{Q}$ the unobservable set of pair (A, C). A conditioned invariant S is internally stabilizable if and only if subspace $S \cap \mathcal{Q}$, which is an A-invariant, is internally stable, i.e., if and only if $A|_{S \cap \mathcal{Q}}$ is stable.

It follows that the intersection of two conditioned invariants is internally stabilizable if any one of them is.

Property 4.1-18.
A conditioned invariant S is both internally and externally stabilizable if and only if there exists at least one real matrix G such that $(A+GC)S \subseteq S$ with $A+GC$ stable.

Property 4.1-19.
If pair (A, C) is detectable, all (A, C)-conditioned invariants are internally stabilizable.

Let $\mathcal{V}$ be an (A, B)-controlled invariant. Figure 4-2(a) specifies the eigenvalues assignability of $A+BF$ subject to the constraint $(A+BF)\mathcal{V} \subseteq \mathcal{V}$. For instance, spectrum $\sigma((A+BF)|_{\mathcal{X}/(\mathcal{V}+\mathcal{R})})$ is fixed, while $\sigma((A+BF)|_{(\mathcal{V}+\mathcal{R})/\mathcal{V}})$ is assignable, and so on. Figure 4-2(b) concerns the similar diagram for matrix

Sect. 4.1 Controlled and Conditioned Invariants

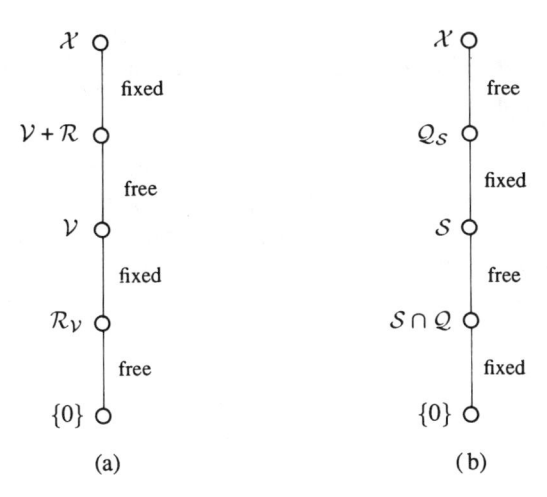

```
   𝒳 ○                              𝒳 ○
       │ fixed                          │ free
 𝒱+ℛ ○                             𝒬_S ○
       │ free                           │ fixed
    𝒱 ○                             𝒮 ○
       │ fixed                          │ free
   ℛ_𝒱 ○                          𝒮∩𝒬 ○
       │ free                           │ fixed
   {0} ○                           {0} ○

    (a)                              (b)
```

Figure 4-2 Assignability of the spectrum of $A+BF$ in connection with the controlled invariant $\mathcal{V}$ and of $A+GC$ in connection with the conditioned invariant $\mathcal{S}$.

$A+GC$ such that $(A+GC)\mathcal{S} \subseteq \mathcal{S}$, where $\mathcal{S}$ is an $(A,\mathcal{C})$-conditioned invariant.

Self-bounded controlled and self-hidden conditioned invariants have particular stabilizability features. Refer to triple $(A,\mathcal{B},\mathcal{C})$ and consider the *fundamental lattices*

$$\Phi(\mathcal{B},\mathcal{C}) := \{\mathcal{V} : A\mathcal{V} \subseteq \mathcal{V}+\mathcal{B}, \mathcal{V} \subseteq \mathcal{C}, \mathcal{V} \supseteq \mathcal{V}_0^* \cap \mathcal{B}\} \tag{38}$$

$$\Psi(\mathcal{C},\mathcal{B}) := \{\mathcal{S} : A(\mathcal{S} \cap \mathcal{C}) \subseteq \mathcal{S}, \mathcal{S} \supseteq \mathcal{B}, \mathcal{S} \subseteq \mathcal{S}_0^* + \mathcal{C}\} \tag{39}$$

with

$$\mathcal{V}_0^* := \max \mathcal{V}(A,\mathcal{B},\mathcal{C}) \tag{40}$$

$$\mathcal{S}_0^* := \min \mathcal{S}(A,\mathcal{C},\mathcal{B}) \tag{41}$$

Structure and stabilizability properties of these lattices will be pointed out through a change of basis. Consider the similarity transformation $T := [T_1\ T_2\ T_3\ T_4]$, with $\mathrm{im}\,T_1 = \mathcal{V}_0^* \cap \mathcal{S}_0^*$, $\mathrm{im}[T_1\ T_2] = \mathcal{V}_0^*$, $\mathrm{im}[T_1\ T_3] = \mathcal{S}_0^*$. Matrices $A' := T^{-1}AT$, $B' := T^{-1}B$ and $C' := CT$ have the structures

$$A' = \begin{bmatrix} A'_{11} & A'_{12} & A'_{13} & A'_{14} \\ O & A'_{22} & A'_{23} & A'_{24} \\ A'_{31} & A'_{32} & A'_{33} & A'_{34} \\ O & O & A'_{43} & A'_{44} \end{bmatrix} \quad B' = \begin{bmatrix} B'_1 \\ O \\ B'_3 \\ O \end{bmatrix} \tag{42}$$

$$C' = [O\ \ O\ \ C'_3\ \ C'_4]$$

The zero submatrices in B' and C' depend on $\mathcal{V}_0^*$ containing $\mathcal{B}$, and $\mathcal{V}_0^*$ being contained in $\mathcal{C}$, while the zero submatrix in the second row of A' is due to the particular structure of B' and to $\mathcal{V}_0^* \cap \mathcal{S}_0^*$ being a controlled invariant (it is the reachable set on $\mathcal{V}_0^*$), those in the fourth row to the whole $\mathcal{V}_0^*$ being a controlled invariant.

Furthermore, $\mathcal{V}_0^*$ being a controlled invariant, submatrices A'_{31} and A'_{32} can be zeroed by means of a suitable state feedback F'; similarly, $\mathcal{S}_0^*$ being a conditioned invariant, A'_{23} and A'_{43} can be zeroed by means of a suitable output injection G'. Clearly these feedbacks cause $\mathcal{V}_0^*$ to be an $(A+BF)$-invariant, with $F := F'T$ and $\mathcal{S}_0^*$ an $(A+GC)$-invariant, with $G := T^{-1}G'$. It is known (Properties 4.1-7 and 4.1-9) that these feedbacks transform any element of $\Phi(\mathcal{B}, \mathcal{C})$ into an $(A+BF)$-invariant and any element of $\Psi(\mathcal{C}, \mathcal{B})$ into an $(A+GC)$-invariant.

The unassignable internal eigenvalues of $\mathcal{V}_0^*$ are those of A'_{22}: since they clearly coincide with the unassignable external eigenvalues of $\mathcal{S}_0^*$, the following property holds.

Property 4.1-20.
$\mathcal{V}_0^*$ is internally stabilizable if and only if $\mathcal{S}_0^*$ is externally stabilizable.

The preceding argument reveals the existence of two interesting one-to-one correspondences between the elements of lattices $\Phi(\mathcal{B}, \mathcal{C})$ and $\Psi(\mathcal{C}, \mathcal{B})$ and the invariants of the linear transformation corresponding to A'_{22} (which, as remarked, expresses $(A+BF)|_{\mathcal{V}_0^*/(\mathcal{V}_0^* \cap \mathcal{S}_0^*)}$ or $(A+GC)|_{(\mathcal{V}_0^* + \mathcal{S}_0^*)/\mathcal{S}_0^*}$). More precisely, the two one-to-one correspondences are set as follows: let $r := \dim(\mathcal{V}_0^* \cap \mathcal{S}_0^*)$, $k := \dim \mathcal{S}_0^*$, and X' be a basis matrix of a generic A'_{22}-invariant. Subspaces

$$\mathcal{V} := \operatorname{im}\left(T \begin{bmatrix} I_r & O \\ O & X' \\ O & O \\ O & O \end{bmatrix}\right) \qquad \mathcal{S} := \operatorname{im}\left(T \begin{bmatrix} I_r & O & O \\ O & X' & O \\ O & O & I_{k-r} \\ O & O & O \end{bmatrix}\right) \qquad (43)$$

are generic elements of $\Phi(\mathcal{B}, \mathcal{C})$ and $\Psi(\mathcal{C}, \mathcal{B})$ respectively.

We shall now consider extension of the concept of complementability, introduced for simple invariants in Subsection 3.2-5, to controlled and conditioned invariants.

Definition 4.1-9. Complementable Controlled Invariant
Let $\mathcal{V}$, $\mathcal{V}_1$, and $\mathcal{V}_2$ be three controlled invariants such that $\mathcal{V}_1 \subseteq \mathcal{V} \subseteq \mathcal{V}_2$. $\mathcal{V}$ is said to be *complementable with respect to* $(\mathcal{V}_1, \mathcal{V}_2)$ if there exists at least one controlled invariant $\mathcal{V}_c$ such that

$$\mathcal{V} \cap \mathcal{V}_c = \mathcal{V}_1$$
$$\mathcal{V} + \mathcal{V}_c = \mathcal{V}_2$$

Definition 4.1-10. Complementable Conditioned Invariant
Let $\mathcal{S}$, $\mathcal{S}_1$, and $\mathcal{S}_2$ be three conditioned invariants such that $\mathcal{S}_1 \subseteq \mathcal{S} \subseteq \mathcal{S}_2$. $\mathcal{S}$ is said to be *complementable with respect to* $(\mathcal{S}_1, \mathcal{S}_2)$ if there exists at least one conditioned invariant $\mathcal{S}_c$ such that

$$\mathcal{S} \cap \mathcal{S}_c = \mathcal{S}_1$$
$$\mathcal{S} + \mathcal{S}_c = \mathcal{S}_2$$

In the particular case of self-bounded controlled and self-hidden conditioned invariants, the complementability condition can still be checked by means of the Sylvester equation. In fact, they correspond to simple A'_{22}-invariants in structure (42).

The Sylvester equation can also be used in the general case. It is worth noting that the complementability condition can be influenced by the feedback matrices which transform controlled and conditioned invariants into simple $(A+BF)$-invariants or $(A+GC)$-invariants.

The one-to-one correspondences between the elements of suitable lattices of self-bounded controlled and self-hidden conditioned invariants and the invariants of related linear transformations are the basis to derive constructive solutions in the framework of the geometric approach, for the most important compensator and regulator synthesis problems.

4.2 DISTURBANCE LOCALIZATION AND UNKNOWN-INPUT STATE ESTIMATION

The disturbance localization problem is one of the first examples of synthesis through the geometric approach.[1]

It is presented in this chapter, which concerns analysis problems, because it is very elementary and completes, by introducing a well-defined structural constraint, the pole assignability problem with state feedback, considered in Section 3.4. Furthermore, it can be considered a basic preliminary approach to numerous more sophisticated regulation problems.

Consider the system

$$\dot{x}(t) = A\,x(t) + B\,u(t) + D\,d(t) \qquad (1)$$
$$e(t) = E\,x(t) \qquad (2)$$

where u denotes the manipulable input, d the nonmanipulable input, which at the moment is assumed to be also completely unaccessible for measurement,

---

[1] See Basile and Marro [6], and Wonham and Morse [44].

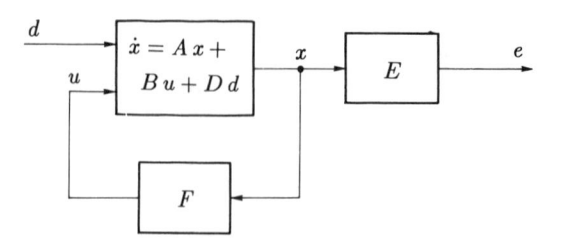

Figure 4-3 The unaccessible disturbance localization problem.

and set the problem of realizing, if possible, a state feedback of the type shown in Fig. 4-3 such that, starting at the zero state, $e(\cdot)=0$ results for all admissible $d(\cdot)$. This is called the *disturbance localization problem* or, more precisely, the *unaccessible disturbance localization problem*. The system with state feedback is described by

$$\dot{x}(t) = (A + B F)\, x(t) + D\, d(t) \tag{3}$$
$$e(t) = E\, x(t) \tag{4}$$

and presents the requested behavior if and only if its reachable set by d, i.e., the minimal $(A+BF)$-invariant containing $\mathcal{D} := \mathrm{im}\, D$, is contained in $\mathcal{E} := \ker E$. Since, by virtue of Theorem 4.1-2, any $(A+BF)$-invariant is an $(A,\mathcal{B})$-controlled invariant, the unaccessible disturbance localization problem admits a solution if and only if the following *structural condition* holds:

$$\mathcal{D} \subseteq \mathcal{V}^* \tag{5}$$

where $\mathcal{V}^* := \max \mathcal{V}(A, \mathcal{B}, \mathcal{E})$ is the same as (4.1.13)

Checking disturbance localization feasibility for a system whose matrices A, B, D, E are known, reduces to few subspace computations: determination of $\mathcal{V}^*$ by means of Algorithm 4.1-2 and checking (5) by using the algorithms described at the end of Subsection 3.1.1 and implemented in Appendix B. For instance, a simple dimensionality check on basis matrices can prove the equalities $\mathcal{V}^* + \mathcal{D} = \mathcal{V}^*$ and $\mathcal{V}^* \cap \mathcal{D} = \mathcal{D}$, clearly equivalent to (5).

A matrix F which makes $\mathcal{V}^*$ an $(A+BF)$-invariant can be determined by means of the algorithm described in Subsection 4.1.1.

On the other hand it is worth noting that:

1. state-to-input feedback in practice is not feasible since in most cases state is not completely accessible for measurement;
2. for the problem to be technically sound it is also necessary to impose the stability requirement, i.e., that matrix F, besides disturbance localization, achieves stability of the overall system matrix $A+BF$.

Point 1 will be overcome in the next chapter, where the more general problem of disturbance localization by dynamic output-to-input feedback will

Sect. 4.2 Disturbance Localization and Unknown-Input State Estimation

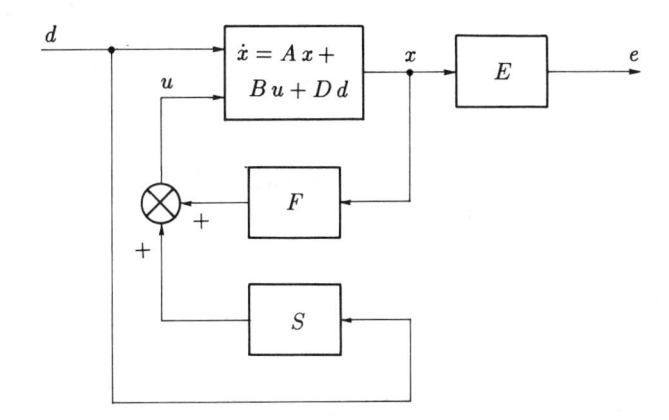

Figure 4-4 The accessible disturbance localization problem.

be considered. Point 2 leads to the unaccessible disturbance localization problem *with stability*, which will be solved later.

We now consider the disturbance localization problem with d accessible: our aim is to make e insensitive to disturbance d by using a linear algebraic regulator that determines control u as a function of state x and disturbance d itself, as shown in Fig. 4-4. In this case (5) is replaced by the less restrictive structural condition

$$\mathcal{D} \subseteq \mathcal{V}^* + \mathcal{B} \tag{6}$$

which implies the existence of an $(A, \mathcal{B})$-controlled invariant $\mathcal{V}$ contained in $\mathcal{E}$ and such that $\mathcal{D} \subseteq \mathcal{V} + \mathcal{B}$ (nonconstructive necessary and sufficient structural condition). Let V be a basis matrix of $\mathcal{V}$; the linear algebraic equation

$$V\alpha - Bu = Dd \tag{7}$$

admits at least one solution in α, u for all d. Express a solution with respect to u as $u = Sd$: it is clear that the total forcing action due to the disturbance, which is $(D+BS)d$ belongs to $\mathcal{V}$, and its effect can be maintained on $\mathcal{V}$, hence on $\mathcal{E}$, by a state feedback F such that $(A+BF)\mathcal{V} \subseteq \mathcal{V}$.

In order to take into account the stability requirement, consider the lattice $\Phi(\mathcal{B}+\mathcal{D}, \mathcal{E})$ of all $(A, \mathcal{B}+\mathcal{D})$-controlled invariants self-bounded with respect to $\mathcal{E}$. The following properties hold.

Property 4.2-1.
Let $\mathcal{V}^* := \max \mathcal{V}(A, \mathcal{B}, \mathcal{E})$. If $\mathcal{D} \subseteq \mathcal{V}^*$ or $\mathcal{D} \subseteq \mathcal{V}^* + \mathcal{B}$ subspace $\max \mathcal{V}(A, \mathcal{B}+\mathcal{D}, \mathcal{E})$ coincides with $\mathcal{V}^*$.

Proof. Apply Algorithm 4.1-2 with $\mathcal{B}+\mathcal{D}$ instead of $\mathcal{B}$ and note that the inclusion $\mathcal{D} \subseteq \mathcal{Z}_i + \mathcal{B}$ holds for all terms of the sequence, which clearly does not change if $\mathcal{B}$ is replaced with $\mathcal{B}+\mathcal{D}$. □

Property 4.2-2.
If $\mathcal{D}\subseteq \mathcal{V}^*$ ($\mathcal{D}\subseteq \mathcal{V}^*+\mathcal{B}$) any element of $\Phi(\mathcal{B}+\mathcal{D},\mathcal{E})$ satisfies $\mathcal{D}\subseteq \mathcal{V}$ ($\mathcal{D}\subseteq \mathcal{V}+\mathcal{B}$).

Proof. By the self-boundedness property $\mathcal{V}^*\cap(\mathcal{B}+\mathcal{D})\subseteq \mathcal{V}$. If $\mathcal{D}\subseteq \mathcal{V}^*$, the intersection is distributive with respect to the sum, so that $\mathcal{V}^*\cap\mathcal{B}+\mathcal{D}\subseteq \mathcal{V}$, hence $\mathcal{D}\subseteq \mathcal{V}$. If $\mathcal{D}\subseteq \mathcal{V}^*+\mathcal{B}$, add $\mathcal{B}$ to both members, thus obtaining $(\mathcal{V}^*+\mathcal{B})\cap(\mathcal{B}+\mathcal{D})\subseteq \mathcal{V}+\mathcal{B}$ and note that $\mathcal{D}$ is contained in both terms of the intersection on the left. $\square$

Denote by

$$\mathcal{V}_m := \mathcal{V}^* \cap \mathcal{S}_1^* \quad \text{with} \quad \mathcal{S}_1^* := \min \mathcal{S}(A,\mathcal{E},\mathcal{B}+\mathcal{D}) \tag{8}$$

the infimum of $\Phi(\mathcal{B}+\mathcal{D},\mathcal{E})$. By Property 4.2-2 it satisfies $\mathcal{D}\subseteq \mathcal{V}_m$ if (5) holds or $\mathcal{D}\subseteq \mathcal{V}_m+\mathcal{B}$ if (6) holds. The following lemma is basic to derive a constructive solution to numerous problems with stability.[2]

Lemma 4.2-1.
Let $\mathcal{D}\subseteq \mathcal{V}^*$ ($\mathcal{D}\subseteq \mathcal{V}^*+\mathcal{B}$). If $\mathcal{V}_m$, the minimal $(A,\mathcal{B}+\mathcal{D})$-controlled invariant self-bounded with respect to $\mathcal{E}$ is not internally stabilizable, no internally stabilizable $(A,\mathcal{B})$-controlled invariant $\mathcal{V}$ exists that satisfies both $\mathcal{V}\subseteq \mathcal{E}$ and $\mathcal{D}\subseteq \mathcal{V}$ ($\mathcal{D}\subseteq \mathcal{V}+\mathcal{B}$).

Proof. Let $\mathcal{V}$ be any $(A,\mathcal{B})$-controlled invariant satisfying all requirements in the statement. Consider the subspace

$$\bar{\mathcal{V}} := \mathcal{V} + \mathcal{R}_{\mathcal{V}^*} \tag{9}$$

which is a controlled invariant as the sum of two controlled invariants and satisfies the inclusions $\mathcal{D}\subseteq \bar{\mathcal{V}}$ ($\mathcal{D}\subseteq \bar{\mathcal{V}}+\mathcal{B}$) and $\mathcal{V}^*\cap\mathcal{B}\subseteq \bar{\mathcal{V}}$ since $\mathcal{D}\subseteq \mathcal{V}$ ($\mathcal{D}\subseteq \mathcal{V}+\mathcal{B}$) and $\mathcal{V}^*\cap\mathcal{B}\subseteq \mathcal{R}_{\mathcal{V}^*}$; by summing $\mathcal{B}$ to both members of the former inclusion we obtain $\mathcal{B}+\mathcal{D}\subseteq \bar{\mathcal{V}}+\mathcal{B}$. By intersecting with $\mathcal{V}^*$ it follows that $\mathcal{V}^*\cap(\mathcal{B}+\mathcal{D})\subseteq \bar{\mathcal{V}}$, hence $\bar{\mathcal{V}}\in \Phi(\mathcal{B}+\mathcal{D},\mathcal{E})$. Furthermore, $\bar{\mathcal{V}}$ is internally stabilizable, being the sum of two internally stabilizable controlled invariants (in particular, the internal eigenvalues of $\mathcal{R}_{\mathcal{V}^*}$ are actually all assignable). Then, there exists an F such that $\bar{\mathcal{V}}$ is an internally stable $(A+BF)$-invariant: all the elements of $\Phi(\mathcal{B}+\mathcal{D},\mathcal{E})$ contained in $\bar{\mathcal{V}}$, in particular $\mathcal{V}_m$, are internally stable $(A+BF)$-invariants, hence internally stabilizable $(A,\mathcal{B})$-controlled invariants. $\square$

We also state, obviously without proof, the dual lemma. Refer to lattice $\Psi(\mathcal{C}\cap\mathcal{E},\mathcal{D})$, whose infimum is $\mathcal{S}^* := \min \mathcal{S}(A,\mathcal{C},\mathcal{B})$, provided that $\mathcal{E}\supseteq \mathcal{S}^*$ or $\mathcal{E}\supseteq \mathcal{S}^*\cap\mathcal{C}$, and whose supremum is

$$\mathcal{S}_M := \mathcal{S}^* + \mathcal{V}_1^* \quad \text{with} \quad \mathcal{V}_1^* := \max \mathcal{V}(A,\mathcal{D},\mathcal{C}\cap\mathcal{E}) \tag{10}$$

If one of the preceding inclusions regarding $\mathcal{S}^*$ holds, any element $\mathcal{S}$ of $\Psi(\mathcal{C}\cap\mathcal{E},\mathcal{D})$ satisfies the similar inclusion $\mathcal{E}\supseteq \mathcal{S}$ or $\mathcal{E}\supseteq \mathcal{S}\cap\mathcal{C}$.

---

[2] See Basile and Marro [9] and Schumacher [39].

Lemma 4.2-2.
Let $\mathcal{E} \supseteq \mathcal{S}^*$ ($\mathcal{E} \supseteq \mathcal{S}^* \cap \mathcal{C}$). If $\mathcal{S}_M$, the maximal $(A, \mathcal{C} \cap \mathcal{E})$-conditioned invariant self-hidden with respect to $\mathcal{D}$, is not externally stabilizable, no externally stabilizable $(A, \mathcal{C})$-conditioned invariant $\mathcal{V}$ exists that satisfies both $\mathcal{D} \supseteq \mathcal{S}$ and $\mathcal{E} \supseteq \mathcal{S}$ ($\mathcal{E} \supseteq \mathcal{S} \cap \mathcal{C}$).

These results are basic to solving both the unaccessible and accessible disturbance localization problem with stability.

Theorem 4.2-1. Unaccessible Disturbance Localization
Consider system (1,2) and assume that (A, B) is stabilizable. The unaccessible disturbance localization problem with stability has a solution if and only if

1. $\mathcal{D} \subseteq \mathcal{V}^*$; (11)
2. $\mathcal{V}_m$ is internally stabilizable. (12)

Proof. *Only if.* Suppose that the problem has a solution, so that there exists an F such that $A+BF$ is stable and $\mathcal{V} := \min \mathcal{J}(A+BF, \mathcal{D})$ is contained in $\mathcal{E}$. Hence $\mathcal{V}$ is an (A, B)-controlled invariant both internally and externally stabilizable. Condition (9) follows from $\mathcal{V}$ being contained in $\mathcal{E}$ and containing $\mathcal{D}$. In turn, (9) implies that the supremum of $\Phi(B+\mathcal{D}, \mathcal{E})$ is $\mathcal{V}^*$ and all elements contain $\mathcal{D}$ (Properties 4.2-1 and 4.2-2). External stabilizability of $\mathcal{V}$ does not correspond to any particular condition, since stabilizability of (A, B) involves external stabilizability of all controlled invariants (Property 4.1-16). Internal stabilizability implies (10) by virtue of Lemma 4.2-1.

If. If (9,10) hold, there exists an F such that $A+BF$ is stable and $(A+BF)\mathcal{V}_m \subseteq \mathcal{V}_m$, so that the problem admits a solution. □

Note that the necessary and sufficient conditions stated in Theorem 4.2-1 are *constructive*, in the sense that they provide a procedure to solve the problem. In general, in the geometric approach to synthesis problems it is possible to state nonconstructive necessary and sufficient conditions, simple and intuitive, and constructive conditions, more involved, but easily checkable with standard algorithms. For the problem considered here, the nonconstructive structural condition consists simply of the existence of a controlled invariant contained in $\mathcal{E}$ and containing $\mathcal{D}$, while the condition with stability requires moreover that this controlled invariant is internally stabilizable. The structural constructive condition is expressed by (5), that with stability by (9,10).

For the accessible disturbance localization problem with stability, the nonconstructive condition differs from the structural one only in the requirement that $\mathcal{V}$ is internally stabilizable, while the constructive one is stated as follows.

Theorem 4.2-2. Accessible Disturbance Localization
Consider system (1,2) and assume that (A, B) is stabilizable. The accessible disturbance localization problem with stability has a solution if and only if

1. $\mathcal{D} \subseteq \mathcal{V}^* + \mathcal{B}$; (13)
2. $\mathcal{V}_m$ is internally stabilizable. (14)

Proof. The statement is proved similarly to Theorem 4.2-1, by again using Lemma 4.2-1. □

We shall now consider the dual problem, which is the asymptotic estimation of a linear function of the state (possibly the whole state) in the presence of an unaccessible disturbance input.[3]

Consider the behavior of an identity observer when the observed system has, besides the accessible input u, an unaccessible input d; referring to the manipulations reported in Subsection 3.4.1, subtract (1) from

$$\dot{z}(t) = (A + GC)\, z(t) + B\, u(t) - G\, y(t)$$

thus obtaining the differential equation

$$\dot{\epsilon}(t) = (A + GC)\, \epsilon(t) - D\, d(t) \tag{15}$$

which shows that the estimate error does not converge asymptotically to zero, even if $A+GC$ is a stable matrix, but converges asymptotically to the subspace min $\mathcal{J}(A+GC, \mathcal{D})$, i.e., to the reachable set of system (15). It follows that, in order to obtain the maximal state estimate, it is convenient to choose G to make this subspace of minimal dimension: since it is an $(A, \mathcal{C})$-conditioned invariant by Theorem 4.1-3, the best choice of G corresponds to transforming into an $(A+GC)$-invariant the minimal $(A, \mathcal{C})$-conditioned invariant containing $\mathcal{D}$ (this is the structural requirement, which refers to the possibility of estimating the state if initial states of both system and observer are congruent) or the minimal externally stabilizable $(A, \mathcal{C})$-conditioned invariant containing $\mathcal{D}$ (this is the stability requirement, which guarantees the convergence of estimate to actual state even if the initial states are not congruent). In the latter case the internal stabilizability of the conditioned invariant is implied on the assumption that (A, C) is detectable (Property 4.1-19), which is clearly necessary if a full-order or identity observer is considered.

Let $\mathcal{S}$ be the minimal $(A+GC)$-invariant containing $\mathcal{D}$ and assume that $A+GC$ is stable. The observer provides an asymptotic estimate of the state "modulo" $\mathcal{S}$ or, in more precise terms, an asymptotic estimate of the state canonical projection on $\mathcal{X}/\mathcal{S}$ (similarly, the direct use of output without any dynamic observer would provide knowledge of state modulo $\mathcal{C}$, or its canonical projection on $\mathcal{X}/\mathcal{C}$). This incomplete estimate may be fully satisfactory if, for instance, it is not necessary to know the whole state, but only a given linear function of it: in this case the asymptotic estimate of this function is complete if and only if $\mathcal{S}$ is contained in its kernel.

These arguments lead to the following statement of the problem of asymptotic estimation of a linear function of state in the presence of an unaccessible input: given the time-invariant linear system

$$\dot{x}(t) = A\, x(t) + D\, d(t) \tag{16}$$

$$y(t) = C\, x(t) \tag{17}$$

---

[3] See Marro [26] and Bhattacharyya [13].

determine an identity observer linear system which, by using y as input, provides an asymptotic estimate of the linear function

$$e(t) = E\,x(t) \tag{18}$$

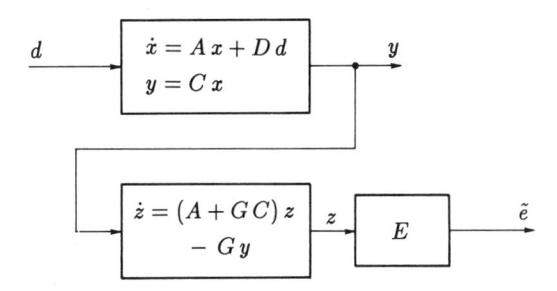

Figure 4-5 Unaccessible input asymptotic state estimation: purely dynamic asymptotic observer.

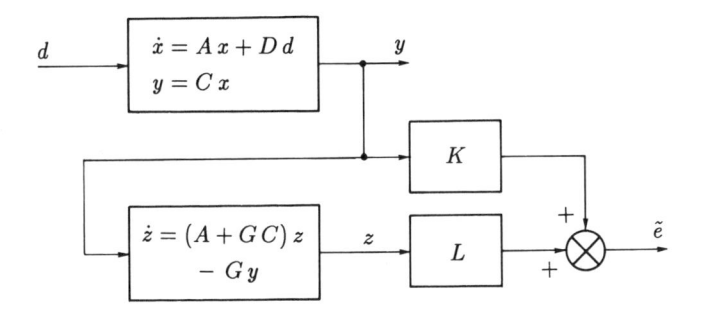

Figure 4-6 Unaccessible input asymptotic state estimation: nonpurely dynamic asymptotic observer.

For the sake of simplicity, accessible input u has not been considered in (16): in fact, if present, it can be applied also to the asymptotic observer. Connections to the system, in cases of both a purely dynamic and a nonpurely dynamic asymptotic observer, are shown respectively in Figs. 4-5 and 4-6.

In conclusion, in geometric terms the synthesis of an asymptotically stable, full-order, purely dynamic state observer reduces to deriving an externally stabilizable (A,C)-conditioned invariant S such that $\mathcal{D} \subseteq \mathcal{S}$ and $\mathcal{S} \subseteq \mathcal{E}$ while, if the estimator is allowed to be nonpurely dynamic, the last condition is replaced by $\mathcal{S} \cap \mathcal{C} \subseteq \mathcal{E}$. Pair (A,C) is assumed to be detectable so $\mathcal{S}$ is internally stabilizable and the full-order observer can be stabilized. However, it is not required if only the state coordinates corresponding to the state canonical projection on $\mathcal{X}/\mathcal{S}$ are reproduced in the observer (see change of basis (4.3.9) in the following Subsection 4.3.1).

To solve the problem we refer to the following results, which can be derived by duality from Theorems 4.2-1 and 4.2-2.

Theorem 4.2-3. Unknown-Input Purely Dynamic Asymptotic Observer

Consider system (16, 17) and assume that (A, C) is detectable. The problem of asymptotically estimating linear function (18) in the presence of the unknown input d with a full-order purely dynamic observer has a solution if and only if

1. $\mathcal{E} \supseteq \mathcal{S}^*$; (19)
2. $\mathcal{S}_M$ is externally stabilizable. (20)

Theorem 4.2-4. Unknown-Input Nonpurely Dynamic Asymptotic Observer

Consider system (16, 17) and assume that (A, C) is detectable. The problem of asymptotically estimating linear function (18) in the presence of the unknown input d with a full-order nonpurely dynamic observer has a solution if and only if

1. $\mathcal{E} \supseteq \mathcal{S}^* \cap \mathcal{C}$; (21)
2. $\mathcal{S}_M$ is externally stabilizable. (22)

In (19, 21) $\mathcal{S}^*$ is the subspace defined by (4.1.24), whereas in (20, 22) $\mathcal{S}_M$ is the subspace defined by (10), which is the maximal $(A, \mathcal{C} \cap \mathcal{E})$-conditioned invariant self-hidden with respect to $\mathcal{D}$ provided (19) or (21) holds.

When $\mathcal{E} = \{0\}$, i.e., when an estimate of the whole state is sought, (20) cannot hold (of course, the trivial case $\mathcal{D} = \{0\}$ is excluded), so a nonpurely dynamic observer must be used. This is possible if the conditions stated in the following corollary, immediately deductable from Theorem 4.2-4, are satisfied.

Corollary 4.2-1.

Consider system (16, 17) and suppose that (A, C) is detectable. The problem of asymptotically estimating the whole state in the presence of the unknown input d has a solution with a nonpurely dynamic observer if and only if

1. $\mathcal{S}^* \cap \mathcal{C} = \{0\}$; (23)
2. $\mathcal{S}^*$ is externally stabilizable. (24)

Computational "recipes" for matrices G, K, L of the observers represented in Figs. 4-5 and 4-6 will be considered in the next chapter (Subsection 5.1.2).

4.3 UNKNOWN-INPUT RECONSTRUCTABILITY, INVERTIBILITY, AND FUNCTIONAL CONTROLLABILITY

The problem of observing the state in the presence of unaccessible inputs by means of a suitable asymptotically stable dynamic system (the estimator or dynamic observer) has been considered in the previous section as a basic application of conditioned invariance. In this section the same problem will be considered in a more extended way and it will be shown that to obtain the maximal information on state in the presence of unaccessible inputs it is necessary to use *differentiators*, so the most general observers are not included in the class of dynamic systems. In other words, the mathematical problem of obtaining final state from input and output functions when some of the inputs are unknown has solvability conditions more extended than the problem of estimating the state through a dynamic observer.

The need to use differentiators, which are linear operators, but not belonging to the class of linear dynamic systems considered in the first two chapters of this book and not susceptible to any ISO representation, is pointed out by a simple example. Consider a dynamic system consisting of n cascaded integrators, with input u to the first and output y from the last: it can be represented by a triple (A, b, c), but its state (which consists of the integrator outputs) can be determined only by means of $n-1$ cascaded differentiators connected to output y. The technique described in the previous section, which uses a dynamic system, clearly cannot be applied in this case.

From a strictly mathematical viewpoint, unknown-input observability is introduced as follows. It is well known that the response of triple (A, B, C) is related to initial state $x(0)$ and control function $u(\cdot)$ by

$$y(t) = C\, e^{At}\, x(0) + C \int_0^t e^{A(t-\tau)}\, B\, u(\tau)\, d\tau \tag{1}$$

where the first term on the right is the free response and the second is the forced response. In order to simplify notation, refer to a fixed time interval $[0, T]$. Hence

$$y|_{[0,T]} = \gamma\big(x(0), u|_{[0,T]}\big) = \gamma_1\big(x(0)\big) + \gamma_2\big(u|_{[0,T]}\big) \tag{2}$$

We recall that (A, C) is *observable* or *reconstructable* (in the continuous-time case these properties are equivalent) if γ_1 is invertible, i.e., $\ker \gamma_1 = \{0\}$. In this case it is possible to derive the initial or the final state from input and output functions. The following definitions extend the reconstructability concept to the case where the input function is unknown and introduce the concept of *system invertibility*, which will be proved to be equivalent to it (Theorem 4.3-1).

Definition 4.3-1.
Triple (A, B, C) is said to be *unknown-state, unknown-input reconstructable* or *unknown-state, unknown-input invertible* if γ is invertible, i.e., $\ker \gamma = \{0\}$.

Definition 4.3-2.
Triple (A, B, C) is said to be *zero-state, unknown-input reconstructable* or *zero-state, unknown-input invertible* if γ_2 is invertible, i.e., $\ker \gamma_2 = \{0\}$.

When (A, C) is not observable or reconstructable, the initial or final state can be determined modulo the subspace

$$\ker \gamma_1 = \mathcal{Q} := \max \mathcal{J}(A,C) \tag{3}$$

which is called the *unobservability subspace* or the *unreconstructability subspace*. This means that the state canonical projection on $\mathcal{X}/\mathcal{Q}$ can be determined from the output function. $\mathcal{Q}$ is the locus of the free motions corresponding to the output function identically zero.

Unknown-input reconstructability in the cases of the above definitions is approached in a similar way: by linearity, when reconstructability is not complete, only the canonical projection of the final state on $\mathcal{X}/\mathcal{Q}_1$ or on $\mathcal{X}/\mathcal{Q}_2$ can be determined, where $\mathcal{Q}_1$ is called the *unknown-state, unknown-input unreconstructability subspace* and $\mathcal{Q}_2$ the *zero-state, unknown-input unreconstructability subspace*. Clearly $\mathcal{Q}_2 \subseteq \mathcal{Q}_1$. Geometric expressions for these subspaces are provided in the following properties.

Property 4.3-1.
Refer to triple (A, B, C). The unknown-state, unknown-input unreconstructability subspace is

$$\mathcal{Q}_1 = \mathcal{V}_0^* := \max \mathcal{V}(A,B,C) \tag{4}$$

Proof. The statement is an immediate consequence of Theorem 4.1-1. □

Property 4.3-2.
Refer to triple (A, B, C). The zero-state, unknown-input unreconstructability subspace is

$$\mathcal{Q}_2 = \mathcal{R}_{\mathcal{V}_0^*} = \mathcal{V}_0^* \cap \mathcal{S}_0^* \quad \text{with} \quad \mathcal{S}_0^* := \min \mathcal{S}(A,C,B) \tag{5}$$

Proof. The statement is an immediate consequence of Theorem 4.1-6. □

4.3.1 A General Unknown-Input Reconstructor

We shall here describe how to implement a general unknown-input state reconstructor. First, we show that the current state is derivable modulo $\mathcal{Q}_1$ by means of an algebraic system with differentiators connected only to the system output. The starting point is the relation

$$y(t) = C\, x(t) \tag{6}$$

which, to emphasize the iterative character of the procedure, is written as

$$q_0(t) = Y_0\, x(t) \tag{7}$$

where $q_0 := y$ is a known continuous function in $[0, T]$ and $Y_0 := C$ is a known constant matrix. The state modulo $\ker Y_0$ can be derived from (7). Differentiating (7) and using the system equation $\dot{x}(t) = A\, x(t) + B\, u(t)$ yields

$$\dot{q}_0(t) - Y_0\, B\, u(t) = Y_0\, A\, x(t)$$

Let P_0 be a projection matrix along $\mathrm{im}(Y_0 B)$, so that $\mathrm{im}(Y_0 B) = \ker P_0$ and

$$P_0 \dot{q}_0(t) = P_0 Y_0 A x(t) \tag{8}$$

Equations (7, 8) can be written together as

$$q_1(t) = Y_1 x(t)$$

where q_1 denotes a known linear function of the output and its first derivative, and

$$Y_1 := \begin{bmatrix} Y_0 \\ P_0 Y_0 A \end{bmatrix}$$

Simple manipulations provide

$$\begin{aligned} \ker Y_1 &= \ker Y_0 \cap \ker(P_0 Y_0 A) \\ &= \ker Y_0 \cap A^{-1} Y_0^{-1} \ker P_0 \\ &= \ker Y_0 \cap A^{-1} Y_0^{-1} Y_0 \, \mathrm{im} B \\ &= \ker Y_0 \cap A^{-1}(\ker Y_0 + \mathrm{im} B) \end{aligned}$$

Iterating k times the procedure yields

$$q_k(t) = Y_k x(t)$$

where q_k denotes a known linear function of the output and its derivatives up to the k-th, and Y_k a matrix such that

$$\begin{aligned} \ker Y_k &= \ker Y_{k-1} \cap A^{-1}(\ker Y_{k-1} + \mathrm{im} B) \\ &= \ker Y_0 \cap A^{-1}(\ker Y_{k-1} + \mathrm{im} B) \end{aligned}$$

where the last equality can be derived with an argument similar to that used in the proof of Algorithm 4.1-1. Sequence $\ker Y_k$ ($k = 0, 1, \ldots$) converges to $\mathcal{V}_0^*$, since it coincides with the sequence provided by Algorithm 4.1-2 to derive $\max \mathcal{V}(A, \mathcal{B}, \mathcal{C})$.

Note that the length of the observation interval $[0, T]$ has not been considered in the preceding argument: since the described technique is based on differentiators, it is only required to be nonzero: functions $q_k(\cdot)$ are continuous, hence differentiable, because each of them is obtained by projecting the previous one along its possible discontinuity directions. Furthermore, from the argument it follows that the maximum order of the involved derivatives is $n-1$.

We shall now prove that a dynamic device exists which, connected to the system output and with initial state suitably set as a linear function of the system state (which is assumed to be known), provides tracking of the system state modulo S_0^*. This device is quite similar to the unknown-input asymptotic estimators considered in the previous section, but it is not necessarily stable. Consider the identity observer shown in Fig. 3-12 and choose matrix G such that $(A+GC)S_0^* \subseteq S_0^*$. The observer equations, expressed in the new basis corresponding to the similarity transformation $T := [T_1 \; T_2]$, with $\text{im} T_1 = S_0^*$, are

$$\begin{bmatrix} \dot{\eta}_1(t) \\ \dot{\eta}_2(t) \end{bmatrix} = \begin{bmatrix} A'_{11} & A'_{12} \\ O & A'_{22} \end{bmatrix} \begin{bmatrix} \eta_1(t) \\ \eta_2(t) \end{bmatrix} + \begin{bmatrix} B'_1 \\ O \end{bmatrix} u(t) + \begin{bmatrix} G'_1 \\ G'_2 \end{bmatrix} y(t) \qquad (9)$$

In (9), η denotes the new state, related to z by $z = T\eta$. Zero submatrices are due to S_0^* being an $(A+GC)$-invariant containing B.

Note that only the second matrix differential equation of (9) (that corresponding to $\dot{\eta}_2(t)$ at the left), has to be reproduced in the observer, since η_2 is not influenced by η_1 or u. If the observer initial state is set according to $\eta(0) = T^{-1} x(0)$, through

$$z_2(t) = T_2 \, \eta_2(t) \qquad (10)$$

a state estimate modulo S_0^* is derived.

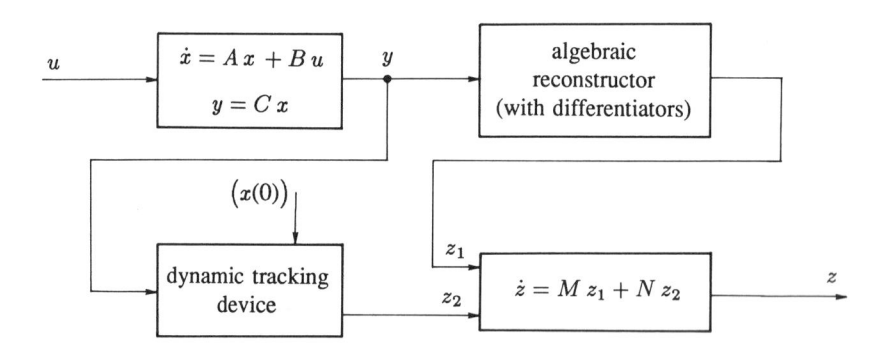

Figure 4-7 The general block diagram of an unknown-input state reconstructor.

The estimator scheme shown in Fig. 4-7 is based on both of the preceding techniques: the algebraic reconstructor with differentiators provides as output z_1 a state estimate modulo Q_1 and works if neither the initial state nor the input function is known, while the dynamic tracking device provides as z_2 a state estimate modulo S_0^*, but requires the initial state to be known. A state estimate modulo Q_2 is obtained as a linear function of the outputs of both devices. Note that the enlarged knowledge of the state obtained by algebraic reconstructor with differentiators is not useful in the dynamic device since if S_0^* is replaced by $\min S(A, V_0^*, B)$, the intersection in equation (5) does not change by virtue of Corollary 4.1-1. The following properties, concerning *complete* unknown-input reconstructability, are particular cases of previous Properties 4.3-1 and 4.3-2.

Property 4.3-3.
Triple (A, B, C) is unknown-state, unknown-input completely reconstructable in any finite time interval $[0, T]$ if and only if

$$V_0^* := \max \mathcal{V}(A, B, C) = \{0\} \tag{11}$$

Note that, if triple (A, B, C) is unknown-state, unknown-input completely reconstructable, pair (A, C) is observable because of the inclusion $\max \mathcal{J}(A, C) \subseteq \max \mathcal{V}(A, B, C)$.

Property 4.3-4.
Triple (A, B, C) is zero-state, unknown-input completely reconstructable in any finite time interval $[0, T]$ if and only if

$$V_0^* \cap \mathcal{B} = \{0\} \tag{12}$$

Proof. An alternative expression for $\mathcal{Q}_2 = \mathcal{R}_{\mathcal{V}^*}$ defined in (5) is $\mathcal{Q}_2 = \min \mathcal{J}(A + BF, \mathcal{V}^* \cap \mathcal{B})$, with F any matrix such that $(A + BF)\mathcal{V}^* \subseteq \mathcal{V}^*$. Therefore $\mathcal{Q}_2 = \{0\}$ if and only if (12) holds. □

The state reconstructor shown in Fig. 4-7 provides the maximal information on the system state when the input function is unknown and the initial state known, by observing the output in any nonzero time interval $[0, T]$. The same scheme can also be used to provide an asymptotic state estimate when the initial state is unknown, provided $\mathcal{S}_0^*$ is externally stabilizable. Since in this case matrix A'_{22} in (9) can be made stable by a suitable choice of G, function $z_2(t)$ in (10) asymptotically converges to a state estimate modulo $\mathcal{S}_0^*$ also if the initial state of the dynamic tracking device is not congruent with that of the system. Necessary and sufficient conditions for complete state asymptotic estimation by means of a device including differentiators are stated in the following property.

Property 4.3-5.
Triple (A, B, C) is unknown-state, unknown-input completely asymptotically observable if and only if

1. $V_0^* \cap \mathcal{B} = \{0\}$; (13)
2. $\mathcal{S}_0^*$ is externally stabilizable. (14)

Proof. Recall Corollary 4.2-1 and note that processing the output through the algebraic reconstructor with differentiators provides the state modulo V_0^* instead of modulo C. Hence (13) follows from (4.2.21) and from the proof of Property 4.3-4. □

Point 2 of Property 4.3-5 can also be expressed in terms of invariant zeros. The external unassignable eigenvalues of $\mathcal{S}_0^*$ or the internal unassignable eigenvalues of $\mathcal{V}_0^*$, which are equal to each other by virtue of Property 4.1-20, are called the *invariant zeros* of triple (A, B, C) (see next section). Therefore, a general unknown-state, unknown-input asymptotic estimator exists if and only if $\mathcal{V}_0^* \cap \mathcal{B}$ reduces to the origin and all invariant zeros of the system are stable.

4.3.2 System Invertibility and Functional Controllability

Refer to a triple (A, B, C). The term *system invertibility* denotes the possibility of reconstructing the input from the output function. Although for the sake of precision it is possible to define both the unknown-state, unknown-input invertibility and the zero-state, unknown-input invertibility (see Definitions 4.3-1 and 4.3-2), the term invertibility "tout court" is referred to the latter, i.e., to the invertibility of map γ_2 in (2).

The term *functional controllability* denotes the possibility of imposing any sufficiently smooth output function by a suitable input function, starting at the zero state. Here "sufficiently smooth" means piecewise differentiable at least n times. It is often also called *right invertibility*, from the identity $y(\cdot) = \gamma_2(u(\cdot)) \circ \gamma_2^{-1}(y(\cdot))$, while simple invertibility is also called *left invertibility*, from $u(\cdot) = \gamma_2^{-1}(y(\cdot)) \circ \gamma_2(u(\cdot))$.

Theorem 4.3-1.

Triple (A, B, C), with B having maximal rank, is unknown-state (zero-state) invertible if and only if it is unknown-state, unknown-input (zero-state, unknown-input) completely reconstructable.

Proof. If. From the system differential equation $\dot{x}(t) = A\,x(t) + B\,u(t)$ it follows that

$$u(t) = (B^T B)^{-1} B^T \left(\dot{x}(t) - A\,x(t) \right) \qquad (15)$$

which provides $u(t)$ almost everywhere in $[0, T]$ if from $y|_{[0,T]}$ (and $x(0)$) it is possible to derive $x|_{[0,T]}$, hence $\dot{x}|_{[0,T]}$ almost everywhere.

Only if. Let the considered system be invertible, i.e., from the output function $y|_{[0,T]}$ it is possible to derive input $u|_{[0,T]}$. By subtracting the forced response from the total response we derive the free response, i.e., the output function of the free system

$$\dot{x}(t) = A\,x(t)$$
$$y(t) = C\,x(t)$$

whose current state $x(t)$ can be determined by repeatedly differentiating the output function if the initial state is unknown (recall that in this case complete reconstructability implies complete observability) or by an identity observer if known. □

From now on, the term "invertibility" will be strictly referred to zero-state invertibility. The following statement is immediately derived from Theorem 4.3-1 and Property 4.3-4.

Sect. 4.3 Unknown-Input Reconstructability, Invertibility, and Functional Controllability **237**

Property 4.3-6. **(Left) Invertibility of a Triple**
Triple (A, B, C) is invertible (left-invertible) if and only if (12) holds.

The device whose block diagram is represented in Fig. 4-7 can easily be extended so as to be a realization of the *inverse system* of triple (A, B, C): connect a further differentiator stage on output z_1 (the time derivative of z_2 can be directly computed as a linear function of z_2 and y): a linear algebraic block implementing (15) will provide input u.

By virtue of Property 4.3-5 and related discussion, the inverse system is asymptotically stable if and only if all the invariant zeros of (A, B, C) are stable (or, equivalently, $\mathcal{V}_0^*$ is internally stabilizable or $\mathcal{S}_0^*$ is externally stabilizable).

We shall now consider the dual concept and prove a theorem that is dual to Property 4.3-6.

Theorem 4.3-2. **Functional Controllability of a Triple**
Triple (A, B, C) is functionally output controllable (or right-invertible) if and only if
$$\mathcal{S}_0^* + \mathcal{C} = \mathcal{X} \qquad (16)$$

Proof. Consider the linear operator γ_2 in (2), which is left invertible if and only if (12) holds. Its adjoint operator
$$u|_{[0,T]} = \gamma_2^T \left(u|_{[0,T]} \right)$$
is defined by
$$u(t) = B^T \int_0^t e^{A^T(t-\tau)} C^T u(\tau) \, d\tau \qquad t \in [0, T]$$
From
$$(\gamma_2^T)^{-1} \circ \gamma_2^T = i$$
where i denotes the identity operator, by taking the adjoint of both members it follows that
$$\gamma_2 \circ \gamma_2^{-1} = i$$
Hence γ_2 admits a right inverse if and only if γ_2^T admits a left inverse, i.e., if and only if
$$\max \mathcal{V}(A^T, \operatorname{im} C^T, \ker B^T) \cap \operatorname{im} C^T = \{0\}$$
from which (16) follows by orthogonal complementation. □

The functional controller is realizable in exactly the same way as the inverse system, i.e., by a state reconstructor of the type shown in Fig. 4-3 completed with a further differentiator stage and an algebraic part. Its dynamic part is asymptotically stable if and only if all the invariant zeros of (A, B, C) are stable (or, equivalently, $\mathcal{S}_0^*$ is externally stabilizable or $\mathcal{V}_0^*$

is internally stabilizable). However, since in this case the system is not necessarily invertible, input function $u(\cdot)$ corresponding to the desired output function is not in general unique. On the other hand, the difference between any two admissible input functions corresponds to a zero-state motion on $\mathcal{R}_{\mathcal{V}_0^*}$ which does not affect the output function, so that the functional controller can be realized to provide any one of the admissible input functions, for instance by setting to zero input components which, expressed in a suitable basis, correspond to forcing actions belonging to $\mathcal{V}_0^* \cap \mathcal{B}$.

4.4 INVARIANT ZEROS AND THE INVARIANT ZERO STRUCTURE

Consider a triple (A, B, C). The concept of "zero," which is a natural counterpart to the concept of "pole" in IO descriptions, is introduced in geometric approach terms through the following definition.[1]

Definition 4.4-1. Invariant Zeros

The *invariant zeros* of triple (A, B, C) are the internal unassignable eigenvalues of $\mathcal{V}_0^* := \max \mathcal{V}(A, B, C)$ or, equivalently by virtue of Property 4.1-20, the external unassignable eigenvalues of $\mathcal{S}_0^* := \min \mathcal{S}(A, C, B)$.

Note that a system whose state or forcing action is completely accessible, i.e., with $\mathcal{B} = \mathcal{X}$ or $\mathcal{C} = \{0\}$, has no invariant zeros. Invariant zeros are easily computable by using the specific geometric approach algorithms.

A more complete definition, which includes the previous one as a particular case, refers to the internal unassignable eigenstructure of $\mathcal{V}_0^*$ or the external unassignable eigenstructure of $\mathcal{S}_0^*$. The eigenstructure of a linear transformation is complete information on its real or complex Jordan form (eigenvalues, number and dimensions of the corresponding Jordan blocks, or orders of corresponding elementary divisors). As for the unassignable eigenvalues, matrix A'_{22} in (4.1.34) is referred to for the eigenstructures here considered.

---

[1] In the literature concerning the matrix polynomial approach, *transmission zeros* are defined as the zeros of the Smith-Macmillan form of the transfer matrix, while those introduced in Definition 4.4-1 are usually named invariant zeros. These definitions of zeros are equivalent to each other if (A, B, C) is minimal (i.e., completely controllable and observable). In this case transmission or invariant zeros have a precise physical meaning, i.e., they "block" transmission of certain frequencies from input to output. Definition 4.4-1 is implicit in an early work by Morse [33] and specifically investigated by Molinari [31]. Extensive reviews on definitions and meanings of multivariable zeros are reported by Francis and Wonham [19], MacFarlane and Karcanias [25], and Schrader and Sain [36].

Definition 4.4-2. Invariant Zero Structure

The *invariant zero structure* of triple (A,B,C) is the internal unassignable eigenstructure of $\mathcal{V}_0^* := \max \mathcal{V}(A,B,C)$ or, equivalently, the external unassignable eigenstructure of $\mathcal{S}_0^* := \min \mathcal{S}(A,C,B)$.

A quite common physical justification of the word "zero" is related to the property to block frequencies. It is worth discussing this property by referring to a suitable extension of the frequency response.

4.4.1 The Generalized Frequency Response

Refer to the block diagram of Fig. 2-4, which represents quadruple (A,B,C,D), cascaded to an *exosystem*, described by the equations

$$\dot{v}(t) = W\,v(t) \tag{1}$$

$$u(t) = L\,v(t) \tag{2}$$

We temporarily assume that A is asymptotically stable, while W, the exosystem matrix, is assumed to have all the eigenvalues with the real parts zero or positive. For instance, the exosystem output could be one of those represented in Fig. 2-5 or any linear combinations of unstable modes.

The exosystem output is

$$u(t) = L\,e^{Wt}\,v_0 \tag{3}$$

Our aim is to search for conditions that ensure that the state evolution of the system, when the possible transient condition is finished, can be expressed as a function of the sole exogenous modes or, in other words, conditions for the existence of a matrix X such that

$$\lim_{t\to\infty} x(t) = X\,e^{Wt}\,v_0$$

for any exosystem initial state v_0. Function

$$x_s(t) := X\,e^{Wt}\,v_0 \tag{4}$$

where x_s is the *state in the steady condition*, is necessarily a solution of the overall system differential equation and can be defined also when the system matrix A is unstable. By substituting it in the system differential equation $\dot{x}(t) = A\,x(t) + B\,u(t)$ and taking into account (3) we get

$$X W\,e^{Wt}\,v_0 = A X\,e^{Wt}\,v_0 + B L\,e^{Wt}\,v_0$$

Since v_0 is arbitrary and the matrix exponential nonsingular, it follows that

$$A X - X W = -B L \tag{5}$$

Matrix X is called *state generalized frequency response*. In general it is a function of matrices W, L of the exosystem.

Relation (5) is a Sylvester equation: if for a given W it admits no solution, the system is said to present a *resonance* at W; by virtue of Theorem 2.5-10 this can occur only if the system and exosystem have common eigenvalues. Let

$$x(t) = x_t(t) + x_s(t)$$

Component x_t is the *state in the transient condition*; since both $x(t)$ and $x_r(t)$ satisfy the system differential equation, by difference we obtain

$$\dot{x}_t(t) = A\, x_t(t)$$

whence

$$x_t(t) = e^{At}\, x_{0t} = e^{At}\,(x_0 - x_{0s}) \tag{6}$$

where, according to (4), $x_{0s} = X\, v_0$ is the particular value of the initial condition that makes the transient motion vanish, i.e., such that the equality $x(t) = x_r(t)$ holds identically in time, not only as t approaches infinity.

As far as the output is concerned, by substituting (3) and (4) in the output equation $y(t) = C\, x(t) + D\, u(t)$ it follows that

$$y_s(t) = (C\, X + D\, L)\, e^{Wt}\, v_0 = Y\, e^{Wt}\, v_0 \tag{7}$$

where

$$Y := C\, X + D\, L \tag{8}$$

is called the *output generalized frequency response*, which is also a function of matrices W, L of the exosystem.

Particular, interesting cases are those of an exosystem with a single real eigenvalue ρ and of an exosystem with a pair of complex conjugate eigenvalues $\sigma \pm j\omega$, i.e.

$$W := \rho \qquad W := \begin{bmatrix} \sigma & \omega \\ -\omega & \sigma \end{bmatrix} \tag{9}$$

or, to consider multiple eigenvalues, the corresponding real Jordan blocks of order k. For instance, in the cases of a double real eigenvalue and a pair of complex conjugate eigenvalues with multiplicity two, we assume respectively

$$W := \begin{bmatrix} \rho & 1 \\ 0 & \rho \end{bmatrix} \qquad W := \begin{bmatrix} \sigma & \omega & & \\ -\omega & \sigma & & I_2 \\ & & \sigma & \omega \\ O & & -\omega & \sigma \end{bmatrix} \tag{10}$$

In these cases matrix L is $p \times k$ or $p \times 2k$ and produces the distribution on the system inputs of the exogenous modes corresponding to the considered eigenvalues or Jordan blocks.

Referring to the generalized frequency response, it is possible to introduce the concepts of blocking zero and blocking structure as follows.

Definition 4.4-3. Blocking Zero and Blocking Structure

A *blocking zero* of quadruple (A, B, C, D) is a value of ρ or $\sigma \pm j\omega$ such that for W defined as in (9) there exists at least one input distribution matrix L corresponding to a state generalized frequency response X such that pair (W, X) is observable and the output generalized frequency response is zero. A *blocking structure* is defined by extending the above to the case of an arbitrary W in real Jordan form.

In other words, a blocking zero or a blocking structure is an exosystem matrix W such that there exists at least one zero-output state trajectory that is a function of all the corresponding modes. Hence, for such a W there exist matrices X, L with (W, X) observable such that

$$AX - XW = -BL \tag{11}$$
$$CX + DL = O \tag{12}$$

Equations (11, 12) are linear in X, L for any W. In the case of a purely dynamic system (i.e., for $D = O$) blocking zeros or blocking structures are not affected by state feedback or output injection. In fact, X_0, W_0, and L_0 satisfy (11, 12) with $D = O$. Then

$$(A + BF)X_0 - X_0 W_0 = -BL_1, \quad \text{with} \quad L_1 := L_0 - FX_0$$
$$(A + GC)X_0 - X_0 W_0 = -BL_0, \quad \text{since} \quad CX_0 = O$$

Invariant zeros and the invariant zero structure are related to blocking zeros and blocking structures, as the following properties state.

Property 4.4-1.

Consider a triple (A, B, C). Its invariant zeros and the invariant zero structure are also blocking zeros and a blocking structure.

Proof. Let F be such that $(A+BF)\mathcal{V}_0^* \subseteq \mathcal{V}_0^*$: among all possible state feedback matrices, this corresponds to the maximal unobservability subspace, since $\mathcal{V}_0^*$ is the maximal controlled invariant contained in $\mathcal{C}$. Furthermore, it allows all the eigenvalues to be arbitrarily assigned, except the internal unassignable eigenvalues of $\mathcal{V}_0^*$, so we can assume that no other eigenvalue of $A+BF$ is equal to them. On this assumption $\mathcal{R}_{\mathcal{V}_0^*}$ as an $(A+BF)$-invariant is complementable with respect to $(\{0\}, \mathcal{V}_0^*)$: this means that there exists a $\mathcal{V}$ such that

$$\mathcal{R}_{\mathcal{V}_0^*} \oplus \mathcal{V} = \mathcal{V}_0^*$$
$$(A + BF)\mathcal{V} \subseteq \mathcal{V}$$

Consider the change of basis defined by transformation $T := [T_1\ T_2\ T_3]$, with $\mathrm{im}\, T_1 = \mathcal{R}_{\mathcal{V}_0^*}$, $\mathrm{im}\, T_2 = \mathcal{V}$. Thus

$$T^{-1}(A+BF)T = \begin{bmatrix} A'_{11} & O & A'_{13} \\ O & A'_{22} & A'_{23} \\ O & O & A'_{33} \end{bmatrix} \tag{13}$$

clearly the invariant zeros are the eigenvalues of A'_{22} (the invariant zero structure is the eigenstructure of A'_{22}). The statement follows by assuming

$$W := A'_{22} \quad X := T_2 \quad L := FT_2 \qquad (14)$$

In fact, it will be shown that the above matrices are such that
1. (X, W) is observable;
2. $CX = O$;
3. $AX - XW = -BL$.

Property 1 is due to the rank of X being maximal and equal to the dimension of W, relation 2 follows from $\text{im} X = \mathcal{V} \subseteq \mathcal{V}_0^*$, while 3 is equivalent to

$$AT_2 - T_2 A'_{22} = -BFT_2$$

i.e.

$$(A + BF)T_2 = T_2 A'_{22}$$

which directly follows from (13). □

Property 4.4-2.

Let triple (A, B, C) be completely controllable and (left) invertible. Its blocking zeros and blocking structures are invariant zeros and parts of the invariant zero structure.

Proof. Let (W, X) be the Jordan block and the state frequency response corresponding to a blocking zero, so that

$$AX - XW = -BL \qquad (15)$$
$$CX = O \qquad (16)$$

This means that the extended free system

$$\dot{\hat{x}}(t) = \hat{A}\,\hat{x}(t) \quad \text{with} \quad \hat{A} := \begin{bmatrix} A & BL \\ O & W \end{bmatrix}$$

admits solutions of the type

$$x(t) = X\,e^{Wt}\,v_0 \quad \text{with} \quad \text{im} X \subseteq \mathcal{V}_0^* \subseteq \mathcal{C}$$

Let F be a state feedback matrix such that $(A+BF)\mathcal{V}_0^* \subseteq \mathcal{V}_0^*$; by adding BFX to both members of (15) it follows that

$$(A+BF)X - XW = -BL + BFX$$

The image of the matrix on the right must belong to $\mathcal{V}_0^*$ since those of both matrices on the left do. On the other hand, provided $\mathcal{V}_0^* \cap \mathcal{B} = \{0\}$ by virtue of the invertibility assumption, matrix on the right is zero. Hence

$$(A+BF)X - XW = O$$

and, consequently

$$x(t) = e^{(A+BF)t}\,X\,v_0 = X\,e^{Wt}\,v_0$$

Pair (W, X) is observable by assumption. Since the modes corresponding to both matrix exponentials in the preceding relations must be identical in function $x(t)$, there exists a Jordan block in $A+BF$ internal to $\mathcal{V}_0^*$ (in this case nonassignable, like all the eigenvalues internal to $\mathcal{V}_0^*$), equal to W. □

Note that, while Definitions 4.4-1 and 4.4-2 (invariant zero and invariant zero structure) refer to a quadruple, Definition 4.4-3 refers to a triple. On the other hand, its extension to quadruples can be achieved by using an artifice, i.e., by cascading to the system a stage of integrators, which does not present either invariant or blocking zeros, so that the invariant zeros and the invariant zero structure of quadruple can be assumed to be equal to those of the augmented system. This topic will be reconsidered in the next section.

4.4.2 The Role of Zeros in Feedback Systems

The concept of zero is of paramount importance in connection with stabilizability of feedback systems. It is well known from the automatic control systems analysis, which is normally developed by using transfer functions, that the presence of zeros in the right-half s plane, i.e., the nonminimum phase condition, generally causes serious stabilizability problems. The preceding multivariable extension of the concept of zero strictly developed in the framework of the geometric approach is similarly connected with stabilizability in the presence of feedback and plays a basic role in synthesis problems.

To clarify this by means of an example, we shall show here that the stabilizability condition for the disturbance localization problem and its dual (see Section 4.2) is susceptible to a quite simple and elegant reformulation in terms of invariant zeros.

Referring to system (4.2.1, 2), denote by $\mathcal{Z}(u;e)$ the set of all invariant zeros between input u and output e, and by $\mathcal{Z}(u,d;e)$ that between inputs u,d (considered as a whole), and output e. The basic result is set in the following theorem.

Theorem 4.4-1.
Let $\mathcal{D} \subseteq \mathcal{V}^*$ or $\mathcal{D} \subseteq \mathcal{V}^* + \mathcal{B}$, with $\mathcal{V}^* := \max \mathcal{V}(A,B,\mathcal{E})$. Then $\mathcal{V}_m := \mathcal{V}^* \cap \mathcal{S}_1^*$, with $\mathcal{S}_1^* := \min \mathcal{S}(A, \mathcal{E}, B+\mathcal{D})$, is internally stabilizable if and only if all the elements of $\mathcal{Z}(u;e) \dot{-} \mathcal{Z}(u,d;e)$ are stable (recall that $\dot{-}$ denotes difference with repetition count).

Proof. It has been proved in Section 4.2 that the assumption regarding $\mathcal{D}$ implies $\max \mathcal{V}(A, B, \mathcal{E}) = \max \mathcal{V}(A, B+\mathcal{D}, \mathcal{E})$, so that the reachable set on $\mathcal{E}$ by the only input u is $\mathcal{R}_\mathcal{E} := \mathcal{V}^* \cap \min \mathcal{S}(A, \mathcal{E}, B)$, while the reachable set by both inputs u,d used together is $\mathcal{V}_m := \mathcal{V}^* \cap \min \mathcal{S}(A, \mathcal{E}, B+\mathcal{D})$.

Assume a coordinate transformation $T := [T_1\, T_2\, T_3\, T_4]$ with $\text{im} T_1 = \mathcal{R}_\mathcal{E}$, $\text{im}[T_1\, T_2] = \mathcal{V}_m$, $\text{im}[T_1\, T_2\, T_3] = \mathcal{V}^*$, $\text{im}[T_1\, T_4] \supseteq \mathcal{B}$, $\text{im}[T_1\, T_2\, T_4] \supseteq \mathcal{S}_1^* \supseteq \mathcal{D}$. Matrices $A' := T^{-1}AT$, $B' := T^{-1}B$ and $D' := T^{-1}D$ have the structures

$$A' = \begin{bmatrix} A'_{11} & A'_{12} & A'_{13} & A'_{14} \\ O & A'_{22} & A'_{23} & A'_{24} \\ O & O & A'_{33} & A'_{34} \\ A'_{41} & A'_{42} & A'_{43} & A'_{44} \end{bmatrix} \qquad B' = \begin{bmatrix} B'_1 \\ O \\ O \\ B'_4 \end{bmatrix} \qquad D' = \begin{bmatrix} D'_1 \\ D'_2 \\ O \\ D'_4 \end{bmatrix}$$

where the zero submatrices of A' are due to $\mathcal{R}_\mathcal{E}$ and $\mathcal{V}_m$ being respectively an $(A,\mathcal{B})$- and an $(A, \mathcal{B}+\mathcal{D})$-controlled invariant and to the structures of B' and D'. The elements of $Z(u;e)$ are the union of the eigenvalues of A'_{22} and A'_{33}, while those of $Z(u,d;e)$ are the eigenvalues of A'_{33}. □

The main results of disturbance localization and unknown-input asymptotic state estimation can be reformulated as follows.

Corollary 4.4-1. Unaccessible Disturbance Localization

Consider system (4.2.1,2) and assume that (A,B) is stabilizable. The unaccessible disturbance localization problem with stability has a solution if and only if

1. $\mathcal{D} \subseteq \mathcal{V}^*$; (17)
2. $Z(u;e) \dot{-} Z(u,d;e)$ has all its elements stable. (18)

Corollary 4.4-2. Accessible Disturbance Localization

Consider system (4.2.1,2) and assume that (A,B) is stabilizable. The accessible disturbance localization problem with stability has a solution if and only if

1. $\mathcal{D} \subseteq \mathcal{V}^* + \mathcal{B}$; (19)
2. $Z(u;e) \dot{-} Z(u,d;e)$ has all its elements stable. (20)

The dual results are stated as follows. Consider system (4.2.16–18) and denote by $Z(d;e)$ the set of all invariant zeros between input u and output e and by $Z(d;y,e)$ that between input d and outputs y,e (considered as a whole).

Theorem 4.4-2.

Let $\mathcal{E} \supseteq \mathcal{S}^*$ or $\mathcal{D} \supseteq \mathcal{S}^* \cap \mathcal{C}$, with $\mathcal{S}^* := \min \mathcal{S}(A,\mathcal{C},\mathcal{D})$. Then $\mathcal{S}_M := \mathcal{S}^* + \max \mathcal{V}(A,\mathcal{D},\mathcal{C} \cap \mathcal{E})$ is externally stabilizable if and only if all the elements of $Z(d;y) \dot{-} Z(d;y,e)$ are stable.

Corollary 4.4-3. Unknown-Input Purely Dynamic Asymptotic Observer

Consider system (4.2.16,17) and assume that (A,C) is detectable. The problem of asymptotically estimating the linear function $e(t) = E x(t)$ in the presence of the unknown input d with a full-order purely dynamic observer, has a solution if and only if

1. $\mathcal{E} \supseteq \mathcal{S}^*$; (21)
2. $Z(d;y) \dot{-} Z(d;y,e)$ has all its elements stable. (22)

Corollary 4.4-4. Unknown-Input Nonpurely Dynamic Asymptotic Observer

Consider system (4.2.16,17) and assume that (A,C) is detectable. The problem of asymptotically estimating the linear function $e(t) = E x(t)$ in the presence of the unknown input d with a full-order purely dynamic observer has a solution if and only if

1. $\mathcal{E} \supseteq \mathcal{S}^* \cap \mathcal{C}$; (23)
2. $Z(d;y) \dot{-} Z(d;y,e)$ has all its elements stable. (24)

4.5 EXTENSIONS TO QUADRUPLES

Most of the previously considered problems were referred to purely dynamic systems of the type (A, B, C) instead of nonpurely dynamic systems of the type (A, B, C, D), which are more general. There are good reasons for this: triples are quite frequent in practice, referring to triples greatly simplifies arguments, and extension to quadruples can often be achieved by using some simple, standard artifices.

A very common artifice that can be adopted for many analysis problems is to connect an integrator stage in cascade to the considered quadruple, at the input or at the output: in this way an extended system is obtained that is modeled by a triple. Problems in which smoothness of the input or output function is a standard assumption, like unknown-input reconstructability, invertibility, functional controllability, introduction of the concepts of transmission zero and zero structure, can thus be extended to quadruples without any loss of generality.

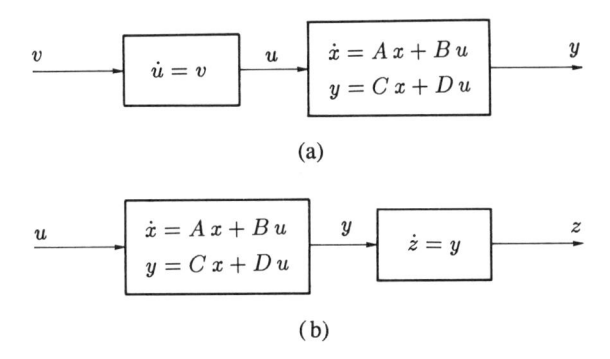

Figure 4-8 Artifices to reduce a quadruple to a triple.

Consider the connections shown in Fig. 4-8: in the case of Fig. 4-8(a) the overall system is modeled by

$$\dot{\hat{x}}(t) = \hat{A}\,\hat{x}(t) + \hat{B}\,v(t) \tag{1}$$
$$y(t) = \hat{C}\,\hat{x}(t) \tag{2}$$

with

$$\hat{x} := \begin{bmatrix} x \\ u \end{bmatrix} \quad \hat{A} := \begin{bmatrix} A & B \\ O & O \end{bmatrix} \quad \hat{B} := \begin{bmatrix} O \\ I_p \end{bmatrix} \quad \hat{C} := [\,C \quad D\,] \tag{3}$$

while in the case of Fig. 4-8(b) the system is described by

$$\dot{\hat{x}}(t) = \hat{A}\,\hat{x}(t) + \hat{B}\,u(t) \tag{4}$$
$$y(t) = \hat{C}\,\hat{x}(t) \tag{5}$$

with

$$\hat{x} := \begin{bmatrix} x \\ z \end{bmatrix} \quad \hat{A} := \begin{bmatrix} A & O \\ C & O \end{bmatrix} \quad \hat{B} := \begin{bmatrix} B \\ D \end{bmatrix} \quad \hat{C} := [\, O \;\; I_q \,] \qquad (6)$$

To approach unknown-input reconstructability and system invertibility it is quite natural to refer to the extended system shown in Fig. 4-8(a), while for functional controllability, that of Fig. 4-8(b) is preferable. For invariant zeros and the invariant zero structure, any one of the extended systems can be used, since in both cases the integrator stage has no influence on zeros.

We shall now consider in greater detail the extension of the concept of zero, referring to (4–6). A controlled invariant contained in $\hat{\mathcal{C}} := \ker \hat{C}$ can be expressed as

$$\hat{\mathcal{V}} = \left\{ \begin{bmatrix} x \\ z \end{bmatrix} : x \in \mathcal{V}, \; z = 0 \right\} \qquad (7)$$

because of the particular structure of $\hat{C}$. The definition property

$$\hat{A}\hat{\mathcal{V}} \subseteq \hat{\mathcal{V}} + \hat{\mathcal{B}} \qquad (8)$$

implies

$$A\mathcal{V} \subseteq \mathcal{V} + \mathcal{B} \qquad (9)$$
$$C\mathcal{V} \subseteq \mathrm{im}\, D \qquad (10)$$

From (9) it follows that $\mathcal{V}$ is an $(A, \mathcal{B})$-controlled invariant. Since any motion on the "extended" controlled invariant $\hat{\mathcal{V}}$ satisfies $z(\cdot) = 0$, then $y(\cdot) = 0$, any state feedback matrix such that $(\hat{A} + \hat{B}\hat{F})\hat{\mathcal{V}} \subseteq \hat{\mathcal{V}}$ has the structure $\hat{F} = [F\; O]$ with

$$(A + BF)\mathcal{V} \subseteq \mathcal{V} \quad \mathcal{V} \subseteq \ker(C + DF) \qquad (11)$$

Relations (7, 8) and (11) can be considered respectively the definition and main property of a geometric tool similar to the controlled invariant, which is called in the literature *output-nulling controlled invariant*.[1]

Of course, being an extension of the regular controlled invariant, it satisfies all its properties, including the semilattice structure. A special algorithm to derive the maximal output-nulling controlled invariant is not needed, since it is possible to use the standard algorithm for the maximal controlled invariant referring to the extended system (4–6).

---

[1] The output-nulling (controlled) invariants were introduced and investigated by Anderson [2,3]. Deep analysis of their properties, definition of their duals, and a complete bibliography are due to Aling and Schumacher [1].

The dual object, the *input-containing conditioned invariant* can also be defined, referring to (1–3) instead of (4–6).

A conditioned invariant of the extended system containing $\hat{\mathcal{B}} := \operatorname{im} \hat{B}$ can be expressed as

$$\hat{\mathcal{S}} = \left\{ \begin{bmatrix} x \\ u \end{bmatrix} : x \in \mathcal{S}, \, u \in \mathbf{R}^p \right\} \tag{12}$$

because of the particular structure of $\hat{B}$. Relation

$$\hat{A}(\hat{\mathcal{S}} \cap \hat{\mathcal{C}}) \subseteq \hat{\mathcal{S}} \tag{13}$$

together with (12) implies

$$A(\mathcal{S} \cap \mathcal{C}) \subseteq \mathcal{S} \tag{14}$$
$$B^{-1} \mathcal{S} \supseteq \ker D \tag{15}$$

It follows that a conditioned invariant $\mathcal{S}$ is input-containing if and only if there exists an extended output injection $\hat{G}^T = [G^T \ I_p]$ such that

$$(A + GC)\mathcal{S} \subseteq \mathcal{S} \quad \mathcal{S} \supseteq \operatorname{im}(B + GD) \tag{16}$$

The minimal input-containing conditioned invariant is the minimal zero-state unknown-input unreconstructability subspace of quadruple (A, B, C, D) by means of a device of the type shown in Fig. 4-7.

The maximal output-nulling controlled invariant and the minimal input-containing conditioned invariant can be determined by means of the standard algorithms referring respectively to the extended system (4–6) and (1–3). Denote them by $\mathcal{V}_0^*$ and $\mathcal{S}_0^*$: it can easily be checked that the reachable set on $\mathcal{V}_0^*$ is $\mathcal{V}_0^* \cap \mathcal{S}_0^*$ and the unobservable set containing $\mathcal{S}_0^*$ is $\mathcal{V}_0^* + \mathcal{S}_0^*$. The invariant zeros of (A, B, C, D) are the elements of $\sigma((A+BF)|_{\mathcal{V}_0^*/(\mathcal{V}_0^* \cap \mathcal{S}_0^*)})$ or those of $\sigma((A+GC)|_{(\mathcal{V}_0^* + \mathcal{S}_0^*)/\mathcal{S}_0^*})$: it is easy to show that these two spectra are identical. The invariant zero structure of (A, B, C, D) coincides with any of the eigenstructures of the corresponding induced maps.

Feedback Connections. When a quadruple, which is a nonpurely dynamic system and hence has an algebraic signal path directly from input to output, is subject to nonpurely dynamic or simply algebraic feedback, like state feedback and output injection, an algebraic closed loop is present and the stability problem cannot be properly approached. In these cases, in general, the assumed mathematical model is not correct for the problem concerned and some neglected dynamics have to be considered.

Nevertheless, there exist artifices that allow standard feedback connections approached for triples to be extended to quadruples. One of these is shown in Fig. 4-9(a): signal Du is subtracted from the output, thus obtaining a new output y_1 which

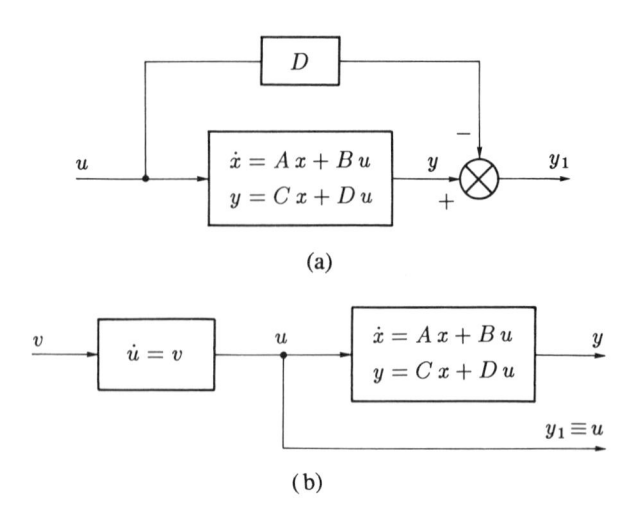

Figure 4-9 Other artifices to deal with quadruples.

can be used for a possible feedback connection. In this way, for instance, quadruple (A, B, C, D) can be stabilized through an observer exactly in the same way as triple (A, B, C).

A drawback of this procedure is that the system feature of being purely dynamic is not robust with respect to uncertainty in D, so that it is often preferable to use one of the artifices shown in Fig. 4-8 again, but providing the maximal observability in the first case and the maximal controllability in the second. For instance, in the first case this is obtained as shown in Fig. 4-9(b), i.e., by including also the auxiliary state variables, which are accessible, in the output distribution matrix $\hat{C}$, which becomes

$$\hat{C} := \begin{bmatrix} C & D \\ O & I_p \end{bmatrix}$$

4.5.1 On Zero Assigment

In some multivariable synthesis problems it is necessary to assign invariant zeros through a convenient choice of the system matrices. Refer to a controllable pair (A, B) and consider the problem of deriving suitable matrices C, D such that quadruple (A, B, C, D) has as many as possible zeros arbitrarily assigned, or, by duality, given the observable pair (A, C) derive B, D such that again (A, B, C, D) has as many as possible zeros arbitrarily assigned. These problems are both reduced to standard pole assignment by state feedback. Denote by p and q respectively the number of the inputs and that of the outputs of the quadruple to be synthesized.

Algorithm 4.5-1. Zero Assignment for a Quadruple (A,B,C,D)

Let (A, B) be controllable and $p \leq q$. It is possible to assign n invariant zeros of the quadruple (A, B, C, D) by the following procedure:

1. Choose D arbitrary of maximal rank;
2. Derive F such that $A+BF$ has the zeros to be assigned as eigenvalues;
3. Assume $C := -DF$.

Proof. Refer to extended system (6) and assume

$$\hat{V}^* := \ker \hat{C} = \begin{bmatrix} I_n \\ O \end{bmatrix} \qquad (17)$$

Being $p \leq q$ and D of maximal rank, clearly

$$\hat{V}^* \cap \hat{B} = \{0\} \qquad (18)$$

Due to the particular choice of C, $\hat{F} := [F \; O]$ is such that

$$\hat{A} + \hat{B}\hat{F} = \begin{bmatrix} A+BF & O \\ O & O \end{bmatrix}$$

This means that $\hat{V}^*$ is an $(\hat{A} + \hat{B}\hat{F})$-invariant (hence the maximal $(\hat{A}, \hat{B})$-controlled invariant contained in $\ker \hat{C}$) and its internal eigenvalues are those of $A+BF$. Due to (18) all these eigenvalues are unassignable, hence they coincide with the invariant zeros of (A, B, C, D). □

If $p \geq q$, the preceding algorithm can be used to derive B, D instead of C, D, provided that (A, C) is observable. In fact, the invariant zeros of (A, B, C, D) coincide with those of (A^T, C^T, B^T, D^T).

REFERENCES

1. ALING, H., and SCHUMACHER, J.M., "A nine-fold canonical decomposition for linear systems," *Int. J. Control*, vol. 39, no. 4, pp. 779–805, 1984.
2. ANDERSON, B.D.O., "Output nulling invariant and controllability subspaces," *Proceedings of the 6th IFAC World Congress*, paper no. 43.6, August 1975.
3. —, "A note on transmission zeros of a transfer function matrix," *IEEE Trans. Autom. Contr.*, vol. AC-21, no. 3, pp. 589–591, 1976.
4. BASILE, G., and MARRO, G., "Controlled and conditioned invariant subspaces in linear system theory," *J. Optimiz. Th. Applic.*, vol. 3, no. 5, pp. 305–315, 1969.
5. —, "On the observability of linear time-invariant systems with unknown inputs," *J. of Optimiz. Th. Applic.*, vol. 3, no. 6, pp. 410–415, 1969.
6. —, "L'invarianza rispetto ai disturbi studiata nello spazio degli stati," *Rendiconti della LXX Riunione Annuale AEI*, paper 1-4-01, Rimini, Italy, 1969.
7. —, "A new characterization of some properties of linear systems: unknown-input observability, invertibility and functional controllability," *Int. J. Control*, vol. 17, no. 5, pp. 931–943, 1973.

8. —, "Some new results on unknown-input observability," *Proceedings of the 8th IFAC Congress*, paper no. 2.1, 1981.

9. —, "Self-bounded controlled invariant subspaces: a straightforward approach to constrained controllability," *J. Optimiz. Th. Applic.*, vol. 38, no. 1, pp. 71–81, 1982.

10. —, "Self-bounded controlled invariants versus stabilizability," *J. Optimiz. Th. Applic.*, vol. 48, no. 2, pp. 245–263, 1986.

11. BASILE, G., MARRO, G., and PIAZZI, A., "A new solution to the disturbance localization problem with stability and its dual," *Proceedings of the '84 International AMSE Conference on Modelling and Simulation*, vol. 1.2, pp. 19–27, Athens, 1984.

12. BHATTACHARYYA, S.P., "On calculating maximal (A, B)-invariant subspaces," *IEEE Trans. Autom. Contr.*, vol. AC-20, pp. 264–265, 1975.

13. —, "Observers design for linear systems with unknown inputs," *IEEE Trans. on Aut. Contr.*, vol. AC-23, no. 3, pp. 483–484, 1978.

14. DAVISON, E.J., and WANG, S.H., "Properties and calculation of transmission zeros of linear multivariable systems," *Automatica*, vol. 10, pp. 643–658, 1974.

15. —, "Remark on multiple transmission zeros" (correspondence item), *Automatica*, vol. 10, pp. 643–658, 1974.

16. DESOER, C.A., and SCHULMAN, J.D., "Zeros and poles of matrix transfer functions and their dynamical interpretation," *IEEE Trans. Circ. Syst.*, vol. CAS-21, no. 1, pp. 3–8, 1974.

17. DORATO, P., "On the inverse of linear dynamical systems," *IEEE Trans. System Sc. Cybern.*, vol. SSC-5, no. 1, pp. 43–48, 1969.

18. FABIAN, E., and WONHAM, W.M., "Decoupling and disturbance rejection," *IEEE Trans. Autom. Contr.*, vol. AC-19, pp. 399–401, 1974.

19. FRANCIS, B.A., and WONHAM, W.M., "The role of transmission zeros in linear multivariable regulators," *Int. J. Control*, vol. 22, no. 5, pp. 657–681, 1975.

20. FUHRMANN, P.A., and WILLEMS, J.C., "A study of (A, B)-invariant subspaces via polynomial models," *Int. J. Control*, vol. 31, no. 3, pp. 467–494, 1980.

21. HAUTUS, M.L.J., "(A, B)-invariant and stabilizability subspaces, a frequency domain description," *Automatica*, no. 16, pp. 703–707, 1980.

22. KOUVARITAKIS, B., and MACFARLANE, A.G.J., "Geometric approach to analysis and synthesis of system zeros - Part 1. Square systems," *Int. J. Control*, vol. 23, no. 2, pp. 149–166, 1976.

23. —, "Geometric approach to analysis and synthesis of system zeros - Part 2. Non-square systems," *Int. J. Control*, vol. 23, no. 2, pp. 167–181, 1976.

24. MACFARLANE, A.G.J., "System matrices," *Proc. IEE*, vol. 115, no. 5, pp. 749–754, 1968.

25. MACFARLANE, A.G.J., and KARCANIAS, N., "Poles and zeros of linear multivariable systems: a survey of the algebraic, geometric and complex-variable theory," *Int. J. Control*, vol. 24, no. 1, pp. 33–74, 1976.

26. MARRO, G., "Controlled and conditioned invariants in the synthesis of unknown-input observers and inverse systems," *Control and Cybernetics* (Poland), vol. 2, no. 3/4, pp. 81–98, 1973.
27. —, *Fondamenti di Teoria dei Sistemi*, Pàtron, Bologna, Italy, 1975.
28. MEDITCH, J.S., and HOSTETTER, G.H., "Observers for systems with unknown and inaccessible inputs," *Int. J. Control*, vol. 19, pp. 473–480, 1974.
29. MOLINARI, B.P., "Extended controllability and observability for linear systems," *IEEE Trans. Autom. Contr.*, vol. AC-21, pp. 136–137, 1976.
30. —, "A strong controllability and observability in linear multivariable control," *IEEE Trans. Autom. Contr.*, vol. AC-21, pp. 761–763, 1976.
31. —, "Zeros of the system matrix," *IEEE Trans. Autom. Contr.*, vol. AC-21, pp. 795–797, 1976.
32. MORSE, A.S., "Output controllability and system synthesis," *SIAM J. Control*, vol. 9, pp. 143–148, 1971.
33. —, "Structural invariants of linear multivariable systems," *SIAM J. Control*, vol. 11, pp. 446–465, 1973.
34. PUGH, A.C., and RATCLIFFE, P.A., "On the zeros and poles of a rational matrix," *Int. J. Control*, vol. 30, no. 2, pp. 213–226, 1979.
35. SAIN, M.K., and MASSEY, J.L., "Invertibility of linear time-invariant dynamical systems," *IEEE Trans. Autom. Contr.*, vol. AC-14, no. 2, pp. 141–149, 1969.
36. SCHRADER, C.B., and SAIN, M.K., "Research on system zeros: a survey," *Int. J. Control*, vol. 50, no. 4, pp. 1407–1433, 1989.
37. SCHUMACHER, J.M., "(C,A)-invariant subspaces: some facts and uses," Vrije Universiteit, Amsterdam, Holland, Report no. 110, 1979.
38. —, "Complement on pole placement," *IEEE Trans. Autom. Contr.*, vol. AC-25, no. 2, pp. 281–282, 1980.
39. —, "On a conjecture of Basile and Marro," *J. Optimiz. Th. Applicat.*, vol. 41, no. 2, pp. 371–376, 1983.
40. SILVERMAN, L.M., "Inversion of multivariable linear systems," *IEEE Trans. Autom. Contr.*, vol. AC-14, no. 3, pp. 270–276, 1969.
41. WEN, J.T., "Time domain and frequency domain conditions for strict positive realness," *IEEE Trans. Autom. Contr.*, vol. 33, no. 10, pp. 988–992, 1988.
42. WONHAM, W.M., "Algebraic methods in linear multivariable control," *System Structure*, ed. A. Morse, IEEE Cat. n. 71C61, New York, 1971.
43. —, *Linear Multivariable Control: A Geometric Approach*, Springer-Verlag, New York, 1974.
44. WONHAM, W.M., and MORSE, A.S., "Decoupling and pole assignment in linear multivariable systems: a geometric approach," *SIAM J. Control*, vol. 8, no. 1, pp. 1–18, 1970.
45. —, "Feedback invariants of linear multivariable systems," *Automatica*, vol. 8, pp. 93–100, 1972.

5

THE GEOMETRIC APPROACH: SYNTHESIS

5.1 THE FIVE-MAP SYSTEM

In this chapter the features of the most general feedback connection, the output-to-input feedback through a dynamic system (or, simply, the output dynamic feedback) are investigated and discussed. Two particular problems are presented as basic applications of output dynamic feedback, namely the disturbance localization problem by means of a dynamic compensator and the regulator problem, which is the most interesting and complete application of the geometric approach.

For either problem, a five-map system (A, B, C, D, E) is needed, modeled by

$$\dot{x}(t) = A\,x(t) + B\,u(t) + D\,d(t) \tag{1}$$
$$y(t) = C\,x(t) \tag{2}$$
$$e(t) = E\,x(t) \tag{3}$$

It is called the *controlled system* and is connected as shown in Fig. 5-1 to a *controller* (compensator or regulator) described by

$$\dot{z}(t) = N\,z(t) + M\,y(t) + R\,r(t) \tag{4}$$
$$u(t) = L\,z(t) + K\,y(t) + S\,r(t) \tag{5}$$

Sect. 5.1 The Five-Map System

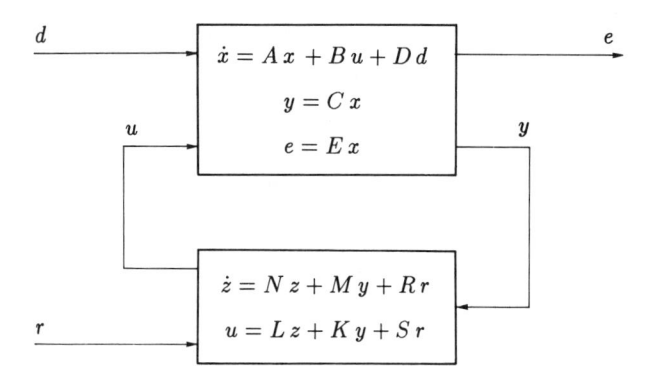

```
d ─────────────┌─────────────────┐──────────── e
               │ ẋ = A x + B u + D d │
               │                 │
   u           │    y = C x      │    y
    ───────────│                 │───────────
               │    e = E x      │
               └─────────────────┘


               ┌─────────────────┐
               │ ż = N z + M y + R r │
   r           │                 │
    ───────────│ u = L z + K y + S r │
               └─────────────────┘
```

Figure 5-1 Controlled system and controller.

The compensator is a device that influences the structural features of the controlled system to which it is connected, while the regulator influences both the system structural features and asymptotic behavior. The *manipulable input* u is separate from the *nonmanipulable input* d, and the *informative output* y is separate from the *regulated output* e. When d is completely unaccessible for measurement, it is also called *disturbance*. Therefore, two distinct input distribution matrices, B and D, and two distinct output distribution matrices, C and E, are considered. The compensator or regulator is a nonpurely dynamic system with an input y (which coincides with the informative output of the controlled system), a *reference input* r, which provides information on the control tasks and possibly includes a part of d (when the nonmanipulable input is accessible for measurement). In the overall system there are two separate state vectors, $x \in \mathcal{X} := \mathbf{R}^n$, the *controlled system state*, and $z \in \mathcal{Z} := \mathbf{R}^m$, the *controller state*.

The overall system considered is very general and versatile: by setting equal to zero some of its matrices it can reproduce in practice all control situations: with dynamic or algebraic output feedback, with dynamic or algebraic precompensation (feedforward), or with mixed feedback and feedforward.

The overall system inputs d and r are assumed to be completely general, i.e., to belong to the class of piecewise continuous functions. In solving control system synthesis problems such a generality may be superfluous and too restrictive: it may be convenient, for instance, to assume that all these inputs or a part of them are generated by a linear time-invariant exosystem. Since in synthesis the exosystem features directly influence some of the obtained regulator features (for instance order and structure), it is convenient to embed the exosystem matrix in that of the controlled system. The controlled system state will be partitioned as

$$x = \begin{bmatrix} x_1 \\ x_2 \end{bmatrix} \tag{6}$$

where x_1 denotes the state of the *plant* and x_2 that of the exosystem. Matrices A, B, C, D, E are accordingly partitioned as

$$A = \begin{bmatrix} A_1 & A_3 \\ O & A_2 \end{bmatrix} \quad B = \begin{bmatrix} B_1 \\ O \end{bmatrix} \quad D = \begin{bmatrix} D_1 \\ O \end{bmatrix} \quad (7)$$

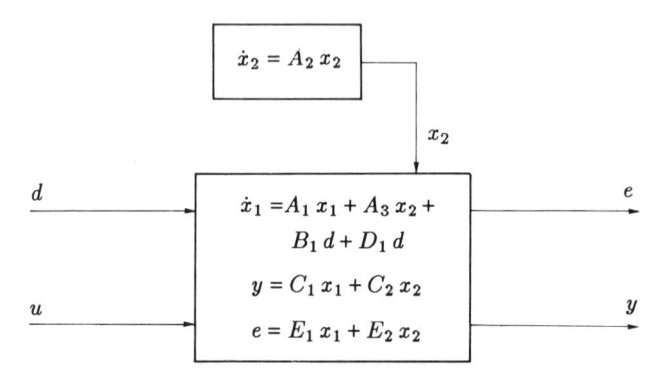

$$C = [\, C_1 \quad C_2 \,] \qquad E = [\, E_1 \quad E_2 \,]$$

Figure 5-2 Controlled system including an exosystem.

Note that the exosystem cannot be influenced by either input, but directly influences both outputs. The controlled system structure is shown in Fig. 5-2. The system is not completely controllable, since inputs act only on the plant, but it is assumed to be completely observable (or, at least, reconstructable) through the informative output. In fact the regulator must receive information, direct or indirect, on all the exogenous modes to counterbalance their effects on the regulated output.

Summing up, the following assumptions are introduced:
1. pair (A_1, B_1) is stabilizable
2. pair (A, C) is detectable

Note that the plant is a well-defined geometric object, namely the A-invariant defined by

$$\mathcal{P} := \{ x \,:\, x_2 = 0 \} \quad (8)$$

The overall system represented in Fig. 5-1 is purely dynamic with two inputs, d and r, and one output, e. In fact, by denoting with

$$\hat{x} := \begin{bmatrix} x \\ z \end{bmatrix} \quad (9)$$

Sect. 5.1 The Five-Map System

the *extended state* (controlled system and regulator state), the overall system equations can be written in compact form as

$$\dot{\hat{x}}(t) = \hat{A}\,\hat{x}(t) + \hat{D}\,d(t) + \hat{R}\,r(t) \qquad (10)$$

$$e(t) = \hat{E}\,\hat{x}(t) \qquad (11)$$

where

$$\hat{A} := \begin{bmatrix} A+BKC & BL \\ MC & N \end{bmatrix} \qquad \hat{D} := \begin{bmatrix} D \\ O \end{bmatrix}$$

$$\hat{R} := \begin{bmatrix} BS \\ R \end{bmatrix} \qquad \hat{E} := [\,E \ \ O\,] \qquad (12)$$

Note that, while the quintuple (A, B, C, D, E) that defines the controlled system is given, the order m of the regulator and matrices K, L, M, N, R, S are a priori unknown: the object of synthesis is precisely to derive them. Thus, the overall system matrices $\hat{A}, \hat{R}$ are also a priori unknown.

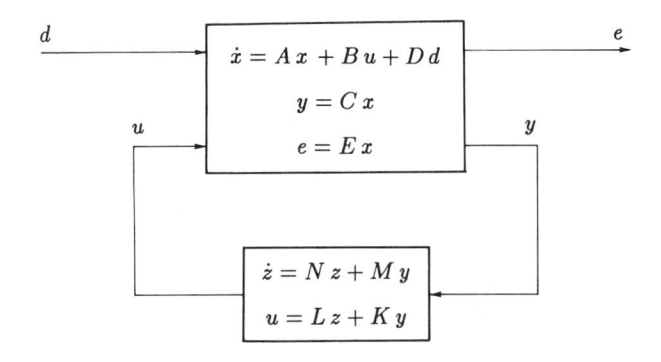

Figure 5-3 Reference block diagram for the disturbance localization problem by dynamic compensator, and the regulator problem.

In some important synthesis problems, like the disturbance localization by dynamic compensator, and the regulator problem (which will both be stated in the next section), input r is not present, so that the reference block diagram simplifies as in Fig. 5-3.

5.1.1 Some Properties of the Extended State Space

We shall now show that geometric properties referring to $\hat{A}$-invariants in the extended state space reflect into properties of $(A, \mathcal{B})$-controlled and $(A, \mathcal{C})$-conditioned invariants, regarding the controlled system alone. This makes it possible to state necessary and sufficient conditions for solvability of the most important synthesis problems in terms of the given quintuple (A, B, C, D, E).

The following property, concerning algebraic output-to-input feedback, is useful to derive the basic necessary structural condition for dynamic compensator and regulator design.[1]

Property 5.1-1.
Refer to triple (A, B, C). There exists a matrix K such that a given subspace $\mathcal{V}$ is an $(A+BKC)$-invariant if and only if $\mathcal{V}$ is both an $(A, \mathcal{B})$-controlled and an $(A, \mathcal{C})$-conditioned invariant.

Proof. Only if. This part of the proof is trivial because if there exists a matrix K such that $(A+BKC)\mathcal{V} \subseteq \mathcal{V}$ clearly there exist matrices $F := KC$ and $G := BK$ such that $\mathcal{V}$ is both an $(A+BF)$-invariant and an $(A+GC)$-invariant, hence an $(A, \mathcal{B})$-controlled and an $(A, \mathcal{C})$-conditioned invariant.

If. Consider a nonsingular matrix $T := [T_1\, T_2\, T_3\, T_4]$, with $\text{im}\, T_1 = \mathcal{V} \cap \mathcal{C}$, $\text{im}\,[T_1\, T_2] = \mathcal{V}$, $\text{im}\,[T_1\, T_3] = \mathcal{C}$, and set the following equation in K:

$$KC[T_2\ T_4] = F[T_2\ T_4] \tag{13}$$

Assume that C has maximal rank (if not, it is possible to ignore some output variables to meet this requirement and insert corresponding zero columns in the derived matrix). On this assumption $C[T_2\, T_4]$ is clearly a nonsingular square matrix, so that equation (13) admits a solution K for all F. Since $\mathcal{V}$ is an $(A, \mathcal{C})$-conditioned invariant

$$(A + BKC)\,\text{im}\, T_1 = A(\mathcal{V} \cap \mathcal{C}) \subseteq \mathcal{V}$$

On the other hand, with $KCT_2 = FT_2$ by virtue of (9), it follows that

$$(A + BKC)\,\text{im}\, T_2 = (A + BF)\,\text{im}\, T_2 \subseteq (A + BF)\mathcal{V} \subseteq \mathcal{V} \quad \square$$

It is now convenient to introduce a new formulation of the problem, where the synthesis of a dynamic regulator is precisely reduced to the derivation of an algebraic output feedback for a new extended system, still of order $n+m$.[2]

The overall system of Fig. 5-3 is equivalent to that shown in Fig. 5-4, where the extended system

$$\dot{\hat{x}}(t) = \hat{A}_0\, \hat{x}(t) + \hat{B}_0\, \hat{u}(t) + \hat{D}\, d(t) \tag{14}$$

$$\hat{y}(t) = \hat{C}_0\, \hat{x}(t) \tag{15}$$

$$e(t) = \hat{E}\, \hat{x}(t) \tag{16}$$

with state, input, and output defined as

$$\hat{x} := \begin{bmatrix} x \\ z \end{bmatrix} \qquad \hat{u} := \begin{bmatrix} u \\ v \end{bmatrix} \qquad \hat{y} := \begin{bmatrix} y \\ w \end{bmatrix} \tag{17}$$

---

[1] See Basile and Marro [4.6], Hamano and Furuta [18].

[2] This artifice is due to Willems and Commault [37].

Sect. 5.1 The Five-Map System

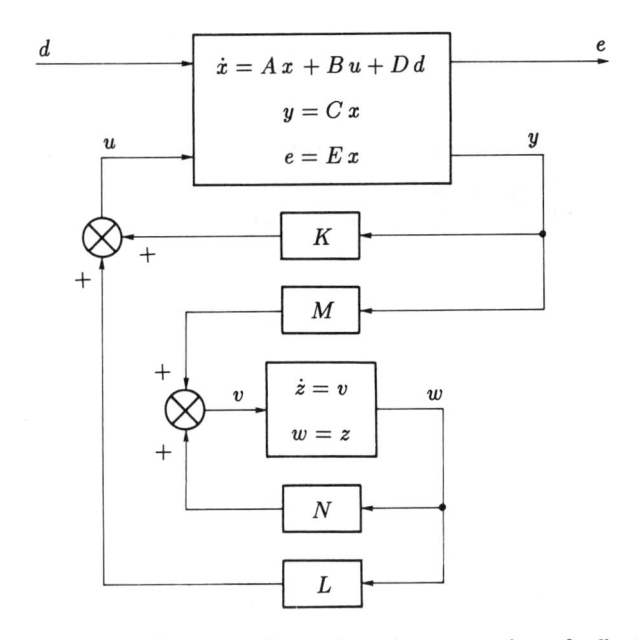

Figure 5-4 Artifice to transform a dynamic output-to-input feedback into an algebraic one.

and matrices

$$\hat{A}_0 := \begin{bmatrix} A & O \\ O & O \end{bmatrix} \quad \hat{B}_0 := \begin{bmatrix} B & O \\ O & I_m \end{bmatrix} \quad \hat{D} := \begin{bmatrix} D \\ O \end{bmatrix}$$

$$\hat{C}_0 := \begin{bmatrix} C & O \\ O & I_m \end{bmatrix} \quad \hat{E} := [E \quad O] \tag{18}$$

is subject to the algebraic output feedback

$$\hat{K} := \begin{bmatrix} K & L \\ M & N \end{bmatrix} \tag{19}$$

Note that in (18) $\hat{D}$ and $\hat{E}$ are defined as in (12).

The equivalence between the block diagrams of Figs. 5-3 and 5-4 and Property 5.1-1 immediately leads to the following statement.

Property 5.1-2.

Any extended subspace $\hat{\mathcal{W}}$ that is an $\hat{A}$-invariant is both an $(\hat{A}_0, \hat{B}_0)$-controlled and an $(\hat{A}_0, \hat{C}_0)$-conditioned invariant.

Consider the following subspaces of $\mathcal{X}$ (the controlled system state space):

$$P(\hat{\mathcal{W}}) := \left\{ x : \begin{bmatrix} x \\ z \end{bmatrix} \in \hat{\mathcal{W}} \right\} \tag{20}$$

$$I(\hat{W}) := \left\{ x : \begin{bmatrix} x \\ 0 \end{bmatrix} \in \hat{W} \right\} \tag{21}$$

which are called, respectively, the *projection* of $\hat{W}$ on $\mathcal{X}$ and the *intersection* of $\hat{W}$ with $\mathcal{X}$. They are effective tools to deal with extended systems.

In the extended state space the controlled system and controller state spaces are respectively

$$\hat{\mathcal{X}} := \left\{ \begin{bmatrix} x \\ z \end{bmatrix} : z = 0 \right\} \qquad \hat{\mathcal{Z}} := \left\{ \begin{bmatrix} x \\ z \end{bmatrix} : x = 0 \right\} \tag{22}$$

They are orthogonal to each other and satisfy the following, easily derivable relations:

$$P(\hat{W}) = I(\hat{W} + \hat{\mathcal{Z}}) \tag{23}$$
$$I(\hat{W}) = P(\hat{W} \cap \hat{\mathcal{X}}) \tag{24}$$
$$I((\hat{W} \cap \hat{\mathcal{X}})^\perp) = P((\hat{W} \cap \hat{\mathcal{X}})^\perp) = I(\hat{W})^\perp \tag{25}$$
$$P((\hat{W} + \hat{\mathcal{Z}})^\perp) = I((\hat{W} + \hat{\mathcal{Z}})^\perp) = P(\hat{W})^\perp \tag{26}$$

Property 5.1-3.

The projection of the orthogonal complement of any extended subspace is equal to the orthogonal complement of its intersection:

$$P(\hat{W}^\perp) = I(\hat{W})^\perp \tag{27}$$

Proof. Equalities (23) and (25) lead to

$$P(\hat{W}^\perp) = I(\hat{W}^\perp + \hat{\mathcal{Z}}) = I((\hat{W} \cap \hat{\mathcal{X}})^\perp) = I(\hat{W})^\perp \quad \square$$

Consider now the following lemma, which will be used in the next section to prove the nonconstructive necessary and sufficient conditions.

Lemma 5.1-1.

Subspace $\hat{\mathcal{V}}$ is an internally and/or externally stabilizable $(\hat{A}_0, \hat{B}_0)$-controlled invariant if and only if $P(\hat{\mathcal{V}})$ is an internally and/or externally stabilizable (A, B)-controlled invariant.

Proof. *Only if.* This part of the proof is an immediate consequence of Definitions 4.1-5 and 4.1-6 (internal and external stabilizability of a controlled invariant).

If. Let $\mathcal{V} := P(\hat{\mathcal{V}})$: it is easily seen that $\hat{\mathcal{V}}$ can be expressed as

$$\hat{\mathcal{V}} := \left\{ \begin{bmatrix} x \\ z \end{bmatrix} : x \in \mathcal{V}, \; z = W x + \eta, \; \eta \in \mathcal{L} \right\} \tag{28}$$

Sect. 5.1 The Five-Map System

where W denotes a suitable $m \times n$ matrix and $\mathcal{L}$ a suitable subspace of the regulator state space $\mathcal{Z}$. From (28) it clearly follows that $\dim \hat{\mathcal{V}} = \dim \mathcal{V} + \dim \mathcal{L}$. Assume a basis matrix $\hat{V}$ of $\hat{\mathcal{V}}$ and, if necessary, reorder its columns in such a way that in the partition

$$\hat{V} = \begin{bmatrix} V_1 & V_2 \\ V_3 & V_4 \end{bmatrix}$$

V_1 is a basis matrix of $\mathcal{V}$. Since all columns of V_2 are linear combinations of those of V_1, by subtracting these linear combinations a new basis matrix of $\hat{\mathcal{V}}$ can be obtained with the structure

$$\hat{V}' = \begin{bmatrix} V_1 & O \\ V_3 & V_4' \end{bmatrix}$$

where V_1 and V_4' have maximal rank. Any $\hat{x} \in \hat{\mathcal{V}}$ can be expressed as

$$\hat{x} = \begin{bmatrix} V_1 & O \\ V_3 & V_4' \end{bmatrix} \begin{bmatrix} \alpha_1 \\ \alpha_2 \end{bmatrix}$$

with α_1, α_2 properly dimensioned arbitrary real vectors, i.e., as

$$x = V_1 \alpha_1$$
$$z = V_3 \alpha_1 + V_4' \alpha_2$$

By eliminating α_1 we finally obtain

$$z = V_3 (V_1^T V_1)^{-1} V_1^T x + V_4' \alpha_2$$

which proves (28). Since $\mathcal{V}$ is an internally and/or externally stabilizable (A, B)-controlled invariant, there exists at least one matrix F such that $\mathcal{V}$ is an internally and/or externally stable $(A+BF)$-invariant. We choose the following state feedback matrix for the extended system:

$$\hat{F} := \begin{bmatrix} F & O \\ W(A+BF)+W & -I_m \end{bmatrix}$$

so that

$$\hat{A}_0 + \hat{B}_0 \hat{F} = \begin{bmatrix} A+BF & O \\ W(A+BF)+W & -I_m \end{bmatrix}$$

Referring to the new state coordinates ρ, η corresponding to the transformation

$$\hat{T} := \begin{bmatrix} I_n & O \\ W & I_m \end{bmatrix}$$

one obtains

$$\hat{T}^{-1}(\hat{A}_0 + \hat{B}_0 \hat{F}) \hat{T} = \begin{bmatrix} A+BF & O \\ O & -I_m \end{bmatrix}$$

and

$$\hat{\mathcal{V}} = \left\{ \begin{bmatrix} \rho \\ \eta \end{bmatrix} : \rho \in \mathcal{V}, \eta \in \mathcal{L} \right\}$$

The above change of coordinates clarifies that $\hat{\mathcal{V}}$ is an internally and/or externally stabilizable $(\hat{A}_0 + \hat{B}_0 \hat{F})$-invariant, hence an internally and/or externally stabilizable $(\hat{A}_0, \hat{B}_0)$-controlled invariant. □

Lemma 5.1-2.
Subspace $\hat{S}$ is an externally and/or internally stabilizable $(\hat{A}_0, \hat{C}_0)$-conditioned invariant if and only if $I(\hat{S})$ is an externally and/or internally stabilizable (A, C)-conditioned invariant.

Proof. Recall that any (A, C)-conditioned invariant is externally and/or internally stabilizable if its orthogonal complement, as an $(A^T, C^\perp)$-controlled invariant, is internally and/or externally stabilizable. Therefore, $\hat{S}$ is externally and/or internally stabilizable if and only if $\hat{S}^\perp$, as an $(\hat{A}_0^T, \text{im}\hat{C}_0^T)$-controlled invariant, is internally and/or externally stabilizable or, by virtue of Lemma 5.1-1, if and only if $P(\hat{S}^\perp)$, as an $(A^T, \text{im}C^T)$-controlled invariant, is internally and/or externally stabilizable. Hence, the statement directly follows from Property 5.1-3. □

5.1.2 Some Computational Aspects

The synthesis procedures that will be presented and used in the next section can be considered extensions of dynamic systems stabilization by means of observers and dual observers, already discussed in Section 3.4.

In general, as a first step it is necessary to derive a matrix F such that a given internally and externally stabilizable controlled invariant $\mathcal{V}$ is an $(A+BF)$-invariant with $A+BF$ stable or a matrix G such that a given externally and internally stabilizable conditioned invariant S is an $(A+GC)$-invariant with $A+GC$ stable. The structure requirement can be imposed independently of the stability requirement: for instance, as far as matrix F is concerned, first derive an F_1 such that $(A+BF_1)\mathcal{V} \subseteq \mathcal{V}$ by means of Algorithm 4.1-3, then express matrices $A+BF_1$ and B in a basis whose vectors span $\mathcal{R}_\mathcal{V}, \mathcal{V}, \mathcal{X}$ (which are $(A+BF_1)$-invariants). Then apply an eigenvalue assignment procedure to the controllable pairs of submatrices (A'_{ij}, B'_i) corresponding to $\mathcal{R}_\mathcal{V}$ and $\mathcal{X}/\mathcal{V}$, which are respectively controllable by construction and stabilizable by assumption. In this way a matrix F_2 is determined which, added to F_1, solves the problem. This procedure can be dualized for matrix G in connection with conditioned invariants.

Refer to the block diagram of Fig. 3-13, where an identity observer is used to indirectly perform state feedback: the purely algebraic block F can be considered as connected between output z of the asymptotic observer shown in Fig. 3-14(b) and system input u. Note that the same result is obtained referring to the dual observer of Fig. 3-14(c): a purely algebraic block G is connected between a summing junction providing the difference $\eta - y$ (of the model and system outputs) and the model forcing action φ.

In the former case, information on the system state to perform state feedback is *completely* derived from the asymptotic observer, and the direct partial information provided by the system output is not taken into account. A more general way to realize state feedback, which includes the complete direct state feedback (which would be possible if C were square and nonsingular)

Sect. 5.1 The Five-Map System

and the complete indirect feedback through the observer as particular cases, is that shown in Fig. 5-4(a). Information on state is there derived as a linear combination of both the system output and the observer state (algebraic blocks L_1 and L_2 and summing junction), then applied to the system input through the algebraic block F.

Let L_1 and L_2 satisfy

$$L_1 C + L_2 = I_n \tag{29}$$

and apply to the extended system

$$\begin{bmatrix} \dot{x}(t) \\ \dot{z}(t) \end{bmatrix} = \begin{bmatrix} A+BFL_1C & BFL_2 \\ BFL_1C-GC & A+GC+BFL_2 \end{bmatrix} \begin{bmatrix} x(t) \\ z(t) \end{bmatrix} + \begin{bmatrix} D \\ O \end{bmatrix} d(t) \tag{30}$$

the coordinate transformation expressed by

$$\begin{bmatrix} x \\ z \end{bmatrix} = \begin{bmatrix} I_n & O \\ I_n & -I_n \end{bmatrix} \begin{bmatrix} \rho \\ \eta \end{bmatrix} \tag{31}$$

i.e., $\rho := x$, $\eta := x - z$. The equivalent system

$$\begin{bmatrix} \dot{\rho}(t) \\ \dot{\eta}(t) \end{bmatrix} = \begin{bmatrix} A+BF & -BFL_2 \\ O & A+GC \end{bmatrix} \begin{bmatrix} \rho(t) \\ \eta(t) \end{bmatrix} + \begin{bmatrix} D \\ D \end{bmatrix} d(t) \tag{32}$$

is obtained. Note, in particular, that the separation property expressed by Theorem 3.4-6 still holds.

Figure 5-5(b) shows the dual connection: difference $\eta - y$ is processed through the algebraic block G, then applied both to the system input and the dual observer forcing action through the algebraic blocks L_1 and L_2 and summing junctions.

Let

$$B L_1 + L_2 = I_n \tag{33}$$

From the extended system, described by

$$\begin{bmatrix} \dot{x}(t) \\ \dot{z}(t) \end{bmatrix} = \begin{bmatrix} A+BL_1GC & BF-BL_1GC \\ -L_2GC & A+BF+L_2GC \end{bmatrix} \begin{bmatrix} x(t) \\ z(t) \end{bmatrix} + \begin{bmatrix} D \\ O \end{bmatrix} d(t) \tag{34}$$

through the coordinate transformation

$$\begin{bmatrix} x \\ z \end{bmatrix} = \begin{bmatrix} -I_n & I_n \\ O & I_n \end{bmatrix} \begin{bmatrix} \rho \\ \eta \end{bmatrix} \tag{35}$$

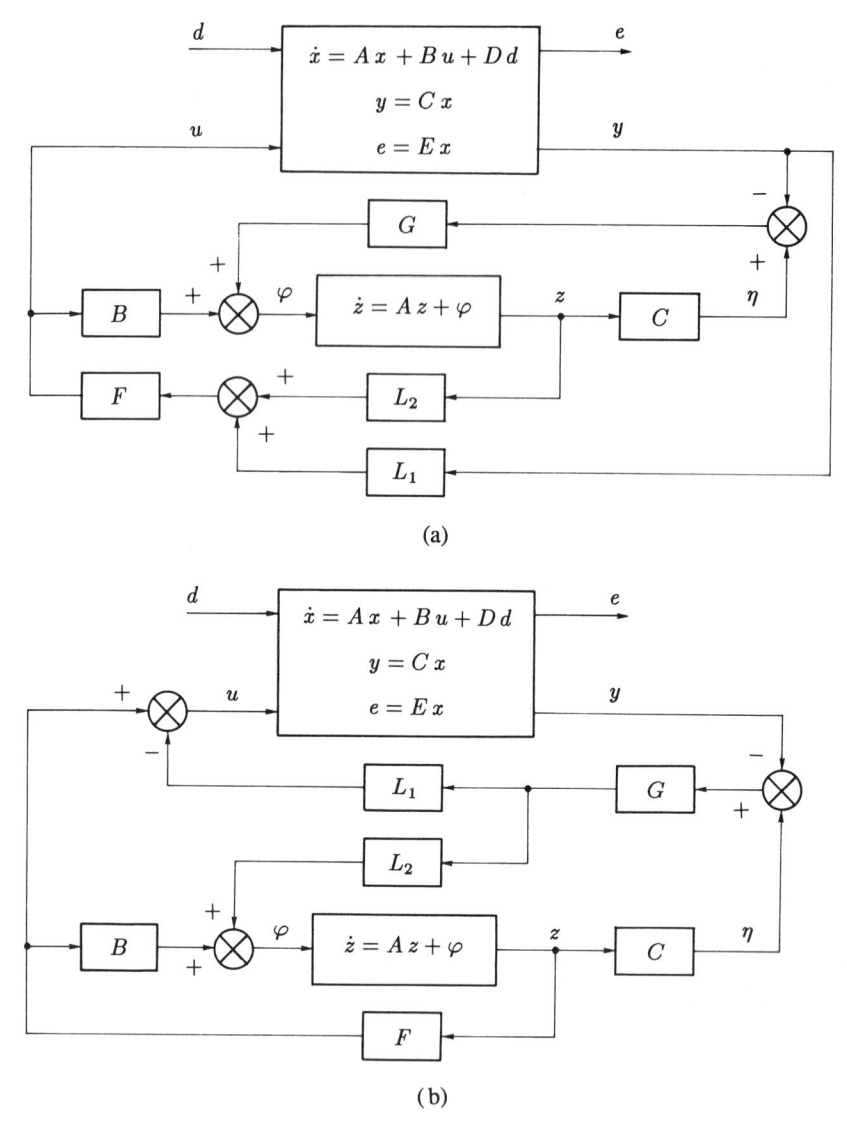

Figure 5-5 Controllers based respectively on the identity observer and the identity dual observer.

i.e., $\rho := x - z$, $\eta := z$, the equivalent system

$$\begin{bmatrix} \dot{\rho}(t) \\ \dot{\eta}(t) \end{bmatrix} = \begin{bmatrix} A+GC & O \\ L_2 GC & A+BF \end{bmatrix} \begin{bmatrix} \rho(t) \\ \eta(t) \end{bmatrix} + \begin{bmatrix} -D \\ O \end{bmatrix} d(t) \qquad (36)$$

is obtained. The separation property also clearly holds in this case.

The crucial point of these procedures is the choice of matrices L_1 and L_2: in fact, while respecting the constraints expressed by (29) and (33), it may be

Sect. 5.1 The Five-Map System

possible to impose further conditions that imply special structural properties for the overall system. The following lemmas provide a useful link between geometric-type conditions and computational support for this problem.

Lemma 5.1-3.
Let C be any $q \times n$ matrix and $\mathcal{L}$ a subspace of $\mathcal{X}$ such that $\mathcal{L} \cap \mathcal{C} = \{0\}$, with $\mathcal{C} := \ker C$. There exist two matrices L_1, L_2 such that

$$L_1 C + L_2 = I_n \qquad \ker L_2 = \mathcal{L} \tag{37}$$

Proof. Let $\mathcal{L}_c$ be any subspace that satisfies

$$\mathcal{L} \oplus \mathcal{L}_c = \mathcal{X} \qquad \mathcal{L}_c \supseteq \mathcal{C} \tag{38}$$

Define L_2 as the projecting matrix on $\mathcal{L}_c$ along $\mathcal{L}$, so that $I_n - L_2$ is the complementary projecting matrix and $\ker(I_n - L_2) = \mathcal{L}_c$. Hence, the equation

$$L_1 C = I_n - L_2$$

is solvable in L_1 by virtue of the second of (38). In fact, recall that the generic linear system $AX = B$ or $X^T A^T = B^T$ is solvable in X if $\operatorname{im} A \supseteq \operatorname{im} B$ or $\ker A^T \subseteq \ker B^T$. □

Note that this proof is constructive, i.e., it provides a procedure to derive L_1, L_2. The dual result, which is useful for synthesis based on the dual observer, is stated without proof as follows.

Lemma 5.1-4.
Let B be any $n \times p$ matrix and $\mathcal{L}$ a subspace of $\mathcal{X}$ such that $\mathcal{L} + \mathcal{B} = \mathcal{X}$, with $\mathcal{B} := \operatorname{im} B$. There exist two matrices L_1, L_2 such that

$$B L_1 + L_2 = I_n \qquad \operatorname{im} L_2 = \mathcal{L} \tag{39}$$

Two Simple Applications. To show how the preceding lemmas can be used in synthesis procedures, we shall look at two simple computational problems. First, consider again the unknown-input asymptotic observers whose block diagrams are shown in Figs. 4-5 and 4-6 or, in more compact form, in Fig. 5-6.

Let S be our *resolvent*, i.e., an internally and externally stabilizable (A, C)-conditioned invariant such that $S \supseteq \mathcal{D}$ and $S \cap \mathcal{C} \subseteq \mathcal{E}$. First, determine a matrix G such that $(A+GC)S \subseteq S$ with $A+GC$ stable. Assume $N := A+GC$, $M := -G$, and $K = O$ in the case of a purely dynamic observer. Our aim is to derive L in the purely dynamic case (it may be different from E if a reduced order device is sought) and K, L in the other case. To this end, derive a subspace $\mathcal{L}$ that satisfies $\mathcal{L} \oplus S \cap \mathcal{C} = S$ (or $\mathcal{L} + S \cap \mathcal{C} = S$ and $\mathcal{L} \cap (S \cap \mathcal{C}) = \{0\}$), i.e., a complement of $S \cap \mathcal{C}$ to S. Clearly $\mathcal{C} \cap \mathcal{L} = \{0\}$. By virtue of Lemma 5.1-3 there exist two matrices L_1, L_2 such that $L_1 C + L_2 = I_n$, $\ker L_2 = \mathcal{L}$. Premultiplying by E yields $E L_1 C + E L_2 = E$, which, by assuming $K := E L_1$, $L := E L_2$, can also be written as

$$KC + L = E \quad \text{with} \quad \ker L \supseteq S \tag{40}$$

To compute L_1, L_2, first derive a matrix $X := [X_1\ X_2]$ such that $\mathrm{im} X_1 = \mathcal{S} \cap \mathcal{C}$, $\mathrm{im}[X_1\ X_2] = \mathcal{S}$, and assume $\mathcal{L} := \mathrm{im} X_2$: clearly $\mathcal{L} \cap \mathcal{C} = \{0\}$, $\mathcal{L} + \mathcal{S} \cap \mathcal{C} = \mathcal{S}$. Then apply the constructive procedure outlined in the proof of Lemma 5.1-3. The inclusion on the right of (40) follows from both subspaces, whose direct sum is $\mathcal{S}$, being contained in $\ker L$: in fact $\ker L_2 = \mathcal{L}$ by construction (so that $\ker L \supseteq \mathcal{L}$) and from $\mathcal{S} \cap \mathcal{C} \subseteq \mathcal{E}$ (a property of $\mathcal{S}$ which can also be written as $E(\mathcal{S} \cap \mathcal{C}) = \{0\}$), and $KC(\mathcal{S} \cap \mathcal{C}) = \{0\}$ (by definition of C), by virtue of (40) it follows that $L(\mathcal{S} \cap \mathcal{C}) = \{0\}$.

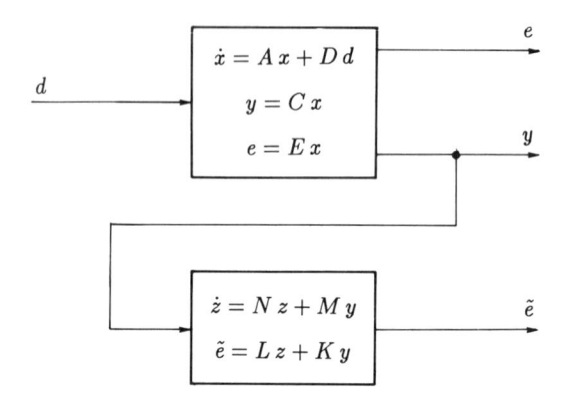

Figure 5-6 Unknown-input nonpurely dynamic asymptotic observer.

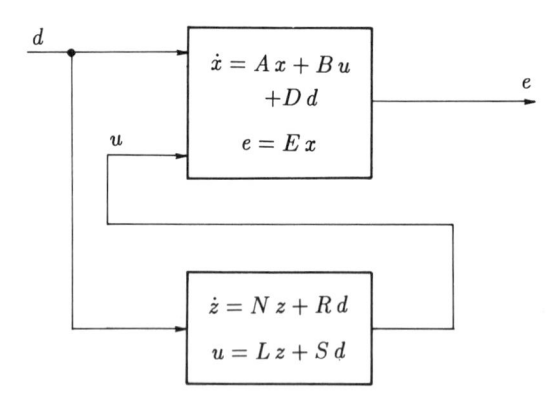

Figure 5-7 Dynamic accessible disturbance localizing unit.

Furthermore, the observer order can be reduced to $n - \dim \mathcal{S}$ and the stability requirement restricted to $\mathcal{S}$ being externally stabilizable. For this, perform in the observer state space the change of basis corresponding to $T := [T_1\ T_2]$ with $\mathrm{im} T_1 = \mathcal{S}$: in practice the first group of coordinates is not needed since it corresponds to an $(A+GC)$-invariant contained in $\ker E$ or in $\ker L$ so that it does not influence the other coordinates and the observer output. Let

$$Q = \begin{bmatrix} Q_1 \\ Q_2 \end{bmatrix} := T^{-1}$$

Sect. 5.1 The Five-Map System

For the purely dynamic observer, we set the equations

$$\dot{z}(t) = N_1 \, z(t) + M_1 \, y(t) \qquad \tilde{e}(t) = E_1 \, z(t)$$

with $N_1 := Q_2(A+GC)T_2$, $M_1 := -Q_2 G$, $E_1 := ET_2$, while for the nonpurely dynamic one we derive

$$\dot{z}(t) = N_1 \, z(t) + M_1 \, y(t) \qquad \tilde{e}(t) = L_1 \, z(t) + K \, y(t)$$

with $N_1 := Q_2(A+GC)T_2$, $M_1 := -Q_2 G$, $L_1 := LT_2$.

We shall now consider the dual problem, i.e., the synthesis of a dynamic precompensator or dual observer which realizes localization of an accessible input according to the block diagram shown in Fig. 5-7. The geometric starting point for solution is to have again a resolvent, which in this case is an internally and externally stabilizable (A, B)-controlled invariant $\mathcal{V}$ such that $\mathcal{V} \subseteq \mathcal{E}$ and $\mathcal{V} + \mathcal{B} \supseteq \mathcal{D}$; hence it is possible to determine a matrix F such that $(A+BF)\mathcal{V} \subseteq \mathcal{V}$ with $A+BF$ stable. Assume $N := A + BF$ and $L := F$. Then, determine a subspace $\mathcal{L}$ that satisfies $\mathcal{L} \cap \mathcal{V} + \mathcal{B} = \mathcal{V}$ and $\mathcal{L} + (\mathcal{V} + \mathcal{B}) = \mathcal{X}$. Clearly $\mathcal{B} + \mathcal{L} = \mathcal{X}$. By virtue of Lemma 5.1-3 there exist two matrices L_1, L_2 such that $BL_1 + L_2 = I_n$, $\text{im}\, L_2 = \mathcal{L}$. Postmultiplying by D yields $BL_1 D + L_2 D = D$ which, by assuming $S := -L_1 D$, $R := L_2 D$, can also be written as

$$-BS + R = D \quad \text{with} \quad \text{im}\, R \subseteq \mathcal{V} \tag{41}$$

The last condition follows from both subspaces whose intersection is $\mathcal{S}$ containing $\text{im}\, R$: in fact $\text{im}\, L_2 = \mathcal{L}$ by construction (so that $\text{im}\, R \subseteq \mathcal{L}$) and from $\mathcal{V} + \mathcal{B} \supseteq \mathcal{D}$ (a property of $\mathcal{V}$ which can also be written as $D^{-1}(\mathcal{V}+\mathcal{B})=\mathcal{X}$), and $(BS)^{-1}(\mathcal{V}+\mathcal{B})=\mathcal{X}$ (by definition of $\mathcal{B}$), by virtue of (41) it follows that $R^{-1}(\mathcal{V}+\mathcal{B})=\mathcal{X}$ or $\mathcal{V}+\mathcal{B} \supseteq \text{im}\, R$.

Furthermore, the dual observer order can be reduced to $\dim \mathcal{V}$ and the stability requirement restricted to $\mathcal{V}$ being internally stabilizable. For this, perform in the dual observer state space the change of basis corresponding to $T := [T_1 \, T_2]$ with $\text{im}\, T_1 = \mathcal{V}$: in practice the second group of coordinates is not needed since all the zero-state admissible trajectories are restricted to the first group, which is an $(A+BF)$-invariant containing $\text{im}\, R$ so that it coincides with the reachable subspace of the dual observer. The recipe for the localizing unit is stated as follows: let

$$Q = \begin{bmatrix} Q_1 \\ Q_2 \end{bmatrix} := T^{-1}$$

and set the equations

$$\dot{z}(t) = N_1 \, z(t) + R_1 \, d(t) \qquad y(t) = L_1 \, z(t) + S \, d(t)$$

with $N := Q_1(A+BF)T_1$, $R_1 := Q_1 R$, $L_1 := FT_1$.

5.1.3 The Dual-Lattice Structures

Theorems 4.1-4 and 4.1-5 point out an interesting connection between controlled and conditioned invariants, which no longer appear as separate objects, connected only by duality relations, but as elements that are both necessary to derive remarkably simple and elegant algebraic expressions: see, for instance, expressions (4.1.30, 32), which provide the constrained reachable set and its dual.

In this subsection the algebraic basic structures of lattices $\Phi(*,*)$ and $\Psi(*,*)$, introduced in Section 4.1, are presented and investigated as a convenient background to their use in solving synthesis problems. Structures will be graphically represented by means of Hasse diagrams referring to the inclusion, which allow a simple representation of relations between the elements that contribute to problem solution, some a priori known and some available through suitable algorithms.

First, we refer to triple (A, B, C) and consider the *fundamental lattices* $\Phi(\mathcal{B},\mathcal{C})$ and $\Psi(\mathcal{C},\mathcal{B})$, with $\mathcal{B} := \text{im} B$, $\mathcal{C} := \text{ker} C$. This particular case will be used as a reference to derive more complex structures, like those that are used in connection with quintuple (A, B, C, D, E) to solve synthesis problems. The basic property that sets a one-to-one correspondence between the lattice of all $(A,\mathcal{B})$-controlled invariants self-bounded with respect to $\mathcal{C}$ and that of $(A,\mathcal{C})$-conditioned invariants self-hidden with respect to $\mathcal{B}$, is stated as follows.

Property 5.1-4.

Let $\mathcal{V}$ be any $(A,\mathcal{B})$-controlled invariant contained in $\mathcal{C}$, and $\mathcal{S}$ any $(A,\mathcal{C})$-conditioned invariant containing $\mathcal{B}$: then

1. $\mathcal{V} \cap \mathcal{S}$ is an $(A,\mathcal{B})$-controlled invariant;
2. $\mathcal{V}+\mathcal{S}$ is an $(A,\mathcal{C})$-conditioned invariant.

Proof. From

$$A(\mathcal{S} \cap \mathcal{C})\mathcal{S} \subseteq \mathcal{S} \qquad \mathcal{S} \supseteq \mathcal{B} \qquad (42)$$

$$A\mathcal{V} \subseteq \mathcal{V}+\mathcal{B} \qquad \mathcal{V} \subseteq \mathcal{C} \qquad (43)$$

it follows that

$$A(\mathcal{V} \cap \mathcal{S}) = A(\mathcal{V} \cap \mathcal{S} \cap \mathcal{C}) \subseteq A\mathcal{V} \cap A(\mathcal{S} \cap \mathcal{C}) \subseteq (\mathcal{V}+\mathcal{B}) \cap \mathcal{S} = \mathcal{V} \cap \mathcal{S} + \mathcal{B}$$

$$A\big((\mathcal{V}+\mathcal{S}) \cap \mathcal{C}\big) = A(\mathcal{V}+\mathcal{S} \cap \mathcal{C}) = A\mathcal{V} + A(\mathcal{S} \cap \mathcal{C}) \subseteq \mathcal{V}+\mathcal{B}+\mathcal{S} = \mathcal{V}+\mathcal{S} \quad \square$$

The fundamental lattices are defined as

$$\Phi(\mathcal{B},\mathcal{C}) := \{\mathcal{V} : A\mathcal{V} \subseteq \mathcal{V}+\mathcal{B}, \mathcal{V} \subseteq \mathcal{C}, \mathcal{V} \supseteq \mathcal{V}_0^* \cap \mathcal{B}\} \qquad (44)$$

$$\Psi(\mathcal{C},\mathcal{B}) := \{\mathcal{S} : A(\mathcal{S} \cap \mathcal{C}) \subseteq \mathcal{S}, \mathcal{S} \supseteq \mathcal{B}, \mathcal{S} \subseteq \mathcal{S}_0^* + \mathcal{C}\} \qquad (45)$$

with

$$V_0^* := \max \mathcal{V}(\mathcal{A}, \mathcal{B}, \mathcal{C}) \tag{46}$$

$$S_0^* := \min \mathcal{S}(\mathcal{A}, \mathcal{C}, \mathcal{B}) \tag{47}$$

Referring to these elements, we can state the following basic theorem.

Theorem 5.1-1.
Relations

$$\mathcal{S} = \mathcal{V} + S_0^* \tag{48}$$
$$\mathcal{V} = \mathcal{S} \cap V_0^* \tag{49}$$

state a one-to-one function and its inverse between $\Phi(\mathcal{B},\mathcal{C})$ and $\Psi(\mathcal{C},\mathcal{B})$. Sums and intersections are preserved in these functions.

Proof. $\mathcal{V} + S_0^*$ is an (A,C)-conditioned invariant by virtue of Property 5.1-4, self-hidden with respect to $\mathcal{B}$ since it is contained in $S_0^* + \mathcal{C}$. Furthermore

$$(\mathcal{V} + S_0^*) \cap V_0^* = \mathcal{V} + S_0^* \cap V_0^* = \mathcal{V}$$

because $\mathcal{V}$, being self-bounded with respect to $\mathcal{C}$, contains the infimum of $\Phi(\mathcal{B},\mathcal{C})$, $V_0^* \cap S_0^*$. By duality, $\mathcal{S} \cap V_0^*$ is an (A,B)-controlled invariant again by Property 5.1-4, self-bounded with respect to $\mathcal{E}$ because it contains $V_0^* \cap \mathcal{B}$. Furthermore

$$(\mathcal{S} \cap V_0^*) + S_0^* = \mathcal{S} \cap V_0^* + S_0^* = \mathcal{S}$$

because $\mathcal{S}$, being self-hidden with respect to $\mathcal{B}$, is contained in the supremum of $\Psi(\mathcal{B},\mathcal{C})$, $V_0^* + S_0^*$. Functions defined by (48,49) are one-to-one because, as just proved, their product is the identity in $\Phi(\mathcal{B},\mathcal{C})$ and their inverse product is the identity in $\Psi(\mathcal{C},\mathcal{B})$. Since (48) preserves sums and (49) intersections and both are one-to-one, sums and intersections are preserved in both functions. □

Figure 5-8 shows the Hasse diagrams of the subspace sets that are referred to in the definitions of the fundamental lattices. Thicker lines denote the parts of the diagrams corresponding to lattices. Note that the eigenvalue assignability is also pointed out and the "zones" corresponding to invariant zeros of triple (A,B,C) are specified in both lattices.

We shall now show that the above one-to-one correspondence can be extended to other lattices, which are more directly connected with the *search for resolvents* for synthesis problems, which usually concerns the quintuple (A,B,C,D,E). Let $\mathcal{D} := \mathrm{im}\, D$, $\mathcal{E} := \ker E$, and assume

$$\mathcal{D} \subseteq \mathcal{V}^* \tag{50}$$

By Property 4.2-1, on this assumption

$$\mathcal{V}^* := \max \mathcal{V}(A, \mathcal{B}, \mathcal{E}) = \max \mathcal{V}(A, \mathcal{B}+\mathcal{D}, \mathcal{E}) \tag{51}$$

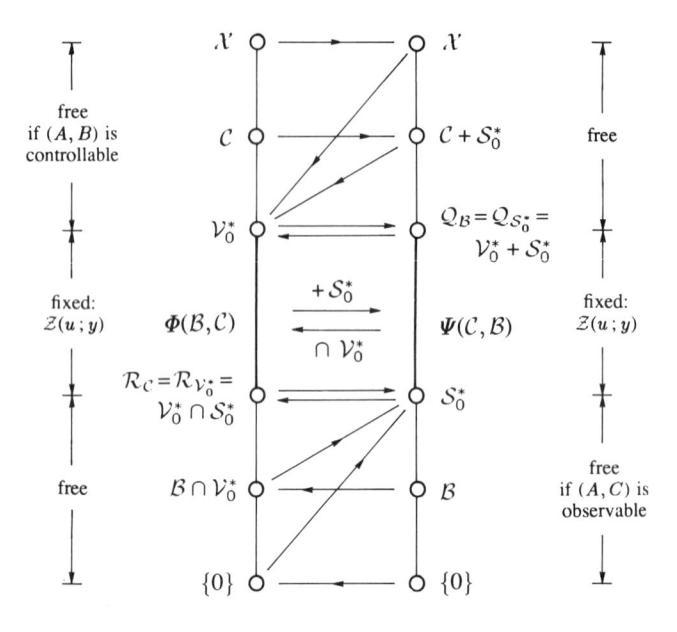

Figure 5-8 The fundamental lattices $\Phi(\mathcal{B},\mathcal{C})$ and $\Psi(\mathcal{C},\mathcal{B})$.

Consider the lattices

$$\Phi(\mathcal{B},\mathcal{E}) \quad \text{and} \quad \Phi(\mathcal{B}+\mathcal{D},\mathcal{E}) \tag{52}$$

which, the latter being a part of the former (see Property 4.4-2), can be represented in the same Hasse diagram, as shown in Fig. 5-9. In the figure the following notations have been introduced:

$$\mathcal{S}_1^* := \min \mathcal{S}(A,\mathcal{E},\mathcal{B}+\mathcal{D}) \tag{53}$$

$$\mathcal{S}_2^* := \min \mathcal{S}(A,\mathcal{E},\mathcal{B}) \tag{54}$$

Their auxiliary dual lattices are

$$\Psi(\mathcal{E},\mathcal{B}) \quad \text{and} \quad \Psi(\mathcal{E},\mathcal{B}+\mathcal{D}) \tag{55}$$

i.e., the lattices of all $(A,\mathcal{E})$-conditioned invariants self-hidden with respect to $\mathcal{B}$ and $\mathcal{B}+\mathcal{D}$, which can also be represented in the same Hasse diagram. Note that the elements of the second auxiliary lattice can be obtained by summing $\mathcal{S}_2^*$ instead of $\mathcal{S}_1^*$ to the corresponding controlled invariants, since all these controlled invariants contain $\mathcal{D}$. The dual-lattice diagram represented in Fig. 5-9 is obtained from the fundamental one simply by replacing $\mathcal{B}$ with $\mathcal{B}+\mathcal{D}$ and $\mathcal{C}$ with $\mathcal{E}$. Also note that invariant zeros are related to lattices, being the unassignable internal or external eigenvalues of suitable well-defined sublattices of controlled or conditioned invariants.

Sect. 5.1 The Five-Map System 269

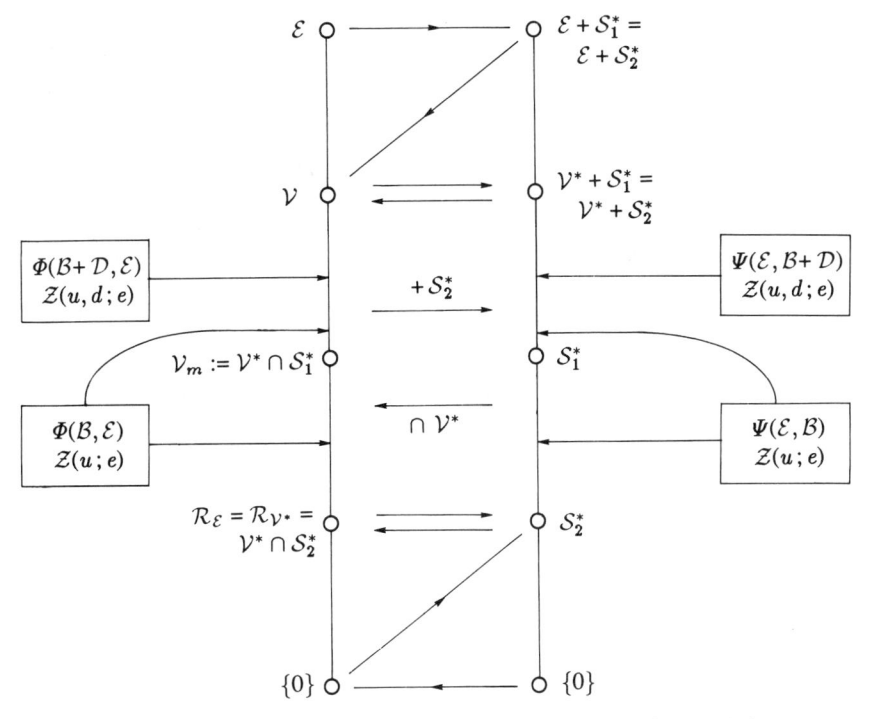

Figure 5-9 Lattices $\Phi(\mathcal{B}, \mathcal{E})$ and $\Phi(\mathcal{B}+\mathcal{D}, \mathcal{E})$ (on the left) and their auxiliary duals.

All of the preceding is dualized as follows. Let

$$\mathcal{S}^* \supseteq \mathcal{E} \tag{56}$$

By the dual of Property 4.2-1, on this assumption

$$\mathcal{S}^* := \min \mathcal{S}(A, \mathcal{C}, \mathcal{D}) = \min \mathcal{S}(A, \mathcal{C} \cap \mathcal{E}, \mathcal{D}) \tag{57}$$

Consider the lattices

$$\Psi(\mathcal{C}, \mathcal{D}) \quad \text{and} \quad \Psi(\mathcal{C} \cap \mathcal{E}, \mathcal{D}) \tag{58}$$

which, the latter being a part of the former (see Property 4.4-2), can be represented in the same Hasse diagram, as shown in Fig. 5-10. In the figure, the following notations have been introduced:

$$\mathcal{V}_1^* := \max \mathcal{V}(A, \mathcal{D}, \mathcal{C} \cap \mathcal{E}) \tag{59}$$

$$\mathcal{S}_2^* := \max \mathcal{V}(A, \mathcal{D}, \mathcal{C}) \tag{60}$$

Their auxiliary dual lattices are

$$\Phi(\mathcal{D}, \mathcal{C}) \quad \text{and} \quad \Phi(\mathcal{D}, \mathcal{C} \cap \mathcal{E}) \tag{61}$$

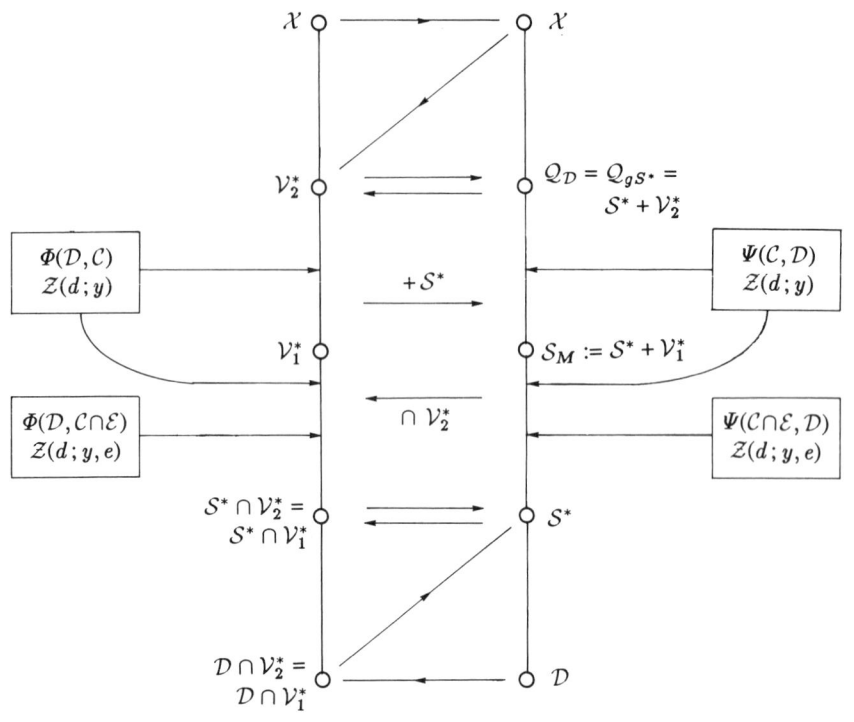

Figure 5-10 Lattices $\Psi(\mathcal{C},\mathcal{D})$ and $\Psi(\mathcal{C}\cap\mathcal{E},\mathcal{D})$ (on the right) and their auxiliary duals.

i.e., the lattices of all $(A,\mathcal{D})$-controlled invariants self-bounded with respect to $\mathcal{C}$ and $\mathcal{C}\cap\mathcal{E}$, which can also be represented in the same Hasse diagram.

The search for resolvents in connection with the most important synthesis problems concerns the elements of lattices $\Phi(\mathcal{B}+\mathcal{D},\mathcal{E})$ and $\Psi(\mathcal{C}\cap\mathcal{E},\mathcal{D})$ (the lattice on the left in Fig. 5-9 and that on the right in Fig. 5-10): in fact, resolvents are, in general, an $(A,\mathcal{B})$-controlled invariant and an $(A,\mathcal{C})$-conditioned invariant both contained in $\mathcal{E}$ and containing $\mathcal{D}$. It will be proved that restricting the choice of resolvents to self-bounded controlled and self-hidden conditioned invariants does not prejudice generality. A question now arises: is it possible to set a one-to-one correspondence directly between these lattices, so that stabilizability features can be comparatively considered? The answer is affirmative: it can be induced by a one-to-one correspondence between subsets of the auxiliary lattices, which are themselves lattices.

The elements of auxiliary lattices $\Phi(\mathcal{D},\mathcal{C}\cap\mathcal{E})$ and $\Psi(\mathcal{C},\mathcal{B}+\mathcal{D})$ are respectively $(A,\mathcal{D})$-controlled invariants contained in $\mathcal{C}\cap\mathcal{E}$ and $(A,\mathcal{E})$-conditioned invariants containing $\mathcal{B}+\mathcal{D}$. On the other hand, note that

$$A\mathcal{V}\subseteq\mathcal{V}+\mathcal{D} \quad\Rightarrow\quad A\mathcal{V}\subseteq\mathcal{V}+\mathcal{B}+\mathcal{D} \tag{62}$$

$$A(\mathcal{S}\cap\mathcal{E})\subseteq\mathcal{S} \quad\Rightarrow\quad A(\mathcal{S}\cap\mathcal{C}\cap\mathcal{E})\subseteq\mathcal{S} \tag{63}$$

i.e., any $(A,\mathcal{D})$-controlled invariant is also an $(A,\mathcal{B}+\mathcal{D})$-controlled invariant and any $(A,\mathcal{E})$-conditioned invariant is also an $(A,\mathcal{C}\cap\mathcal{E})$-conditioned invariant. Unfortunately, not all the elements of $\Phi(\mathcal{D},\mathcal{C}\cap\mathcal{E})$ are self-bounded with respect to $\mathcal{C}\cap\mathcal{E}$ as $(A,\mathcal{B}+\mathcal{D})$-controlled invariants, and not all the elements of $\Psi(\mathcal{C},\mathcal{B}+\mathcal{D})$ are self-hidden with respect to $\mathcal{B}+\mathcal{D}$ as $(A,\mathcal{C}\cap\mathcal{E})$-conditioned invariants; the elements that meet this requirement belong to the lattices

$$\Phi(\mathcal{B}+\mathcal{D},\mathcal{C}\cap\mathcal{E}) := \{\mathcal{V} : \mathcal{V} \in \Phi(\mathcal{D},\mathcal{C}\cap\mathcal{E}), \mathcal{V} \supseteq \mathcal{V}_1^* \cap (\mathcal{B}+\mathcal{D})\} \quad (64)$$

$$\Psi(\mathcal{C}\cap\mathcal{E},\mathcal{B}+\mathcal{D}) := \{\mathcal{S} : \mathcal{S} \in \Psi(\mathcal{C},\mathcal{B}+\mathcal{D}), \mathcal{S} \subseteq \mathcal{S}_1^* + (\mathcal{C}\cap\mathcal{E})\} \quad (65)$$

to which Theorem 5.1-1 can still be applied. By virtue of Theorems 4.1-4 and 4.1-5, the previous lattices can be defined also by the relations

$$\Phi(\mathcal{B}+\mathcal{D},\mathcal{C}\cap\mathcal{E}) := \{\mathcal{V} : \mathcal{V} \in \Phi(\mathcal{D},\mathcal{C}\cap\mathcal{E}), \mathcal{V} \supseteq \mathcal{V}_1^* \cap \mathcal{S}_1^*\} \quad (66)$$

$$\Psi(\mathcal{C}\cap\mathcal{E},\mathcal{B}+\mathcal{D}) := \{\mathcal{S} : \mathcal{S} \in \Psi(\mathcal{C},\mathcal{B}+\mathcal{D}), \mathcal{S} \subseteq \mathcal{V}_1^* + \mathcal{S}_1^*\} \quad (67)$$

which point out the new infimum and supremum, which are different from those of $\Phi(\mathcal{D},\mathcal{C}\cap\mathcal{E})$ and $\Psi(\mathcal{C},\mathcal{B}+\mathcal{D})$.

The sublattices of $\Phi(\mathcal{B}+\mathcal{D},\mathcal{E})$ and $\Psi(\mathcal{C}\cap\mathcal{E},\mathcal{D})$ defined by the one-to-one correspondences shown in Figs. 5-9 and 5-10 with the auxiliary sublattices (67,66) are defined by

$$\Phi_R := \{\mathcal{V} : \mathcal{V} \in \Phi(\mathcal{B}+\mathcal{D},\mathcal{E}), \mathcal{V}_m \subseteq \mathcal{V} \subseteq \mathcal{V}_M\} \quad (68)$$

$$\Psi_R := \{\mathcal{S} : \mathcal{S} \in \Psi(\mathcal{C}\cap\mathcal{E},\mathcal{D}), \mathcal{S}_m \subseteq \mathcal{S} \subseteq \mathcal{S}_M\} \quad (69)$$

with

$$\mathcal{V}_m := \mathcal{V}^* \cap \mathcal{S}_1^* \quad (70)$$

$$\mathcal{V}_M := \mathcal{V}^* \cap (\mathcal{V}_1^* + \mathcal{S}_1^*) = \mathcal{V}^* \cap \mathcal{S}_1^* + \mathcal{V}_1^* \quad (71)$$

$$\mathcal{S}_m := \mathcal{S}^* + \mathcal{V}_1^* \cap \mathcal{S}_1^* = (\mathcal{S}^* + \mathcal{V}_1^*) \cap \mathcal{S}_1^* \quad (72)$$

$$\mathcal{S}_M := \mathcal{S}^* + \mathcal{V}_1^* \quad (73)$$

The overall dual-lattice layout is represented in Fig. 5-11. The identities expressed in (71,72) follow from $\mathcal{V}_1^* \subseteq \mathcal{V}^*$ and $\mathcal{S}_1^* \supseteq \mathcal{S}^*$. The former derives from

$$\mathcal{V}_1^* := \max \mathcal{V}(A,\mathcal{D},\mathcal{C}\cap\mathcal{E}) \subseteq \max \mathcal{V}(A,\mathcal{B}+\mathcal{D},\mathcal{C}\cap\mathcal{E}) \subseteq \max \mathcal{V}(A,\mathcal{B}+\mathcal{D},\mathcal{E}) = \mathcal{V}^*$$

where the first inclusion is related to the procedure for the computation of $\max \mathcal{V}(*,*)$ and the last equality from $\mathcal{D} \subseteq \mathcal{V}^*$. Relation $\mathcal{S}_1^* \supseteq \mathcal{S}^*$ can be proved

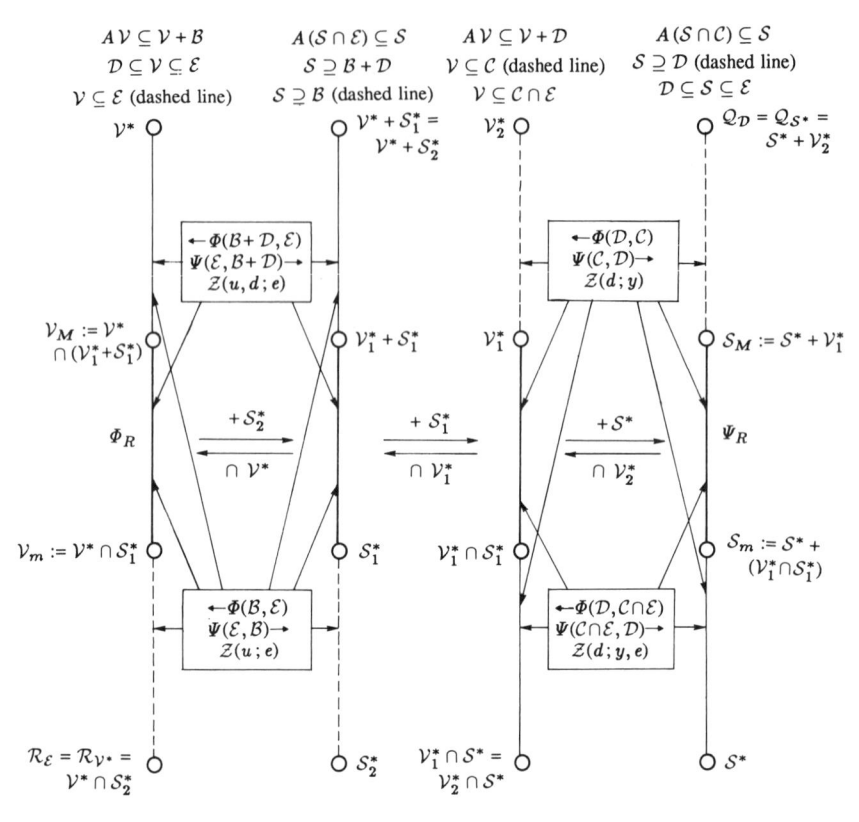

Figure 5-11 Induced one-to-one correspondence between suitable sublattices of $\Phi(\mathcal{B}+\mathcal{D}, \mathcal{E})$ and $\Psi(\mathcal{C}\cap\mathcal{E}, \mathcal{D})$ (which are denoted by Φ_R and Ψ_R).

by duality. The one-to-one correspondences between sublattices (68, 69) are defined by

$$S = \big((\mathcal{V}+\mathcal{S}_1^*) \cap \mathcal{V}_1^*\big) + \mathcal{S}^* = (\mathcal{V}+\mathcal{S}_1^*) \cap (\mathcal{V}_1^*+\mathcal{S}^*) = (\mathcal{V}+\mathcal{S}_1^*) \cap \mathcal{S}_M$$

$$\mathcal{V} = \big((\mathcal{S}\cap\mathcal{V}_1^*) + \mathcal{S}_1^*\big) \cap \mathcal{V}^* = (\mathcal{S}\cap\mathcal{V}_1^*) + (\mathcal{S}_1^*+\mathcal{V}^*) = (\mathcal{S}\cap\mathcal{V}_1^*) + \mathcal{V}_m$$

Note, in particular, that $\mathcal{V}_m$ and $\mathcal{V}_M$ are $(A, \mathcal{B})$-controlled invariants self-bounded with respect to $\mathcal{E}$, and that $\mathcal{S}_m$ and $\mathcal{S}_M$ are $(A, \mathcal{C})$-conditioned invariants self-hidden with respect to $\mathcal{D}$. These particular elements of $\Phi(\mathcal{B}, \mathcal{E})$ and $\Psi(\mathcal{C}, \mathcal{D})$ are very useful in the regulator and compensator synthesis procedures, which will be approached and thoroughly investigated in the next section.

5.2 THE DYNAMIC DISTURBANCE LOCALIZATION AND THE REGULATOR PROBLEM

The solution of two basic problems, where the power of the geometric approach is particularly stressed, will now be discussed. They are the *disturbance localization by dynamic compensator* and the *regulator problem*. First, nonconstructive but very simple and intuitive necessary and sufficient conditions, will be derived. Then constructive necessary and sufficient conditions that directly provide resolvents - and so can be directly used for synthesis - will be stated for the solvability of both problems.

For the disturbance localization by dynamic compensator we refer to the block diagram of Fig. 5-3 and assume that the controlled system consists only of the plant, without any exosystem; thus it is completely stabilizable and detectable.

Problem 5.2-1. Disturbance Localization by Dynamic Compensator

Refer to the block diagram of Fig. 5-3 and assume that (A, B) is stabilizable and (A, C) detectable. Determine, if possible, a feedback compensator of the type shown in the figure such that:

1. $e(t)=0$, $t \geq 0$, for all admissible $d(\cdot)$ and for $x(0)=0$, $z(0)=0$;
2. $\lim_{t \to \infty} x(t)=0$, $\lim_{t \to \infty} z(t)=0$ for all $x(0), z(0)$ and for $d(\cdot)=0$.

Condition 1 is the *structure requirement* and 2 the *stability requirement*. Problem 5.2-1 can be stated also in geometric terms referring to the extended system (5.1.9–12), obviously with $\hat{R}=O$.

Problem 5.2-1/G. Geometric Formulation of Problem 5.2-1

Refer to the block diagram of Fig. 5-3 and assume that (A, B) is stabilizable and (A, C) detectable. Determine, if possible, a feedback dynamic compensator of the type shown in the figure such that:

1. the overall system has an $\hat{A}$-invariant $\hat{\mathcal{W}}$ that satisfies

$$\hat{\mathcal{D}} \subseteq \hat{\mathcal{W}} \subseteq \hat{\mathcal{E}} \quad \text{with} \quad \hat{\mathcal{D}} := \operatorname{im} \hat{D}, \ \hat{\mathcal{E}} := \ker \hat{E};$$

2. $\hat{A}$ is stable.

Necessary and sufficient conditions for the solvability of Problem 5.2-1 are given in the following theorem in geometric terms regarding (A, B, C, D, E).

Theorem 5.2-1.[1]

The disturbance localization problem by a dynamic compensator admits a solution if and only if there exist both an (A, B)-controlled invariant $\mathcal{V}$ and an (A, C)-conditioned invariant $\mathcal{S}$ such that:

1. $\mathcal{D} \subseteq \mathcal{S} \subseteq \mathcal{V} \subseteq \mathcal{E}$; (1)

---

[1] This theorem is due to Willems and Commault [37].

2. S is externally stabilizable; (2)
3. $\mathcal{V}$ is internally stabilizable. (3)

Conditions stated in Theorem 5.2-1 are nonconstructive, since they refer to a resolvent pair $(S, \mathcal{V})$ which is not defined. Equivalent constructive conditions are stated as follows. They are formulated in terms of subspaces $S^*, \mathcal{V}^*, S_M$, and $\mathcal{V}_M$ defined in (5.1.57, 51, 73, 71).[2]

Theorem 5.2-2.
The disturbance localization problem by a dynamic compensator admits a solution if and only if:
1. $S^* \subseteq \mathcal{V}^*$; (4)
2. S_M is externally stabilizable; (5)
3. $\mathcal{V}_M$ is internally stabilizable. (6)

We shall now consider the regulator problem. The formulation herein presented is very general and includes all feedback connections examined so far as particular cases (disturbance localization, unknown-input asymptotic estimation, the above dynamic compensator). Moreover, it will be used as a reference for further developments of the theory, like approach to reduced-order devices and robust regulation. We still refer to the block diagram of Fig. 5-3, assuming in this case that an exosystem is included as part of the controlled system, as in Fig. 5-2.

Problem 5.2-2. The Regulator Problem[3]
Refer to the block diagram of Fig. 5-3, where the controlled system is assumed to have the structure of Fig. 5-2 with (A_{11}, B_1) stabilizable and (A, C) detectable. Determine, if possible, a feedback regulator of the type shown in Fig. 5-3 such that:
1. $e(t)=0$, $t \geq 0$, for all admissible $d(\cdot)$ and for $x_1(0)=0$, $x_2(0)=0$, $z(0)=0$;
2. $\lim_{t \to \infty} e(t)=0$ for all $x_1(0), x_2(0), z(0)$ and for $d(\cdot)=0$;
3. $\lim_{t \to \infty} x_1(t)=0$, $\lim_{t \to \infty} z(t)=0$ for all $x_1(0), z(0)$ and for $x_2(0)=0$, $d(\cdot)=0$.

Condition 1 is the *structure requirement*, 2 the *regulation requirement*, and 3 the *stability requirement*. Problem 5.2-2 can also be stated in geometric

---

[2] These constructive conditions without eigenspaces have been introduced by Basile, Marro, and Piazzi [6].

[3] The regulator problem has been the object of very intensive research. The most important contributions to its solution in the framework of the geometric approach are due to Wonham [39] (problem without the stability requirement), Wonham and Pearson [40], and Francis [17] (problem with the stability requirement). The statement reported here, which includes disturbance localization as a particular case, is due to Schumacher [31]. Theorem 5.2-3, where the plant is explicitly introduced as a geometric object, is due to Basile, Marro, and Piazzi [8], as well as Theorem 5.2-4, where the stability requirement is handled without any use of eigenspaces [9].

Sect. 5.2 The Dynamic Disturbance Localization and the Regulator Problem

terms: first, define the *extended plant* as the $\hat{A}$-invariant

$$\hat{\mathcal{P}} := \left\{ \begin{bmatrix} x_1 \\ x_2 \\ z \end{bmatrix} : x_2 = 0 \right\} \tag{7}$$

and refer again to the extended system (9–12) with $\hat{R} = O$. The three points in the statement of Problem 5.2-2 can be reformulated as follows:

1. There exists an $\hat{A}$-invariant $\hat{\mathcal{W}}_1$ such that $\hat{\mathcal{D}} \subseteq \hat{\mathcal{W}}_1 \subseteq \hat{\mathcal{E}}$;
2. There exists an externally stable $\hat{A}$-invariant $\hat{\mathcal{W}}_2$ such that $\hat{\mathcal{W}}_2 \subseteq \hat{\mathcal{E}}$;
3. $\hat{\mathcal{P}}$, as an $\hat{A}$-invariant, is internally stable.

Note that $\hat{\mathcal{W}} := \hat{\mathcal{W}}_1 + \hat{\mathcal{W}}_2$ is an externally stable $\hat{A}$-invariant as the sum of two $\hat{A}$-invariants, one of which is externally stable, so that $\hat{\mathcal{P}}$ is internally stable if and only if $\hat{\mathcal{W}} \cap \hat{\mathcal{P}}$ is so. In fact, if $\hat{\mathcal{P}}$ is internally stable, the invariant $\hat{\mathcal{W}} \cap \hat{\mathcal{P}}$ is also internally stable, being contained in $\hat{\mathcal{P}}$. To prove the converse, consider the similarity transformation $\hat{T} := [T_1\ T_2\ T_3\ T_4]$ with $\text{im}\, T_1 = \hat{\mathcal{W}} \cap \hat{\mathcal{P}}$, $\text{im}\,[T_1\ T_2] = \hat{\mathcal{P}}$, $\text{im}\,[T_1\ T_3] = \hat{\mathcal{W}}$, which leads to

$$\hat{T}^{-1} \hat{A} \hat{T} = \begin{bmatrix} A'_{11} & A'_{12} & A'_{13} & A'_{14} \\ O & A'_{22} & O & A'_{24} \\ O & O & A'_{33} & A'_{34} \\ O & O & O & A'_{44} \end{bmatrix}$$

Submatrix A'_{11} is stable since $\hat{\mathcal{W}} \cap \hat{\mathcal{P}}$ is internally stable, A'_{22} and A'_{44} are stable since $\hat{\mathcal{W}}$ is externally stable, so that $\hat{\mathcal{P}}$ is internally stable as a consequence of A'_{11} and A'_{22} being stable. Now Problem 5.2-2 can be stated in geometric terms as follows.

Problem 5.2-2/G. Geometric Formulation of Problem 5.2-2

Refer to the block diagram of Fig. 5-3, where the controlled system is assumed to have the structure of Fig. 5-2 with (A_1, B_1) stabilizable and (A, C) detectable. Determine, if possible, a feedback regulator of the type shown in Fig. 5-3 such that:

1. the overall system has an $\hat{A}$-invariant $\hat{\mathcal{W}}$ that satisfies

$$\hat{\mathcal{D}} \subseteq \hat{\mathcal{W}} \subseteq \hat{\mathcal{E}}, \quad \text{with} \quad \hat{\mathcal{D}} := \text{im}\,\hat{D}, \ \hat{\mathcal{E}} := \ker \hat{E};$$

2. $\hat{\mathcal{W}}$ is externally stable;
3. $\hat{\mathcal{W}} \cap \hat{\mathcal{P}}$ (which is an $\hat{A}$-invariant) is internally stable.

Necessary and sufficient conditions for solvability of the regulator problem are stated in the following theorem, which can be considered as an extension of Theorem 5.2-1.

Theorem 5.2-3.
The regulator problem admits a solution if and only if there exist both an (A,B)-controlled invariant $\mathcal{V}$ and an (A,C)-conditioned invariant $\mathcal{S}$ such that:[4]

1. $\mathcal{D} \subseteq \mathcal{S} \subseteq \mathcal{V} \subseteq \mathcal{E}$; (8)
2. $\mathcal{S}$ is externally stabilizable; (9)
3. $\mathcal{V}$ is externally stabilizable; (10)
4. $\mathcal{V} \cap \mathcal{P}$ is internally stabilizable. (11)

The corresponding constructive conditions are stated in the following theorem, which extends Theorem 5.2-2.

Theorem 5.2-4.
Let all the exogenous modes be unstable. The regulator problem admits a solution if and only if:

1. $\mathcal{S}^* \subseteq \mathcal{V}^*$; (12)
2. $\mathcal{V}^*$ is externally stabilizable; (13)
3. $\mathcal{S}_M$ is externally stabilizable; (14)
4. $\mathcal{V}_M \cap \mathcal{P}$ is internally stabilizable; (15)
5. $\mathcal{V}_M + \mathcal{V}^* \cap \mathcal{P}$ is complementable with respect to $(\mathcal{V}_M, \mathcal{V}^*)$. (16)

On stabilizability and complementability of controlled invariants, see Subsection 4.1.4. The assumption that all the exogenous modes are unstable in practice does not affect generality. In fact, any possible asymptotically stable exogenous mode can be eliminated in the mathematical model as it does not influence the asymptotic behavior of the overall system, since the extended plant is required to be asymptotically stable.

5.2.1 Proof of the Nonconstructive Conditions

This subsection reports the proofs of Theorems 5.2-1 and 5.2-3. Of course, they are related to each other, since the second theorem extends the first.

Proof of Theorem 5.2-1. *Only if.* Assume that conditions 1 and 2 stated in Problem 5.2-1/G are satisfied. Hence, $\hat{\mathcal{W}}$ is an $\hat{A}$-invariant, so that, by virtue of Property 5.1-1, it is also an $(\hat{A}_0, \hat{B}_0)$-controlled and an $(\hat{A}_0, \hat{C}_0)$-conditioned invariant, both internally and externally stabilizable since $\hat{\mathcal{W}}$ is both an internally and externally stable $\hat{A}$-invariant. Thus, by Lemma 5.1-1 $\mathcal{V} := P(\hat{\mathcal{W}})$ is an internally and externally stabilizable (A,B)-controlled invariant, and by Lemma 5.1-2 $\mathcal{S} := I(\hat{\mathcal{W}})$ is an internally and externally stabilizable (A,C)-conditioned invariant. Inclusions (1) follow from

$$I(\hat{\mathcal{W}}) \subseteq P(\hat{\mathcal{W}}) \text{ for all } \hat{\mathcal{W}}$$
$$\mathcal{D} = I(\hat{\mathcal{D}}) = P(\hat{\mathcal{D}})$$
$$\mathcal{E} = P(\hat{\mathcal{E}}) = I(\hat{\mathcal{E}})$$

---

[4] Note that $\mathcal{V} \cap \mathcal{P}$ is an (A,B)-controlled invariant as the intersection of an (A,B)-controlled invariant and an A-invariant containing $\mathcal{B}$. In fact, $A(\mathcal{V} \cap \mathcal{P}) \subseteq A\mathcal{V} \cap A\mathcal{P} \subseteq (\mathcal{V}+\mathcal{B}) \cap \mathcal{P} = \mathcal{V} \cap \mathcal{P} + \mathcal{B}$.

Sect. 5.2 The Dynamic Disturbance Localization and the Regulator Problem

If. Assume that the conditions reported in the statement are satisfied. Recall that, if pair (A, B) is stabilizable, any $(A, \mathcal{B})$-controlled invariant is externally stabilizable and, if (A, C) is detectable, any $(A, \mathcal{C})$-conditioned invariant is internally stabilizable. In the compensator synthesis assume $m = n$ and

$$\hat{\mathcal{W}} := \left\{ \begin{bmatrix} x \\ z \end{bmatrix} : x \in \mathcal{V}, \ z = x - \eta, \ \eta \in \mathcal{S} \right\} \tag{17}$$

so that, clearly, $P(\hat{\mathcal{W}}) = \mathcal{V}$ and $I(\hat{\mathcal{W}}) = \mathcal{S}$; hence, by Lemmas 5.1-1 and 5.1-2, $\hat{\mathcal{W}}$ is both an internally and externally stabilizable $(\hat{A}_0, \hat{B}_0)$-controlled invariant and an internally and externally stabilizable $(\hat{A}_0, \hat{C}_0)$-conditioned invariant. Note that the internal and external stabilizability of any $\hat{\mathcal{W}}$ as an $(\hat{A}_0, \hat{B}_0)$-controlled and an $(\hat{A}_0, \hat{C}_0)$-conditioned invariant in general is not sufficient for the existence of an algebraic output-to-input feedback such that $\hat{\mathcal{W}}$ is an $\hat{A}$-invariant internally and externally stable, i.e., Property 5.1-1 cannot be extended to include the stability requirement. However, in the particular case of (17) it will be proved by a direct check that such an input-to-output feedback exists. Define matrices L_1 and L_2 which satisfy

$$L_1 C + L_2 = I_n \quad \text{with} \quad \ker L_2 = \mathcal{L} \tag{18}$$

and $\mathcal{L}$ such that

$$\mathcal{L} \oplus \mathcal{S} \cap \mathcal{C} = \mathcal{S} \tag{19}$$

These matrices exist by virtue of Lemma 5.1-3. It follows that, $\mathcal{L}$ being contained in $\mathcal{S}$ but having zero intersection with $\mathcal{S} \cap \mathcal{C}$

$$\mathcal{L} \cap \mathcal{C} = \{0\} \tag{20}$$

Also, derive F and G such that

$$(A + BF)\mathcal{V} \subseteq \mathcal{V} \tag{21}$$
$$(A + GC)\mathcal{S} \subseteq \mathcal{S} \tag{22}$$

with both $A + BF$ and $A + GC$ stable. The extended system (5.1.30) solves the problem. In fact, consider the equivalent system (5.1.32): in terms of the new coordinates (ρ, η) $\hat{\mathcal{W}}$ is expressed as

$$\hat{\mathcal{W}} = \left\{ \begin{bmatrix} \rho \\ \eta \end{bmatrix} : \rho \in \mathcal{V}, \ \eta \in \mathcal{S} \right\} \tag{23}$$

and from (18) it follows that

$$\bigl(A + BF(L_1 C + L_2)\bigr)(\mathcal{S} \cap \mathcal{C}) = (A + BF)(\mathcal{S} \cap \mathcal{C}) \subseteq \mathcal{V}$$

From $A(\mathcal{S} \cap \mathcal{C}) \subseteq \mathcal{S}$ ($\mathcal{S}$ is an $(A, \mathcal{C})$-conditioned invariant) and $BFL_1 C(\mathcal{S} \cap \mathcal{C}) = \{0\}$ (by definition of $\mathcal{C}$) we derive

$$BFL_2(\mathcal{S} \cap \mathcal{C}) \subseteq \mathcal{V}$$

As $BFL_2\mathcal{L}=\{0\}$, (19) yields

$$BFL_2\mathcal{S} \subseteq \mathcal{V}$$

so that $\hat{\mathcal{W}}$ is an $\hat{A}$-invariant. Clearly $\hat{\mathcal{D}} \subseteq \hat{\mathcal{W}} \subseteq \hat{\mathcal{E}}$ and matrix $\hat{A}$ is stable by virtue of its particular structure in the new basis. □

Proof of Theorem 5.2-3. *Only if.* Refer to Problem 5.2-2/G. By virtue of Property 5.1-1, $\hat{\mathcal{W}}$ is both an $(\hat{A}_0, \hat{B}_0)$-controlled and an $(\hat{A}_0, \hat{C}_0)$-conditioned invariant. $\hat{\mathcal{W}}$ is externally stabilizable, both as a controlled invariant and as a conditioned invariant because, as an $\hat{A}$-invariant it is externally stable (recall that $\hat{A}$ can be expressed as $\hat{A}_0 + \hat{B}\hat{K}\hat{C}$). Furthermore, $\hat{\mathcal{W}} \cap \hat{\mathcal{P}}$ as an internally stable $\hat{A}$-invariant is an internally stabilizable $(\hat{A}_0, \hat{B}_0)$-controlled invariant. Consider the subspaces $\mathcal{V} := P(\hat{\mathcal{W}})$, $\mathcal{S} := I(\hat{\mathcal{W}})$: inclusions (8) are proved as in Theorem 5.2-1, while, by virtue of Lemmas 5.1-1 and 5.1-2, $\mathcal{V}$ and $\mathcal{S}$ are externally stabilizable and $P(\hat{\mathcal{W}} \cap \hat{\mathcal{P}})$ is internally stabilizable. Since

$$P(\hat{\mathcal{W}} \cap \hat{\mathcal{P}}) = P(\hat{\mathcal{W}}) \cap P(\hat{\mathcal{P}})$$

because $\hat{\mathcal{P}} \supseteq \hat{\mathcal{Z}}$,[5] and

$$P(\hat{\mathcal{P}}) = \mathcal{P}$$

it follows that $\mathcal{V} \cap \mathcal{P}$ is internally stabilizable. Note that by means of a similar argument it would be possible to prove that

$$I(\hat{\mathcal{W}} \cap \hat{\mathcal{P}}) = I(\hat{\mathcal{W}}) \cap I(\hat{\mathcal{P}}) = \mathcal{S} \cap \mathcal{P}$$

is an internally stabilizable (A, C)-conditioned invariant. Nevertheless, this property is implied by the detectability of the plant, since the intersection of any two conditioned invariants is internally stabilizable if any one of them is.

If. Assume that all the conditions reported in the statement are met; define matrices L_1 and L_2 satisfying (18,19), a matrix F such that (21) holds with both $(A+BF)|_{\mathcal{R}_\mathcal{V}}$ and $(A+BF)|_{\mathcal{X}/\mathcal{V}}$ stable, and a matrix G such that (22) holds with $A+GC$ stable: this is possible since $\mathcal{S}$ is externally stabilizable and the controlled system is detectable. It is easily proved that this choice of F also makes $(A+BF)|_{\mathcal{P}}$ stable. In fact, consider the similarity transformation $T := [T_1\, T_2\, T_3\, T_4]$, with $\mathrm{im}\, T_1 = \mathcal{V} \cap \mathcal{P}$, $\mathrm{im}\,[T_1\, T_2] = \mathcal{V}$, $\mathrm{im}\,[T_1\, T_3] = \mathcal{P}$. Matrix $A+BF$ in the new basis has the structure

$$T^{-1}(A+BF)T = \begin{bmatrix} A'_{11} & A'_{12} & A'_{13} & A'_{14} \\ O & A'_{22} & O & A'_{24} \\ O & O & A'_{33} & A'_{34} \\ O & O & O & A'_{44} \end{bmatrix} \quad (24)$$

due to both $\mathcal{V}$ and $\mathcal{P}$ being $(A+BF)$-invariants. $\mathcal{V} \cap \mathcal{P}$ is internally stable, being internally stabilizable by assumption and having, as constrained reachable set, $\mathcal{R}_\mathcal{V}$,

---

[5] In fact, the following equalities hold:

$$P(\hat{\mathcal{W}} \cap \hat{\mathcal{P}}) = I(\hat{\mathcal{W}} \cap \hat{\mathcal{P}} + \hat{\mathcal{Z}}) = I\big((\hat{\mathcal{W}} + \hat{\mathcal{Z}}) \cap (\hat{\mathcal{P}} + \hat{\mathcal{Z}})\big)$$
$$= I(\hat{\mathcal{W}} + \hat{\mathcal{Z}}) \cap I(\hat{\mathcal{P}} + \hat{\mathcal{Z}}) = P(\hat{\mathcal{W}}) \cap P(\hat{\mathcal{P}})$$

Sect. 5.2 The Dynamic Disturbance Localization and the Regulator Problem

which has been stabilized by the particular choice of F: hence A'_{11} is stable. On the other hand, A'_{33} and A'_{44} are stable because $\mathcal{V}$ is externally stable: it follows that $\mathcal{P}$ is internally stable. Having determined matrices L_1, L_2, F, and G, the regulator synthesis can be performed as in the previous proof. Define again $\hat{\mathcal{W}}$ as in (23): it is immediately verified that $\hat{\mathcal{W}}$ is an $\hat{A}$-invariant and satisfies $\hat{\mathcal{D}} \subseteq \hat{\mathcal{W}} \subseteq \hat{\mathcal{E}}$.

It still has to be proved that the regulation and plant stability requirements are met, i.e., in geometric terms, that $\hat{\mathcal{W}}$ is externally and $\hat{\mathcal{P}}$ internally stable. Regarding the first requirement, let us make the change of basis (5.1.31) a little finer by defining new coordinates (ρ', η') according to

$$\begin{bmatrix} \rho \\ \eta \end{bmatrix} = \begin{bmatrix} P & O \\ O & Q \end{bmatrix} \begin{bmatrix} \rho' \\ \eta' \end{bmatrix}$$

with $P := [P_1\ P_2]$, $\mathrm{im}\, P_1 = \mathcal{V}$ and $Q := [Q_1\ Q_2]$, $\mathrm{im}\, Q_1 = \mathcal{S}$. In this new basis the system matrix (5.1.32) assumes the structure

$$\begin{bmatrix} \times & \times & \times & \times \\ O & S & O & \times \\ \hline O & O & S & \times \\ O & O & O & S \end{bmatrix}$$

where $\times$ denotes a generic and S a stable submatrix. Stability of the submatrices on the main diagonal depends on $\mathcal{V}$ being internally stable and $A+GC$ stable. In the new basis

$$\hat{\mathcal{W}} = \mathrm{im}\, \left(\begin{bmatrix} I_1 & O \\ O & O \\ O & I_2 \\ O & O \end{bmatrix} \right)$$

where I_1, I_2 denote properly dimensioned identity matrices: it is immediately verified that $\hat{\mathcal{W}}$ is externally stable. To prove that $\hat{\mathcal{P}}$ is internally stable, note that all the extended system eigenvalues, the exogenous excepted, are stable, because maps $(A+BF)|_{\mathcal{P}}$ and $A+GC$ are stable. Their eigenvalues are all and only those internal of $\hat{\mathcal{P}}$. $\square$

5.2.2 Proof of the Constructive Conditions

The constructive conditions stated in Theorems 5.2-2 and 5.2-4 are expressed in terms of the subspaces defined and analyzed in the previous section. First of all, note that the previously proved necessary and sufficient structural conditions (1) and (8) clearly imply the necessity of (4) and (12); then, since $\mathcal{D} \subseteq \mathcal{V}^*$, $\mathcal{V}^*$, the maximal $(A, \mathcal{B})$-controlled invariant contained in $\mathcal{E}$ coincides with the maximal $(A, \mathcal{B}+\mathcal{D})$-controlled invariant contained in $\mathcal{E}$ and, dually, since $\mathcal{S}^* \supseteq \mathcal{E}$, $\mathcal{S}^*$, the minimal $(A, \mathcal{C})$-conditioned invariant containing $\mathcal{D}$ coincides with the minimal $(A, \mathcal{C} \cap \mathcal{E})$-conditioned invariant containing $\mathcal{D}$. Also recall that $\mathcal{V}_m$ denotes the infimum of lattice $\Phi(\mathcal{B}+\mathcal{D}, \mathcal{E})$ of all $(A, \mathcal{B}+\mathcal{D})$-controlled invariants self-bounded with respect to $\mathcal{E}$, $\mathcal{S}_M$ the supremum of the lattice $\Psi(\mathcal{C}\cap\mathcal{E},\mathcal{D})$ of all $(A,\mathcal{C}\cap\mathcal{E})$-conditioned invariants self-hidden with respect to $\mathcal{D}$. $\mathcal{V}_M$ and $\mathcal{S}_m$ denote respectively the supremum and the infimum of restricted lattices Φ_R and Ψ_R. If the necessary inclusion $\mathcal{S}^* \subseteq \mathcal{V}^*$ is satisfied, the latter subspaces (which in the general case are defined by (5.1.71) and (5.1.72)) can be expressed as functions of $\mathcal{V}_m$ and $\mathcal{S}_M$ by

$$\mathcal{V}_M = \mathcal{V}_m + \mathcal{S}_M \tag{25}$$

$$\mathcal{S}_m = \mathcal{V}_m \cap \mathcal{S}_M \tag{26}$$

which are proved by the following manipulations:

$$\mathcal{V}_M = \mathcal{V}_m + \mathcal{S}_M = (\mathcal{V}^* \cap \mathcal{S}_1^*) + (\mathcal{S}^* + \mathcal{V}_1^*) = \big((\mathcal{V}^* + \mathcal{S}^*) \cap \mathcal{S}_1^*\big) + \mathcal{V}_1^* = (\mathcal{V}^* \cap \mathcal{S}_1^*) + \mathcal{V}_1^*$$

$$\mathcal{S}_m = \mathcal{S}_M \cap \mathcal{V}_m = (\mathcal{S}^* + \mathcal{V}_1^*) \cap (\mathcal{V}^* \cap \mathcal{S}_1^*) = \big((\mathcal{S}^* \cap \mathcal{V}^*) + \mathcal{V}_1^*\big) \cap \mathcal{S}_1^* = (\mathcal{S}^* + \mathcal{V}_1^*) \cap \mathcal{S}_1^*$$

In the expression of $\mathcal{V}_M$ the third equality derives from the distributivity of the sum with $\mathcal{S}^*$ with respect to the intersection $\mathcal{V}^* \cap \mathcal{S}_1^*$ and the subsequent one from inclusion $\mathcal{S}^* \subseteq \mathcal{V}^*$. The equalities in the expression of $\mathcal{S}_m$ can be proved by duality.

The proof of the constructive conditions will be developed by using the nonconstructive ones as a starting point. In particular, for necessity it will be proved that the existence of a resolvent pair $(\mathcal{S},\mathcal{V})$, i.e., of a pair of subspaces satisfying the conditions stated in Theorems 5.2-2 and 5.2-3, implies the existence of a resolvent pair with the conditioned invariant self-hidden and the controlled invariant self-bounded. This property leads to conditions for the bounds of suitable sublattices of $\Psi(\mathcal{C}\cap\mathcal{E},\mathcal{D})$ and $\Phi(\mathcal{B}+\mathcal{D},\mathcal{E})$ to which the elements of this second resolvent pair must belong.

Proof of Theorem 5.2-2. *Only if.* Assume that the problem has a solution; hence, by virtue of Theorem 5.2-1 there exists a resolvent pair $(\mathcal{S},\mathcal{V})$, where $\mathcal{S}$ is an externally stabilizable $(A,\mathcal{C})$-conditioned invariant and $\mathcal{V}$ an internally stabilizable $(A,\mathcal{B})$-controlled invariant satisfying the inclusions $\mathcal{D}\subseteq\mathcal{S}\subseteq\mathcal{V}\subseteq\mathcal{E}$. Then, as already pointed out, the structural property $\mathcal{S}^*\subseteq\mathcal{V}^*$ holds. First we prove that in this case there also exists an externally stabilizable $(A,\mathcal{C})$-conditioned invariant $\bar{\mathcal{S}}$, self-hidden with respect to $\mathcal{D}$, such that

$$\mathcal{S}_m \subseteq \bar{\mathcal{S}} \subseteq \mathcal{S}_M \tag{27}$$

and an internally stabilizable $(A,\mathcal{B})$-controlled invariant $\bar{\mathcal{V}}$, self-bounded with respect to $\mathcal{E}$, such that

$$\mathcal{V}_m \subseteq \bar{\mathcal{V}} \subseteq \mathcal{V}_M \tag{28}$$

which satisfy the same inclusions that, since $\mathcal{D}\subseteq\bar{\mathcal{S}}$ and $\bar{\mathcal{V}}\subseteq\mathcal{E}$, reduce to

$$\bar{\mathcal{S}} \subseteq \bar{\mathcal{V}} \tag{29}$$

Assume

$$\bar{\mathcal{S}} := (\mathcal{S}+\mathcal{S}_m) \cap \mathcal{S}_M = \mathcal{S}\cap\mathcal{S}_M + \mathcal{S}_m \tag{30}$$

$$\bar{\mathcal{V}} := \mathcal{V}\cap\mathcal{V}_M + \mathcal{S}_M = (\mathcal{V}+\mathcal{V}_m)\cap\mathcal{V}_M \tag{31}$$

Sect. 5.2 The Dynamic Disturbance Localization and the Regulator Problem

Note that, by virtue of Lemmas 4.2-1 and 4.2-2, V_m is internally stabilizable and S_M externally stabilizable. $S \cap S_M$ is externally stabilizable and self-hidden with respect to $\mathcal{D}$, i.e., it belongs to $\Psi(\mathcal{C} \cap \mathcal{E}, \mathcal{D})$, so that $\bar{S}$, as the sum of two elements of $\Psi(\mathcal{C} \cap \mathcal{E}, \mathcal{D})$ one of which is externally stabilizable, is externally stabilizable and belongs to $\Psi(\mathcal{C} \cap \mathcal{E}, \mathcal{D})$. The dual argument proves that $\bar{V}$ is internally stabilizable and belongs to $\Phi(\mathcal{B}+\mathcal{D}, \mathcal{E})$. Relations (30) and (31) are equivalent to $\bar{S} \in \bar{\Psi}_R$ and $\bar{V} \in \bar{\Phi}_R$. Inclusion (29) follows from

$$S \cap S_M \subseteq S \subseteq V \subseteq V + V_m$$
$$S \cap S_M \subseteq S_M \subseteq S_M + V_m = V_M$$
$$S_m = S_M \cap V_m \subseteq V_m \subseteq V + V_m$$
$$S_m = S_M \cap V_m \subseteq V_m \subseteq V_M$$

We now introduce a change of basis that will lead to the proof. This type of approach has already been used in Subsection 4.1.3 to point out connections between stabilizability features of self-bounded controlled invariants and related self-hidden conditioned invariants. Let us assume the similarity transformation $T := [T_1\, T_2\, T_3\, T_4]$, with $\mathrm{im}\, T_1 = S_m = S_M \cap V_m$, $\mathrm{im}\, [T_1\, T_2] = V_m$ and $\mathrm{im}\, [T_1\, T_3] = S_M$. Since $S_M \subseteq S^* \cap \mathcal{C} \subseteq S_m + \mathcal{C}$, it is possible to choose T_3 in such a way that

$$\mathrm{im}\, T_3 \subseteq \mathcal{C} \tag{32}$$

By duality, since $V_m \supseteq V^* \cap \mathcal{B} \supseteq V_m \cap \mathcal{B}$, matrix T_4 can be chosen in such a way that

$$\mathrm{im}\, [T_1\, T_2\, T_4] \supseteq \mathcal{B} \tag{33}$$

In the new basis matrices $A' := T^{-1} A T$, $B' := T^{-1} B$ and $C' := C T$ are expressed as

$$A' = \begin{bmatrix} A'_{11} & A'_{12} & A'_{13} & A'_{14} \\ A'_{21} & A'_{22} & O & A'_{24} \\ O & O & A'_{33} & A'_{34} \\ A'_{41} & A'_{42} & O & A'_{44} \end{bmatrix} \quad B' = \begin{bmatrix} B'_1 \\ B'_2 \\ O \\ B'_4 \end{bmatrix} \tag{34}$$

$$C' = [\, C'_1 \quad C'_2 \quad O \quad C'_4 \,]$$

Conditions (32) and (33) imply the particular structures of matrices B' and C'. As far as the structure of A' is concerned, note that the zero submatrices in the third row are due to the particular structure of B' and to V_m being an $(A, \mathcal{B})$-controlled invariant, while those in the third column are due to the particular structure of C' and to S_M being an $(A, \mathcal{C})$-conditioned invariant. If the structural zeros in A', B', and C' are taken into account, from (27, 28) it follows that all possible pairs $\bar{S}, \bar{V}$ can be expressed as

$$\bar{S} = S_m + \mathrm{im}\,(T_3 X_S) \quad \text{and} \quad \bar{V} = V_m + \mathrm{im}\,(T_3 X_V) \tag{35}$$

where X_S, X_V are basis matrices of an externally stable and an internally stable A'_{33}-invariant subspace respectively. These stability properties follow from $\bar{S}$ being externally stabilizable and $\bar{\mathcal{V}}$ internally stabilizable. Condition (29) clearly implies $\text{im} X_S \subseteq \text{im} X_V$, so that A'_{33} is stable. Since, as has been previously pointed out, $\mathcal{S}_M$ is externally stabilizable and $\mathcal{V}_m$ internally stabilizable, the stability of A'_{33} implies the external stabilizability of $\mathcal{S}_m$ and the internal stabilizability of $\mathcal{V}_M$.[6]

If. The problem admits a solution by virtue of Theorem 5.2-1 with $\mathcal{S}:=\mathcal{S}_M$ and $\mathcal{V}:=\mathcal{V}_M$. □

Proof of Theorem 5.2-4. *Only if.* We shall first present some general properties and remarks on which the proof will be based.

(a) The existence of a resolvent pair $(\mathcal{S},\mathcal{V})$ induces the existence of a second pair $(\bar{\mathcal{S}},\bar{\mathcal{V}})$ whose elements, respectively self-hidden and self-bounded, satisfy $\bar{\mathcal{S}} \in \Psi_R$ and $\bar{\mathcal{V}} \in \Phi_E$, with Ψ_R defined by (5.1.69) and

$$\Phi_E := \{\mathcal{V} : \mathcal{V}_m \subseteq \mathcal{V} \subseteq \mathcal{V}^*\} \qquad (36)$$

Assume

$$\bar{\mathcal{S}} := (\mathcal{S}+\mathcal{S}_m) \cap \mathcal{S}_M = \mathcal{S} \cap \mathcal{S}_M + \mathcal{S}_m \qquad (37)$$
$$\bar{\mathcal{V}} := \mathcal{V} + \mathcal{V}_m \qquad (38)$$

Note that $\mathcal{S} \cap \mathcal{S}_M$ is externally stabilizable since $\mathcal{S}$ and $\mathcal{S}_M$ are, respectively by assumption and by virtue of Lemma 4.2-1, so that $\bar{\mathcal{S}}$ is externally stabilizable as the sum of two self-hidden conditioned invariants, one of which is externally stabilizable. $\mathcal{V} \cap \mathcal{P}$ contains $\mathcal{D}$ and is internally stabilizable by assumption and $\mathcal{V}_m$ is internally stabilizable by virtue of Lemma 4.2-1. Furthermore, $\mathcal{V}_m \subseteq \mathcal{P}$: in fact, refer to the defining expression of $\mathcal{V}_m$ - (5.1.70) - and note that

$$\mathcal{S}_1^* := \min \mathcal{S}(A,\mathcal{E},\mathcal{B}+\mathcal{D}) \subseteq \min \mathcal{J}(A,\mathcal{B}+\mathcal{D}) \subseteq \mathcal{P}$$

The intersection $\bar{\mathcal{V}} \cap \mathcal{P} = \mathcal{V} \cap \mathcal{P} + \mathcal{V}_m$ is internally stabilizable since both controlled invariants on the right are. On the other hand, $\mathcal{V}$ being externally stabilizable implies that also $\bar{\mathcal{V}}$ is so because of the inclusion $\mathcal{V} \subseteq \bar{\mathcal{V}}$.

To emphasize some interesting properties of lattices Ψ_R, Φ_R, and Φ_E, let us now introduce a suitable change of basis, which extends that introduced in the proof of Theorem 5.2-2: consider the similarity transformation $T := [T_1\, T_2\, T_3\, T_4\, T_5]$, with $\text{im} T_1 = \mathcal{S}_m = \mathcal{S}_M \cap \mathcal{V}_m$, $\text{im}[T_1\, T_2] = \mathcal{V}_m$, $\text{im}[T_1\, T_3] = \mathcal{S}_M$ and $\text{im}[T_1\, T_2\, T_3\, T_4] = \mathcal{V}^*$. Furthermore, we can choose T_3 and T_5 in such a way that the further conditions

$$\text{im} T_3 \subseteq \mathcal{C} \qquad (39)$$
$$\text{im}[T_1\, T_2\, T_5] \supseteq \mathcal{B} \qquad (40)$$

are satisfied. This is possible since, $\mathcal{S}_M$ being self-hidden and $\mathcal{V}_m$ self-bounded, the inclusions $\mathcal{S}_M \subseteq \mathcal{S}^* + \mathcal{C} \subseteq \mathcal{S}_m + \mathcal{C}$ and $\mathcal{V}_m \supseteq \mathcal{V}^* \cap \mathcal{B}$ hold. From $\mathcal{V}_M = \mathcal{S}_M + \mathcal{V}_m$ it

---

[6] External stabilizability of $\mathcal{S}_m$, which is more restrictive than that of $\mathcal{S}_M$, is not considered in the statement, since it is a consequence of the other conditions.

Sect. 5.2 The Dynamic Disturbance Localization and the Regulator Problem

follows that $\text{im}\,[T_1\ T_2\ T_3] = \mathcal{V}_m$. In this basis the structures of matrices $A' := T^{-1}AT$, $B' := T^{-1}B$, and $C' := CT$, partitioned accordingly, are

$$A' = \begin{bmatrix} \times & \times & \times & \times & \times \\ \times & \times & O & \times & \times \\ \hline O & O & P & R & \times \\ O & O & O & Q & \times \\ \hline \times & \times & O & \times & \times \end{bmatrix} \qquad B' = \begin{bmatrix} \times \\ \times \\ O \\ O \\ \times \end{bmatrix} \qquad (41)$$

$$C' = \begin{bmatrix} \times & \times & | & O & \times & | & \times \end{bmatrix}$$

The zeros in B' and C' are due respectively to inclusions (40) and (39), those in the first and second column of A' are due to $\mathcal{V}_m$ being a controlled invariant and to the particular structure of B', those in the third column to $\mathcal{S}_M$ being a conditioned invariant and the structure of C'. Note that the zero in the third column and fourth row also depends on $\mathcal{V}_M$ being a controlled invariant. The displayed subpartitioning of matrices (41) stresses the particular submatrix

$$V := \begin{bmatrix} P & R \\ O & Q \end{bmatrix} \qquad (42)$$

which will play a key role in the search for resolvents. This change of basis emphasizes a particular structure of matrices A', B', and C', which immediately leads to the following statements.

(b) Any element $\mathcal{S}$ of Ψ_R can be expressed as

$$\mathcal{S} = \mathcal{S}_m + \text{im}\,(T_3 X_S) \qquad (43)$$

where X_S is the basis matrix of a P-invariant. On the assumption that $\mathcal{S}_M$ is externally stabilizable, this invariant is externally stable if and only if $\mathcal{S}$ is externally stabilizable.

(c) Any element $\mathcal{V}$ of Φ_R can be expressed as

$$\mathcal{V} = \mathcal{V}_m + \text{im}\,(T_3 X_V) \qquad (44)$$

where X_V is the basis matrix of a P-invariant. On the assumption that $\mathcal{V}_m$ is internally stabilizable, this invariant is internally stable if and only if $\mathcal{V}$ is internally stabilizable.

On the ground of b and c, if $\mathcal{S} \in \Psi_R$ and $\mathcal{V} \in \Phi_R$ are such that $\mathcal{S} \subseteq \mathcal{V}$, clearly also $\text{im}\,X_S \subseteq \text{im}\,X_V$. In other words, inclusions involving elements of Ψ_R and Φ_R imply inclusions of the corresponding P-invariants. In the sequel, we will refer also to the lattice of self-bounded controlled invariants

$$\Phi_L := \{\mathcal{V} : \mathcal{V}_M \subseteq \mathcal{V} \subseteq \mathcal{V}^*\} \qquad (45)$$

which enjoy the following feature, similar to c.

(d) Any element V of Φ_L can be expressed as

$$V = V_M + \text{im}\,(T_4 X_q) \qquad (46)$$

where X_q is the basis matrix of a Q-invariant.
Other useful statements are the following:

(e) Let V be an (A, B)-controlled invariant, self-bounded with respect to V^*. The internal unassignable eigenvalues in between $V \cap P$ and V are all exogenous.

(f) Let $\mathcal{R} := \min \mathcal{J}(A, B)$ be the reachable set of the controlled system and V any controlled invariant. The following assertions are equivalent:

V is externally stabilizable;
$V + \mathcal{R}$ (which is an A-invariant) is externally stabilizable;
$V + P$ (which is an A-invariant) is externally stabilizable.

To prove e, consider a change of basis defined by the transformation matrix $T = [T_1\ T_2\ T_3\ T_4]$, with $\text{im}\,T_1 = V \cap P$, $\text{im}[T_1\ T_2] = V$, $\text{im}[T_1\ T_3] = P$. We obtain the following structures of matrices $A' := T^{-1}AT$ and $B' := T^{-1}B$:

$$A' = \begin{bmatrix} \times & \times & \times & \times \\ O & A'_{22} & O & A'_{24} \\ \times & \times & \times & \times \\ O & O & O & A'_{44} \end{bmatrix} \qquad B' = \begin{bmatrix} B'_1 \\ O \\ B'_3 \\ O \end{bmatrix}$$

the eigenvalues referred to in the statement are those of A'_{22}. The zeros in B' are due to $P \supseteq B$, whereas those in A' are due to the invariance of P and the controlled invariance of V. The eigenvalues external with respect to P (the exogenous ones) are those of

$$\begin{bmatrix} A'_{22} & A'_{24} \\ O & A'_{44} \end{bmatrix}$$

hence the eigenvalues of A'_{22} are all exogenous. Also, f is easily proved by means of an appropriate change of basis, taking into account the external stability of $\mathcal{R}$ with respect to P.

We shall now review all the points in the statement, and prove the necessity of the given conditions. Condition (12) is implied by Theorem 5.2-2, in particular by the existence of a resolvent pair which satisfies $\mathcal{D} \subseteq \mathcal{S} \subseteq V \subseteq \mathcal{E}$. Let V be a resolvent, i.e., an externally stabilizable (A, B)-controlled invariant: by virtue of Property 4.1-13 $V + \mathcal{R}$ is externally stable; since $V \subseteq V^*$, $V^* + \mathcal{R}$ is also externally stable, hence V^* is externally stabilizable and (13) holds. If there exists a resolvent S, i.e., an (A, C)-conditioned invariant contained in $\mathcal{E}$, containing $\mathcal{D}$ and externally stabilizable, $\mathcal{S}_M$ is externally stabilizable by virtue of Lemma 4.2-2. Thus, the necessity of (14) is proved. To prove the necessity of (15), consider a resolvent pair $(\bar{S}, \bar{V})$ with $\bar{S} \in \Psi_R$ and $\bar{V} \in \Phi_E$, whose existence has been previously proved in point a. Clearly, $V_L := \bar{V} \cap P$ belongs to Φ_R. On the other hand, $\bar{S} \cap P$, which is a conditioned invariant as the intersection of two conditioned invariants, belongs to Ψ_R (remember that $\mathcal{S}_m \subseteq V_m$ and $V_m \subseteq P$); furthermore, subspaces $V_L \cap P$ and $V_M \cap P$ clearly belong to Φ_R. From $\bar{S} \subseteq \bar{V}$ it follows that $\bar{S} \cap P \subseteq V_L \cap P$; moreover, clearly $V_L \cap P \subseteq V_M \cap P$. By virtue of points b and c, $\bar{S} \cap P$, $V_L \cap P$, $V_M \cap P$, $\bar{S}$, correspond to invariants $\mathcal{J}_1, \mathcal{J}_2, \mathcal{J}_3, \mathcal{J}_4$ of matrix P such that

$$\mathcal{J}_1 \subseteq \mathcal{J}_2 \subseteq \mathcal{J}_3 \quad \text{and} \quad \mathcal{J}_1 \subseteq \mathcal{J}_4$$

Sect. 5.2 The Dynamic Disturbance Localization and the Regulator Problem 285

$\bar{S}$ being externally stabilizable $\mathcal{J}_4$ and, consequently, $\mathcal{J}_3 + \mathcal{J}_4$, is externally stable. Therefore, considering that all the eigenvalues external to $\mathcal{J}_3$ are the unassignable ones between $\mathcal{V}_M \cap \mathcal{P}$ and $\mathcal{V}_M$, $\mathcal{J}_3 + \mathcal{J}_4$ must be the whole space upon which the linear transformation expressed by P is defined. This feature can also be pointed out with the relation

$$\mathcal{V}_M \cap \mathcal{P} + \bar{S} = \mathcal{V}_M$$

Matrix P is similar to

$$\begin{bmatrix} P'_{11} & \times & \times & \times \\ O & P'_{22} & \times & O \\ O & O & P'_{33} & O \\ O & O & O & P'_{44} \end{bmatrix}$$

where the partitioning is inferred by a change of coordinates such that the first group corresponds to $\mathcal{J}_1$, the first and second to $\mathcal{J}_2$, the first three to $\mathcal{J}_3$ and the first and fourth to $\mathcal{J}_4$. The external stabilizability of $\bar{S}$ implies the stability of P'_{22} and P'_{33}, while the internal stabilizability of $\mathcal{V}_L \cap \mathcal{P}$ implies the stability of P'_{11} and P'_{22}: hence $\mathcal{J}_3$ is internally stable, that is to say, $\mathcal{V}_M \cap \mathcal{P}$ is internally stabilizable. A first step toward the proof of complementability condition (16) is to show that any resolvent $\bar{\mathcal{V}} \in \Phi_E$ is such that $\mathcal{V}_p := \bar{\mathcal{V}} + \mathcal{V}_M \cap \mathcal{P}$ is also a resolvent and contains $\mathcal{V}_M$. Indeed, $\mathcal{V}_p$ is externally stabilizable because of the external stabilizability of $\bar{\mathcal{V}}$; furthermore $\mathcal{V}_p \cap \mathcal{P}$ is internally stabilizable since

$$(\bar{\mathcal{V}} + \mathcal{V}_M \cap \mathcal{P}) \cap \mathcal{P} = \bar{\mathcal{V}} \cap \mathcal{P} + \mathcal{V}_m \cap \mathcal{P}$$

From $\bar{S} \subseteq \bar{\mathcal{V}}$ and (46) it follows that

$$\mathcal{V}_M = \bar{S} + \mathcal{V}_M \cap \mathcal{P} \subseteq \bar{\mathcal{V}} + \mathcal{V}_M \cap \mathcal{P}$$

By virtue of point f, considering that all the exogenous modes are unstable, and $\mathcal{V}_p$ and $\mathcal{V}^*$ are externally stabilizable, it follows that

$$\mathcal{V}_p + \mathcal{P} = \mathcal{V}^* + \mathcal{P} = \mathcal{X}$$

where $\mathcal{X}$ denotes the whole state space of the controlled system, hence

$$\mathcal{V}_p + \mathcal{V}^* \cap \mathcal{P} = \mathcal{V}^* \qquad (47)$$

By virtue of d, the controlled invariants $\mathcal{V}_M + \mathcal{V}_p \cap \mathcal{P}$, $\mathcal{V}_M + \mathcal{V}^* \cap \mathcal{P}$ and $\mathcal{V}_p$ correspond to Q-invariants $\mathcal{K}_1, \mathcal{K}_2, \mathcal{K}_3$, such that $\mathcal{K}_1 \subseteq \mathcal{K}_2$, $\mathcal{K}_1 \subseteq \mathcal{K}_3$. Note also that, by virtue of (47), $\mathcal{K}_2 + \mathcal{K}_3$ is the whole space on which the linear transformation expressed by Q is defined. Therefore, matrix Q is similar to

$$\begin{bmatrix} Q'_{11} & \times & Q'_{13} \\ O & \times & O \\ O & O & Q'_{33} \end{bmatrix}$$

where the partitioning is inferred by a change of coordinates such that the first group corresponds to $\mathcal{K}_1$, the first and second to $\mathcal{K}_2$, the first and third to $\mathcal{K}_3$. Submatrix Q'_{11} is stable since its eigenvalues are the unassignable ones internal to $\mathcal{V}_p \cap \mathcal{P}$, while Q'_{33} has all its eigenvalues unstable since, by the above point e, they correspond to the unassignable ones between $\mathcal{V}^* \cap \mathcal{P}$ and $\mathcal{V}^*$. Therefore, $\mathcal{K}_1$ is complementable with respect to $(\{0\}, \mathcal{K}_3)$; this clearly implies that $\mathcal{K}_2$ is complementable with respect to $(\{0\}, \mathcal{K}_2 + \mathcal{K}_3)$, hence (16) holds for the corresponding controlled invariants.

If. By virtue of the complementability condition (16), there exists a controlled invariant $\mathcal{V}_c$ satisfying

$$\mathcal{V}_c \cap (\mathcal{V}_M + \mathcal{V}^* \cap \mathcal{P}) = \mathcal{V}_M \qquad (48)$$
$$\mathcal{V}_c + (\mathcal{V}_M + \mathcal{V}^* \cap \mathcal{P}) = \mathcal{V}^* \qquad (49)$$

We will show that $(\mathcal{S}_M, \mathcal{V}_c)$ is a resolvent pair. Indeed, by (49) $\mathcal{V}_M \subseteq \mathcal{V}_c$, and by (12) and (49) $\mathcal{D} \subseteq \mathcal{S}_M \subseteq \mathcal{V}_c \subseteq \mathcal{E}$, since $\mathcal{S}_M \subseteq \mathcal{V}_M$. Adding $\mathcal{P}$ to both members of (48) yields $\mathcal{V}_c + \mathcal{P} = \mathcal{V}^* + \mathcal{P}$: by virtue of f, (13) implies the external stabilizability of $\mathcal{V}_c$. By intersecting both members of (48) with $\mathcal{P}$ and considering that $\mathcal{V}_c \subseteq \mathcal{V}^*$, it follows that $\mathcal{V}_c \cap \mathcal{P} = \mathcal{V}_M \cap \mathcal{P}$, hence by (15) $\mathcal{V}_c \cap \mathcal{P}$ is internally stabilizable. □

5.2.3 General Remarks and Computational Recipes

The preceding results are the most general state-space formulations on the regulation of multivariable linear systems. Their statements are quite simple and elegant, completely in coordinate-free form. The constructive conditions make an automatic feasibility analysis possible by means of a computer, having the five matrices of the controlled system as the only data for both problems considered (disturbance localization by means of a dynamic output feedback and asymptotic regulation). This automatic check is particularly interesting in the multivariable case, where loss of structural features for implementability of a given control action may arise from parameter or structure changes (a structure change may be due, for instance, to interruption of communication channels in the overall system).

However, a completely automatic synthesis procedure based on the aforementioned constructive conditions is not in general satisfactory for the following reasons:

1. When the existence conditions are met, in general the problem admits several solutions, which are not equivalent to each other; for instance, if the plant is stable, regulation can be obtained both by means of a feedforward or a feedback controller since either device satisfies the conditions of Problem 5.2-2 but, in general, feedback is preferable since it is more *robust* against parameter variation or uncertainty;

2. The order of the regulator derived in the constructive proofs of Theorems 5.2-2 and 5.2-4 is quite high (the plant plus the exosystem order); however, it is worth noting that this is the maximal order that may be needed, since the

Sect. 5.2 The Dynamic Disturbance Localization and the Regulator Problem

regulator has both the asymptotic tracking and stabilization functions, and is actually needed only if the controlled plant is strongly intrinsically unstable.

Both points 1 and 2 will be reconsidered in the next chapter, where a new formulation and solution for the regulator problem will be presented. It is a particular case of that discussed in this chapter, but specifically oriented toward the achievement of robustness and order reduction, hence more similar to the formulations of synthesis problems for standard single-input, single-output automatic control systems.

An Example. To illustrate these arguments, a simple example is in order. Consider the single time constant plant described by

$$G(s) = \frac{K_1}{1 + \tau_1 s} \tag{50}$$

and suppose that a controller is to be designed such that the output of the plant asymptotically tracks a reference signal r consisting of an arbitrary step plus an arbitrary ramp. Thus, the exosystem is modeled by two integrators in cascade having arbitrary initial conditions, and the tracking error, which is required to converge to zero as t approaches infinity, is defined as

$$e(t) = r(t) - y(t) \tag{51}$$

The controlled system (plant plus exosystem) is described by equations (5.1.1–3) with

$$A := \begin{bmatrix} -\frac{1}{\tau_1} & 0 & 0 \\ 0 & 0 & 0 \\ 0 & 1 & 0 \end{bmatrix} \quad B := \begin{bmatrix} \frac{K_1}{\tau_1} \\ 0 \\ 0 \end{bmatrix} \quad D := \begin{bmatrix} 0 \\ 0 \\ 0 \end{bmatrix}$$

$$C := \begin{bmatrix} 1 & 0 & 0 \\ 0 & 0 & 1 \end{bmatrix} \quad E := \begin{bmatrix} -1 & 0 & 0 \end{bmatrix}$$

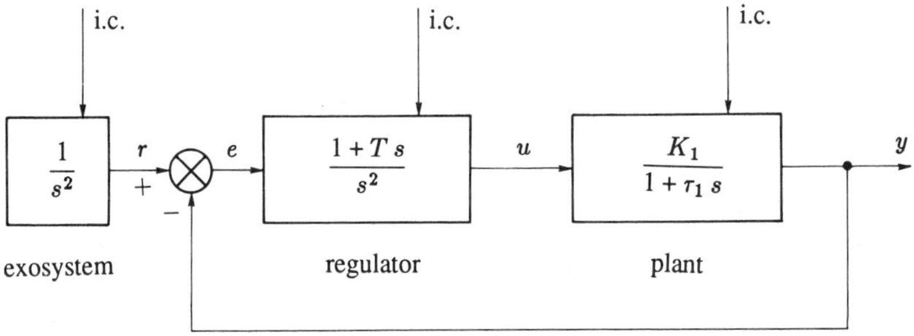

Figure 5-12 A typical feedback control.

Two possible solutions are represented in Figs. 5-12 and 5-13. The first is a typical *feedback* control system and is susceptible to the state-space representation shown in Fig. 5-3 with

$$N := \begin{bmatrix} 0 & 0 \\ 1 & 0 \end{bmatrix} \quad M := \begin{bmatrix} -1 & 1 \\ 0 & 0 \end{bmatrix}$$

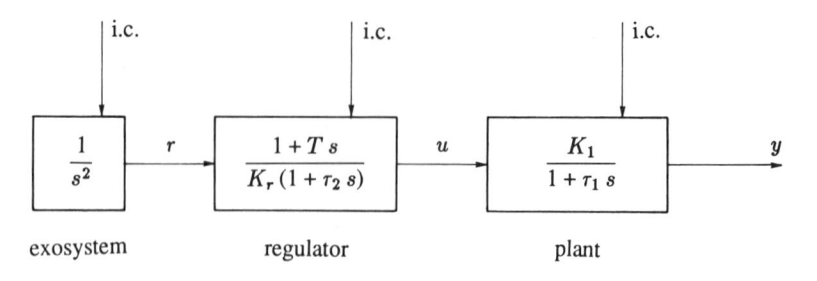

Figure 5-13 A typical feedforward control.

$$L := [T \quad 1] \qquad K := [0 \quad 0]$$

In this case, the only requirement to achieve the prescribed behavior is the *stability condition*

$$T > \tau_1 \qquad (52)$$

The solution presented in Fig. 5-13 is a typical *feedforward* control system and corresponds to

$$N := \frac{1}{\tau_2} \qquad M := \left[-1 \quad \frac{1}{\tau_2}\right]$$

$$L := \frac{1}{K_v}\left(1 - \frac{T}{\tau_2}\right) \qquad K := \left[0 \quad \frac{T}{K_r \tau_2}\right]$$

In this case, to achieve the prescribed behavior the *structural conditions*

$$K_r = K_1 \quad \text{and} \quad T = \tau_1 + \tau_2 \qquad (53)$$

must be satisfied. Note that (52) is expressed by an inequality so that, if the choice of T is sufficiently conservative, it continues to be satisfied also in the presence of small parameter changes, and the feedback scheme maintains the regulation property or is *robust* with respect to parameter variation or uncertainty. On the other hand, the strict equalities in (53) are both necessary for the regulation requirement to be met, so that the feedforward scheme is not robust. In the feedback case robustness is achieved through two significant, basic features: the controlled variable e (i.e., the tracking error which must be asymptotically nulled) is measured or computed without any error (the summing junction is assumed to be perfect), and an *internal model* of the exosystem is embodied in the regulator (the double pole at the origin). In this way a replica of the signal to be asymptotically tracked is internally generated in the extended plant and "automatically trimmed" in the presence of stability of the overall system to null the effects of any parameter variation on the asymptotic behavior of e and exactly track the reference input r. In both schemes the regulator has a relatively low order (two in the feedback case, one in feedforward), but the plant is per se stable. Of course, a purely feedforward control cannot be implemented when the plant is unstable: in this case a dynamic feedback must be added to it, at least to achieve stability, independently of regulation.

Sect. 5.2 The Dynamic Disturbance Localization and the Regulator Problem

We shall now consider compensator and regulator synthesis in greater detail and show that the synthesis procedures presented to prove sufficiency of the conditions stated in Theorems 5.2-1 and 5.2-3 can be framed in a more general context, which will be used in the next section for complete treatment of reduced-order observers, compensators, and regulators.

The procedures used in the proofs of the preceding theorems lead to synthesis of compensators and regulators called *full-order* since their state dimension coincides with that of the controlled system (including the exosystem) and *observer-based* since they are realized according to the scheme of Fig. 5-5(a), in which state feedback is obtained through an observer. They are exactly dualizable: thus full-order compensators and regulators *dual observer-based* are obtained. These are realized according to the scheme of Fig. 5-5(b), in which output injection is obtained through a dual observer. We shall now briefly present the algebraic manipulations concerning this dualization.

Dualizing the Constructive Synthesis Procedures. Instead of (18, 19) we use

$$B L_1 + L_2 = I_n \quad \text{with} \quad \text{im} L_2 = \mathcal{L} \tag{54}$$

and $\mathcal{L}$ such that

$$\mathcal{L} \cap (\mathcal{V} + \mathcal{B}) = \mathcal{V} \quad \mathcal{L} + \mathcal{B} = \mathcal{X} \tag{55}$$

and show that in this case

$$L_2 G C \mathcal{S} \subseteq \mathcal{V} \tag{56}$$

In fact

$$\bigl(A + (BL_1 + L_2) G C\bigr) \mathcal{S} = (A + G C) \mathcal{S} \subseteq \mathcal{V} \subseteq \mathcal{V} + \mathcal{B}$$

and, from $A\mathcal{S} \subseteq A\mathcal{V} \subseteq \mathcal{V} + \mathcal{B}$ (since $\mathcal{S} \subseteq \mathcal{V}$ and $\mathcal{V}$ is a controlled invariant) and $BL_1 G C \mathcal{S} \subseteq \mathcal{B} \subseteq \mathcal{V} + \mathcal{B}$ (by definition of $\mathcal{B}$), it follows that

$$L_2 G C \mathcal{S} \subseteq \mathcal{V} + \mathcal{B}$$

On the other hand, since $L_2 G C \mathcal{S} \subseteq \mathcal{L}$, (54) implies (56). The extended subspace

$$\hat{\mathcal{W}} := \left\{ \begin{bmatrix} x \\ z \end{bmatrix} : z \in \mathcal{V}, \ x = z - \rho, \ \rho \in \mathcal{S} \right\} \tag{57}$$

which, in the coordinates (ρ, η) of system (5.1.36), is expressed by

$$\hat{\mathcal{W}} = \left\{ \begin{bmatrix} \rho \\ \eta \end{bmatrix} : \rho \in \mathcal{S}, \ \eta \in \mathcal{V} \right\} \tag{58}$$

is clearly an $\hat{A}$-invariant. It is easy to verify that it satisfies all the structure and stability requirements stated in Problems 5.2-1/G and 5.2-2/G.

We now consider recipes to derive solutions to the compensator and regulator problems. The computational aspects on which they are based have already been considered in Subsection 5.1.2 (see the two simple applications therein presented), so that they will be reported here in a very schematic way.

Observer-Based Full-Order Compensator. Given the resolvent pair $(\mathcal{S}, \mathcal{V})$, determine L_1, L_2, F, G such that
1. $L_1 C + L_2 = I_n$, $\ker L_2 = \mathcal{L}$, with $\mathcal{L} \cap \mathcal{C} = \{0\}$, $\mathcal{L} + \mathcal{S} \cap \mathcal{C} = \mathcal{S}$; (59)
2. $(A + BF)\mathcal{V} \subseteq \mathcal{V}$, $A + BF$ is stable; (60)
3. $(A + GC)\mathcal{S} \subseteq \mathcal{S}$, $A + GC$ is stable; (61)

then realize the compensator according to

$$N := A + GC + BFL_2 \tag{62}$$
$$M := BFL_1 - G \tag{63}$$
$$L := FL_2 \tag{64}$$
$$K := FL_1 \tag{65}$$

Dual Observer-Based Full-Order Compensator. Given the resolvent pair $(\mathcal{S}, \mathcal{V})$, determine L_1, L_2, F, G such that
1. $BL_1 + L_2 = I_n$, $\operatorname{im} L_2 = \mathcal{L}$, with $\mathcal{L} + \mathcal{B} = \mathcal{X}$, $\mathcal{L} \cap (\mathcal{V} + \mathcal{B}) = \mathcal{V}$; (66)
2. $(A + BF)\mathcal{V} \subseteq \mathcal{V}$, $A + BF$ is stable; (67)
3. $(A + GC)\mathcal{S} \subseteq \mathcal{S}$, $A + GC$ is stable; (68)

then realize the compensator according to

$$N := A + BF + L_2 GC \tag{69}$$
$$M := -L_2 G \tag{70}$$
$$L := F + L_1 GC \tag{71}$$
$$K := L_1 G \tag{72}$$

Observer-Based Full-Order Regulator. Given the resolvent pair $(\mathcal{S}, \mathcal{V})$, determine L_1, L_2 still according to (59) while, as the second group of conditions (those regarding F), instead of (60) consider:

2. $(A + BF)\mathcal{V} \subseteq \mathcal{V}$, $(A+BF)|_{\mathcal{R}_\mathcal{V}}$ is stable, $(A+BF)|_{\mathcal{X}/\mathcal{V}}$ is stable; (73)

and derive G still according to (61). The regulator is defined by (62–65).

Dual Observer-Based Full-Order Regulator. Given the resolvent pair $(\mathcal{S}, \mathcal{V})$, determine L_1, L_2 still according to (66), F according to (73), and G according to (68). The regulator is defined by (69–72).

5.2.4 Sufficient Conditions in Terms of Zeros

We can easily derive conditions expressed in terms of invariant zeros which imply the stabilizability conditions of Theorems 5.2-2 and 5.2-4; hence, joined to the structural condition, they are sufficient for the solvability of the corresponding problems. They are straightforward extensions of the necessary and sufficient conditions in terms of invariant zeros considered in Subsection

4.4.2 for disturbance localization and unknown-input asymptotic observation problems.[7]

Corollary 5.2-1.
The disturbance localization problem by a dynamic compensator admits a solution if:

1. $\mathcal{S}^* \subseteq \mathcal{V}^*$; (74)
2. $\mathcal{Z}(d;y) \dot{-} \mathcal{Z}(d;y,e)$ has all its elements stable; (75)
3. $\mathcal{Z}(u;e) \dot{-} \mathcal{Z}(u,d;e)$ has all its elements stable; (76)
4. $\mathcal{Z}(u,d;e) \cap \mathcal{Z}(d;y,e)$ has all its elements stable. (77)

Proof. (hint) Condition (75) is equivalent to the external stabilizability of $\mathcal{S}_M$, (76) to the internal stabilizability of $\mathcal{V}_m$, (77) implies that the internal unassignable eigenvalues between $\mathcal{V}_m$ and $\mathcal{V}_M$ (those corresponding to lattices Φ_R or Φ_R in Fig. 5-11) are stable. □

Corollary 5.2-2.
Let all the exogenous modes be unstable. The regulator problem admits a solution if:

1. $\mathcal{S}^* \subseteq \mathcal{V}^*$; (78)
2. $\mathcal{Z}(d;y) \dot{-} \mathcal{Z}(d;y,e)$ has all its elements stable; (79)
3. $\mathcal{Z}(u;e) \dot{-} \mathcal{Z}(u,d;e)$ has all its elements stable; (80)
4. $\mathcal{Z}_P(u,d;e) \cap \mathcal{Z}(d;y,e)$ has all its elements stable; (81)
5. $\mathcal{Z}(u;e)$ contains all the eigenvalues of the exosystem; (82)
6. $\mathcal{Z}_P(u;e)$ has no element equal to an eigenvalue of the exosystem. (83)

In (81) and (83) $\mathcal{Z}_P(*;*)$ denotes a set of invariant zeros referred only to the plant, i.e., with the exosystem not present.

Proof. (hint) Relation (79) insures that $\mathcal{S}_M$ is externally stabilizable, (80) and (81) that $\mathcal{V}_M \cap \mathcal{P}$ is internally stabilizable, and (82, 83) that $\mathcal{V}^*$ is externally stabilizable and $\mathcal{V}_M + \mathcal{V}^* \cap \mathcal{P}$ complementable with respect to $(\mathcal{V}_M, \mathcal{V}^*)$. □

5.3 REDUCED-ORDER DEVICES

In this section we shall state and prove a general theorem for order reduction, which allows unitary treatment of all reduced-order devices (observers, compensators, and regulators).

We refer to triple (A, B, C) with the aim of investigating correlation between structural features and eigenvalue assignability. In reduced form, this problem has already been approached in Subsection 4.1-3, where two basic problems for synthesis have been considered: pole assignability with state feedback under the constraint that feedback also transforms a given controlled

---
[7] For a more extended treatment of this topic, see Piazzi and Marro [29].

invariant into a simple invariant and its dual, pole assignability by output injection and contemporary transformation of a conditioned invariant into a simple invariant. The results of these approaches to pole assignability under structural constraints are presented in schematic form in Fig. 4-2 and will now be extended by the following theorem.

Theorem 5.3-1. The Basic Theorem for Order Reduction

Given an (A,C)-conditioned invariant S, there exist both an (A,C)-conditioned invariant S_1 and an output injection matrix G such that the following structural features are satisfied:[1]

1. $C \oplus S_1 = \mathcal{X}$; (1)
2. $S = S \cap C \oplus S \cap S_1$; (2)
3. $(A + GC)S_1 \subseteq S_1$; (3)
4. $(A + GC)S \subseteq S$; (4)

The corresponding spectra assignability are specified by:

5. $\sigma\big((A+GC)|_{\mathcal{Q} \cap S}\big)$ is fixed; (5)
6. $\sigma\big((A+GC)|_{S/(\mathcal{Q} \cap S)}\big)$ is free; (6)
7. $\sigma\big((A+GC)|_{\mathcal{Q}_S/S}\big)$ is fixed; (7)
8. $\sigma\big((A+GC)|_{\mathcal{X}/\mathcal{Q}_S}\big)$ is free. (8)

Theorem 5.3-1 is dualized as follows.

Theorem 5.3-2. The Dual Basic Theorem for Order Reduction

Given an (A,B)-controlled invariant $\mathcal{V}$, there exist both an (A,B)-controlled invariant $\mathcal{V}_1$ and a state feedback matrix F such that the following structural features are satisfied:

1. $B \oplus \mathcal{V}_1 = \mathcal{X}$; (9)
2. $\mathcal{V} = \mathcal{V} \cap B \oplus \mathcal{V} \cap \mathcal{V}_1$; (10)
3. $(A + BF)\mathcal{V}_1 \subseteq \mathcal{V}_1$; (11)
4. $(A + BF)\mathcal{V} \subseteq \mathcal{V}$; (12)

The corresponding spectra assignability are specified by:

5. $\sigma\big((A+BF)|_{\mathcal{R}_\mathcal{V}}\big)$ is free; (13)
6. $\sigma\big((A+BF)|_{\mathcal{V}/\mathcal{R}_\mathcal{V}}\big)$ is fixed; (14)
7. $\sigma\big((A+BF)|_{(\mathcal{V}+\mathcal{R})/\mathcal{V}}\big)$ is free; (15)
8. $\sigma\big((A+BF)|_{\mathcal{X}/(\mathcal{V}+\mathcal{R})}\big)$ is fixed. (16)

Proof of Theorem 5.3-2. Perform the change of basis corresponding to $x = Tz$, with $T := [T_1\, T_2\, T_3\, T_4\, T_5\, T_6]$ such that

$$\mathrm{im}\, T_1 = \mathcal{V} \cap B$$

---

[1] Theorem 5.3-1 and its dual, Theorem 5.3-2, are due to Piazzi [27].

Sect. 5.3 Reduced-Order Devices

$$\text{im}\,[T_1\,T_2] = \mathcal{R}_\mathcal{V} = \mathcal{V} \cap \min \mathcal{S}(A,\mathcal{V},\mathcal{B})$$
$$\text{im}\,[T_1\,T_2\,T_3] = \mathcal{V}$$
$$\text{im}\,[T_1\,T_4] = \mathcal{B}$$
$$\text{im}\,[T_1\,T_2\,T_3\,T_4\,T_5] = \mathcal{V} + \mathcal{R}$$

in the new basis matrices $A':=T^{-1}AT$ and $B':=T^{-1}B$ present the structures

$$A' = \begin{bmatrix} A'_{11} & A'_{12} & A'_{13} & A'_{14} & A'_{15} & A'_{16} \\ A'_{21} & A'_{22} & A'_{23} & A'_{24} & A'_{25} & A'_{26} \\ O & O & A'_{33} & A'_{34} & A'_{35} & A'_{36} \\ A'_{41} & A'_{42} & A'_{43} & A'_{44} & A'_{45} & A'_{46} \\ O & O & O & A'_{54} & A'_{55} & A'_{56} \\ O & O & O & O & O & A'_{66} \end{bmatrix} \qquad B' = \begin{bmatrix} B'_1 \\ O \\ O \\ B'_4 \\ O \\ O \end{bmatrix}$$

The structural zeros in A' are due to $\mathcal{V}+\mathcal{R}$ being an invariant and $\mathcal{R}_\mathcal{V}$, $\mathcal{V}$ controlled invariants. Then, perform in the input space the change of basis defined by $u = Nv$ with

$$N := \begin{bmatrix} B'_1 \\ B'_4 \end{bmatrix}^{-1}$$

which transforms the input distribution matrix as follows:

$$B'' := B'N = \begin{bmatrix} I_1 & O \\ O & O \\ O & O \\ O & I_4 \\ O & O \\ O & O \end{bmatrix}$$

with identity matrices I_1 and I_4 having dimensions $\dim(\mathcal{B}\cap\mathcal{V})$ and $\dim((\mathcal{V}+\mathcal{B})/\mathcal{V})$ respectively. Note that, due to the properties of $\mathcal{R}_\mathcal{V}$ and $\mathcal{R}$, pairs

$$\left(\begin{bmatrix} A'_{11} & A'_{12} \\ A'_{21} & A'_{22} \end{bmatrix},\begin{bmatrix} I_1 \\ O \end{bmatrix}\right) \qquad \left(\begin{bmatrix} A'_{44} & A'_{45} \\ A'_{54} & A'_{55} \end{bmatrix},\begin{bmatrix} I_4 \\ O \end{bmatrix}\right)$$

are controllable; by virtue of the particular structure of the input distribution matrices, pairs (A'_{22}, A'_{21}) and (A'_{55}, A'_{54}) are also controllable; hence there exist matrices F'_{12} and F'_{45} which allow arbitrary assignment of spectra $\sigma(A'_{22}+A'_{21}F'_{12})$ and $\sigma(A'_{55}+A'_{54}F'_{45})$.

We now perform in the state space the further change of basis defined by $z=\tilde{T}\tilde{z}$ with

$$\tilde{T} := \begin{bmatrix} I_1 & F'_{12} & O & O & O & O \\ O & I_2 & O & O & O & O \\ O & O & I_3 & O & O & O \\ O & O & O & I_4 & F'_{45} & O \\ O & O & O & O & I_5 & O \\ O & O & O & O & O & I_6 \end{bmatrix}$$

Thus, system matrix $\tilde{A} := \tilde{T}^{-1} A' \tilde{T}$ and input distribution matrix $\tilde{B} := \tilde{T}^{-1} B''$ assume the structures

$$\tilde{A} = \begin{bmatrix} \tilde{A}_{11} & \tilde{A}_{12} & \tilde{A}_{13} & \tilde{A}_{14} & \tilde{A}_{15} & \tilde{A}_{16} \\ A'_{21} & A'_{22}+A'_{21}F'_{12} & A'_{23} & A'_{24} & \tilde{A}_{25} & A'_{26} \\ O & O & A'_{33} & A'_{34} & \tilde{A}_{35} & A'_{36} \\ A'_{41} & \tilde{A}_{42} & A'_{43} & \tilde{A}_{44} & \tilde{A}_{45} & \tilde{A}_{46} \\ O & O & O & A'_{54} & A'_{55}+A'_{54}F'_{45} & A'_{56} \\ O & O & O & O & O & A'_{66} \end{bmatrix} \quad \tilde{B} = \begin{bmatrix} I_1 & O \\ O & O \\ O & O \\ O & I_4 \\ O & O \\ O & O \end{bmatrix}$$

In the actual state and input bases (i.e., in the coordinates $\tilde{z}$ and v) define subspace $\mathcal{V}_1$ and state feedback matrix $\tilde{F}$ as

$$\mathcal{V}_1 := \text{im}\left(\begin{bmatrix} O & O & O & O \\ I_2 & O & O & O \\ O & I_3 & O & O \\ O & O & O & O \\ O & O & I_5 & O \\ O & O & O & I_6 \end{bmatrix}\right)$$

$$\tilde{F} := \begin{bmatrix} -\tilde{A}_{11}+P & -\tilde{A}_{12} & -\tilde{A}_{13} & R & -\tilde{A}_{15} & -\tilde{A}_{16} \\ -A'_{41} & -\tilde{A}_{42} & -A'_{43} & -\tilde{A}_{44}+Q & -\tilde{A}_{45} & -\tilde{A}_{46} \end{bmatrix}$$

where P and Q are free matrices with arbitrary eigenvalues and R is a further free matrix. In the new basis, checking all stated conditions is quite easy. Structural conditions (9, 10) are clear; to verify the other ones, consider the matrix

$$\tilde{A} + \tilde{B}\tilde{F} = \begin{bmatrix} P & O & O & \tilde{A}_{14}+R & O & O \\ A'_{24} & A'_{22}+A'_{21}F'_{12} & A'_{23} & A'_{24} & \tilde{A}_{25} & A'_{26} \\ O & O & A'_{33} & A'_{34} & \tilde{A}_{35} & A'_{36} \\ O & O & O & Q & O & O \\ O & O & O & A'_{54} & A'_{55}+A'_{54}F'_{45} & A'_{56} \\ O & O & O & O & O & A'_{66} \end{bmatrix} \quad (17)$$

The structural zeros in (17) prove that $\mathcal{V}_1$ and $\mathcal{V}$ are $(A+BF)$-invariants [relations (11, 12)]. The remaining stated relations derive from the following properties of the spectra of some submatrices of (17):

$$\sigma\big((A+BF)|_{\mathcal{R}_\mathcal{V}}\big) = \sigma(P) \uplus \sigma(A'_{22}+A'_{21}F'_{12})$$
$$\sigma\big((A+BF)|_{\mathcal{V}/\mathcal{R}_\mathcal{V}}\big) = \sigma(A'_{33})$$
$$\sigma\big((A+BF)|_{(\mathcal{V}+\mathcal{R})/\mathcal{V}}\big) = \sigma(Q) \uplus \sigma(A'_{55}+A'_{54}F'_{45})$$
$$\sigma\big((A+BF)|_{\mathcal{X}/(\mathcal{V}+\mathcal{R})}\big) = \sigma(A'_{66})$$

Some straightforward manipulations enable the controlled invariant $\mathcal{V}_1$ and the state feedback matrix F to be expressed in the originary basis, i.e., with respect to coordinates x, u. They lead to:

$$\mathcal{V}_1 = \text{im}\,[T_1 \mid F'_{12}+T_2 \mid T_3 \mid T_4 \mid F'_{45}+T_5 \mid T_6]$$
$$F = N\tilde{F}\tilde{T}^{-1}T^{-1}$$

where T_i ($i = 1,\ldots,6$) are the submatrices of T defined at the beginning of this proof. □

5.3.1 Reduced-Order Observers

Consider a triple (A, B, C). The following properties hold.[2]

Property 5.3-1.
If (A, C) is detectable there exists an (A, C)-conditioned invariant S_1 such that:
1. $C \oplus S_1 = \mathcal{X}$; (18)
2. S_1 is externally stabilizable. (19)

Property 5.3-2.
If (A, B) is stabilizable there exists an (A, B)-controlled invariant V_1 such that:
1. $B \oplus V_1 = \mathcal{X}$; (20)
2. V_1 is internally stabilizable. (21)

Proof of Property 5.3-2. Apply Theorem 5.3-2 with $V := \mathcal{X}$: relations (9) and (10) are both equivalent to (20), while (11) and (13) together with the assumption on stabilizability of (A, B) guarantee the existence of F and V_1 such that $(A+BF)V_1 \subseteq V_1$ with $A+BF$ stable (since the internal eigenvalues of $\mathcal{R} := \min \mathcal{J}(A, B) = \min \mathcal{J}(A+BF, B)$ are assignable, while the external ones are stable by assumption). Hence, V_1 is an internally stabilizable (A, B)-controlled invariant. □

Properties 5.3-1 and 5.3-2 can be used to derive reduced-order observers and dual observers, i.e., asymptotic state observers of order $n - q$, where q denotes the number of linearly independent output variables, and stable precompensators of order $n - p$, where p denotes the number of linearly independent input variables.

The Synthesis Procedure. Consider the block diagram of Fig. 5-5(a) and suppose that block F is not present and that input u is applied from the outside: the remainder of the device, up to the summing junction where the signals from L_1 and L_2 converge, is a whole state asymptotic observer which derives the state estimate in part from the system output. From this observer, which is still full-order, we can derive a reduced-order one by using a computational procedure similar to that described at the end of Subsection 5.1.2.

According to Property 5.3-1, choose matrix G so that $(A+GC)S_1 \subseteq S_1$ and $A+GC$ is stable. In this way the matrix of the dynamic part of the device, corresponding to the differential equation

$$\dot{z}(t) = (A + GC) z(t) - G y(t) + B u(t) \qquad (22)$$

has an externally stable invariant, i.e., if the state is expressed in a suitable basis some components do not influence the remaining ones: by eliminating them a stable system is obtained which provides a state estimate modulo S_1. By virtue of (18) it is possible to determine L_1, L_2 in such a way that the corresponding linear combination of the output and this partial estimate is a complete estimate $\tilde{x}$ of state x.

We shall now give a complete recipe for the reduced-order observer: first of all apply Lemma 5.1-3 with $\mathcal{L} := S_1$ and derive L_1 and L_2 by means of the

---

[2] Property 5.3-1 and its dual, Property 5.3-2, are due to Wonham [38].

$$\dot{x} = Ax + Bu$$
$$y = Cx$$

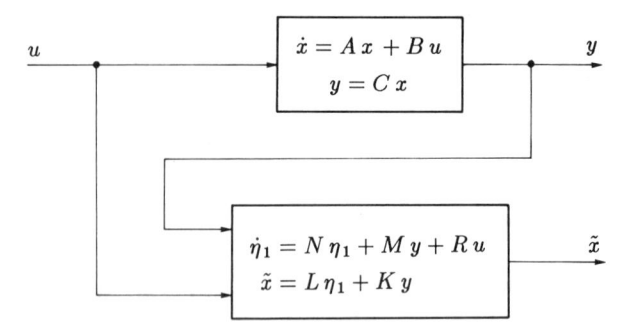

$$\dot{\eta}_1 = N\eta_1 + My + Ru$$
$$\tilde{x} = L\eta_1 + Ky$$

Figure 5-14 Reduced-order observer.

constructive procedure used in the proof. Then, compute a similarity transformation matrix $T := [T_1\ T_2]$ with $\mathrm{im}\,T_1 = \mathcal{C}$, $\mathrm{im}\,T_2 = \mathcal{S}_1$ and assume

$$Q = \begin{bmatrix} Q_1 \\ Q_2 \end{bmatrix} := T^{-1}$$

In the new coordinate η defined by $z = T\eta$ equation (22) becomes

$$\begin{bmatrix} \dot{\eta}_1(t) \\ \dot{\eta}_2(t) \end{bmatrix} = \begin{bmatrix} Q_1(A+GC)T_1 & O \\ Q_2(A+CG)T_1 & Q_2(A+GC)T_2 \end{bmatrix} \begin{bmatrix} \eta_1(t) \\ \eta_2(t) \end{bmatrix} - \begin{bmatrix} Q_1 G \\ Q_2 G \end{bmatrix} y(t) + \begin{bmatrix} Q_1 B \\ Q_2 B \end{bmatrix} u(t)$$

Implement the observer as in the block diagram of Fig. 5-14 with:

$$N := Q_1(A+GC)T_1 \tag{23}$$
$$M := -Q_1 G \tag{24}$$
$$R := Q_1 B \tag{25}$$
$$L := L_2 T_1 \tag{26}$$
$$K := L_1 \tag{27}$$

Perfectly dual arguments can be developed for dynamic precompensators. Refer to the block diagram of Fig. 5-5(b) and suppose that block G is not present: the remainder of the device is a dynamic precompensator (or dual observer) where the control action, applied to L_1 and L_2 in parallel, is suitably distributed to the inputs of the controlled system and precompensator itself. By applying Property 5.3-2, the precompensator order can be reduced to $n - p$ while preserving stability.

5.3.2 Reduced-Order Compensators and Regulators

Consider a triple (A, B, C). The following properties hold.[3]

---

[3] Property 5.3-3 and its dual, Property 5.3-4, are due to Imai and Akashi [19].

Property 5.3-3.
Let (A,C) be detectable. Given any externally stabilizable (A,C)-conditioned invariant S, there exists another (A,C)-conditioned invariant S_1 such that:

1. $C \oplus S_1 = \mathcal{X}$; (28)
2. $S = S \cap S_1 \oplus S \cap C$; (29)
3. $S+S_1$ is an (A,C)-conditioned invariant; (30)
4. S_1 is externally stabilizable. (31)

Property 5.3-4.
Let (A,B) be stabilizable. Given any internally stabilizable (A,B)-controlled invariant $\mathcal{V}$, there exists another (A,B)-controlled invariant $\mathcal{V}_1$ such that:

1. $B \oplus \mathcal{V}_1 = \mathcal{X}$; (32)
2. $\mathcal{V} = \mathcal{V} \cap \mathcal{V}_1 \oplus \mathcal{V} \cap B$; (33)
3. $\mathcal{V} \cap \mathcal{V}_1$ is an (A,B)-controlled invariant; (34)
4. $\mathcal{V}_1$ is internally stabilizable. (35)

Proof of Property 5.3-4. Also this property is reconducted to Theorem 5.3-2. The existence of a state feedback matrix F such that $(A+BF)\mathcal{V}_1 \subseteq \mathcal{V}_1$ and $(A+BF)\mathcal{V} \subseteq \mathcal{V}$ by virtue of Property 4.1-5 is equivalent to (34), while conditions on spectra assignability, added to the assumptions that $\mathcal{V}$ is internally stabilizable and (A,B) stabilizable, imply the possibility of $A+BF$ being stable, hence $\mathcal{V}_1$ internally stabilizable. □

It is now possible to state recipes for reduced-order compensators, assuming as the starting point those for full-order compensators presented in the previous section. The only difference in computations is that, when deriving matrices L_1, L_2 satisfying (5.2.18) in the case of the observer-based compensator or (5.2.54) in that of the dual observer-based compensator, it is assumed that $\mathcal{L} := S_1$ or $\mathcal{L} := \mathcal{V}_1$ (S_1 and $\mathcal{V}_1$ are defined in Properties 5.3-3 and 5.3-4) and G, F are determined in such a way that $(A+GC)S_1 \subseteq S_1$, $(A+GC)S \subseteq S$ and $A+GC$ is stable in the former case, $(A+BF)\mathcal{V}_1 \subseteq \mathcal{V}_1$, $(A+BF)\mathcal{V} \subseteq \mathcal{V}$ and $A+BF$ is stable in the latter. In the proofs of the synthesis procedures only a few changes are needed: relations (5.2.19, 20) are replaced by (29, 28) above in the direct case, and (5.2.55) by (33, 32) in the dual one. Suitable changes of basis, like that presented in the previous section for the reduced-order observer, allow elimination of $n-q$ equations in the direct case and $n-p$ equations in the dual one.

Observer-Based Reduced-Order Compensator. Given the resolvent pair $(S, \mathcal{V})$, determine L_1, L_2, F, G such that

1. $L_1 C + L_2 = I_n$, $\ker L_2 = S_1$; (36)
2. $(A+BF)\mathcal{V} \subseteq \mathcal{V}$, $A+BF$ is stable; (37)
3. $(A+GC)S_1 \subseteq S_1$, $(A+GC)S \subseteq S$, $A+GC$ is stable; (38)

then derive T_1 and Q_1 from $T:=[T_1\ T_2]$, $\operatorname{im}T_1=\mathcal{C}$, $\operatorname{im}T_2=\mathcal{S}_1$ and

$$Q = \begin{bmatrix} Q_1 \\ Q_2 \end{bmatrix} := T^{-1}$$

and realize the compensator according to

$$N := Q_1(A + GC + BFL_2)T_1 \qquad (39)$$
$$M := Q_1(BFL_1 - G) \qquad (40)$$
$$L := FL_2 T_1 \qquad (41)$$
$$K := FL_1 \qquad (42)$$

Dual Observer-Based Reduced-Order Compensator. Given the resolvent pair $(\mathcal{S},\mathcal{V})$, determine L_1, L_2, F, G such that
1. $BL_1 + L_2 = I_n$, $\operatorname{im}L_2 = \mathcal{V}_1$; $\qquad (43)$
2. $(A+BF)\mathcal{V}_1 \subseteq \mathcal{V}_1$, $(A+BF)\mathcal{V} \subseteq \mathcal{V}$, $A+BF$ is stable; $\qquad (44)$
3. $(A+GC)\mathcal{S} \subseteq \mathcal{S}$, $A+GC$ is stable. $\qquad (45)$

Then, derive T_1 and Q_1 from $T:=[T_1\ T_2]$, $\operatorname{im}T_1=\mathcal{V}_1$, $\operatorname{im}T_2=\mathcal{B}$, and

$$Q = \begin{bmatrix} Q_1 \\ Q_2 \end{bmatrix} := T^{-1}$$

and realize the compensator according to

$$N := Q_1(A + BF + L_2 GC)T_1 \qquad (46)$$
$$M := -Q_1 L_2 G \qquad (47)$$
$$L := (F + L_1 GC)T_1 \qquad (48)$$
$$K := L_1 G \qquad (49)$$

We shall now consider the synthesis of reduced-order regulators. The only difference with respect to the compensators is that pair (A, B) is not stabilizable; hence Theorem 3.5-2 is considered in its most general form through the following corollary.

Corollary 5.3-1.
Consider a generic pair (A, B). Given an internally stable A-invariant $\mathcal{P}$ and an (A, B)-controlled invariant $\mathcal{V}$ such that $\mathcal{V} \cap \mathcal{P}$ is internally stabilizable, there exist both another (A, B)-controlled invariant $\mathcal{V}_1$ and a matrix F such that:
1. $\mathcal{B} \oplus \mathcal{V}_1 = \mathcal{X}$; $\qquad (50)$
2. $\mathcal{V} = \mathcal{V} \cap \mathcal{V}_1 \oplus \mathcal{V} \cap \mathcal{B}$; $\qquad (51)$
3. $(A+BF)\mathcal{V}_1 \subseteq \mathcal{V}_1$; $\qquad (52)$
4. $(A+BF)\mathcal{V} \subseteq \mathcal{V}$; $\qquad (53)$
5. $(A+BF)|_\mathcal{P}$ is stable; $\qquad (54)$
6. $(A+BF)|_{\mathcal{X}/\mathcal{V}}$ is stable. $\qquad (55)$

Proof. Consider Theorem 5.3-2 and determine $\mathcal{V}_1$ and F in such a way that (50–53) hold and all the free spectra are stabilized. Relation (55) holds since, $\mathcal{V}$ being externally stabilizable, $\mathcal{V}+\mathcal{R}$ is externally stable both as an A-invariant and an $(A+BF)$-invariant. $\mathcal{V} \cap \mathcal{P}$ is an (A,B)-controlled invariant (as the intersection of an (A,B)-controlled invariant and an A-invariant containing $\mathcal{B}$), self-bounded with respect to $\mathcal{V}$ since $\mathcal{V} \cap \mathcal{B} \subseteq \mathcal{V} \cap \mathcal{P}$ (as $\mathcal{B} \subseteq \mathcal{P}$). Restriction $(A+BF)|_{(\mathcal{V} \cap \mathcal{P})/\mathcal{R}_\mathcal{V}}$ is stable because $\mathcal{V} \cap \mathcal{P}$ is internally stabilizable and restriction $(A+BF)|_{\mathcal{R}_\mathcal{V}}$ is stable because of the above choice of F; hence $(A+BF)|_{\mathcal{V} \cap \mathcal{P}}$ is stable. Since $\mathcal{V}$ is externally stable as an $(A+BF)$-invariant, it follows that $(A+BF)|_\mathcal{P}$ is stable. $\square$

We now present the recipes for regulators.

Observer-Based Reduced-Order Regulator. Given the resolvent pair $(\mathcal{S}, \mathcal{V})$, determine L_1, L_2 still according to (36), while for F instead of (37) consider the conditions:

2. $(A+BF)\mathcal{V} \subseteq \mathcal{V}$, $(A+BF)|_\mathcal{P}$ is stable, $(A+BF)|_{\mathcal{X}/\mathcal{V}}$ is stable; (56)

and derive G still according to (38). Definitions of T_1 and Q_1 are the same as in the compensator case. The regulator is defined again by (39–42).

Dual Observer-Based Reduced-Order Regulator. Given the resolvent pair $(\mathcal{S}, \mathcal{V})$, determine L_1, L_2 still according to (43), while for F, instead of (44), consider the conditions:

2. $(A+BF)\mathcal{V}_1 \subseteq \mathcal{V}_1$, $(A+BF)\mathcal{V} \subseteq \mathcal{V}$, $(A+BF)|_\mathcal{P}$ is stable,
$(A+BF)|_{\mathcal{X}/\mathcal{V}}$ is stable; (57)

and derive G still according to (45). Definitions of T_1 and Q_1 are the same as in the compensator case. The regulator is defined again by (46–49).

5.4 ACCESSIBLE DISTURBANCE LOCALIZATION AND MODEL-FOLLOWING CONTROL

We go back to the disturbance localization problem by dynamic compensator, already treated in Section 5.2, to give it a more general formulation. Refer to the block diagram of Fig. 5-15, where disturbances (or nonmanipulable inputs) entering the controlled plant are two: an *unaccessible disturbance d*, and an *accessible disturbance* d_1.

The overall system, with extended state $\hat{x} := (x, z)$, is described by

$$\dot{\hat{x}}(t) = \hat{A}\hat{x}(t) + \hat{D} d(t) + \hat{R} d_1(t) \quad (1)$$
$$e(t) = \hat{E}\hat{x}(t) \quad (2)$$

with

$$\hat{A} := \begin{bmatrix} A+BKC & BL \\ MC & N \end{bmatrix} \quad \hat{D} := \begin{bmatrix} D \\ O \end{bmatrix}$$
$$\hat{R} := \begin{bmatrix} D_1+BS \\ R \end{bmatrix} \quad \hat{E} := [\, E \quad O \,] \quad (3)$$

```
                    │ d
                    ▼
  d₁       ┌──────────────────┐    e
 ──────●──▶│ ẋ = Ax + Bu + Dd │──────▶
          │     + D₁d₁        │
          │                   │
          │    y = Cx         │
       u  │                   │ y
       ┌─▶│    e = Ex         │──┐
       │  └──────────────────┘  │
       │                         │
       │  ┌──────────────────┐  │
       │  │ ż = Nz + My + Rd₁│  │
       └──│ u = Lz + Ky + Sd₁│◀─┘
          └──────────────────┘
```

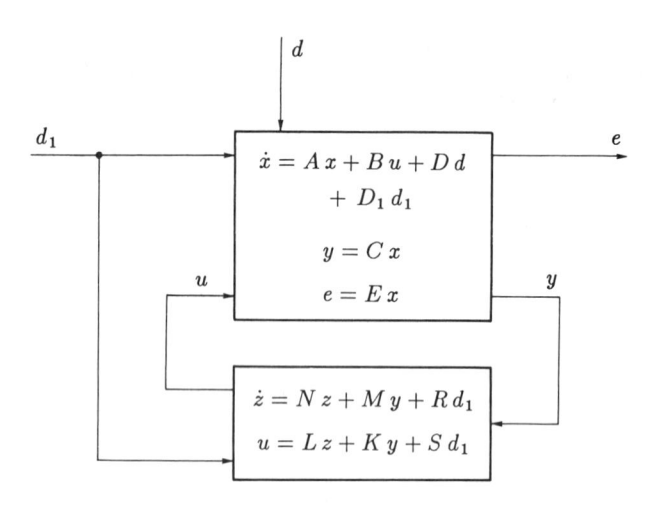

Figure 5-15 Disturbance localization by dynamic compensation: disturbances in part accessible.

Recalling the results of Section 4.2 on unaccessible disturbance localization by dynamic output feedback and those of Section 5.2 on accessible disturbance localization by algebraic state feedback, we can give the problem the following geometric formulation.

Problem 5.4-1. General Disturbance Localization by Dynamic Feedback

Refer to the block diagram of Fig. 5-15 and assume that (A, B) is stabilizable and (A, C) detectable. Determine, if possible, a feedback compensator of the type shown in the figure such that:

1. the overall system has an $\hat{A}$-invariant $\hat{\mathcal{W}}$ that satisfies

$$\hat{\mathcal{D}} \subseteq \hat{\mathcal{W}} \subseteq \hat{\mathcal{E}}, \quad \hat{\mathcal{R}} \subseteq \hat{\mathcal{W}} \quad (\text{with } \hat{\mathcal{R}} := \operatorname{im} \hat{R})$$

2. $\hat{A}$ is stable.

Theorems 5.2-1 (nonconstructive conditions) and 5.2-2 (constructive conditions) can be extended to solve this more general problem. In the following the notation $\mathcal{D}_1 := \operatorname{im} D_1$ will be used.

Theorem 5.4-1.

The disturbance localization problem by a dynamic compensator with disturbance in part accessible admits a solution if and only if there exist both an (A, B)-controlled invariant $\mathcal{V}$ and an (A, C)-conditioned invariant $\mathcal{S}$ such that:

1. $\mathcal{D} \subseteq \mathcal{S} \subseteq \mathcal{V} \subseteq \mathcal{E}$; (4)
2. $\mathcal{D}_1 \subseteq \mathcal{V} + \mathcal{B}$; (5)
3. $\mathcal{S}$ is externally stabilizable; (6)
4. $\mathcal{V}$ is internally stabilizable. (7)

Theorem 5.4-2.
The disturbance localization problem by a dynamic compensator with disturbance in part accessible admits a solution if and only if

1. $S^* \subseteq V^*$; (8)
2. $\mathcal{D}_1 \subseteq V^* + \mathcal{B}$; (9)
3. S_M is externally stabilizable; (10)
4. V_M is internally stabilizable; (11)
5. $V'_m := V^* \cap \min S(A, \mathcal{E}, \mathcal{B}+\mathcal{D}_1)$ is internally stabilizable. (12)

Subspaces S^*, V^*, S_M, V_M are defined by (5.1.57, 51, 73, 71).

Proof of Both Theorems. (hint) Necessity of (4–7) is proved as for Theorem 5.2-1, by projection of the conditions stated in the geometric formulation of the problem (in this case Problem 5.4-1); sufficiency of (4) and (6,7) still as for Theorem 5.2-1, while sufficiency of (5) is proved as follows: determine a dynamic compensator that realizes the unaccessible disturbance localization and denote by $\hat{\mathcal{W}}$ the corresponding extended invariant. Since

$$\hat{\mathcal{D}}_1 \subseteq \hat{\mathcal{W}} + \hat{\mathcal{B}}_0 \quad \text{with} \quad \hat{D}_1 := \begin{bmatrix} D_1 \\ O \end{bmatrix} \text{ and } \hat{B}_0 := \begin{bmatrix} B & O \\ O & I_m \end{bmatrix} \quad (13)$$

where, of course, $\hat{\mathcal{D}}_1 := \text{im}\,\hat{D}_1$, $\hat{\mathcal{B}}_0 := \text{im}\,\hat{B}_0$, is clearly possible to derive matrices R and S such that the corresponding $\hat{R}$, defined in (3), satisfies the requirement stated in Problem 5.4-1. As far as (12) is concerned, necessity follows from Property 4.1-5, while sufficiency is reduced to Theorem 5.4-1 by assuming $S := S_M$, $V := V_M + V'_m$ as a resolvent pair. □

We shall now show that another well-known problem of control theory, the *model-following control*, can be formally reduced to the accessible disturbance localization problem.[1]

This problem is stated in the following terms: refer to a completely controllable and observable three-map system, described by

$$\dot{x}_s(t) = A_s\, x_s(t) + B_s\, u_s(t) \quad (14)$$
$$y_s(t) = C_s(t)\, x_s(t) \quad (15)$$

and suppose that a *model*, completely controllable, observable, and stable, is given as

$$\dot{x}_m(t) = A_m\, x_m(t) + B_m\, u_m(t) \quad (16)$$
$$y_m(t) = C_m(t)\, x_m(t) \quad (17)$$

---

[1] Model-following systems were geometrically approached first by Morse [23].

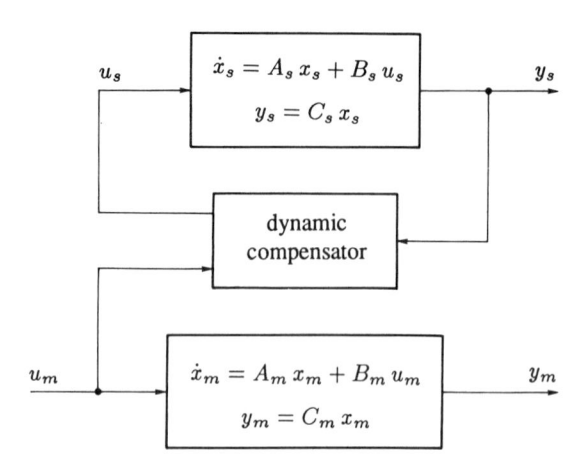

```
                    ┌──────────────────┐
         uₛ         │ ẋₛ = Aₛ xₛ + Bₛ uₛ │         yₛ
       ─────┬──────▶│                  ├──────┬─────▶
            │       │ yₛ = Cₛ xₛ       │      │
            │       └──────────────────┘      │
            │                                 │
            │       ┌──────────────┐          │
            └──────▶│   dynamic    │◀─────────┤
                    │ compensator  │          │
            ┌──────▶│              │          │
            │       └──────────────┘          │
            │                                 │
         uₘ │       ┌──────────────────┐      │ yₘ
       ─────┴──────▶│ ẋₘ = Aₘ xₘ + Bₘ uₘ ├──────┴─────▶
                    │                  │
                    │ yₘ = Cₘ xₘ       │
                    └──────────────────┘
```

Figure 5-16 Model-following compensator.

The dimensions of system and model output spaces are assumed to be equal.

A control device is sought which, connected to the system and model as shown in Fig. 5-16, corresponds to a stable overall system and realizes the tracking of the model output, i.e., is such that, starting at the zero state and for all admissible input functions $u_m(\cdot)$, automatically provides a control function $u_s(\cdot)$ which realizes equality $y_s(\cdot) = y_m(\cdot)$.

This problem can be reduced to accessible disturbance localization by dynamic feedback and solved according to the block diagram shown in Fig. 5-15. To this end, assume

$$x := \begin{bmatrix} x_s \\ x_m \end{bmatrix} \qquad u := u_s \qquad d_1 := u_m \tag{18}$$

and, for matrices

$$A := \begin{bmatrix} A_s & O \\ O & A_m \end{bmatrix} \qquad B := \begin{bmatrix} B_s \\ O \end{bmatrix} \qquad C := [\, C_s \quad O \,]$$

$$D := \begin{bmatrix} O \\ O \end{bmatrix} \qquad D_1 := \begin{bmatrix} O \\ B_m \end{bmatrix} \qquad E := [\, C_s \quad -C_m \,] \tag{19}$$

5.5 NONINTERACTING CONTROLLERS

Consider the block diagram of Fig. 5-1 and assume that the controlled system is stabilizable and detectable (i.e., that it coincides with the plant alone, without any exosystem). It is a five-map system (A, B, C, D, E), described by the equations:

$$\dot{x}(t) = A\,x(t) + B\,u(t) + D\,d(t) \tag{1}$$
$$y(t) = C\,x(t) \tag{2}$$
$$e(t) = E\,x(t) \tag{3}$$

Suppose that a partition in k blocks of the controlled output e is given so that by applying a suitable reordering of these components if needed, it is possible to set $e = (e_1, \ldots, e_k)$; denote by $E_1, \ldots, E_k$ the corresponding submatrices of E. A noninteracting controller is defined as follows.[1]

Definition 5.5-1. Noninteracting Controller

The control apparatus of Fig. 5-1 is said to be *noninteracting* with respect to partition $(e_1, \ldots, e_k)$ of output e if there exists a corresponding partition $(r_1, \ldots, r_k)$ of the reference input r such that, starting at the zero state, by acting on a single input r_i (with all the other inputs, d included, identically zero) only the corresponding output e_i is changed, while the others remain identically zero.

The existence of a noninteracting controller for a given output partition is strictly related to the system structure: in fact, noninteraction may involve loss of controllability if the system structure is not favorable. Noninteraction is clearly related to the concept of constrained controllability, introduced and discussed in Subsection 4.2.3. Let

$$\mathcal{E}_i := \bigcap_{j \neq i} \ker E_j \quad (i=1,\ldots,k) \tag{4}$$

It is clear that the reachable set on $\mathcal{E}_i$, i.e.

$$\mathcal{R}_{\mathcal{E}_i} = \mathcal{V}_i^* \cap \min \mathcal{S}(A, \mathcal{E}_i, \mathcal{B}) \quad (i=1,\ldots,k) \tag{5}$$

with

$$\mathcal{V}_i^* := \max \mathcal{V}(A, \mathcal{B}, \mathcal{E}_i) \quad (i=1,\ldots,k) \tag{6}$$

---

[1] The first geometric approaches to noninteracting controllers are due to Wonham and Morse [4.44, 24], and Basile and Marro [3]. We shall report here this latter treatment, which is less general but more elementary and tutorial.

is the maximal subspace on which, starting at the origin, state trajectories can be obtained that affect only output e_i, without influencing the other outputs, which remain identically zero.[2]

Therefore, conditions

$$E_i \mathcal{R}_{\mathcal{E}_i} = \text{im} E_i \quad (i=1,\ldots,k) \tag{7}$$

are necessary to perform a complete noninteracting control. They clearly depend on the system structure, hence are necessary whatever the type of controller used (for instance a nonlinear one). Necessary and sufficient conditions for the existence of a noninteracting control device implemented according to the block diagram of Fig. 5-1 are stated in the following theorem.

Theorem 5.5-1.

Refer to a quintuple (A, B, C, D, E) with (A, B) stabilizable and (A, C) detectable. Given a partition $(e_1, \ldots, e_k)$ of the controlled output variables, there exists a noninteracting control device of the type shown in Fig. 5-17 if and only if conditions (7) hold.

Proof. Only if. This part of the proof is directly implied by the concept of constrained controllability.

If. Consider the extended mathematical model

$$\dot{\hat{x}}(t) = \hat{A}\,\hat{x}(t) + \hat{D}\,d(t) + \hat{R}\,r(t) \tag{8}$$
$$e(t) = \hat{E}\,\hat{x}(t) \tag{9}$$

where matrices are the same as in (5.1.12). By virtue of Theorem 3.3-1 on controllability, system (8–9) is noninteracting if and only if:

$$\hat{\mathcal{R}}_i^* \subseteq \bigcap_{j \neq i} \ker \hat{\mathcal{E}}_j \quad (i=1,\ldots,k) \tag{10}$$

$$\hat{E}_i \hat{\mathcal{R}}_i^* = \text{im} E_i \quad (i=1,\ldots,k) \tag{11}$$

where $\hat{\mathcal{R}}_i^* := \min \mathcal{J}(\hat{A}, \text{im}\,\hat{R}_i)$ and $\hat{R}_i$, $\hat{E}_i$ denote the submatrices of $\hat{R}$ and $\hat{E}$ corresponding, respectively by columns and rows, to the above-mentioned partitions of r and e. To prove sufficiency, we shall perform the synthesis of a linear compensator satisfying (10, 11) and such that matrix $\hat{A}$ is stable.

For this, refer to the simplified scheme of Fig. 5-18 (where d and y have been temporarily ignored) and suppose, for a moment, that A is a stable matrix: this

---

[2] It is possible that, for one or more of the groups of output components required to be noninteracting, something more than simple controllability is preferred, for instance functional controllability. If so, it is sufficient to replace $\mathcal{R}_{\mathcal{E}_i}$ with $\mathcal{F}_{\mathcal{E}_i}$, the functional controllability subspace on $\mathcal{E}_i$, which can be computed by a suitable extension of the arguments presented in Subsection 4.3.2. For specific treatment, see Basile and Marro [4].

Sect. 5.5 Noninteracting Controllers

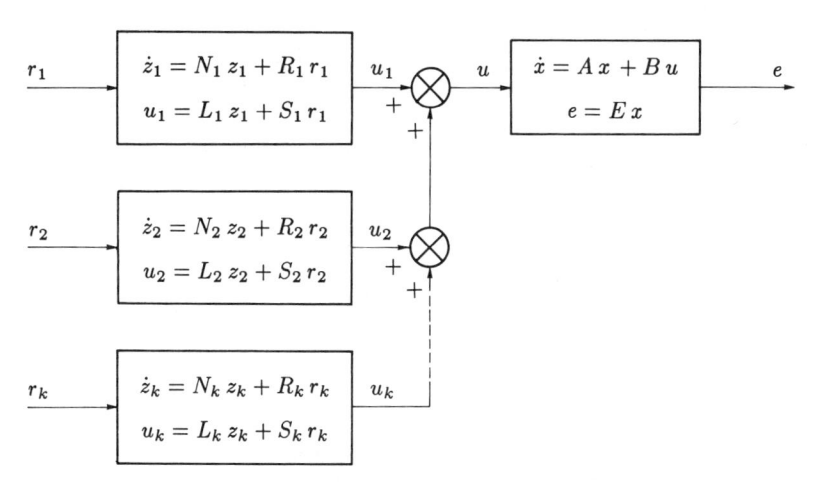

Figure 5-17 Realizing a noninteracting controller by means of k dynamic precompensators.

assumption will be removed later. For each e_i ($i=1,\ldots,k$) realize an identity dynamic precompensator of the type shown in Fig. 3-14(c) and express its state in the basis defined by similarity transformation $T_i := [T_{i,1}\ T_{i,2}]$ with $\mathrm{im}\,T_{i,1} = \mathcal{R}_{\mathcal{E}_i}$, and input in the basis defined by transformation $U_i := [U_{i,1}\ U_{i,2}]$ with $\mathrm{im}(BU_{i,1}) = \mathcal{V}_i^* \cap \mathcal{B}$, $\mathrm{im}(BU_i) = \mathcal{B}$. The system equations become

$$A_i' := T_i^{-1} A T_i = \begin{bmatrix} A_{i,11}' & A_{i,12}' \\ A_{i,21}' & A_{i,22}' \end{bmatrix} \quad B_i' := T_i^{-1} B U_i = \begin{bmatrix} B_{i,11}' & O \\ O & B_{i,22}' \end{bmatrix} \quad (12)$$

with $(A_{i,11}', B_{i,11}')$ controllable. Hence, there exists at least one matrix

$$F_i' := \begin{bmatrix} F_{i,11}' & O \\ F_{i,21}' & O \end{bmatrix} \quad (13)$$

such that $A_{i,11}' + B_{i,11}' F_{i,11}'$ is stable and $A_{i,21}' + B_{i,22}' F_{i,21}'$ a zero matrix. In the new basis the identity dynamic precompensator is described by the equations

$$\begin{bmatrix} \dot{z}_{i,1}(t) \\ \dot{z}_{i,2}(t) \end{bmatrix} = \begin{bmatrix} A_{i,11}' + B_{i,11}' F_{i,11}' & A_{i,12}' \\ O & A_{i,22}' \end{bmatrix} \begin{bmatrix} z_{i,1}(t) \\ z_{i,2}(t) \end{bmatrix} + \begin{bmatrix} B_{i,11}' & O \\ O & B_{i,22}' \end{bmatrix} \begin{bmatrix} r_{i,1}(t) \\ r_{i,2}(t) \end{bmatrix}$$

(14)

Note that the components of $z_{i,2}$ belong to $\ker F_i'$ and that to obtain noninteraction it is in any case necessary to assume $r_{i,2}(\cdot) = 0$, so that it is possible to ignore these components in the actual realization of the device. Hence, we assume $z_i := z_{i,1}$, $r_i := r_{i,1}$ and

$$N_i := A_{i,11}' + B_{i,11}' F_{i,11}' \quad (15)$$

$$R_i := B_{i,11}' \quad (16)$$

$$L_i := U_{i,1} F_{i,11}' \quad (17)$$

$$S_i := U_{i,1} \quad (18)$$

```
         d                                              e
         ─────────┐  ┌──────────────────┐  ┌─────────
                  │  │ ẋ = A x + B u + D d│  │
                  │  │   y = C x          │  │
     v₁    ⊗   u  │  │   e = E x          │  y
     ─────→⊕─────→│  │                    │──┬──────
           ↑ +    │  └──────────────────┘     │
           │ +    │                            │
           │      │  ┌──────────────────┐     │
           │      │  │ ż = N z + M y + v₂│←────┤
     v₂    │      │  │ u = L z + K y     │     │
     ──────┴──────┘  └──────────────────┘     │
```

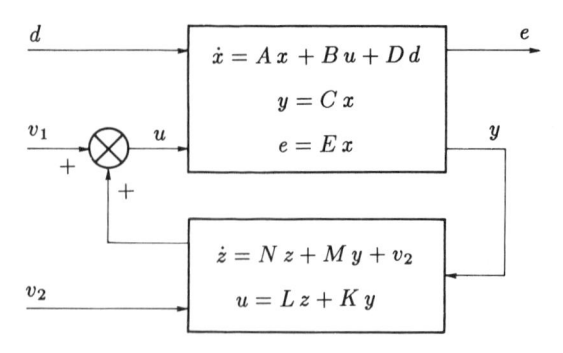

Figure 5-18 Stabilizing the plant.

The order of the obtained precompensator is clearly equal to $\dim \mathcal{R}_{\mathcal{E}_i}$.

We shall now remove the assumption that matrix A is stable. Should A not be stable, there would exist a dynamic feedback of the type shown in Fig. 5-18 so that the corresponding extended system is stable. This is described by

$$\dot{\hat{x}}(t) = (\hat{A}_0 + \hat{B}_0 \hat{K} \hat{C}_0) \hat{x}(t) + \hat{B}_0 \hat{u}(t) + \hat{D} d(t) \tag{19}$$

$$\hat{y}(t) = \hat{C}_0 \hat{x}(t) \tag{20}$$

$$e(t) = \hat{E} \hat{x}(t) \tag{21}$$

where

$$\hat{u} := \begin{bmatrix} v_1 \\ v_2 \end{bmatrix}$$

and $\hat{A}_0, \hat{B}_0, \hat{C}_0, \hat{D}, \hat{E}, \hat{K}$ are the same as in (5.1.18, 5.1.19).
By virtue of Lemma 5.1-1, subspaces

$$\hat{\mathcal{V}}_i^* := \left\{ \begin{bmatrix} x \\ z \end{bmatrix} : x \in \mathcal{V}_i^*, \ z = T x \right\} \quad (i = 1, \ldots, k) \tag{22}$$

where T denotes an arbitrary suitably dimensioned matrix, are $(\hat{A}_0, \hat{B}_0)$-controlled invariants, hence $(\hat{A}_0 + \hat{B}_0 \hat{K} \hat{C}_0, \hat{B}_0)$-controlled invariants, since feedback through input does not influence controlled invariance (the arbitrariness of T depends on the forcing action of the stabilizing unit being completely accessible). The reachable sets on them are

$$\hat{\mathcal{R}}_{\mathcal{E}_i} = \left\{ \begin{bmatrix} x \\ z \end{bmatrix} : x \in \mathcal{R}_{\mathcal{E}_i}, \ z = T x \right\} \quad (i = 1, \ldots, k) \tag{23}$$

and clearly have the same dimensions as the reachable sets $\mathcal{R}_{\mathcal{E}_i}$ $(i = 1, \ldots, k)$ in the nonextended state space.

Therefore, this procedure can be applied referring to the extended system (controlled system and stabilizing feedback unit), without any change in the state dimensions of the dynamic precompensators. A straightforward check shows that the extended system (8,9), which is clearly stable, satisfies the noninteraction conditions (10,11). □

Obtaining Stability, Disturbance Localization, and Noninteraction Simultaneously. Disturbance d has not been considered in the previous proof. Indeed, due to linearity and superposition property, it can be handled completely independently of noninteraction: if the necessary and sufficient conditions stated in Theorems 5.2-1 and 5.2-2 (when disturbance is completely unaccessible), or in Theorems 5.4-1 and 5.4-2 (when disturbance is in part accessible), are satisfied, a stabilizing dynamic unit of the type shown in Fig. 5-18 can be used to provide also disturbance localization. Since controlled invariants of the extended system are preserved in the presence of any through-input feedback, the possibility of simultaneously achieving noninteraction is not affected by the actual values of the dynamic compensator matrices K, L, M, N.

The solution to the noninteracting control problem presented in the proof of Theorem 5.5-1 is completely exhaustive, since it realizes noninteraction whenever all the necessary and sufficient conditions are met. However it is not claimed here to be the most convenient with respect to robustness in the presence of parameter changes and/or uncertainty, and the minimal with respect to the regulator order. On the contrary, its order appears to be relatively high in comparison with other regulation problems (where the order of the controller coincides, at most, with that of the controlled system). This happens because the intersections of controlled invariants $\mathcal{R}_{\mathcal{E}_i}$ in general are not controlled invariants themselves, so that it is not possible to achieve noninteraction by means of state feedback through a unique asymptotic state observer.

REFERENCES

1. AKASHI, H., and IMAI, H., "Disturbance localization and output deadbeat control through an observer in discrete-time linear multivariable systems," *IEEE Trans. Autom. Contr.*, vol. AC-24, pp. 621–627, 1979.
2. ANTOULAS, A.C., "A new approach to synthesis problems in linear system theory," *IEEE Trans. Autom. Contr.*, vol. AC-30, no. 5, pp. 465–473, 1985.
3. BASILE, G., and MARRO, G., "A state space approach to noninteracting controls," *Ricerche di Automatica*, vol. 1, no. 1, pp. 68–77, 1970.
4. —, "On the perfect output controllability of linear dynamic systems," *Ricerche di Automatica*, vol. 2, no. 1, pp. 1–10, 1971.
5. —, "Dual-lattice theorems in the geometric approach," *J. Optimiz. Th. Applic.*, vol. 48, no. 2, pp. 229–244, 1986.
6. BASILE, G., MARRO, G., and PIAZZI, A., "Stability without eigenspaces in the geometric approach: some new results," *Frequency Domain and State Space Methods for Linear Systems*, edited by C. A. Byrnes and A. Lindquist, North-Holland (Elsevier), Amsterdam, pp. 441–450, 1986.
7. —, "Revisiting the regulator problem in the geometric approach. Part I. Disturbance localization by dynamic compensation," *J. Optimiz. Th. Applic.*, vol. 53, no. 1, pp. 9–22, 1987.
8. —, "Revisiting the regulator problem in the geometric approach. Part II.

Asymptotic tracking and regulation in the presence of disturbances," *J. Optimiz. Th. Applic.*, vol. 53, no. 1, pp. 23–36, 1987.

9. —, "Stability without eigenspaces in the geometric approach: the regulator problem," *J. Optimiz. Th. Applic.*, vol. 64, no. 1, pp. 29–42, 1990.

10. BHATTACHARYYA, S.P., "Disturbance rejection in linear systems," *Int. J. Control*, vol. 5, no. 7, pp. 633–637, 1974.

11. —, "Frequency domain condition for disturbance rejection," *IEEE Trans. Autom. Contr.*, vol. AC-25, no. 6, pp. 1211–1213, 1980.

12. —, "Frequency domain condition for output feedback disturbance rejection," *IEEE Trans. Autom. Contr.*, vol. AC-27, no. 4, pp. 974–977, 1982.

13. BHATTACHARYYA, S.P., PEARSON, J.B., and WONHAM, W.M., "On zeroing the output of a linear system," *Information and Control*, vol. 20, no. 2, pp. 135–142, 1972.

14. BRASH, F.M., and PEARSON, J.B., "Pole placement using dynamic compensators," *IEEE Trans. Autom. Contr.*, vol. AC-15, no. 1, pp. 34–43, 1970.

15. CHANG, M.F., and RHODES, I.B., "Disturbance localization in linear systems with simultaneous decoupling, pole assignment, or stabilization," *IEEE Trans. Autom. Contr.*, vol. AC-20, pp. 518–523, 1975.

16. CHENG, L., and PEARSON, J.B., "Frequency domain synthesis of multivariable linear regulators," *IEEE Trans. Autom. Contr.*, vol. AC-23, no. 1, pp. 3–15, 1978.

17. FRANCIS, B.A., "The multivariable servomechanism problem from the input-output viewpoint," *IEEE Trans. Autom. Contr.*, vol. AC-22, no. 3, pp. 322–328, 1977.

18. HAMANO, F., and FURUTA, K., "Localization of disturbances and output decomposition in decentralized linear multivariable systems," *Int. J. Control*, vol. 22, no. 4, pp. 551–562, 1975.

19. IMAI, H., and AKASHI, H., "Disturbance localization and pole shifting by dynamic compensation," *IEEE Trans. Autom. Contr.*, vol. AC-26, no. 1, pp. 226–235, 1981.

20. KUČERA, V., "Discrete linear model following systems," *Kybernetika* (Praga), vol. 13, no. 5, pp. 333–342, 1977.

21. —, "Disturbance rejection: a polynomial approach," *IEEE Trans. Autom. Contr.*, vol. AC-28, no. 4, pp. 508–511, 1983.

22. MARRO, G., and PIAZZI, A., "Duality of reduced-order regulators," *Proceedings of the '88 International AMSE Conference on Modelling and Simulation*, Istambul, 1988.

23. MORSE, A.S., "Structure and design of linear model following systems," *IEEE Trans. Autom. Contr.*, vol. AC-18, no. 4, pp. 346–354, 1973.

24. MORSE, A.S., and WONHAM, W.M., "Decoupling and pole assignment by dynamic compensation," *SIAM J. Control*, vol. 8, no. 3, pp. 317–337, 1970.

25. OHM, D., BHATTACHARYYA, S.P., and HOUZE, J.W., "Transfer matrix conditions for (C, A, B)-pairs," *IEEE Trans. Autom. Contr.*, vol. AC-29, pp. 172–

174, 1984.

26. ÖZGÜLER, A.B., and ELDEM, V., "Disturbance decoupling problems via dynamic output feedback," *IEEE Trans. Autom. Contr.*, vol. AC-30, pp. 756–764, 1986.

27. PIAZZI, A., "Pole placement under structural constraints," *IEEE Trans. on Automatic Control*, vol. 35, no. 6, pp. 759–761, 1990.

28. —, "Geometric aspects of reduced-order compensators for disturbance rejection," *IEEE Trans. on Automatic Control*, vol. 36, no. 1, pp. 102–106, 1991.

29. PIAZZI, A., and MARRO, G., "The role of invariant zeros in multivariable system stability," *Proceedings of the 1991 European Control Conference*, Grenoble, 1991.

30. SCHUMACHER, J.M.H., "Compensator synthesis using (C, A, B)-pairs," *IEEE Trans. Autom. Contr.*, vol. AC-25, no. 6, pp. 1133–1138, 1980.

31. —, "Regulator synthesis using (C, A, B)-pairs," *IEEE Trans. Autom. Contr.*, vol. AC-27, no. 6, pp. 1211–1221, 1982.

32. —, "The algebraic regulator problem from the state-space point of view," *Linear Algebra and its Applications*, vol. 50, pp. 487–520, 1983.

33. —, "Almost stabilizability subspaces and high gain feedback," *IEEE Trans. Autom. Contr.*, vol. AC-29, pp. 620–627, 1984.

34. SHAH, S.L., SEBORG, D.E., and FISHER, D.G., "Disturbance localization in linear systems by eigenvector assignment," *Int. J. Control*, vol. 26, no. 6, pp. 853–869, 1977.

35. WILLEMS, J.C., "Almost invariant subspaces: an approach to high gain feedback design - Part I: Almost controlled invariant subspaces," *IEEE Trans. Autom. Contr.*, vol. AC-26, no. 1, pp. 235–252, 1981.

36. WILLEMS, J.C., "Almost invariant subspaces: an approach to high gain feedback design - Part II: Almost conditionally invariant subspaces," *IEEE Trans. Autom. Contr.*, vol. AC-27, no. 5, pp. 1071–1085, 1982.

37. WILLEMS, J.C., and COMMAULT, C., "Disturbance decoupling by measurement feedback with stability or pole placement," *SIAM J. Contr. Optimiz.*, vol. 19, no. 4, pp. 490–504, 1981.

38. WONHAM, W.M., "Dynamic observers – geometric theory," *IEEE Trans. Autom. Contr.*, vol. AC-15, no. 2, pp. 258–259, 1970.

39. —, "Tracking and regulation in linear multivariable systems," *SIAM J. Control*, vol. 11, no. 3, pp. 424–437, 1973.

40. WONHAM, W.M., and PEARSON, J.B., "Regulation and internal stabilization in linear multivariable systems," *SIAM J. Control*, vol. 12, no. 1, pp. 5–18, 1974.

6

THE ROBUST REGULATOR

6.1 THE SINGLE-VARIABLE FEEDBACK REGULATION SCHEME

In this chapter the robust multivariable regulator will be investigated by using the geometric approach techniques presented in Chapters 3, 4, and 5. A regulator is said to be *robust* if it preserves the regulation property and satisfactory dynamic behavior also in the presence of variations of the parameters of the plant in well-defined neighborhoods of their nominal values, called *uncertainty domains*. These parameter variations are assumed to be "slow" with respect to the most significant time constants of the controlled plant, so that their influence on the controlled output is negligible if the regulation property is maintained for all the parameter values.

Many basic concepts of the standard single-variable regulator design techniques will be revisited with a different language and extended to the multivariable case. For better understanding and framing of this connection in light of the traditional approach, in this first section the most important terms and concepts of automatic control theory will be briefly recalled and a short sketch of the standard synthesis philosophy will be discussed.

Refer to the block diagram of Fig. 6-1. It is well known that in standard single-variable systems robustness is achieved by using feedback, i.e., by feeding back to the controller a measurement of the controlled output (which must be as accurate as possible) or, more exactly, through very accurate direct

Sect. 6.1 The Single-Variable Feedback Regulation Scheme

determination of the tracking error variable, which is a known, simple function of the controlled output.

The meanings of the symbols in the figure are:

r: *reference input*
e: *error variable*
m: *manipulable input*
d: *disturbance input*
c: *controlled output*
y_1, y_2: *informative outputs*

The informative outputs are in general stabilizing signals: a typical example is the tachymetric feedback in position control systems.

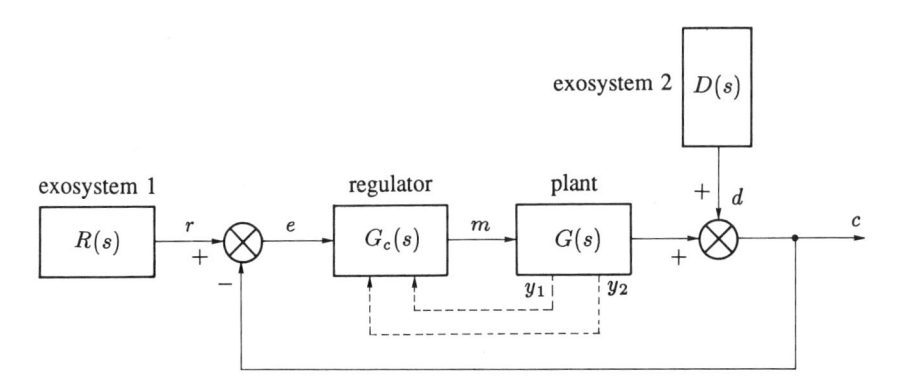

Figure 6-1 A feedback control system.

We shall now consider the particular case

$$G(s) := \frac{1}{(s+a)(s+b)(s+c)} \qquad R(s) = \frac{1}{s^2} \qquad D(s) = \frac{1}{s} \qquad (1)$$

The displayed functions are "generalized transfer functions": in the case of the exosystems the inputs are understood to be identically zero, so that the corresponding outputs are affected only by the initial conditions. Our aim is to determine a $G_c(s)$ such that the *extended plant* (i.e., the plant plus the regulator) is (asymptotically) stable and the *overall system* (i.e., the extended plant plus the exosystems) satisfies the regulation condition $\lim_{t \to \infty} e(t) = 0$. Note that the exosystems are (asymptotically) unstable and introduce into the extended plant signals of the general type

$$\mu + \nu t$$

whose coefficients μ and ν depend on the initial conditions of the exosystems.

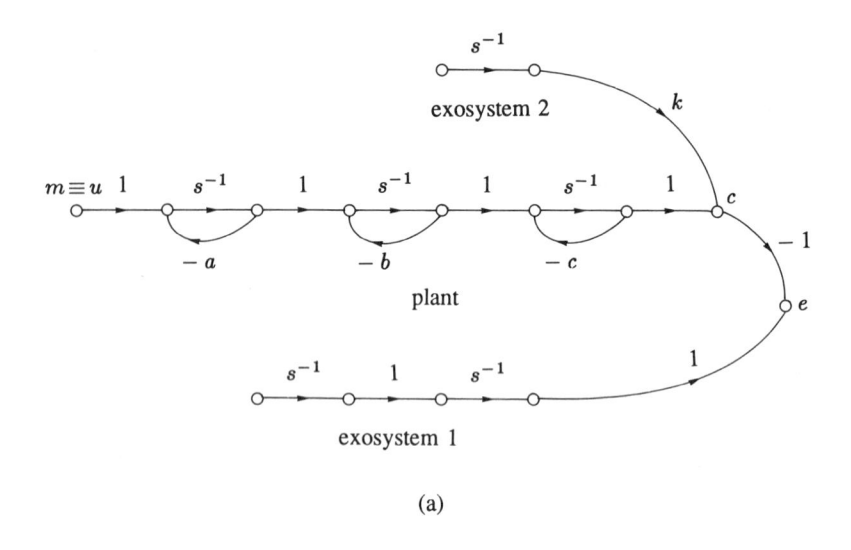

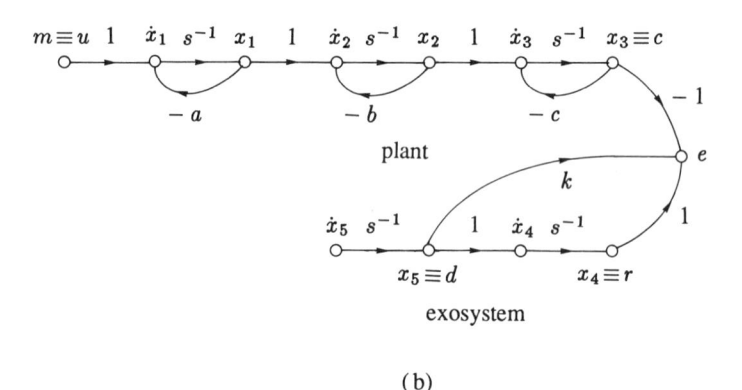

Figure 6-2 Signal-flow graph representations of a controlled system.

This problem can be reformulated in the state space. Consider the signal-flow graph of Fig. 6-2(a), derive the equivalent graph of Fig. 6-2(b) (where the second exosystem, which is not independently observable from output e, is embedded in the first), and define a state variable for every integrator as shown in the figure: we derive the state space representation

$$A = \begin{bmatrix} -a & 0 & 0 & 0 & 0 \\ 1 & -b & 0 & 0 & 0 \\ 0 & 1 & -c & 0 & 0 \\ 0 & 0 & 0 & 0 & 1 \\ 0 & 0 & 0 & 0 & 0 \end{bmatrix} \quad B = \begin{bmatrix} 1 \\ 0 \\ 0 \\ 0 \\ 0 \end{bmatrix} \quad D = \begin{bmatrix} 0 \\ 0 \\ 0 \\ 0 \\ 0 \end{bmatrix} \quad (2)$$

$$C = E = [0 \quad 0 \quad -1 \quad -k \quad 1]$$

Sect. 6.1 The Single-Variable Feedback Regulation Scheme

Note that in this case the informative and controlled outputs coincide. A traditional design is developed according to the following steps: since the exosystem introduces observable unstable modes, the regulator must generate identical modes so that the steady state error is zero. In other words, a model of the exosystem must be contained in the regulator. This expresses the so-called *internal model principle*, which corresponds to a design expedient that is well known in the single-variable case: to achieve zero steady state error in the step response, the regulator must have a pole at the origin, like the exosystem that generates the step, while the same requirement for the ramp response implies a double pole at the origin, and so on. For the sake of generality we assume

$$G_c(s) := \frac{\gamma(s+\alpha)(s+\beta)}{s^2} \quad (3)$$

In this way the maximum number of zeros compatible with the physical realizability or causality condition (the degree of the numerator must be less than, or equal to that of the denominator) is inserted into the regulator: the values of parameters α, β, γ are free and can be chosen to achieve stability and improve the transient behavior of the loop, if possible. If stability is not achievable or the transient behavior cannot be made completely satisfactory, significant improvements can be generally obtained by inserting some further pole-zero pairs into the regulator.

A realization of the regulator described by transfer function (3) is represented in Fig. 6-3. From the signal-flow graph one immediately derives the quadruple

$$K := \gamma \quad L := [\beta \quad \alpha] \quad M := \begin{bmatrix} \gamma \\ \gamma \end{bmatrix} \quad N := \begin{bmatrix} 0 & \alpha \\ 0 & 0 \end{bmatrix} \quad (4)$$

which completes the state space representation of the type shown in Fig. 5-3.

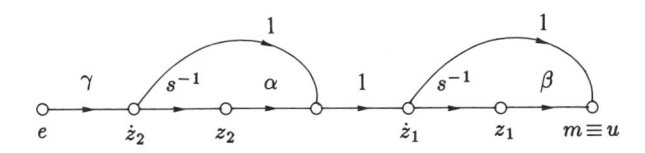

Figure 6-3 Signal-flow graph representation of a possible feedback regulator.

The state-space solution of the multivariable regulator problem will be described in the next section. Let us now list and explain the most significant steps of the traditional synthesis by assuming the previous example as reference but trying to derive general results, which will be extended to the multivariable case.

Robust Stability. In the case of the previous example, the typical situation that can be referred to when coefficients a, b, c are nonnegative and α, β, γ positive is illustrated by the root locus shown in Fig. 6-4: the system is surely stable for a proper choice of the gain γ if the absolute values of a, b, c are large with respect to those of α, β.

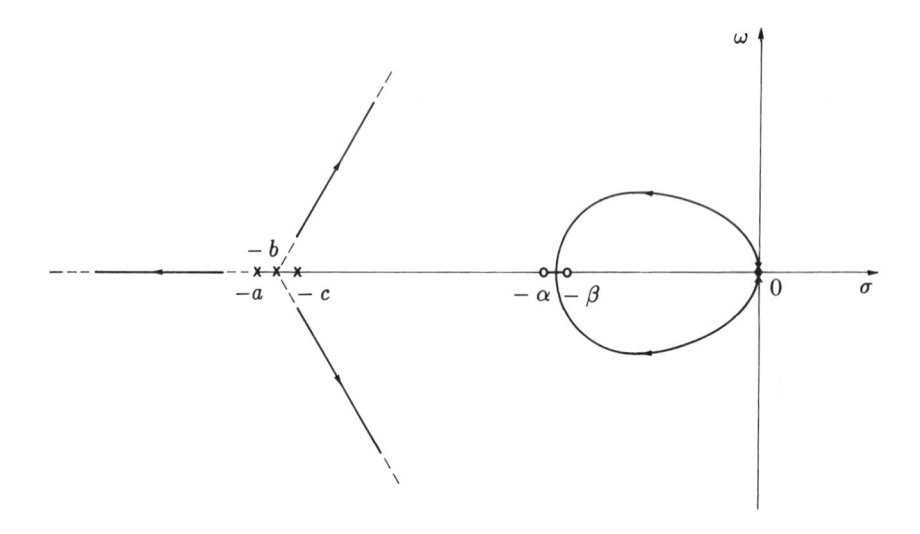

Figure 6-4 Root locus of system (1) and regulator (3).

If all the exosystem poles are located at the origin (as in the most common cases) this condition is met if sufficiently small α and β are chosen; however, by so doing, the poles that originate from the exosystem are maintained very close to the origin, and this causes a very slow transient. In any case it is convenient to translate the poles of the plant toward the left by inserting in the regulator some pole-zero pairs or by means of feedback connections using the informative outputs: since the plant is completely controllable and observable, its poles are all arbitrarily assignable, at least with a proper state observer with arbitrary poles. If the plant is *minimum-phase*, i.e., has all the poles and zeros with the real parts negative, a robust stability is always possible. If the plant has some poles and zeros with the real parts positive it is still pole assignable, hence stabilizable, by means of a suitable dynamic feedback connection, but the root locus corresponding to this feedback, which originates and terminates in the right half-plane is, in general, very sensitive to parameter changes in its stable portion and remains in the left half-plane for a very short gain interval, so that robust stability is very difficult to achieve.

Observability and the Internal Model. In the preceding example the mode generated by the second exosystem (a step) is a part of those generated by the first one (a step and a ramp), so that it is not necessary to reproduce both exosystems in the regulator; in fact a double pole at the origin allows steady state compensation of any set of disturbances consisting of a linear combination of a step and a ramp, independently of the points where they are introduced in the regulation loop. The reason it is possible to use a single internal model in this case is that the exosystems that generate the disturbances are not all separately observable from the controlled output e, so that it is sufficient to reproduce in the regulator only the eigenstructures corresponding to the observable exogenous modes. In the single-variable case it is generally convenient to assume in the mathematical model a single exosystem which is completely observable [as was obtained by replacing the signal-flow graph of Fig. 6-2(a) with that of Fig. 6-2(b)], but this may not apply to the multivariable case where a single exosystem that influences several regulated outputs cannot be compensated by a single internal model if coefficients of influence are subject to changes, resulting in lack of robustness. This point, which represents an important difference between the single-variable and the multivariable case, will be thoroughly investigated in the Section 6.3.

Robust Regulation. In the single-variable case, regulation is robust versus parameter variations provided the plant remains stable and the internal model is preserved when coefficients change. As previously mentioned, if the plant pole-zero configuration is favorable, the plant stability requirement can be robustly satisfied by suitable placement of the free poles and zeros. It is not difficult to build a robust internal model if the exogenous eigenvalues are all zero: fortunately this case is very frequent in practice since specifications on regulator asymptotic behavior usually refer to *test signals* consisting of steps, ramps, parabolas, and so on. However, in the multivariable case it is further necessary that individual asymptotic controllability of regulated outputs from the plant inputs is maintained in the presence of parameter variations. This structural requirement is another important point that distinguishes the multivariable from the single-variable case.

6.2 THE AUTONOMOUS REGULATOR: A GENERAL SYNTHESIS PROCEDURE

In the regulator problem formulated in Section 5.2, two types of input functions were considered: an input function $d(\cdot)$ belonging to a very general class (that of bounded and piecewise continuous functions), representing a nonmanipulable input to be identically localized, i.e., to be made noninfluencing of the controlled output at any time through a suitable change of structure, and an input function $x_2(\cdot)$, generated by an exosystem, of a more particular class (that of the linear combinations of all possible modes of a linear time-invariant system), representing a nonmanipulable input to be asymptotically localized. This formulation of the problem has the advantage of being very general: for instance, it extends in the most natural way a classic geometric approach problem, i.e., disturbance localization through state feedback.

When robustness is a required regulator feature, "rigid" solutions like the standard structural disturbance localization are not acceptable since, in general, they lose efficiency in the presence of parametric variations; therefore it is necessary to have to resort to "elastic" solutions like the asymptotic insensitivity to disturbances modeled by an exosystem. Note that if the controlled output dynamics is sufficiently fast (or even arbitrarily fast, i.e., with arbitrarily assignable modes), this approach is almost equivalent to rigid localization since the localization error corresponding to a disturbance signal with limited bandwidth can be made arbitrarily small.

In light of these considerations we shall first of all see how the statements of Theorems 5.2-2 and 5.2-4 simplify on the assumption that $D = O$, i.e., when an *autonomous* regulator is considered. In this case the overall system is not subject to any external signal, since the only considered perturbations are those of the initial states (of the plant, exosystem, and regulator): it is required that for any set of initial states the regulation condition $\lim_{t \to \infty} e(t) = 0$ is met. This formulation is the direct extension of that presented in the previous section for single-variable systems. It is worth pointing out that the robustness requirement will not be considered in this section. The conditions that are derived as particular cases of those considered in Section 5.2 are necessary and sufficient for the existence of a generic, stable autonomous regulator, but do not guarantee the existence of an autonomous regulator of the feedback type (it may be feedforward). Hence, they are only necessary to extend to multivariable systems the synthesis procedure for robust regulators (based on feedback through an internal model of the exosystem) illustrated in the previous section for the single-variable case.

For the sake of clarity and to make this chapter self-contained, it appears convenient to report first a specific formulation of the autonomous regulator problem. Consider the *overall system* represented in Fig. 6-5, whose state evolution is described by the homogeneous matrix differential equation

$$\begin{bmatrix} \dot{x}_1(t) \\ \dot{x}_2(t) \\ \dot{z}(t) \end{bmatrix} = \begin{bmatrix} A_1+B_1KC_1 & A_3+B_1KC_2 & B_1L \\ O & A_2 & O \\ MC_1 & MC_2 & N \end{bmatrix} \begin{bmatrix} x_1(t) \\ x_2(t) \\ z(t) \end{bmatrix} \quad (1)$$

which, according to the notation introduced in Section 5.1, can also be written, together with the output equation, in the compact form

$$\dot{\hat{x}}(t) = \hat{A}\,\hat{x}(t)$$
$$e(t) = \hat{E}\,\hat{x}(t) \quad \text{with} \quad \hat{A} := \begin{bmatrix} \hat{A}_1 \hat{A}_3 \\ \hat{A}_4 \hat{A}_2 \end{bmatrix} \quad \text{and} \quad \hat{E} := [\hat{E}_1 \; O] \quad (2)$$

Sect. 6.2 The Autonomous Regulator: A General Synthesis Procedure

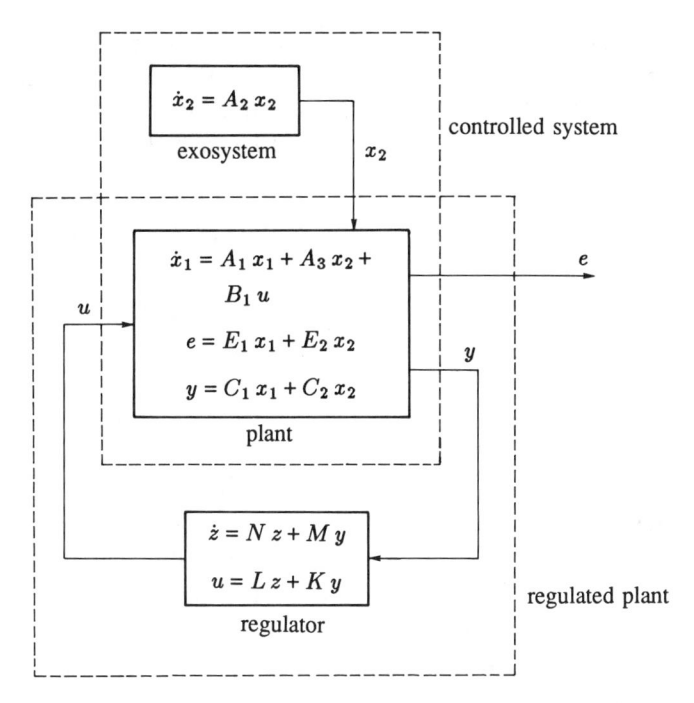

Figure 6-5 The general multivariable autonomous regulator.

i.e., as a unique *extended system* with the submatrices of $\hat{A}$ and $\hat{E}$ defined by

$$\hat{A}_1 := \begin{bmatrix} A_1 + B_1 K C_1 & A_3 + B_1 K C_2 \\ O & A_2 \end{bmatrix} \quad \hat{A}_3 := \begin{bmatrix} B_1 L \\ O \end{bmatrix} \quad (3)$$

$$\hat{A}_4 := [\, M C_1 \quad M C_2 \,] \quad \hat{A}_2 := N \quad \hat{E}_1 := E = [\, E_1 \quad E_2 \,]$$

The part of the overall system driven by the exosystem is called the *regulated plant* (the plant plus the regulator): it has x_2 as the only input and e as the only output and is a nonpurely dynamic system represented by the quadruple (A_p, B_p, C_p, D_d) with

$$A_p := \begin{bmatrix} A_1 + B_1 K C_1 & B_1 L \\ M C_1 & N \end{bmatrix} \quad B_p := \begin{bmatrix} A_3 + B_1 K C_2 \\ M C_2 \end{bmatrix} \quad (4)$$

$$C_p := [\, E_1 \quad O \,] \quad D_p := [\, E_2 \,]$$

As the last step of this review of the new and old notations, recall the four remaining matrices of the five-map controlled system, which are the only data provided to solve our regulation problem:

$$A := \begin{bmatrix} A_1 & A_3 \\ O & A_2 \end{bmatrix} \quad B := \begin{bmatrix} B_1 \\ O \end{bmatrix} \quad C := [\, C_1 \quad C_2 \,] \quad E := [\, E_1 \quad E_2 \,] \quad (5)$$

We shall denote with n_1 the state dimension of the plant, with n_2 that of the exosystem, and with m that of the regulator; the *plant*, and the *extended plant* defined by (5.1.8) and (5.2.7) are respectively an A-invariant and an $\hat{A}$-invariant which in the above considered coordinate systems can be expressed as

$$\mathcal{P} = \operatorname{im} P \quad \text{with} \quad P := \begin{bmatrix} I_{n_1} \\ O \end{bmatrix} \tag{6}$$

and

$$\hat{\mathcal{P}} = \operatorname{im} \hat{P} \quad \text{with} \quad \hat{P} := \begin{bmatrix} I_{n_1} & O \\ O & O \\ O & I_m \end{bmatrix} \tag{7}$$

In the autonomous regulator case the plant stability condition simply means that the regulated plant must be stable, i.e., that A_p must be a stable matrix or that $\hat{\mathcal{P}}$ must be internally stable as an $\hat{A}$-invariant, and the regulation condition means that there exists an externally stable $\hat{A}$-invariant contained in $\ker \hat{E}$. Geometrically the autonomous regulator problem is stated in the following terms, as a particular case of Problem 5.2-2/G.

Problem 6.2-1.
Refer to the block diagram of Fig. 6-5 and assume that (A_1, B_1) is stabilizable and (A, C) detectable. Determine, if possible, a feedback regulator of the type shown in the figure such that:
1. the overall system has an $\hat{A}$-invariant $\hat{\mathcal{W}}$ that satisfies

$$\hat{\mathcal{W}} \subseteq \hat{\mathcal{E}} \quad \text{with} \quad \hat{\mathcal{E}} := \ker \hat{E};$$

2. $\hat{\mathcal{W}}$ is externally stable;
3. $\hat{\mathcal{W}} \cap \hat{\mathcal{P}}$ (which is an $\hat{A}$-invariant) is internally stable.

The elimination of input D significantly simplifies the necessary and sufficient conditions stated in Chapter 5. In fact, from $\mathcal{D} := \operatorname{im} D = \{0\}$ it follows that $\mathcal{S}^* = \mathcal{S}_m = \mathcal{S}_M = \{0\}$; hence

$$\mathcal{V}_m = \mathcal{V}_M = \mathcal{V}^* \cap \min \mathcal{S}(A, \mathcal{E}, B) = \mathcal{R}_{\mathcal{V}^*} = \mathcal{R}_\mathcal{E} \tag{8}$$

so that from Theorems 5.2-2 and 5.2-4 the following corollaries are immediately derived.

Corollary 6.2-1.
The autonomous regulator problem has a solution if and only if there exists an (A, B)-controlled invariant $\mathcal{V}$ such that:
1. $\mathcal{V} \subseteq \mathcal{E}$; (9)
2. $\mathcal{V}$ is externally stabilizable; (10)
3. $\mathcal{V} \cap \mathcal{P}$ is internally stabilizable. (11)

Sect. 6.2 The Autonomous Regulator: A General Synthesis Procedure

Corollary 6.2-2.
Let all the exogenous modes be unstable. The autonomous regulator problem has a solution if and only if:
1. V^* is externally stabilizable; (12)
2. $V^* \cap \mathcal{P}$ is complementable with respect to $(\mathcal{R}_{V^*}, V^*)$. (13)

At this point some remarks are in order.
1. Due to the particular structure of matrix A, the assumption that (A, C) is detectable clearly implies that (A_1, C_1) also is.
2. Let all the exogenous modes be unstable. For the regulator problem to have a solution, as a clear consequence of (12) it is necessary that

$$\operatorname{im} E_2 \subseteq \operatorname{im} E_1 \qquad (14)$$

3. Condition (12) can be reformulated in the very strict geometric way

$$V^* + \mathcal{P} = \mathcal{X} \qquad (15)$$

In fact it has been previously proved (point e of the proof of Theorem 5.2-4) that if V is any (A, B)-controlled invariant self-bounded with respect to V^* the internal unassignable eigenvalues in between $V \cap \mathcal{P}$ and V are all exogenous. This property, applied to V^* itself, clearly implies (15).

These conditions hold also in the general case considered in Chapter 5, but have not been mentioned before, since they are straightforward consequences of the assumptions or of the derived geometric conditions.

We shall now derive an important theorem (Theorem 6.2-1) which sets a further, significant simplification of the necessary and sufficient conditions (15, 13) and provides a basic means to approach structural robustness in the multivariable regulator case. First, we state a property that geometrically characterizes any solution of the problem.

Property 6.2-1.
Let all the exogenous modes be unstable. The regulator corresponding to the overall system matrices $\hat{A}, \hat{E}$ satisfies both plant stability and the regulation condition if and only if there exists an externally stable $\hat{A}$-invariant $\hat{\mathcal{W}}$ such that:
1. $\hat{\mathcal{W}} \subseteq \hat{\mathcal{E}}$ with $\hat{\mathcal{E}} := \ker \hat{E}$; (16)
2. $\hat{\mathcal{W}} \oplus \hat{\mathcal{P}} = \hat{\mathcal{X}}$. (17)

Proof. *Only if.* The plant stability condition means that $\hat{\mathcal{P}}$ is internally stable as an $\hat{A}$-invariant and the regulation condition holds if and only if there exists an externally stable $\hat{A}$-invariant contained in $\hat{\mathcal{E}}$. Denote this by $\hat{\mathcal{W}}_1$. Since the eigenvalues of A_2, i.e., the eigenvalues of the exosystem (which have been assumed to be unstable) are a part of those of $\hat{A}$, they must be internal eigenvalues of $\hat{\mathcal{W}}_1$. By reason of dimensionality the other eigenvalues of $\hat{A}$ coincide with those of A_p so that the $\hat{A}$-invariant $\hat{\mathcal{W}}_1 \cap \hat{\mathcal{P}}$ is internally stable and $\hat{\mathcal{P}}$ is complementable with respect to $(\hat{\mathcal{W}}_1 \cap \hat{\mathcal{P}}, \hat{\mathcal{W}}_1)$. Assume the similarity transformation $T := [T_1 \, T_2 \, T_3]$, with $\text{im}\, T_1 = \hat{\mathcal{W}}_1 \cap \hat{\mathcal{P}}$, $\text{im}\, [T_1 \, T_2] = \hat{\mathcal{W}}_1$, $\text{im}\, [T_1 \, T_3] = \hat{\mathcal{P}}$. In the new basis, $\hat{A}' := T^{-1} \hat{A} T$ and $\hat{E}' := \hat{E} T$ have the structures

$$\hat{A}' = \begin{bmatrix} A'_{11} & A'_{12} & A'_{13} \\ O & A'_{22} & A'_{23} \\ O & O & A'_{33} \end{bmatrix} \qquad \hat{E}' = [\,O \quad O \quad E'_3\,] \qquad (18)$$

Matrices A'_{11} and A'_{33} are stable while A'_{22} is unstable, so that the Sylvester equation

$$A'_{11} X - X A'_{22} = -A'_{12} \qquad (19)$$

admits a unique solution X. Define $\hat{\mathcal{W}}$ as the image of $T_1 X + T_2$. Clearly it satisfies both the conditions in the statement.

If. The regulation condition is satisfied since $\hat{\mathcal{W}}$ attracts all the external motions. Furthermore, $\hat{\mathcal{W}}$ has the eigenvalues of A_2 as the only internal eigenvalues and those of the regulated plant as the only external ones, so that the plant stability condition is satisfied. □

For those who like immediate translation of geometric conditions into matrix equations, we can state the following corollary.

Corollary 6.2-3.
Let all the exogenous modes be unstable and suppose that the regulated plant is stable. The regulation condition is satisfied if and only if the equations

$$A_p X_p - X_p A_2 = -B_p \qquad (20)$$
$$C_p X_p + D_p = O \qquad (21)$$

have a solution in X_p.

Proof. The complementability condition implies the possibility of assuming

$$\hat{\mathcal{W}} := \text{im}\left(\begin{bmatrix} X_1 \\ I_{n_2} \\ Z \end{bmatrix} \right) \quad \text{with} \quad \begin{bmatrix} X_1 \\ Z \end{bmatrix} = X_p \qquad (22)$$

From Property 3.2-1 it follows that equations (20, 21) are satisfied if and only if $\hat{\mathcal{W}}$ is an $\hat{A}$-invariant contained in $\hat{E}$. □

As in the case of the general regulator problem approached in Chapter 5, necessary and sufficient conditions referring to the overall system and expressed in terms of invariants, reflect in necessary and sufficient conditions referring to the sole controlled system and expressed in terms of controlled and/or conditioned invariants. These conditions are directly usable for feasibility checks and for constructive synthesis procedures. The basic result is stated in the following theorem whose proof, although contained in that of Theorem 5.2-4, is developed here in a completely self-contained way.

Theorem 6.2-1. The Fundamental Theorem on the Autonomous Regulator

Let all the exogenous modes be unstable. The autonomous regulator problem admits a solution if and only if there exists an (A, B)-controlled invariant $\mathcal{V}$ such that

1. $\mathcal{V} \subseteq \mathcal{E}$; (23)
2. $\mathcal{V} \oplus \mathcal{P} = \mathcal{X}$. (24)

Proof. Only if. Consider the $\hat{A}$-invariant $\hat{\mathcal{W}}$ defined by (22) and denote by $\hat{W}$ the corresponding basis matrix, also shown in (22): by Property 3.2-1 there exists a matrix X such that $\hat{A}\hat{W} = \hat{W}X$. Straightforward manipulations show that this equation implies the existence of a matrix U such that

$$A \begin{bmatrix} X_1 \\ I_{n_2} \end{bmatrix} = X \begin{bmatrix} X_1 \\ I_{n_2} \end{bmatrix} + BU$$

Hence

$$\mathcal{V} := \begin{bmatrix} X_1 \\ I_{n_2} \end{bmatrix}$$

is an (A, B)-controlled invariant by Property 4.1-4. It clearly satisfies both the stated conditions.

If. We prove that the conditions in the statement imply those of Corollary 6.2-1. Condition (9) holds by assumption. Condition (10) is proved as follows: the A-invariant $\mathcal{V}+\mathcal{P}$, which covers the whole space, is externally stable (hence, in particular, externally stabilizable), so that $\mathcal{V}$ is externally stabilizable (see point f of the proof of Theorem 5.2-4). Finally, (11) is implied by $\mathcal{V} \cap \mathcal{P} = \{0\}$. □

This proof of the fundamental theorem is the most direct in connection with the previously developed theory. Another proof, which has the advantage of being completely constructive, can be based on Algorithm 6.2-1 (for the only if part) and Algorithm 6.2-3 (for the if part).

Note that the controlled invariant $\mathcal{V}$ considered in the statement has the dimension equal to that of the exosystem, so that its unique internal eigenvalues (clearly unassignable) coincide with those of the exosystem. Since any resolvent must have the eigenvalues of the exosystem as internal, since it must be externally stabilizable, $\mathcal{V}$ is a *minimal-dimension resolvent*. It will be shown in Section 6.4 that having a minimal-dimension resolvent is a basic step for the synthesis of a particular multivariable robust feedback regulator that is also minimal-dimension.

Given the data of our problem, i.e., matrices A, B, C, E, the conditions of Corollary 6.2-1 can be checked in a completely automatic way through the computation of $\mathcal{V}^*$ and $\mathcal{R}_{\mathcal{V}^*} = \mathcal{V}^* \cap \mathcal{S}^*$ with $\mathcal{S}^* := \min \mathcal{S}(A, \mathcal{E}, \mathcal{B})$. To check complementability and, if possible, to derive a complement of $\mathcal{V}^*$ that satisfies (23, 24) the following algorithm can be used.

Algorithm 6.2-1. Complementation of the Maximal Controlled Invariant

Let $V^* + P = X$. A controlled invariant V such that $V \oplus P = X$ and $V \subseteq \mathcal{E}$ can be derived as follows. Consider the similarity transformation defined by $T := [T_1\ T_2\ T_3\ T_4]$, with $\mathrm{im}\, T_1 = \mathcal{R}_{V^*}$, $\mathrm{im}\,[T_1\ T_2] = V^* \cap P$, $\mathrm{im}\,[T_1\ T_2\ T_3] = V^*$, and $T_4 = [T_4'\ T_4'']$ with $\mathrm{im}\,[T_1\ T_4'] = S^*$, $\mathrm{im}\,[T_1\ T_2\ T_4''] = P$. The system matrices expressed in the new basis, i.e., $A' := T^{-1}AT$, $B' := T^{-1}BT$, and $E' := ET$, have the structures

$$A' = \begin{bmatrix} A'_{11} & A'_{12} & A'_{13} & A'_{14} \\ O & A'_{22} & A'_{23} & A'_{24} \\ O & O & A'_{33} & O \\ A'_{41} & A'_{42} & A'_{43} & A'_{44} \end{bmatrix} \qquad B' = \begin{bmatrix} B'_1 \\ O \\ O \\ B'_4 \end{bmatrix} \qquad (25)$$

$$E' = [O\ \ O\ \ O\ \ E'_4] \qquad (26)$$

Note that S^*, being the minimal $(A, \mathcal{E})$-conditioned invariant containing B, is contained in the reachable set, which is the minimal A-invariant containing B, hence in P, which is an A-invariant containing B. The structure of B' follows from $B \subseteq S^*$, the zero submatrix in the second row of A' is due to $\mathcal{R}_{V^*}$ being a controlled invariant, the first two in the first row to $\mathcal{J} \cap P$ being a controlled invariant, and the third to P being an A-invariant. The structure of E' is due to $V^* \subseteq \mathcal{E}$. A complement of $V^* \cap P$ with respect to $(\mathcal{R}_{V^*}, V^*)$ exists if and only if the Sylvester equation

$$A'_{22} X - X A'_{33} = -A'_{23} \qquad (27)$$

admits a solution in X. If so, a V satisfying (23,24) is defined by $V := \mathrm{im}(T_2 X + T_3)$. □

This change of basis allows immediate derivation of another interesting result, a sufficient condition that turns out to be very useful for a quick check of the complementability condition.

Theorem 6.2-2. A Sufficient Condition in Terms of Invariant Zeros

Let all the exogenous modes be unstable. The autonomous regulator problem has a solution if:

1. $V^* + P = X$; (28)
2. no invariant zero of the plant [i.e., of the triple (A_1, B_1, E_1)] coincides with an eigenvalue of the exosystem. (29)

Proof. The maximal controlled invariant contained in V^* and in P is $V^* \cap P$ itself, so that the invariant zeros referred to in the statement are clearly the eigenvalues of A'_{22}. Since the eigenvalues of the exosystem are those of A'_{33}, condition (29) implies the solvability of (27). □

Another algorithm that is used in the regulator synthesis provides a suitable state feedback matrix to make V an invariant and stabilize the plant. It can be set as an extension of Algorithm 4.1-3 in the following terms.

Sect. 6.2 The Autonomous Regulator: A General Synthesis Procedure

Algorithm 6.2-2. Computation of the State Feedback Matrix

Assume that B has maximal rank. Given $\mathcal{V}$ such that $\mathcal{V} \oplus \mathcal{P} = \mathcal{X}$ and $\mathcal{V} \subseteq \mathcal{E}$, a state feedback matrix $F = [F_1 \ F_2]$ [partitioned according to (5)] such that $(A+BF)\mathcal{V} \subseteq \mathcal{V}$ and $A_1 + B_1 F_1$ is stable, can be derived as follows. Consider the similarity transformation defined by $T := [T_1 \ T_2 \ T_3]$, with $\operatorname{im} T_1 = \mathcal{B}$, $\operatorname{im}[T_1 \ T_2] = \mathcal{P}$, and $\operatorname{im}[T_1 \ T_3] = \mathcal{V}$, so that matrices $A' := T^{-1}AT$, $B' := T^{-1}BT$, and $E' := ET$ have the structures

$$A' = \begin{bmatrix} A'_{11} & A'_{12} & A'_{13} \\ A'_{21} & A'_{22} & O \\ O & O & A'_{33} \end{bmatrix} \quad B' = \begin{bmatrix} I_p \\ O \\ O \end{bmatrix} \tag{30}$$

$$E' = [E'_1 \ E'_2 \ O] \tag{31}$$

The zero submatrix in the second row of A' is due to $\mathcal{V}$ being a controlled invariant, those in the third row to $\mathcal{P}$ being an invariant, while the zero in E' is due to $\mathcal{V}$ being contained in $\mathcal{E}$. The particular structure of B' follows from B being the image of the first part of the transformation matrix. Assume, in the new basis, a matrix $F' = [F'_1 \ F'_2 \ F'_3]$, accordingly partitioned, with $F'_3 := -A'_{13}$ and F'_1, F'_2 such that

$$\begin{bmatrix} A'_{11} & A'_{12} \\ A'_{21} & A'_{22} \end{bmatrix} + \begin{bmatrix} F'_1 & F'_2 \\ O & O \end{bmatrix}$$

is stable: this is possible because (A_1, B_1) has been assumed to be stabilizable. Then compute $F := F'T^{-1}$. □

We can now set an algorithm for the regulator synthesis. The order of the regulator is n, so that an overall system of order $2n$ is obtained. Although the procedure is a particularization of the recipe for the observer-based full-order regulator reported in Subsection 5.2.3 (with $L_1 := O$, $L_2 := I_n$), in order to keep the contents of this chapter self-contained, it will be proved again.

Algorithm 6.2-3. The Autonomous Regulator Synthesis Algorithm

Given $\mathcal{V}$ such that $\mathcal{V} \oplus \mathcal{P} = \mathcal{X}$ and $\mathcal{V} \subseteq \mathcal{E}$, determine F such that $(A+BF)\mathcal{V} \subseteq \mathcal{V}$ and $A_1 + B_1 F_1$ is stable, G such that $A+GC$ is stable and assume

$$N := A + BF + GC \quad M := -G \quad L := F \quad K := O \tag{32}$$

Proof. The extended system matrices $\hat{A}$ and $\hat{E}$ are partitioned as

$$\hat{A} = \begin{bmatrix} A_1 & A_3 & B_1 F_1 & B_1 F_2 \\ O & A_2 & O & O \\ -G_1 C_1 & -G_1 C_2 & A_1 + B_1 F_1 + G_1 C_1 & A_3 + B_1 F_2 + G_1 C_2 \\ -G_2 C_1 & -G_2 C_2 & G_2 C_1 & A_2 + G_2 C_2 \end{bmatrix} \tag{33}$$

$$\hat{E} = [E_1 \ E_2 \ O \ O] \tag{34}$$

By means of the similarity transformation

$$T = T^{-1} := \begin{bmatrix} I_{n_1} & O & O & O \\ O & I_{n_2} & O & O \\ I_{n_1} & O & -I_{n_1} & O \\ O & I_{n_2} & O & -I_{n_2} \end{bmatrix}$$

we derive as $\hat{A}' := T^{-1}\hat{A}T$ and $\hat{E}' := \hat{E}T$ the matrices

$$\hat{A}' = \begin{bmatrix} A_1+B_1F_1 & A_3+B_1F_2 & -B_1F_1 & -B_1F_2 \\ O & A_2 & O & O \\ O & O & A_1+G_1C_1 & A_3+G_1C_2 \\ O & O & G_2C_1 & A_2+G_2C_2 \end{bmatrix} \tag{35}$$

$$\hat{E}' = [\, E_1 \quad E_2 \quad O \quad O\,] \tag{36}$$

Let X_1 be such that

$$\mathcal{V} = \text{im}\left(\begin{bmatrix} X_1 \\ I_{n_2} \end{bmatrix}\right) \tag{37}$$

From (35,36) it is immediately seen that the regulated plant is stable and that the $\hat{A}$-invariant defined by

$$\hat{\mathcal{W}} := \text{im}\left(\begin{bmatrix} X_1 \\ I_{n_2} \\ X_1 \\ I_{n_2} \end{bmatrix}\right) = \text{im}\left(T \begin{bmatrix} X_1 \\ I_{n_2} \\ O \\ O \end{bmatrix}\right) \tag{38}$$

is externally stable and contained in $\hat{\mathcal{E}} := \ker \hat{E}$. □

6.2.1 On the Separation Property of Regulation

Refer again to the overall system shown in Fig. 6-5 and suppose that the matrices M, N, L, K of the regulator have been determined with Algorithm 6.2-3. Note that the plant stabilizing feedback matrix F_1 has been computed independently of matrix F_2, which causes a change of structure such that $\mathcal{V}$ becomes an $(A+BF)$-invariant. In other words, feedback through F_1 has a stabilizing task, while regulation is due to feedback through F_2.

The block diagram shown in Fig. 6-6 is equivalent, but the regulator has been split into two separate units: a *stabilizing unit* and a *strict regulator*. The matrices shown in the figure are easily derived from (33) as

$$N_{11} := A_1 + B_1F_1 + G_1C_1$$
$$N_{12} := A_3 + B_1F_2 + G_1C_2$$
$$N_{21} := G_2C_1$$
$$N_{22} := A_2 + G_2C_2$$
$$M_1 := -G_1$$
$$M_2 := -G_2$$
$$L_1 := F_1$$
$$L_2 := F_2$$

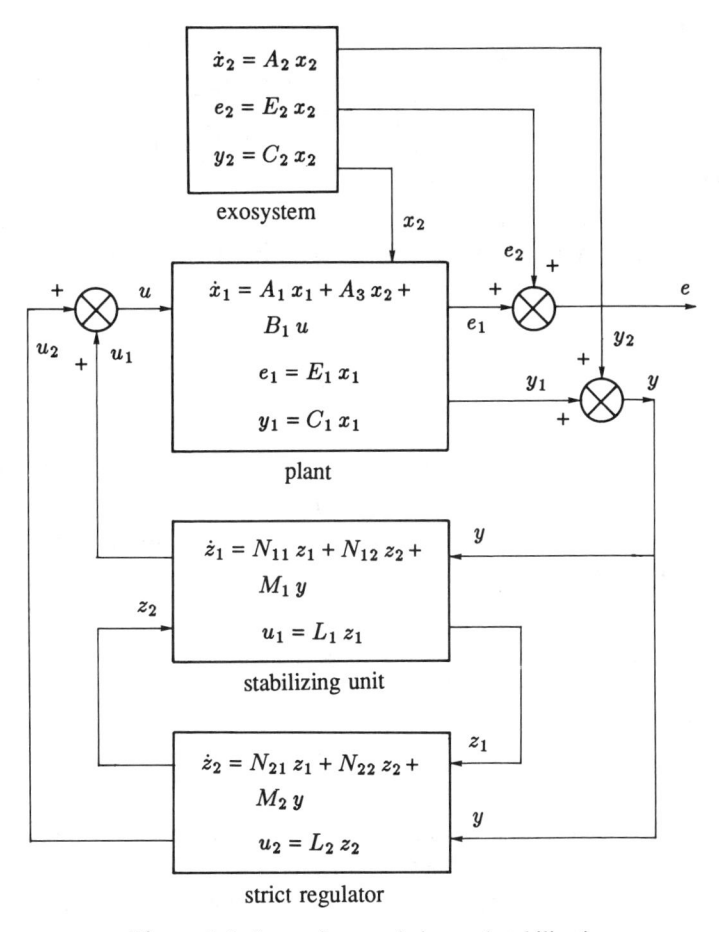

Sect. 6.2 The Autonomous Regulator: A General Synthesis Procedure

Figure 6-6 Separating regulation and stabilization.

[exosystem]
$$\dot{x}_2 = A_2 x_2$$
$$e_2 = E_2 x_2$$
$$y_2 = C_2 x_2$$

[plant]
$$\dot{x}_1 = A_1 x_1 + A_3 x_2 + B_1 u$$
$$e_1 = E_1 x_1$$
$$y_1 = C_1 x_1$$

[stabilizing unit]
$$\dot{z}_1 = N_{11} z_1 + N_{12} z_2 + M_1 y$$
$$u_1 = L_1 z_1$$

[strict regulator]
$$\dot{z}_2 = N_{21} z_1 + N_{22} z_2 + M_2 y$$
$$u_2 = L_2 z_2$$

The interesting question now arises whether or not a preliminary stabilization of the plant compromises solvability of the regulation problem.

Note that, the state of the stabilizing unit being clearly completely accessible both for control and observation, the interconnection of the exosystem, the plant, and the stabilizing unit is defined by the matrices

$$\hat{A}_s := \begin{bmatrix} A & BL_1 \\ M_1 C & N_{11} \end{bmatrix} \quad \hat{B}_s := \begin{bmatrix} B & O \\ O & I_{n_1} \end{bmatrix}$$

$$\hat{C}_s := \begin{bmatrix} C & O \\ O & I_{n_1} \end{bmatrix} \quad \hat{E}_s := [\, E \quad O \,]$$

Clearly, the interconnection of the plant and the stabilizing unit is stabilizable if and only if (A_1, B_1) is, and pair $(\hat{A}_s, \hat{C}_s)$ is detectable if and only if (A, C)

is. By virtue of Lemma 5.1-1 the subspace

$$\hat{\mathcal{V}} := \mathrm{im}\left(\begin{bmatrix} V \\ O \end{bmatrix}\right) \quad \text{with} \quad V := \begin{bmatrix} X_1 \\ I_{n_2} \end{bmatrix}$$

and with X_1 defined as in (37), is an externally stabilizable $(\hat{A}_s, \hat{B}_s)$-controlled invariant if and only if $\mathcal{V}$ is an externally stabilizable (A, B)-controlled invariant. It clearly satisfies

$$\hat{\mathcal{V}} \oplus \hat{\mathcal{P}}_s = \hat{\mathcal{X}}_s \quad \text{with} \quad \hat{\mathcal{P}}_s := \left(\begin{bmatrix} P \\ I_{n_1} \end{bmatrix}\right)$$

and with P defined as in (6). Also the sufficient condition regarding zeros stated by Theorem 6.2-2 still holds since the interconnection of the plant and the stabilizing unit has the same invariant zeros as the plant. Summing up, the following property has been proved.

Property 6.2-2. The Separation Property of Regulation
Interconnecting the plant with any dynamic feedback device having the state completely accessible for control and measurement does not influence solvability of the regulation problem.

It is worth pointing out that the name "stabilizing unit" used earlier is not rigorous. In fact, although the overall system has been stabilized (and all the assignable poles have been arbitrarily defined) by the synthesis algorithm, it is not guaranteed that the interconnection of the plant and the stabilizing unit is stable if the strict regulator is disconnected. However, in many "regular" cases (plant open-loop stable and minimum phase) this usually happens, particularly if the free closed-loop poles have been assigned in a conservative way.

6.2.2 The Internal Model Principle

We shall now consider the generalization of the internal model principle for the multivariable case.[1] Refer to the autonomous regulator shown in Fig. 6-7, whose informative output coincides with the regulated one, hence described by triple (A, B, E).

In this case the overall system is still described by equations (1–3), but with pair (C_1, C_2) replaced by (E_1, E_2). This causes an internal model of the exosystem to be included in the regulator, as is stated in the following theorem.

---

[1] The extension of the internal model principle to multivariable systems is due to Francis, Sebakhy, and Wonham [13, 14].

Sect. 6.2 The Autonomous Regulator: A General Synthesis Procedure

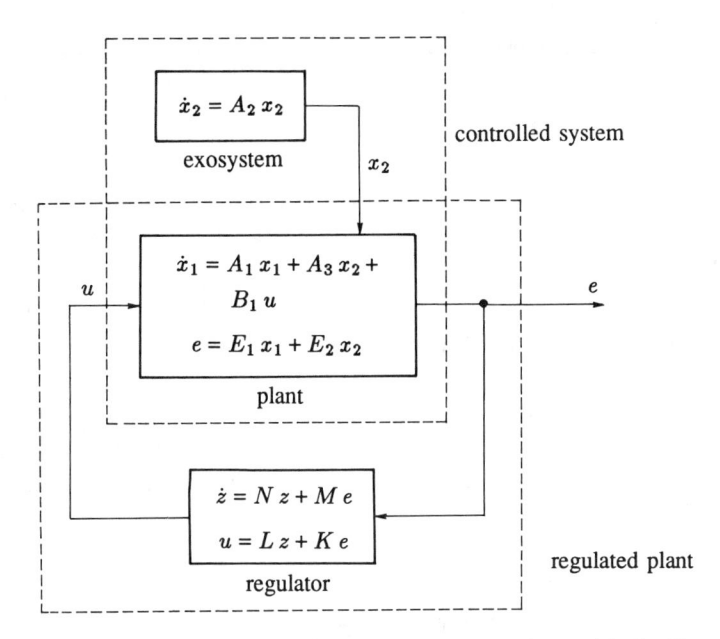

Figure 6-7 The multivariable autonomous regulator with $C = E$.

Theorem 6.2-3. The Internal Model Principle

Refer to the system shown in Fig. 6-7. Assume that (A_1, B_1) is stabilizable, (A, E) is detectable, and all the eigenvalues of A_2 are unstable. In any possible solution of the autonomous regulator problem the eigenstructure of A_2 is repeated in N, i.e., the Jordan blocks (or the elementary divisors) of A_2 are a subset of those of N.

Proof. If the regulator problem is solved by the quadruple (N, M, L, K), i.e., if $\lim_{t \to \infty} e(t) = 0$ for all the initial states, there exists an $\hat{A}$-invariant $\hat{\mathcal{W}}$ contained in $\hat{\mathcal{E}} := \ker \hat{E}$. Assume any initial state (x_{01}, x_{02}, z_0) belonging to $\hat{\mathcal{W}}$; the corresponding trajectory $\hat{x}(t) = (x_1(t), x_2(t), z(t))$ identically belongs to $\hat{\mathcal{W}}$, i.e., is such that

$$e(t) = E_1 x_1(t) + E_2 x_2(t) = 0 \quad \forall \, t \geq 0 \tag{39}$$

and satisfies the matrix differential equation

$$\begin{bmatrix} \dot{x}_1(t) \\ \dot{x}_2(t) \\ \dot{z}(t) \end{bmatrix} = \begin{bmatrix} A_1 & A_3 & B_1 L \\ O & A_2 & O \\ O & O & N \end{bmatrix} \begin{bmatrix} x_1(t) \\ x_2(t) \\ z(t) \end{bmatrix}$$

which is derived from (1) (with E_1, E_2 instead of C_1, C_2) taking into account (39). Relation (39) can also be written as

$$e(t) = [E_1 \; E_2] e^{At} \begin{bmatrix} x'_{01} \\ x_{02} \end{bmatrix} + [E_1 \; O] e^{N_e t} \begin{bmatrix} x''_{01} \\ z_0 \end{bmatrix} = 0 \tag{40}$$

where

$$N_e := \begin{bmatrix} A_1 & B_1 L \\ O & N \end{bmatrix} \tag{41}$$

and x'_{01}, x''_{01} denote any two vectors such that $x_{01} = x'_{01} + x''_{01}$. Since (A, E) is detectable, the time function

$$[\, E_1 \quad E_2 \,]\, e^{At} \begin{bmatrix} x'_{01} \\ x_{02} \end{bmatrix} \tag{42}$$

contains all the unstable modes of A. Then from (40) it follows that all the unstable modes of A are also modes of N_e; this means that the eigenstructure of A_2 is repeated in N. □

The application of the internal model principle is the basic tool to orient the autonomous regulator synthesis procedures towards feedback, hence to achieve robustness. In fact feedback controllers, relying on direct measurement of the controlled variable, directly neutralize any possible parametric change, if some structure and stability requirements are satisfied, while feedforward, being based on a model of the controlled system, is remarkably influenced by parameter changes and, in general, is not robust.

6.3 THE ROBUST REGULATOR: SOME SYNTHESIS PROCEDURES

Refer to the overall system represented in Fig. 6-7 (in which the informative output coincides with the regulated one) and assume that some of the matrices of the plant and regulator are subject to parametric changes, i.e., that their elements are functions of a parameter vector $q \in \mathcal{Q}$.[1] Matrices that are allowed to change are $A_1, A_3, B_1, E, M, L, K$, while A_2 and N are not varied, since the exosystem is a mathematical abstraction and the internal model principle is a necessary condition for regulation in this case. Here, as in all the most elementary approaches to robustness, particular dependence on parameters is not considered, but simply all the elements or all the nonzero elements of the above matrices are assumed to change independently of each other in given neighborhoods of their nominal values. In order to avoid heavy notation, we shall use the symbols $A_1^\circ, \ldots$ instead of $A_1(q), \ldots \; \forall \, q \in \mathcal{Q}$. Our aim is to derive a regulator that works at $A_1^\circ, A_3^\circ, B_1^\circ, E^\circ, M^\circ, L^\circ, K^\circ$.

Clearly, an autonomous regulator is robust if and only if the $\hat{A}^\circ$-invariant $\hat{\mathcal{P}}$ is internally stable and there exists an externally stable $\hat{A}^\circ$-invariant $\hat{\mathcal{W}}^\circ$

---

[1] Multivariable robust regulation was the object of very deep and competitive investigations in the mid-1970s. The most significant contributions are those of Davison [4], with Ferguson [5], Goldemberg [6], with Scherzinger [7], and Francis [12, 5.17], with Sebakhy and Wonham [13, 14]. A different approach is presented by Pearson, Shields, and Staats [27]. The results reported in this section are similar to those of Francis, but presented in a simplified and less rigorous way. The above references are a good basis for a deeper insight into the mathematical implications of robust regulation.

contained in $\hat{\mathcal{E}}^\circ := \ker \hat{E}^\circ$. Hence, a robust version of Property 6.2-1 can be immediately derived in the following terms.

Property 6.3-1.
Let all the exogenous modes be unstable. The regulator corresponding to the overall system matrices $\hat{A}^\circ, \hat{E}^\circ$ is robust if and only if there exists an externally stable $\hat{A}^\circ$-invariant $\hat{\mathcal{W}}^\circ$ such that:

1. $\hat{\mathcal{W}}^\circ \subseteq \hat{\mathcal{E}}^\circ$ with $\hat{\mathcal{E}}^\circ := \ker \hat{E}^\circ$; (1)
2. $\hat{\mathcal{W}}^\circ \oplus \hat{\mathcal{P}} = \hat{\mathcal{X}}$. (2)

The fundamental theorem applies to the robust case as a necessary condition stated as follows.

Property 6.3-2.
Let all the exogenous modes be unstable. The autonomous regulator problem admits a robust solution only if there exists an (A°, B°)-controlled invariant $\mathcal{V}^\circ$ such that

1. $\mathcal{V}^\circ \subseteq \mathcal{E}^\circ$; (3)
2. $\mathcal{V}^\circ \oplus \mathcal{P} = \mathcal{X}$. (4)

Note that, because of uniqueness of the solution of the Sylvester equation in Algorithm 6.2-1, if a $\mathcal{V}^\circ$ exists it is unique. The above necessary conditions are guaranteed by the following property, which is a consequence of Theorem 6.2-2.

Property 6.3-3.
Conditions (3,4) are satisfied if:

1. $\mathcal{V}^{*\circ} + \mathcal{P} = \mathcal{X}$; (5)
2. no invariant zero of the plant [i.e., of the triple $(A_1^\circ, B_1^\circ, E_1^\circ)$] coincides with an eigenvalue of the exosystem. (6)

Clearly, condition (5) is in any case necessary for (4) to hold. If generic robustness is considered (i.e., with all the elements of the system matrices varying anyhow) it is satisfied if and only if $\mathcal{E}^\circ + \mathcal{B}^\circ = \mathcal{X}$ (with $\mathcal{V}^{*\circ} := \mathcal{E}^\circ$). This implies that the input components are not less than the regulated output components.

A multivariable autonomous regulator synthesized by means of Algorithm 6.2-3 is, in general, not robust. To point out the reasons for this and suggest some simple expedients to achieve robustness, refer to the following example.

Example 6.3-1.
Consider the controlled system defined by

$$A_1 := \begin{bmatrix} -.4 & -4.04 & 0 \\ 1 & 0 & 0 \\ 0 & 0 & -1 \end{bmatrix} \quad A_3 := \begin{bmatrix} 0 & 0 \\ .2 & 0 \\ 0 & .3 \end{bmatrix} \quad B_1 := \begin{bmatrix} 2 & 0 \\ .4 & 1 \\ 0 & 1.5 \end{bmatrix} \quad (7)$$

![Figure 6-8: A multivariable controlled system - signal-flow graph showing regulator, plant, and exosystem with state variables x_1, x_2, x_3, x_4, x_5, inputs u_1, u_2, and outputs e_1, e_2.]

Figure 6-8 A multivariable controlled system.

$$E_1 := \begin{bmatrix} 0 & -1 & 0 \\ 0 & 0 & -1 \end{bmatrix} \quad E_2 := \begin{bmatrix} 1 & 0 \\ 0 & 1 \end{bmatrix} \quad A_2 := \begin{bmatrix} 0 & 1 \\ 0 & 0 \end{bmatrix} \tag{8}$$

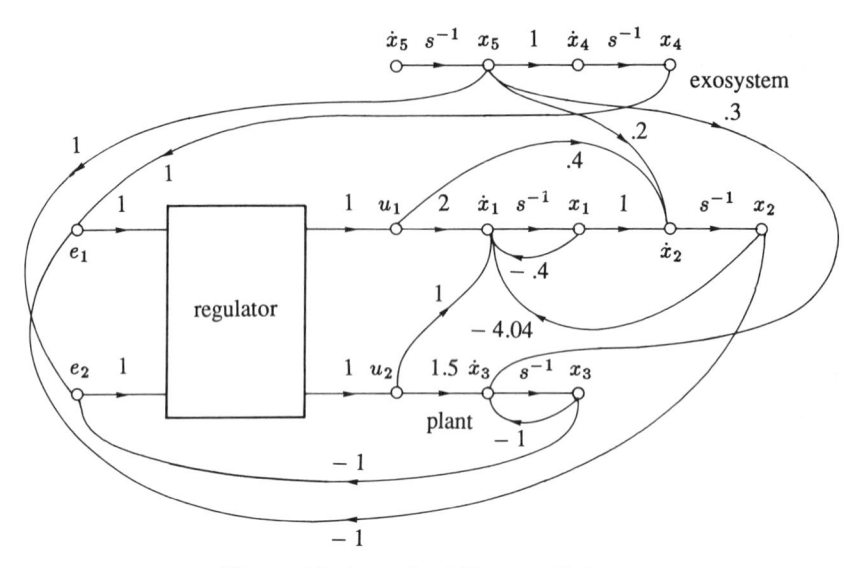

whose signal-flow graph is shown in Fig. 6-8.

The philosophy that underlies the multivariable robust regulator design and distinguishes it from the single-variable case is pointed out by the following remarks.

1. A single exosystem that influences several controlled outputs cannot be robustly neutralized by a single internal model. In fact, if some of the influence coefficients are subject to independent variations, the coefficients of the corresponding compensating actions should also vary accordingly so that regulation is maintained, and this, of course, is not possible. This drawback may be overcome by associating a different exosystem to every controlled output. In the case on hand, regulated output e_1 is influenced by a linear combination of a step and a ramp, while e_2 is affected by a step: in the mathematical model we can consider two different exosystems: a second-order one to generate the exogenous signals affecting e_1 and a first-order one for those affecting e_2, as shown in Fig. 6-9, where a further state variable x_6 has been introduced to this end. Since the design algorithm implies separate reproduction of each exosystem in the internal model, in this way the solution provided by the synthesis algorithm will be forced to have two independent internal models, one for each exosystem.

2. Unfortunately, in general this is not enough. In fact, a ramp introduced at input u_1 to compensate for the action of exosystem 1 on e_1 may also appear at e_2 because of interaction (nominal or due to parametric changes in A_1 and B_1). Since a regulator designed by using Algorithm 6.2-3 tends to distribute the internal model actions on all the input variables, a ramp signal generated by the internal model corresponding to exosystem 1 might appear at u_2, hence at e_2, in consequence of a possible variation of matrix L. Note, incidentally, that in the particular case of this example a design of the multivariable regulator as two single-variable regulators based on exosystems 1

Sect. 6.3 The Robust Regulator: Some Synthesis Procedures

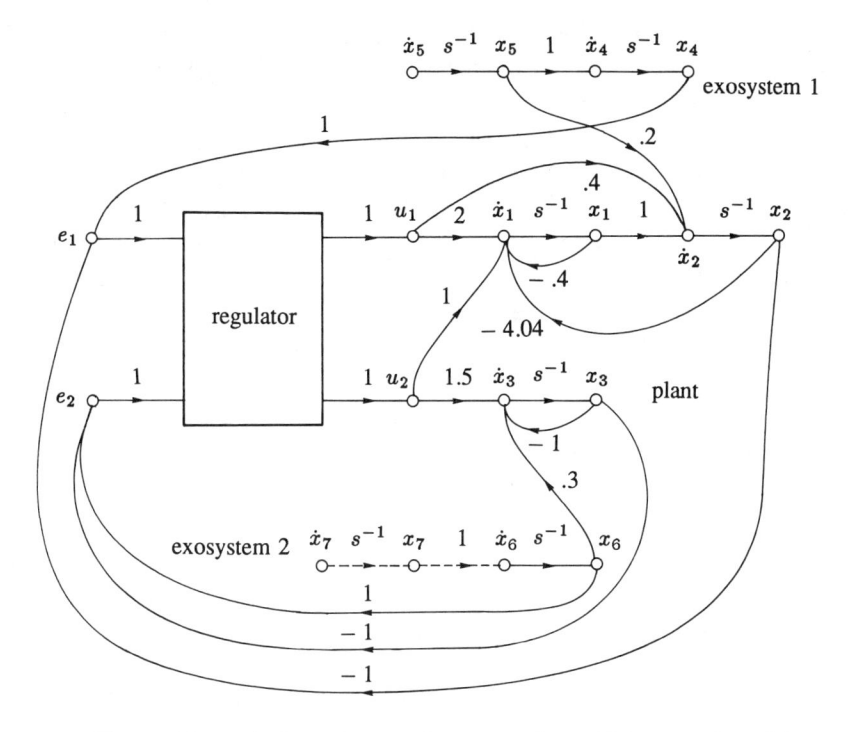

Figure 6-9 Achieving robustness by means of a multiple internal model.

and 2 respectively could be satisfactory, if only the nonzero elements of the involved matrices are subject to change. In general, however, in order to make the regulator robust it is necessary to make the internal model corresponding to exosystem 2 capable of generating ramps as well: this is obtained simply by adding to exosystem 2 a further state variable x_7, as shown in the figure by dashed lines. If referred to the overall system of Fig. 6-9 with this completion, the regulator provided by Algorithm 6.2-3 is robust with respect to variations of all the elements of matrices (not necessarily only the nonzero ones).

Taking into account these changes, the matrices of the controlled system are written as:

$$A_1 := \begin{bmatrix} -.4 & -4.04 & 0 \\ 1 & 0 & 0 \\ 0 & 0 & -1 \end{bmatrix} \quad A_3 := \begin{bmatrix} 0 & 0 & 0 & 0 \\ .2 & 0 & 0 & 0 \\ 0 & 0 & .3 & 0 \end{bmatrix} \quad B_1 := \begin{bmatrix} 2 & 0 \\ .4 & 1 \\ 0 & 1.5 \end{bmatrix} \quad (9)$$

$$E_1 := \begin{bmatrix} 0 & -1 & 0 \\ 0 & 0 & -1 \end{bmatrix} \quad E_2 := \begin{bmatrix} 1 & 0 & 0 & 0 \\ 0 & 0 & 1 & 0 \end{bmatrix} \quad A_2 := \begin{bmatrix} J_1 & O \\ O & J_1 \end{bmatrix} \quad (10)$$

with

$$J_1 := \begin{bmatrix} 0 & 1 \\ 0 & 0 \end{bmatrix} \quad (11)$$

We shall now translate the preceding considerations into a general recipe to achieve robustness in multivariable regulators designed by means of Algorithm

6.2-3, simply by using the expedient of replicating a significant part of the exosystem in the mathematical model of the controlled system. We assume that pair (A_1, E_1) is observable.

1. Suppose at the moment that the exosystem has only one eigenvalue (for instance zero, as in the previous example) and that matrix A_2 is in real Jordan form (as in the example). For every regulated output consider the observable part of the exosystem and determine the dimension of its maximal Jordan block (or, if preferred, the degree of its minimal polynomial). Assign to the considered regulated output a new individual exosystem with the dimension of this Jordan block (so that the observability condition is certainly met). In the single-variable case of Fig. 6-2, this step has been accomplished by replacing the first signal-flow graph shown in the figure with the second one.

2. Extend backwards the Jordan blocks assigned to each individual regulated output to obtain the same dimension for all.

3. If no part of the original exosystem is observable from a particular controlled output, first check if this unobservability property is robustly maintained (for instance, this is very likely to happen if only the nonzero elements of the involved matrices are subject to vary). If so, do nothing. If not, associate also to this output an exosystem of the same type as the others and create for it an observability path as follows: consider any nonzero element in the corresponding row of matrix E_1 (for instance the element having the maximal absolute value) and denote by k its column index; then assume as the new part of matrix A_3 (that which distributes the action of the new exosystem on the plant state variables) one with all elements equal to zero, except that in the first column and k-th row, which can be set equal to one.

4. If the original exosystem has several distinct eigenvalues, repeat the outlined procedure for each of them.

Note that to achieve robustness, an overall exosystem that is in general of greater dimension than the original one is finally obtained. This is formally correct, because the original autonomous regulator problem has been changed into another one, in which the class of the exogenous signals that are allowable for every path from the exosystem has been extended and unified. Also note that, as in the single-variable case, the particular points where the exogenous and internal model signals input the plant, are immaterial provided the observability assumption is satisfied for both.

These remarks are formalized in the following algorithm.

Algorithm 6.3-1. The Autonomous Robust Regulator Synthesis

Redefine, if necessary, the mathematical model of the controlled system in such a way that an independent replica of a unique type of exosystem is observable from every regulated output. Then synthesize the nominal regulator by means of Algorithm 6.2-3 (of course, with $C = E$).

Sect. 6.3 The Robust Regulator: Some Synthesis Procedures

Proof. Consider the robustness-oriented version of (6.2.31,32):

$$\hat{A}^\circ = \begin{bmatrix} A_1^\circ & A_3^\circ & B_1^\circ F_1^\circ & B_1^\circ F_2^\circ \\ O & A_2 & O & O \\ -G_1^\circ E_1^\circ & -G_1^\circ E_2^\circ & A_1+B_1 F_1+G_1 E_1 & A_3+B_1 F_2+G_1 E_2 \\ -G_2^\circ E_1^\circ & -G_2^\circ E_2^\circ & G_2 E_1 & A_2+G_2 E_2 \end{bmatrix} \quad (12)$$

$$\hat{E}^\circ = [\, E_1^\circ \quad E_2^\circ \quad O \quad O \,] \quad (13)$$

or, in more compact form

$$\hat{A}^\circ = \begin{bmatrix} A_1^\circ & A_3^\circ & B_1^\circ L^\circ \\ O & A_2 & O \\ M^\circ E_1^\circ & M^\circ E_2^\circ & N \end{bmatrix} \quad (14)$$

$$\hat{E}^\circ = [\, E_1^\circ \quad E_2^\circ \quad O \,] \quad (15)$$

with

$$N := \begin{bmatrix} A_1+B_1 F_1+G_1 E_1 & A_3+B_1 F_2+G_1 E_2 \\ G_2 E_1 & A_2+G_2 E_2 \end{bmatrix}$$

$$M^\circ := \begin{bmatrix} -G_1^\circ \\ -G_2^\circ \end{bmatrix} \qquad L^\circ := [\, F_1^\circ \quad F_2^\circ \,] \quad (16)$$

Recall that at the nominal values of the parameters there exists an (A,B)-controlled invariant $\mathcal{V} \subseteq \mathcal{E}$, which complements $\mathcal{P}$ (i.e., having the eigenvalues of A_2 as the only internal ones), and whose basis matrix can be expressed as in (6.2.37). It is an $(A+BF)$-invariant by construction, so that

$$(A_1 + B_1 F_1) X_1 + A_3 + B_1 F_2 = X_1 A_2$$

and, since $\mathcal{V} \subseteq \mathcal{E}$ (or $E_1 X_1 + E_2 = O$), it is also an N-invariant, again having the eigenvalues of A_2 as the only internal ones. Note the structure of the $\hat{A}$-invariant defined by (6.2.38): a motion starting from a point of $\mathcal{V}$ (corresponding to a generic initial condition x_{20} of the exosystem and $X_1 x_{20}$ of the plant) can be maintained on $\mathcal{V}$ (in the regulated system state space) if and only if the regulator is given the same initial condition: this corresponds to a "steady state" or "limit" trajectory of the overall system. The feedback connection that causes the state to be maintained on $\mathcal{V}$ (hence on $\mathcal{E}$) is derived from the regulator instead of the exosystem. Since the exosystem is completely observable from e, to ensure that all its modes do not appear in e, the internal model also must be completely observable from e (with the exosystem disconnected): this is automatically guaranteed by the construction procedure of matrix F. In fact, since F is such that $\mathcal{V}$ is an $(A+BF)$-invariant, it provides for the exosystem modes a path $B_1 F$ through the input which cancels their influence at the regulated output through the dynamic connection (A_1, A_3, E_1) and the algebraic connection E_2. In other words, while pair

$$\left(\begin{bmatrix} A_1 & A_3 \\ O & A_2 \end{bmatrix}, [E_1 \; E_2] \right)$$

is observable by assumption, pair

$$\left(\begin{bmatrix} A_1 & B_1 F \\ O & A_2 \end{bmatrix}, [E_1 \ O]\right)$$

is observable by construction, and its observability is implied by that of the former pair. Define the matrices of the regulated plant (the particularization of (6.2.4) for the case on hand) as

$$A_p^\circ := \begin{bmatrix} A_1^\circ & B_1^\circ L^\circ \\ M^\circ E_1^\circ & N \end{bmatrix} \quad B_p^\circ := \begin{bmatrix} A_3^\circ \\ M^\circ E_2^\circ \end{bmatrix}$$

$$C_p^\circ := [E_1^\circ \ O] \qquad D_p^\circ := [E_2^\circ] \tag{17}$$

and consider the "nominal" overall system invariant $\hat{\mathcal{W}}$ defined by (6.2.38) to which the steady state trajectories belong at the nominal value of the parameters. It can be derived by means of the Sylvester equation (6.2.20) with matrices (17) at their nominal values: by uniqueness (recall that A_p is stable and A_2 unstable) the solution must be

$$X_p = \begin{bmatrix} X_1 \\ X_1 \\ I_{n_2} \end{bmatrix} \tag{18}$$

Let us now consider robustness: how does solution (18) modify when parameters change? As long as A_p° remains stable, the solution of the corresponding robustness oriented equation

$$A_p^\circ X_p^\circ - X_p^\circ A_2 = -B_p^\circ \tag{19}$$

remains unique. If the nonnominal solution

$$X_p^\circ = \begin{bmatrix} X_1^\circ \\ X_2^\circ \\ X_3^\circ \end{bmatrix} \tag{20}$$

is such that the $\hat{A}^\circ$-invariant

$$\hat{\mathcal{W}} = \begin{bmatrix} X_1^\circ \\ I_{n_2} \\ X_2^\circ \\ X_3^\circ \end{bmatrix} \tag{21}$$

is contained in ker$\hat{E}$, the regulation property still holds. If for any initial condition x_{20} of the exosystem there exist corresponding initial conditions of the plant and regulator such that e remains zero, by unicity they are respectively $X_1^\circ x_{20}$ and $[X_2^\circ \ X_3^\circ]^T x_{20}$. We claim that (20) is of the general type

$$X_p^\circ = \begin{bmatrix} X_1^\circ \\ X_1 T^\circ \\ T^\circ \end{bmatrix} \tag{22}$$

with X_1° and X_1 defined by

$$\mathcal{V}^\circ = \mathrm{im}\left(\begin{bmatrix} X_1^\circ \\ I_{n_2} \end{bmatrix}\right) \quad \text{and, as before,} \quad \mathcal{V} = \mathrm{im}\left(\begin{bmatrix} X_1 \\ I_{n_2} \end{bmatrix}\right) \tag{23}$$

($\mathcal{V}^\circ$ is the new complement of the plant – see Property 6.3-2) and T° denotes a matrix commuting with A_2, i.e., such that

$$A_2 T^\circ = T^\circ A_2 \tag{24}$$

It is easily checked that (22) is a solution of (19) if (24) holds. In connection with (22) it is possible to define the $\hat{A}^\circ$-invariant

$$\hat{\mathcal{W}}^\circ := \mathrm{im}\left(\begin{bmatrix} X_1^\circ \\ I_{n_2} \\ X_1 T^\circ \\ T^\circ \end{bmatrix}\right) \tag{25}$$

which is clearly externally stable (it is a complement of the regulated plant, which is a stable $\hat{A}^\circ$-invariant) and contained in $\hat{\mathcal{E}}^\circ$. Since observability is locally preserved in the presence of parameter variations, both the exosystem and the internal model remain independently observable from e. However, in this case the initial conditions corresponding to a limit trajectory (on $\mathcal{V}^\circ$ for the controlled system and on $\mathcal{V}$ for the regulator) are not equal, but related by matrix T°, having (24) as the only constraint. Condition (24), on the other hand, provides a class of matrices depending on a large number of parameters and allows independent tuning of every mode generated by the internal model. For instance, in the particular case of (9–11) we have [2]

$$T^\circ = \begin{bmatrix} \alpha & \beta & \gamma & \delta \\ 0 & \alpha & 0 & \gamma \\ \epsilon & \eta & \lambda & \mu \\ 0 & \epsilon & 0 & \lambda \end{bmatrix} \quad \text{and} \quad e^{A_2 t} T^\circ = \begin{bmatrix} \alpha & \beta+\alpha t & \gamma & \delta+\gamma t \\ 0 & \alpha & 0 & \gamma \\ \epsilon & \eta+\epsilon t & \lambda & \mu+\lambda t \\ 0 & \epsilon & 0 & \lambda \end{bmatrix} \tag{26}$$

If the exosystem were not replicated, i.e., if synthesis were based on (7,8) instead of (9–11), we would have

$$T^\circ = \begin{bmatrix} \alpha & \beta \\ 0 & \alpha \end{bmatrix} \quad \text{and} \quad e^{A_2 t} T^\circ = \begin{bmatrix} \alpha & \beta+\alpha t \\ 0 & \alpha \end{bmatrix} \tag{27}$$

and the number of parameters clearly would not be sufficient to guarantee neutralization of an arbitrary step plus an arbitrary ramp at both regulated outputs. On the other hand, if the exosystem is suitably replicated and reproduced in the internal model, an overall state trajectory that does not affect e exists for all the initial conditions of the

---

[2] An extended treatment of matrix equations of the general types $AX=XA$ and $AX=XB$ with discussion of their solutions in terms of parameters is developed in Gantmacher's book [A.8], Chapter 8.

exosystem: structure (22) for the solution of the Sylvester equation and (25) for the corresponding $\hat{W}^\circ$ follow by uniqueness. □

A question now arises: what happens, in mathematical terms, if the exosystem is not replicated for every regulated output? The Sylvester equation (19) still admits a unique solution, which corresponds to the externally stable $\hat{A}^\circ$-invariant $\hat{W}^\circ$ defined by (21), but this is not contained in $\hat{\mathcal{E}}^\circ$, so that the regulation requirement is not met. In this case the regulator in general is not robust because there does not exist any externally stable $\hat{A}^\circ$-invariant contained in $\hat{\mathcal{E}}^\circ$ or, in other words, the maximal $\hat{A}^\circ$-invariant contained in $\hat{\mathcal{E}}^\circ$ is externally unstable.

A more elegant formalization of this procedure, which, in practice, is equivalent to it, but only points out the extension of the internal model (not of the class of the admissible exogenous modes by replicas of a unified exosystem), is set in the following algorithm.

Algorithm 6.3-2. The Francis Robust Regulator Synthesis

Let h be the number of distinct eigenvalues of the exosystem and denote by $J_1, J_2, \ldots, J_h$ the corresponding maximal real Jordan blocks. Define the eigenstructure to be replicated in the internal model as

$$J := \begin{bmatrix} J_1 & O & \cdots & O \\ O & J_2 & \cdots & O \\ \vdots & \vdots & \ddots & \vdots \\ O & O & \cdots & J_h \end{bmatrix} \tag{28}$$

and assume as the matrix of the internal model

$$A_{2e} := \begin{bmatrix} J & O & \cdots & O \\ O & J & \cdots & O \\ \vdots & \vdots & \ddots & \vdots \\ O & O & \cdots & J \end{bmatrix} \tag{29}$$

where J is replicated as many times as there are regulated output components. Let n_e be the dimension of A_{2e}. Then define the extension of the controlled system

$$A_e := \begin{bmatrix} A_1 & A_{3e} \\ O & A_{2e} \end{bmatrix} \quad B_e := \begin{bmatrix} B_1 \\ O \end{bmatrix}$$

$$E_e := [\, E_1 \quad O \,] \quad P_e := \begin{bmatrix} I_{n_1} \\ O \end{bmatrix} \tag{30}$$

where A_{3e} is simply chosen in such a way that (A_{2e}, A_{3e}) is observable and (A_e, E_e) detectable (see the previously outlined procedure to create, if necessary, an observability path for every Jordan block of the internal model). Note that the existence of a $\mathcal{V}^\circ \subseteq \mathcal{E}$ such that $\mathcal{V}^\circ \oplus \mathcal{P} = \mathcal{X}$ means simply that B_1 is such that every steady state influence of the exosystem on the regulated output (from which the exosystem is

Sect. 6.3 The Robust Regulator: Some Synthesis Procedures

completely observable) can be cancelled by a suitable feedback from the state of the exosystem itself. Since this is an intrinsic property of (A_1, B_1, E_1), it is also valid for the new system (30) i.e., $\mathcal{V}^* + \mathcal{P} = \mathcal{X}$ implies $\mathcal{V}_e^* + \mathcal{P}_e = \mathcal{X}_e$ with $\mathcal{V}_e^* := \max \mathcal{V}(A_e, B_e, \mathcal{E}_e)$ ($\mathcal{E}_e := \ker E_e$). Furthermore, the complementation algorithm shows that $\mathcal{V}_e^*$ is complementable if $\mathcal{V}^*$ is. Let $\mathcal{V}_e$ be such that $\mathcal{V}_e \oplus \mathcal{P}_e = \mathcal{X}_e$. Determine $F_e = [F_1 \; F_{2e}]$ such that $(A_e + B_e F_e) \mathcal{V}_e \subseteq \mathcal{V}_e$ and $A_1 + B_1 F_1$ is stable, G_e such that $A_e + G_e E_e$ is stable, and assume

$$N := A_e + B_e F_e + G_e E_e \qquad M := -G_e \qquad L := F_e \qquad K := 0 \qquad (31)$$

Proof. The extended system matrices $\hat{A}$ and $\hat{E}$ can be partitioned as

$$\hat{A} = \begin{bmatrix} A_1 & A_3 & B_1 F_1 & B_1 F_{2e} \\ 0 & A_2 & 0 & 0 \\ -G_{1e} E_1 & -G_{1e} E_2 & A_1 + B_1 F_1 + G_{1e} E_1 & A_{3e} + B_1 F_{2e} \\ -G_{2e} E_1 & -G_{2e} E_2 & G_{2e} E_1 & A_{2e} \end{bmatrix} \qquad (32)$$

$$\hat{E} = [E_1 \; E_2 \; 0 \; 0] \qquad (33)$$

By means of the similarity transformation

$$T = T^{-1} := \begin{bmatrix} I_{n_1} & 0 & 0 & 0 \\ 0 & I_{n_2} & 0 & 0 \\ I_{n_1} & 0 & -I_{n_1} & 0 \\ 0 & 0 & 0 & -I_{n_e} \end{bmatrix}$$

we derive as $\hat{A}' := T^{-1} \hat{A} T$ and $\hat{E}' := \hat{E} T$ the matrices

$$\hat{A}' = \begin{bmatrix} A_1 + B_1 F_1 & A_3 & -B_1 F_1 & -B_1 F_{2e} \\ 0 & A_2 & 0 & 0 \\ 0 & A_3 + G_{1e} E_2 & A_1 + G_{1e} E_1 & A_{3e} \\ 0 & G_{2e} E_2 & G_{2e} E_1 & A_{2e} \end{bmatrix} \qquad (34)$$

$$\hat{E}' = [E_1 \; E_2 \; 0 \; 0] \qquad (35)$$

Let $\mathcal{V}$ be such that $\mathcal{V} \oplus \mathcal{P} = \mathcal{X}$ and define X_1, X_{1e} through

$$\mathcal{V} = \mathrm{im}\left(\begin{bmatrix} X_1 \\ I_{n_2} \end{bmatrix}\right) \qquad \mathcal{V}_e = \mathrm{im}\left(\begin{bmatrix} X_{1e} \\ I_{n_e} \end{bmatrix}\right) \qquad (36)$$

From (34) it is immediately seen that the regulated plant is stable. By an argument similar to that already used to illustrate the internal model operation in connection with Algorithm 6.3-1, it can be shown that at the nominal values of the parameters there exists a matrix T satisfying

$$A_{2e} T = T A_2 \qquad (37)$$

such that the subspace

$$\hat{\mathcal{W}} := \mathrm{im}\left(\begin{bmatrix} X_1 \\ I_{n_2} \\ X_{1e} T \\ T \end{bmatrix}\right) \qquad (38)$$

is an $\hat{A}$-invariant. If so, it is clearly externally stable and contained in $\hat{\mathcal{E}} := \ker\hat{E}$. Matrix T can easily be derived by means of a Sylvester equation of type (19). Again, since the number of the free parameters in a generic T satisfying (37) is large enough to compensate for any influence of the exosystem at the regulated output, there exists at least one T that causes the regulated output to be maintained at zero. By uniqueness, the solution of the Sylvester equation must have the structure

$$X_p = \begin{bmatrix} X_1 \\ X_{1e}T \\ T \end{bmatrix} \tag{39}$$

which reflects into structure (38) for $\hat{\mathcal{W}}$. For instance, in the particular case of (7,8) we have

$$T = \begin{bmatrix} \alpha & \beta \\ 0 & \alpha \\ \gamma & \delta \\ 0 & \gamma \end{bmatrix} \quad \text{and} \quad e^{A_2 t}T = \begin{bmatrix} \alpha & \beta+\alpha t \\ 0 & \alpha \\ \gamma & \delta+\gamma t \\ 0 & \gamma \end{bmatrix} \tag{40}$$

The robustness-oriented version of (32, 33) is

$$\hat{A}^\circ = \begin{bmatrix} A_1^\circ & A_3^\circ & B_1^\circ F_1^\circ & B_1^\circ F_{2e}^\circ \\ O & A_2 & O & O \\ -G_{1e}^\circ E_1^\circ & -G_{1e}^\circ E_2^\circ & A_1+B_1F_1+G_{1e}E_1 & A_{3e}+B_1F_{2e} \\ -G_{2e}^\circ E_1^\circ & -G_{2e}^\circ E_2^\circ & G_{2e}E_1 & A_{2e} \end{bmatrix} \tag{41}$$

$$\hat{E}^\circ = [\, E_1^\circ \quad E_2^\circ \quad O \quad O \,] \tag{42}$$

Let $\mathcal{V}^\circ$ be such that $\mathcal{V}^\circ \oplus \mathcal{P} = \mathcal{X}$ and denote by X_1° the corresponding matrix defined as in the first of (36). Robustness is related to the existence of a T° satisfying (37), i.e., such that the subspace

$$\hat{\mathcal{W}}^\circ := \text{im}\left(\begin{bmatrix} X_1^\circ \\ I_{n_2} \\ X_{1e}T^\circ \\ T^\circ \end{bmatrix} \right) \tag{43}$$

is an $\hat{A}^\circ$-invariant. If so, it is clearly externally stable and contained in $\hat{\mathcal{E}}^\circ := \ker\hat{E}^\circ$. In the case of Example 6.3-1 T° has the same structure as T in (40), hence includes as many parameters as needed to reproduce a step of arbitrary amplitude and a ramp of arbitrary slope at both the regulated outputs. The invariant (43) can easily be derived by means of a Sylvester equation of type (18). □

6.4 THE MINIMAL-ORDER ROBUST REGULATOR

Algorithms 6.3-1 and 6.3-2 result in regulator designs with a very large number of poles assignable in the overall nominal system. If in particular the plant, i.e., the triple (A_1, B_1, E_1), is completely controllable and observable, all the poles

Sect. 6.4 The Minimal-Order Robust Regulator

of the regulated plant ($2n_1 + n_2$ in the first case and $2n_1 + n_{2e}$ in the second) are arbitrarily assignable, and the dynamics of $e(t)$ can be made arbitrarily fast. Structural robustness is achieved by multiple replicas of the exosystem in the internal model, while robustness with respect to stability, not considered so far, is ensured at least locally because the eigenvalues are continuous functions of the parameters. Unfortunately, contrary to what may appear at first glance, pole placement in the far left half-plane is not conservative since it results in wider spreading when parameters vary, hence in less robust design with respect to preservation of both standard behavior and stability.

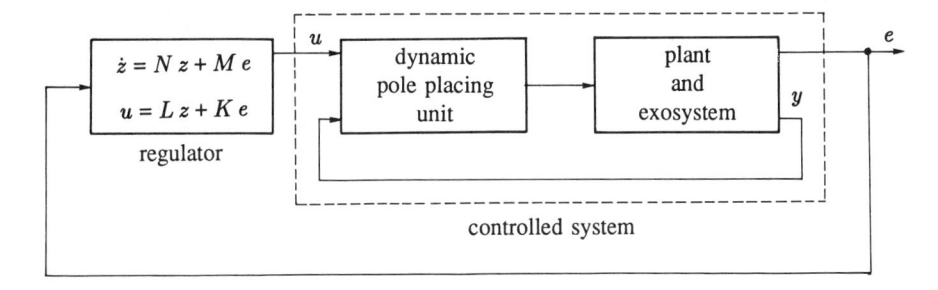

Figure 6-10 The structure of a minimal-order regulator.

We shall now present a completely different design philosophy: to assign zeros instead of poles. This is a more direct extension to multivariable systems of the standard single-variable design technique briefly recalled in Section 6.1. The number of zeros that are freely assignable is equal to the order of the internal model. We shall refer to a plant that is open-loop stable, but this does not imply any loss of generality since stability or arbitrary pole assignment has been shown to be obtainable, where possible, independently of regulation. This can be done by means of a suitable dynamic unit, connected as shown in Fig. 6-10, which may be considered as a part of the plant in the regulator design (separation property).

Refer to the block diagram of Fig. 6-7. A synthesis procedure that assigns as many zeros as needed to match the order of the internal model is set as follows. Note that it is not completely automatic, but may be assumed as a guideline for trial and error or CAD design.

Algorithm 6.4-1. The Minimal-Order Robust Regulator Synthesis

Assume that A_1 is stable and define matrix K of the regulator as

$$K := (E_1 A_1^{-1} B_1)^+ \tag{1}$$

(the pseudoinverse of the static gain of the plant). This has been proved to be a good choice in most practical cases. However, K can be assumed simply of maximal rank and such that $A + B_1 K E_1$ is stable or the K provided by (1) can be varied if needed to satisfy these requirements. Define

$$A_e := \begin{bmatrix} A_1 + B_1 K E_1 & A_{3e} \\ O & A_{2e} \end{bmatrix} \quad B_e := \begin{bmatrix} B_1 \\ O \end{bmatrix}$$

$$E_e := [\, E_1 \quad O \,] \quad P_e := \begin{bmatrix} I_{n_1} \\ O \end{bmatrix} \tag{2}$$

with A_{2e} given by (6.3.24) and A_{3e} chosen to establish an observability path at the regulated output for the complete internal model. Denote by $\mathcal{X}_e$ the corresponding state space and proceed as in Algorithm 6.3-2: let $\mathcal{V}_e$ be such that $\mathcal{V}_e \oplus \mathcal{P}_e = \mathcal{X}_e$, F_{2e} such that $(A_e + B_e F_e) \mathcal{V}_e \subseteq \mathcal{V}_e$, with $F_e := [K E_1 \; F_{2e}]$. Then by Algorithm 4.5-1 determine G_e such that quadruple (A_{2e}, G_e, F_{2e}, K) has the desired invariant zeros and assume

$$N := A_{2e} \quad M := G_e \quad L := F_e \tag{3}$$

The overall system matrices are, in this case

$$\hat{A}^\circ = \begin{bmatrix} A_1^\circ + B_1'^\circ K^\circ E_1^\circ & A_3^\circ & B_1'^\circ F_{2e}^\circ \\ O & A_2 & O \\ G_e^\circ E_1^\circ & G_e^\circ E_2^\circ & A_{2e} \end{bmatrix} \tag{4}$$

$$\hat{E}^\circ = [\, E_1^\circ \quad E_2^\circ \quad O \,] \tag{5}$$

Note that matrix A_{3e} does not appear in (4): however, it has been useful in computing F_{2e} to provide a suitable signal path from the internal model to the regulated output. Matrix $B_1'^\circ$ in (4) is defined as

$$B_1'^\circ := k\, B_1^\circ \tag{6}$$

where k, real positive, is the *gain constant* of the loop. The design consists of the following steps:

1. Consider the initial pole-zero layout of the plant and internal model and choose suitable locations for the invariant zeros of the regulator.

2. Design the regulator by the previously outlined procedure.

3. Trace out the multivariable root locus (the locus of the eigenvalues of (4) versus k) and determine a value of k corresponding to a satisfactory pole location; if this is not possible, go back to step 1 and choose a different set of invariant zeros. When stability is not achievable for any choice of zero locations, it is possible to augment the number of arbitrary zeros as explained in step 4.

4. Add any number of arbitrary stable poles in matrix A_{2e} (as diagonal elements or real Jordan blocks, single or chained anyhow, in conservative left half-plane locations) and make them observable from all the regulated outputs through a suitable choice of the nonzero elements in the corresponding rows of matrix A_{3e}; then go back to step 1 with as many more arbitrarily assignable zeros as the added poles. This procedure extends the well-known pole-zero cancellation technique to the multivariable case, a technique widely used in single-variable design to shift stable poles toward the left.

Sect. 6.4 The Minimal-Order Robust Regulator

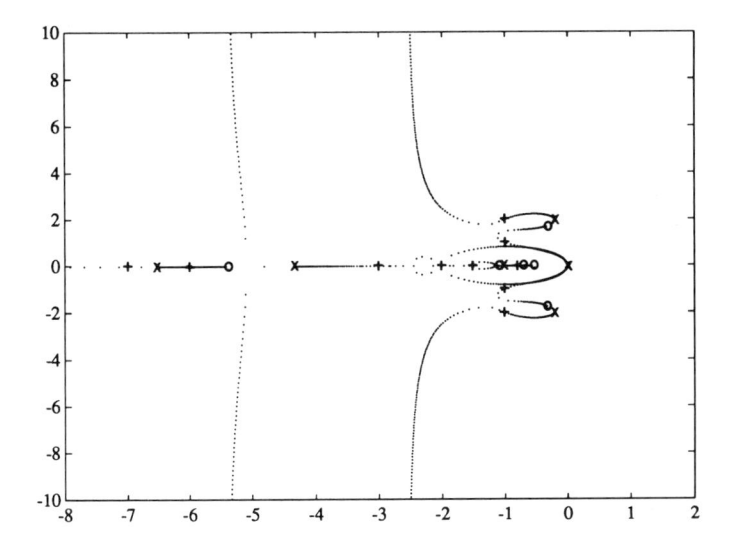

Figure 6-11 Designing with Algorithm 6.3-1: the multivariable root locus.

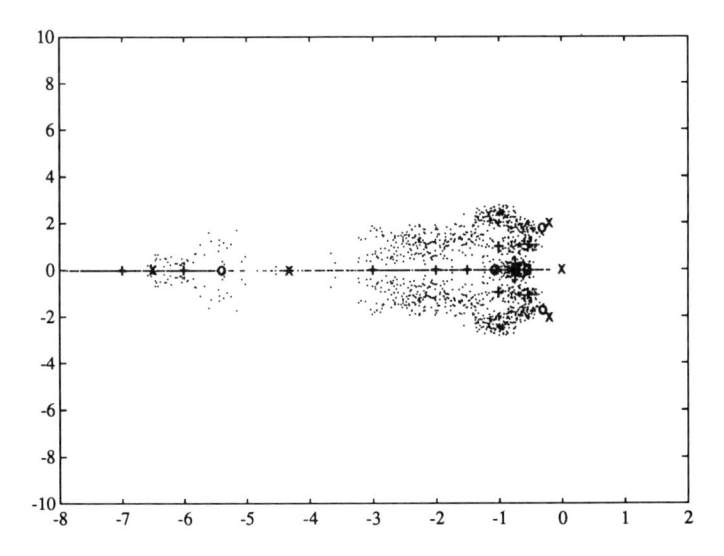

Figure 6-12 Designing with Algorithm 6.3-1: stability robustness check corresponding to ±5% variations of the nonzero parameters.

It is worth noting that if the plant is open-loop stable with a sufficient stability margin and the exosystem has all the eigenvalues at zero (this is the standard situation in technical applications), the regulation problem in a strict sense can be solved by choosing all the invariant zeros in the left half-plane and very close to the origin. Thus, they attract the root locus branches from the exosystem poles while the other branches (those originating from the plant

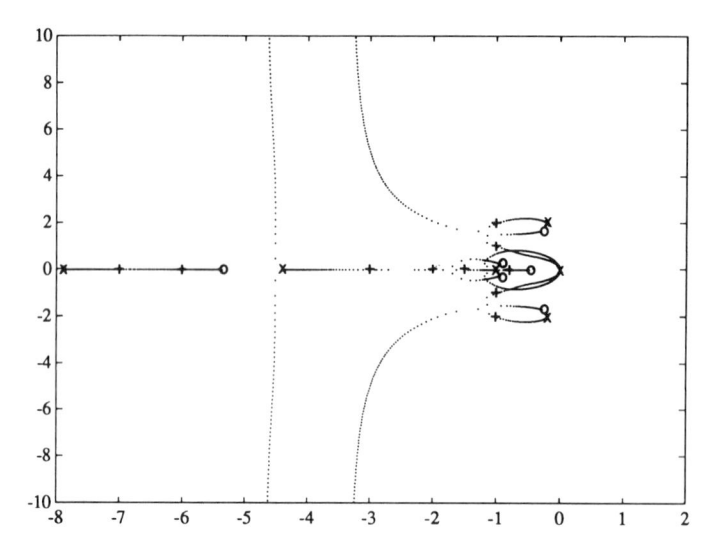

Figure 6-13 Designing with Algorithm 6.3-2: the multivariable root locus.

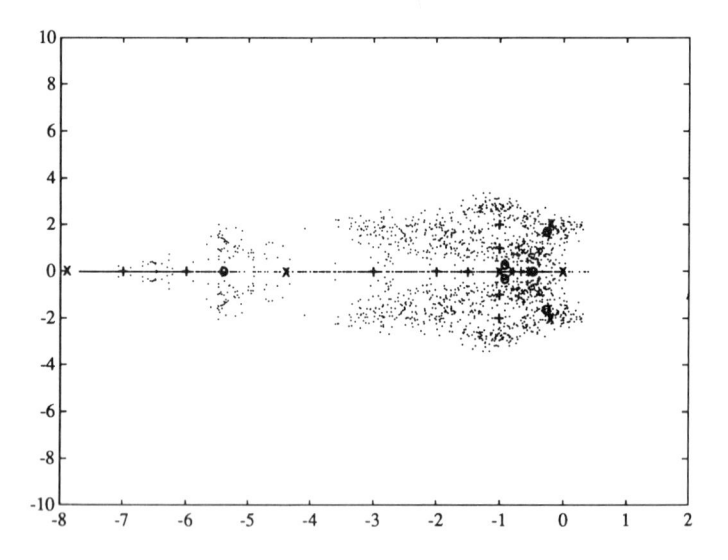

Figure 6-14 Designing with Algorithm 6.3-2: stability robustness check corresponding to ±5 % variations of the nonzero parameters.

open-loop poles) remain in the stability region. Stability is easily achievable by this technique, in particular if the dimensions of the Jordan blocks of the exosystem is not excessive: recall that for every Jordan block the corresponding branches of the locus leave the origin in directions that are angularly equally spaced so that some branches may lie on the right half-plane for small values of k and cross the imaginary axis for finite values of k. However, a solution of

Sect. 6.4 The Minimal-Order Robust Regulator 343

this type is not in general the best due to the relatively long regulation transient caused by poles having small real parts.

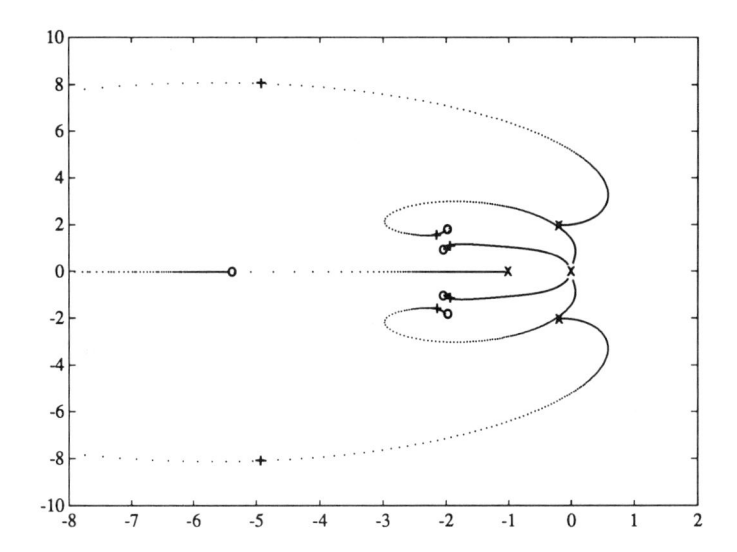

Figure 6-15 Designing with Algorithm 6.4-1: the multivariable root locus.

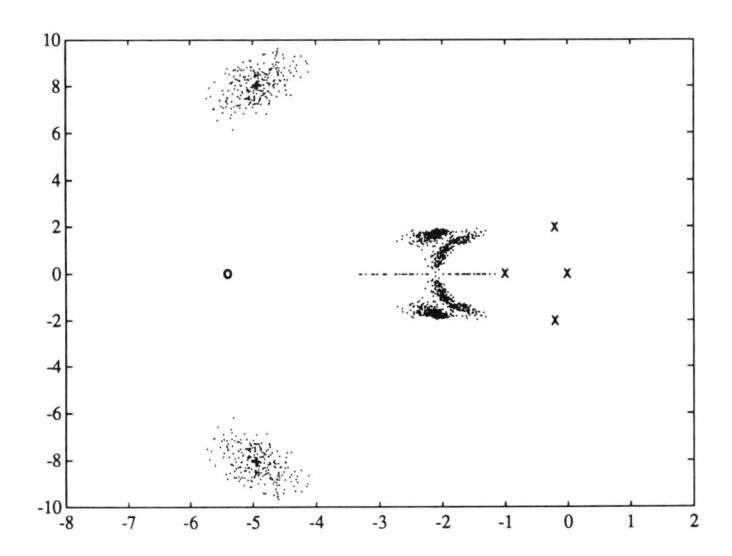

Figure 6-16 Designing with Algorithm 6.4-1: stability robustness check corresponding to ±5% variations of the nonzero parameters.

An Example. To compare the design techniques followed in Algorithms 6.3-1 and 6.3-2 (complete pole placement) with respect to that developed in Algorithm

6.4-1 (regulator zero placement) refer again to Example 6.3-1. In the corresponding mathematical model (6.3.9–11) the exosystem has already been extended to provide structural robustness. The poles to impose (for instance for a good regulation transient) are assumed to be:
- plant: $-1\pm j$, -2;
- regulator: $-1\pm 2j$, $-.8$, -1.5, -3, -6, -7.

In the root locus of Fig. 6-11 the open-loop poles are denoted by **x**, the open-loop zeros by **o**, and the above closed-loop poles by **+**; the plant has open-loop poles at $-.2\pm 2j$, -1, and a unique zero at -5.4, while the regulator has four open-loop poles at the origin (those of the internal model). The locations of the other zeros cannot be influenced in the design (they are imposed by the algorithm). Fig. 6-12 shows the spreading out of poles due to parameter variations (250 random changes of the nonzero coefficients of matrices A_1, A_3, B_1, L, M within $\pm 5\%$ of their nominal values). Similarly, design with the same close-loop pole locations was also performed with the Francis algorithm (Algorithm 6.3-2): the corresponding multivarible root locus and spreading out of poles are represented respectively in Figs. 6-13 and 6-14.

Referring to the same example, the design was repeated with Algorithm 6.4-1 and the four assignable regulator zeros were set at $-2\pm 2j$, $-2\pm 1.8j$. The corresponding root locus, shown in Fig. 6-15, appears to be quite robust with respect to gain changes. The pole locations shown in the figure correspond to $k=20$. Fig. 6-16 shows that the spreading out of poles (due to random changes of the nonzero coefficients of matrices A_1, A_3, B_1, L, M, K within $\pm 5\%$ of their nominal values) is significantly less in this case.

6.5 THE ROBUST CONTROLLED INVARIANT

Previously in this chapter a regulator has been considered robust if its behavior is good for a significantly large class of regulated systems or when parameters of a controlled plant vary "slowly" in time. A question now arises: what happens if parameters vary quickly or even are discontinuous in time like, for instance, when the plant can be "switched" from one configuration to another? The regulator can be robust in the sense that it behaves satisfactorily for all the allowable parameter configurations, but in general sudden change of structure or fast parameter variation causes a significant transient at the regulated output: if it is at steady state value (zero), in general it is subject to a pulse variation, then gradually returns to zero. In many applications it is desirable to eliminate this transient, particularly when the instants of time when configuration changes can be communicated to the regulator (for instance, action on flaps or gear of an airplane may be communicated to the autopilot to avoid any transient in the airplane attitude). If a regulator is robust also in this sense, i.e., with respect to very fast parameter variations or changes of structure, it is said to be *hyper-robust*.

To deal with hyper-robustness some new geometric concepts will be introduced, in particular the *robust controlled invariant* and the *robust self-*

bounded controlled invariant.[1] Refer to the controlled system

$$\dot{x}(t) = A(q)\,x(t) + B(q)\,u(t) \tag{1}$$

where $q \in \mathcal{Q}$ denotes a parameter subject to variation in time. The following definitions and properties extend the concept of controlled invariance to take into account "strong" or "fast" parameter variation.

Definition 6.5-1. Robust Controlled Invariant

Let $\mathcal{B}(p) := \operatorname{im} B(p)$. A subspace $\mathcal{V} \subseteq \mathcal{X}$ is called a *robust $(A(q), B(q))$-controlled invariant relative to $\mathcal{Q}$* if

$$A(q)\mathcal{V} \subseteq \mathcal{V} + \mathcal{B}(q) \quad \forall\, q \in \mathcal{Q} \tag{2}$$

Property 6.5-1.

Given a subspace $\mathcal{V} \in \mathcal{X}$ and any two instants of time t_0, t_1 with $T_1 > t_0$, for any initial state $x(t_0) \in \mathcal{V}$ and any $q \in \mathcal{Q}$ there exists at least one control function $u|_{[t_0, t_1]}$ such that the corresponding state trajectory $x|_{[t_0, t_1]}$ of system (1) completely belongs to $\mathcal{V}$ if and only if $\mathcal{V}$ is a robust controlled invariant.

Property 6.5-2.

Given a subspace $\mathcal{V} \in \mathcal{X}$, for any $q \in \mathcal{Q}$ there exists at least one state feedback matrix $F(q)$ such that

$$\bigl(A(q) + B(q)\,F(q)\bigr)\mathcal{V} \subseteq \mathcal{V} \tag{3}$$

if and only if $\mathcal{V}$ is a robust controlled invariant.

It is easy to check that the sum of any two robust $(A(q), B(q))$-controlled invariants is a robust $(A(q), B(q))$-controlled invariant, so that the set of all robust controlled invariants contained in a given subspace $\mathcal{E}$ is a semilattice with respect to $+, \subseteq$. Denote its supremum by $\max \mathcal{V}_R(A(q), B(q), \mathcal{E})$ (*maximal robust $(A(q), B(q))$-controlled invariant contained in $\mathcal{E}$*). Algorithm 4.1-2 for computation of $\max \mathcal{V}(A, B, \mathcal{E})$ (the maximal (A, B)-controlled invariant contained in $\mathcal{E}$) can be extended as follows to compute the supremum of the new semilattice with robustness.

Algorithm 6.5-1. The Maximal Robust Controlled Invariant

For simpler notation, given a family $\mathcal{W}(q)$ of subspaces of $\mathcal{X}$ depending on the parameter $q \in \mathcal{Q}$, we shall denote with $\overline{\mathcal{W}(q)}$ the intersection of all members of the family, i.e.

$$\overline{\mathcal{W}(q)} := \bigcap_{q \in \mathcal{Q}} \mathcal{W}(q)$$

---

[1] The robust controlled invariant and the robust self-bounded controlled invariant were introduced by Basile and Marro [1]. Algorithm 6.5-1 is due to Conte, Perdon, and Marro [3].

Subspace max $V_R(A(q), B(q), \mathcal{E})$ coincides with the last term of the sequence

$$Z_0 := \mathcal{E} \tag{4}$$
$$Z_i := \mathcal{E} \cap \overline{A^{-1}(q)(Z_{i-1} + B(q))} \qquad (i=1,\ldots,k) \tag{5}$$

where the value of $k \leq n-1$ is determined by condition $Z_{k+1} = Z_k$.

Proof. First note that $Z_i \subseteq Z_{i-1}$ $(i=1,\ldots,k)$. In fact, instead of (5), consider the recursion expression

$$Z'_i := Z'_{i-1} \cap \overline{A^{-1}(q)(Z'_{i-1} + B(q))} \qquad (i=1,\ldots,k) \tag{6}$$

with $Z'_0 := \mathcal{E}$, which defines a sequence such that $Z'_i \subseteq Z'_{i-1}$ $(i=1,\ldots,k)$; hence

$$\overline{A^{-1}(q)(Z'_i + B(q))} \subseteq \overline{A^{-1}(q)(Z'_{i-1} + B(q))} \qquad (i=1,\ldots,k)$$

This sequence is equal to (5): by induction, note that if $Z'_j = Z_j$ $(j=1,\ldots,i-1)$, also

$$Z'_i := \mathcal{E} \cap \overline{A^{-1}(q)(Z_{i-2} + B(q))} \cap \overline{A^{-1}(q)(Z_{i-1} + B(q))} = Z_i$$

being

$$\overline{A^{-1}(q)(Z_{i-2} + B(q))} \supseteq \overline{A^{-1}(q)(Z_{i-1} + B(q))}$$

If $Z_{k+1} = Z_k$, also $Z_j = Z_k$ for all $j > k+1$ and Z_k is a robust controlled invariant contained in $\mathcal{E}$. In fact, in such a case

$$Z_k = \mathcal{E} \cap \overline{A^{-1}(q)(Z_k + B(q))}$$

hence $Z_k \subseteq \mathcal{E}$, $A(q)Z_k \subseteq Z_k + B(q)$ for all $q \in Q$.

Since two subsequent subspaces are equal if and only if they have equal dimensions and the dimension of the first subspace is at most $n-1$, a robust controlled invariant is obtained in at most $n-1$ steps.

The last term of the sequence is the maximal robust $(A(q), B(q))$-controlled invariant contained in $\mathcal{E}$, as can again be proved by induction. Let V be another robust controlled invariant contained in $\mathcal{E}$: if $V \subseteq Z_{i-1}$, it follows that $V \subseteq Z_i$. In fact

$$V \subseteq \mathcal{E} \cap \overline{A^{-1}(q)(V + B(q))}$$
$$\subseteq \mathcal{E} \cap \overline{A^{-1}(q)(Z_{i-1} + B(q))} = Z_i \quad \square$$

If Q is a finite set, the above algorithm can be applied without any difficulty, since it reduces to a sequence of standard manipulations of subspaces (sum, inverse linear transformation, intersection). On the other hand, if Q is a compact set and $A(q), B(q)$ continuous functions of q (for instance, polynomial matrices in q), the most difficult step to overcome (and the only one that requires special procedures) is to compute the intersection of all the elements of a family of subspaces depending on parameter q. For this the following algorithm can be profitably used.

Algorithm 6.5-2. The Intersection Algorithm (Conte and Perdon)

A sequence of subspaces $\{\mathcal{W}_{(i,j)}\}$ converging to

$$\overline{A^{-1}(q)(\mathcal{Z}_{i-1} + \mathcal{B}(q))}$$

is computed as follows:

step 0: choose $q' \in \mathcal{Q}$ and set $\mathcal{W}_{i0} := A^{-1}(q')(\mathcal{Z}_{i-1} + \mathcal{B}(q'))$;

step j: denote respectively by $W_{i(j-1)}$ and Z two matrices whose columns span $\mathcal{W}_{i,j-1}$ and $\mathcal{Z}_{i-1}$, and consider

$$r_j(q) := \rho([A(q)W_{i(j-1)} \mid Z \mid B(q)]) - \rho([Z \mid B(q)])$$

($\rho(M)$ is the rank of matrix M), then:
if $r_j(q) = 0$ for all $q \in \mathcal{Q}$, stop;
if $r_j(q'') \neq 0$ for some $q'' \in \mathcal{Q}$, set $\mathcal{W}_{ij} := A^{-1}(q'')(\mathcal{Z}_{i-1} + \mathcal{B}(q''))$.

Proof. The sequence is decreasing, therefore it converges in a finite number of steps. If $r_j(q) = 0$ for all $q \in \mathcal{Q}$, then clearly $A(q)\mathcal{W}_{i,j-1} \subseteq (\mathcal{Z}_{i-1} + \mathcal{B}(q))$ for all $q \in \mathcal{Q}$ and

$$\mathcal{W}_{i,j-1} = \overline{A^{-1}(q)(\mathcal{Z}_{i-1} + \mathcal{B}(q))} \quad \square$$

The problem of computing the maximal robust controlled invariant is thus reduced to that of checking whether $r_j(q) = 0$ for all $q \in \mathcal{Q}$. One of the possible procedures is to discretize $\mathcal{Q}$. Unfortunately, in this case lack of dimension may occur only at some isolated points of $\mathcal{Q}$, which may not have been considered in the discretization, thus going undetected. However, because of rounding errors, in implementing the algorithm for intersection on digital computers it is necessary to introduce a suitable threshold in the linear dependence test that provides the sequence stop: this causes linear dependence, although occurring only at isolated points of $\mathcal{Q}$, to be detected also in small neighborhoods of these points. Hence, it is sufficient to discretize $\mathcal{Q}$ over a grid fine enough to guarantee detection of any lack of rank.

The assumption that $\mathcal{E}$ is constant is not particularly restrictive in practice: if $\mathcal{E}$ depends on q but has a constant dimension k_e, it is possible to refer matrices $A(q)$ and $B(q)$ to a basis with k_e elements belonging to $\mathcal{E}(q)$: with respect to this basis subspace $\mathcal{E}$ and the robust controlled invariant provided by Algorithm 6.5-1 are clearly constant, although they would depend on q in the original basis.

Also, self-bounded controlled invariants, defined in Subsection 4.1.2 and basic to deal with stabilizability in synthesis problems, can be extended for hyper-robust regulation.

Definition 6.5-2. Robust Self-Bounded Controlled Invariant

Let $\mathcal{V}_R^* := \max \mathcal{V}_R(A(q), B(q), \mathcal{E})$. A robust controlled invariant $\mathcal{V}$ contained in $\mathcal{E}$ (hence in $\mathcal{V}_R^*$) is said to be *self-bounded with respect to* $\mathcal{E}$ if for all initial states belonging to $\mathcal{V}$ and all admissible values of q any state trajectory that completely belongs to $\mathcal{E}$ (hence to $\mathcal{V}_R^*$) lies on $\mathcal{V}$.

The following properties are straightforward extensions of similar ones concerning the nonrobust case.

Property 6.5-3.

A robust controlled invariant $\mathcal{V}$ is self-bounded with respect to $\mathcal{E}$ if and only if

$$\mathcal{V} \supseteq \mathcal{V}_R^* \cap \mathcal{B}(p) \quad \forall q \in \mathcal{Q}$$

Property 6.5-4.

The set of all robust controlled invariants self-bounded with respect to a given subspace $\mathcal{E}$ is a lattice with respect to $\subseteq, +, \cap$.

Proof. Let $\mathcal{V}_1, \mathcal{V}_2$ be any two elements of the set referred to in the statement. Their sum is an element of the set since it is a robust controlled invariant and contains $\mathcal{V}_R^* \cap \mathcal{B}(p)$; it will be shown that their intersection is also an element of the set. Let $\mathcal{V} := \mathcal{V}_1 \cap \mathcal{V}_2$ and denote by $F(q)$ any matrix function of q such that $(A(q) + B(q)F(q))\mathcal{V}_R^* \subseteq \mathcal{V}_R^*$, which exists by virtue of Property 6.5-2: since $\mathcal{V}_1$ and $\mathcal{V}_2$ are self-bounded, they must be invariant under $A(q) + B(q)F(q)$, i.e., they satisfy

$$\bigl(A(q) + B(q)F(q)\bigr)\mathcal{V}_1 \subseteq \mathcal{V}_1, \quad \mathcal{V}_1 \supseteq \mathcal{V}_R^* \cap \mathcal{B}(q) \quad \forall q \in \mathcal{Q}$$

$$\bigl(A(q) + B(q)F(q)\bigr)\mathcal{V}_2 \subseteq \mathcal{V}_2, \quad \mathcal{V}_2 \supseteq \mathcal{V}_R^* \cap \mathcal{B}(q) \quad \forall q \in \mathcal{Q}$$

From these relations, it follows that

$$\bigl(A(q) + B(q)F(q)\bigr)\mathcal{V} \subseteq \mathcal{V}, \quad \mathcal{V} \supseteq \mathcal{V}_R^* \cap \mathcal{B}(q) \quad \forall q \in \mathcal{Q}$$

so that $\mathcal{V}$ is a robust controlled invariant, again by Property 6.5-2, and self-bounded with respect to $\mathcal{E}$ by Property 6.5-3. □

The supremum of the lattice of all robust controlled invariants self-bounded with respect to $\mathcal{E}$ is clearly $\mathcal{V}_R^*$. The following algorithm sets a numerical procedure for computation of the infimum of the lattice. It is a generalization of the well-known algorithm for computation of controllability subspaces [4.43]: this is consistent with the property, stated by Theorem 4.1-6 for constant systems, that the controllability subspace on a given controlled invariant coincides with the minimum self-bounded controlled invariant contained in it. For the sake of completeness, we shall refer to the more general case in which the lattice of all robust controlled invariants self-bounded with respect to $\mathcal{E}$ is constrained to contain a given subspace $\mathcal{D}$ which, of course, must be contained in $\mathcal{V}_R^*$ for the lattice to be nonempty. $\mathcal{D}$ can be assumed to be the origin when such a constraint does not exist.

Sect. 6.5 The Robust Controlled Invariant

Algorithm 6.5-3. The Minimal Robust Self-Bounded Controlled Invariant
Let $\mathcal{D} \subseteq \mathcal{V}_R^*$, with $\mathcal{V}_R^* := \max \mathcal{V}_R(A(q), B(q), \mathcal{E})$. Consider the sequence of subspaces

$$\mathcal{Z}_0 = \sum_{q \in \mathcal{Q}} \mathcal{V}_R^* \cap \bigl(B(q) + \mathcal{D}\bigr) \tag{7}$$

$$\mathcal{Z}_i = \sum_{q \in \mathcal{Q}} \mathcal{V}_R^* \cap \bigl(A(q)\,\mathcal{Z}_{i-1} + B(q) + \mathcal{D}\bigr) \quad (i=1,\ldots,k) \tag{8}$$

When $\mathcal{Z}_{i+1} = \mathcal{Z}_i$, stop. The last term of the sequence is $\mathcal{R}_R$, the infimum of the lattice of all robust controlled invariants self-bounded with respect to $\mathcal{E}$ and containing $\mathcal{D}$.

Proof. Sequence (7,8) converges in at most $n-1$ steps by reason of dimensionality because each term contains the previous one or is equal to it, and in case of equality the sequence is constant. Let

$$B_t(p) := B(q) + \mathcal{D} \quad \forall q \in \mathcal{Q}$$

so that

$$\mathcal{V}_R^* \cap B_t(q) = \mathcal{V}_R^* \cap B(q) + \mathcal{D} \quad \forall q \in \mathcal{Q}$$

since the intersection is distributive with respect to the sum because $\mathcal{D} \subseteq \mathcal{V}_R^*$. At the limit we have

$$\mathcal{R}_R = \sum_{q \in \mathcal{Q}} \mathcal{V}_R^* \cap \bigl(A(q)\,\mathcal{R}_R + B_t(q)\bigr) \tag{9}$$

First we prove that $\mathcal{R}_R$ is a robust controlled invariant. Since it is contained in the robust controlled invariant $\mathcal{V}_R^*$, it follows that

$$A(q)\,\mathcal{R}_R \subseteq \mathcal{V}_R^* + B_t(q) \quad \forall q \in \mathcal{Q}$$

Intersection with the trivial inclusion

$$A(q)\,\mathcal{R}_R \subseteq A(q)\mathcal{R}_R + B_t(q) \quad \forall q \in \mathcal{Q}$$

yields

$$A(q)\,\mathcal{R}_R \subseteq \bigl(\mathcal{V}_R^* + B_t(q)\bigr) \cap \bigl(A(q)\,\mathcal{R}_R + B_t(q)\bigr)$$
$$= \mathcal{V}_R^* \cap \bigl(A(q)\,\mathcal{R}_R + B_t(q)\bigr) + B_t(q) \quad \forall q \in \mathcal{Q}$$

The last equality follows from the intersection with the second sum being distributive with respect to the former (which contains $B_t(q)$) and from $B_t(q)$ being contained in the second sum. By suitably adding terms on the right, we can set the further relation

$$A(q)\,\mathcal{R}_R \subseteq \sum_{q \in \mathcal{Q}} \mathcal{V}_R^* \cap \bigl(A(q)\,\mathcal{R}_R + B_t(q)\bigr) + B_t(q) \quad \forall q \in \mathcal{Q}$$

which, by virtue of equality (9), proves that $\mathcal{R}_R$ is a robust $(A(q), B_t(q))$-controlled invariant, hence a robust $(A(q), B(q))$-controlled invariant, provided it contains $\mathcal{D}$. It is self-bounded too, since it contains $\mathcal{V}_R^* \cap B_t(q)$, hence $\mathcal{V}_R^* \cap B(q)$ for all admissible q.

It remains to prove that $\mathcal{R}_R$ is the minimum robust self-bounded controlled invariant contained in $\mathcal{E}$ and containing $\mathcal{D}$, i.e., it is contained in any other element $\mathcal{V}$ of the lattice. Such a $\mathcal{V}$ satisfies

$$A(q)\mathcal{V} \subseteq \mathcal{V} + \mathcal{B}_t(q) \quad \text{and} \quad \mathcal{V} \supseteq \mathcal{V}_R^* \cap \mathcal{B}_t(q) \quad \forall q \in \mathcal{Q}$$

Refer again to sequence (7, 8). Clearly $\mathcal{Z}_0 \subseteq \mathcal{V}$; by induction, suppose that $\mathcal{Z}_{i-1} \subseteq \mathcal{V}$, so that

$$A(q)\mathcal{Z}_{i-1} \subseteq \mathcal{V} + \mathcal{B}_t(q) \quad \forall q \in \mathcal{Q}$$

or

$$A(q)\mathcal{Z}_{i-1} + \mathcal{B}_t(q) \subseteq \mathcal{V} + \mathcal{B}_t(q) \quad \forall q \in \mathcal{Q}$$

By intersecting both members with $\mathcal{V}_R^*$, we obtain

$$\mathcal{V}_R^* \cap \left(A(q)\mathcal{Z}_{i-1} + \mathcal{B}_t(q)\right) \subseteq \mathcal{V}_R^* \cap \left(\mathcal{V} + \mathcal{B}_t(q)\right) = \mathcal{V} + \mathcal{V}_R^* \cap \mathcal{B}_t(q) = \mathcal{V} \quad \forall q \in \mathcal{Q}$$

and, by summing over q, $\mathcal{Z}_i \subseteq \mathcal{V}$. $\square$

6.5.1 The Hyper-Robust Disturbance Localization Problem

A typical application of the "robust" tools of the geometric approach is the *hyper-robust disturbance localization problem* by state feedback, which is a straightforward extension of the classic disturbance localization problem considered in Section 4.2. Now refer to the disturbed linear system

$$\dot{x}(t) = A(q)\,x(t) + B(q)\,u(t) + D(q)\,d(t) \tag{10}$$

$$e(t) = E(q)\,x(t) \tag{11}$$

and consider the problem of making the controlled output e insensitive to the disturbance input d for any admissible value of q by means of a suitable action on the control input u. Note that both d and q are nonmanipulable inputs but, while d is inaccessible, so that it cannot be used as an input for the controller, q is assumed to be completely accessible, possibly through a suitable identification process. Other information available to the controller is the state vector, possibly through an observer, so that u may be considered a function of both x and q.

The hyper-robust disturbance localization problem is said to have a solution if there exists at least one function $u(x,q)$ such that the zero-state response $e(\cdot)$ of system (10, 11) corresponding to any disturbance function $d(\cdot)$ is identically zero. As a consequence of the following Theorem 6.5-1 and Property 6.5-2, if the problem admits a solution, function $u(x,q)$ can be assumed to be linear in the state without any loss of generality. Let

$$\mathcal{D}(q) := \operatorname{im} D(q) \quad \text{and} \quad \mathcal{E}(q) := \ker E(q) \tag{12}$$

and assume that $\dim \mathcal{D}(q)$ and $\dim \mathcal{E}(q)$ are constant with respect to any parameter change, i.e.

$$\dim \mathcal{D}(q) = k_d, \quad \dim \mathcal{E}(q) = k_e \quad \forall q \in \mathcal{Q} \tag{13}$$

It being clearly necessary that

$$\mathcal{D}(q) \subseteq \mathcal{E}(q) \quad \forall q \in \mathcal{Q} \tag{14}$$

assuming that $\mathcal{D}$ and $\mathcal{E}$ are constant does not cause any loss of generality. In fact, if they are not, consider a new basis in the state space with k_d elements belonging to $\mathcal{D}(q)$, $k_e - k_d$ elements belonging to $\mathcal{E}(q)$, and the remaining elements chosen to be linearly independent of the previous ones: clearly, in this new basis $\mathcal{D}$ and $\mathcal{E}$ are constant. Our result is then stated as follows.

Theorem 6.5-1.
Consider system (10,11) with $\mathcal{D}$ and $\mathcal{E}$ not depending on parameter q. The hyper-robust disturbance localization problem has a solution if and only if

$$\mathcal{D} \subseteq \mathcal{V}_R^* \quad \text{with} \quad \mathcal{V}_R^* := \max \mathcal{V}_R(A(q), B(q), \mathcal{E}) \tag{15}$$

Proof. *Only if.* Recall that the state trajectory of a constant system can be controlled on a subspace starting at any initial state belonging to it only if it is a controlled invariant (Theorem 4.1-1). Clearly, the state trajectory of a system subject to parameter changes can be controlled on a subspace starting at any initial state belonging to it and for any admissible value of the parameter only if it is a robust controlled invariant. For any (fixed) value of the parameter, in order to obtain insensitivity to a general disturbance function $d(\cdot)$, the corresponding state trajectory must be kept on a subspace of $\mathcal{E}$ which necessarily has to be a controlled invariant containing $\mathcal{D}$. Since this controllability feature must also be preserved at any state when the parameter changes, the above controlled invariant must be robust, hence (15) is necessary.
If. By virtue of (2) at any state on a robust controlled invariant $\mathcal{V}$ there exists a control function $u(x,q)$ such that the corresponding state velocity belongs to $\mathcal{V}$. If (15) holds, such a control action derived in connection with $\mathcal{V}_R^*$ clearly solves the problem. □

Note that the "if" part of the proof suggests a practical implementation of the hyper-robust disturbance decoupling controller. However, if condition (12) holds, it is convenient, for every value of q, to use the minimum self-bounded $(A(q), B(q))$-controlled invariant contained in $\mathcal{E}$ and containing $\mathcal{D}$, which has a maximal stabilizability feature.

6.5.2 Some Remarks on Hyper-Robust Regulation

We recall that hyper-robust regulation is a robust regulation such that the steady state condition (regulated output at zero) is maintained also when parameters are subject to fast variations or the structure of the plant is suddenly changed. A necessary condition is stated by the following theorem.

Theorem 6.5-2.
Consider the autonomous regulator of Fig. 6-7 and assume that A_1, A_3, and B_1 depend on a parameter $q \in \mathcal{Q}$ and subspace $\mathcal{E}$ is constant. The hyper-robust autonomous regulation problem admits a solution only if

$$V_R^* + \mathcal{P}(q) = \mathcal{X} \quad \forall q \in \mathcal{Q} \tag{16}$$

where V_R^* denotes the maximal $(A(q), B(q))$-controlled invariant robust with respect to $\mathcal{Q}$ and contained in $\mathcal{E}$.

Proof. By contradiction, let $\bar{q} \in \mathcal{Q}$ be a value of the parameter such that (16) does not hold. Hence

$$\max \mathcal{V}(A(\bar{q}), B(\bar{q}), \mathcal{E}) + \mathcal{P}(\bar{q}) \subset \mathcal{X}$$

(note the strict inclusion) and necessary condition (6.2.15) is not satisfied at $\bar{q}$. □

REFERENCES

1. BASILE, G., and MARRO, G., "On the robust controlled invariant," *Systems & Control Letters*, vol. 9, no. 3, pp. 191–195, 1987.

2. BHATTACHARYYA, S.P., "Generalized controllability, (A, B)-invariant subspaces and parameter invariant control," *SIAM J. on Alg. Disc. Meth.*, no. 4, pp. 529–533, 1983.

3. CONTE, G., PERDON, A.M., and MARRO, G., "Computing the maximum robust controlled invariant subspace," *Systems & Control Letters*, forthcoming, 1991.

4. DAVISON, E.J., "The robust control of a servomechanism problem for linear time-invariant multivariable systems," *IEEE Trans. on Autom. Contr.*, no. 21, pp. 25–33, 1976.

5. DAVISON, E.J., and FERGUSON, I.J., "The design of controllers for the multivariable robust servomechanism problem using parameter optimization methods," *IEEE Trans. on Autom. Contr.*, vol. AC-26, no. 1, pp. 93–110, 1981.

6. DAVISON, E.J., and GOLDEMBERG, A., "Robust control of a general servomechanism problem: the servo compensator," *Automatica*, vol. 11, pp. 461–471, 1975.

7. DAVISON, E.J., and SCHERZINGER, B.M., "Perfect control of the robust servomechanism problem," *IEEE Trans. on Autom. Contr.*, vol. AC-32, no. 8, pp. 689–701, 1987.

8. DICKMAN, A., "On the robustness of multivariable linear feedback systems in state-space representation," *IEEE Trans. on Autom. Contr.*, vol. AC-32, no. 5, pp. 407–410, 1987.

9. DORATO, P., "A historical review of robust control," *IEEE Control Systems Magazine*, April 1987.

10. —, (editor), *Robust Control*, IEEE Press, New York, 1987.

11. DOYLE, J.C., and STEIN, G., "Multivariable feedback design: concepts for a classical/modern synthesis," *IEEE Trans. on Autom. Contr.*, vol. AC-26, no. 1, pp. 4–16, 1981.
12. FRANCIS, B.A., "The linear multivariable regulator problem," *SIAM J. Contr. Optimiz.*, vol. 15, no. 3, pp. 486–505, 1977.
13. FRANCIS, B., SEBAKHY, O.A., and WONHAM, W.M., "Synthesis of multivariable regulators: the internal model principle," *Applied Math. & Optimiz.*, vol. 1, no. 1, pp. 64–86, 1974.
14. FRANCIS, B.A., and WONHAM, W.M., "The internal model principle of control theory," *Automatica*, no. 12, pp. 457–465, 1976.
15. FUHRMANN, P.A., "Duality in polynomial models with some applications to geometric control theory," *IEEE Trans. Autom. Contr.*, vol. AC-26, no. 1, pp. 284–295, 1981.
16. GOLUB, G.H., and WILKINSON, J.H., "Ill-conditioned eigensystems and the computation of the Jordan canonical form," *SIAM Review*, vol. 18, no. 4, pp. 578–619, 1976.
17. GRASSELLI, O.M., "On the output regulation problem and its duality," *Ricerche di Automatica*, vol. 6, pp. 166–177, 1975.
18. —, "Steady-state output insensitivity to step-wise disturbances and parameter variations," *System Science*, vol. 2, pp. 13–28, 1976.
19. GRASSELLI, O.M., and LONGHI, S., "Robust linear multivariable regulators under perturbations of physical parameters," *Proceedings of the Joint Conference on New Trends in System Theory*, Genoa, Italy, July 1990.
20. HA, I.J., and GILBERT, E.G., "Robust tracking in nonlinear systems," *IEEE Trans. on Autom. Contr.*, vol. AC-32, no. 9, pp. 763–771, 1987.
21. ISIDORI, A., and BYRNES, C.I., "Output regulation in nonlinear systems," *IEEE Trans. on Autom. Contr.*, vol. 35, no. 2, pp. 131–140, 1990.
22. KHARGONEKAR, P.P., GEORGIOU, T.T., and PASCOAL, A.M., "On the robust stabilizability of linear time-invariant plants with unstructured uncertainty," *IEEE Trans. on Autom. Contr.*, vol. AC-32, no. 3, pp. 201–207, 1988.
23. LEITMANN, G., "Guaranteed asymptotic stability for some linear system with bounded uncertainties," *ASME J. Dynam. Syst., Meas., Contr.*, vol. 101, pp. 212–215, 1979.
24. —, "On the efficacy of nonlinear control in uncertain linear systems," *ASME J. Dynam. Syst., Meas., Contr.*, vol. 102, pp. 95–102, 1981.
25. MARTIN, J.M., "State-space measures for stability robustness," *IEEE Trans. on Autom. Contr.*, vol. AC-32, no. 6, pp. 509–512, 1987.
26. MOORE, B.C., and LAUB, A.J., "Computation of supremal (A, B)-invariant and controllability subspaces," *IEEE Trans. Autom. Contr.*, vol. AC-23, no. 5, pp. 783–792, 1978.
27. PEARSON, J.B., SHIELDS, R.W., and STAATS, P.W., "Robust solution to linear multivariable control problems," *IEEE Trans. on Autom. Contr.*, vol. AC-19, pp. 508–517, 1974.

28. PERNEBO, L., "An algebraic theory for design of controllers for linear multivariable systems – Part II: feedback realizations and feedback design," *IEEE Trans. Autom. Contr.*, vol. AC-26, no. 2, pp. 183–193, 1981.
29. SAFONOV, M.G., *Stability and Robustness of Multivariable Feedback Systems*, MIT Press, Cambridge, Mass., 1980.
30. SCHMITENDORF, W.E., "Design of observer-based robust stabilizing controllers," *Automatica*, vol. 24, no. 5, pp. 693–696, 1988.
31. SOLAK, M.K., "A direct computational method for determining the maximal (A, B)-invariant subspace contained in Ker$C$," *IEEE Trans. Autom. Contr.*, vol. AC-31, no. 4, pp. 349–352, 1986.
32. TSUI, C.C., "On robust observer compensator design," *Automatica*, vol. 24, no. 5, pp. 687–692, 1988.
33. ZHOU, K., and KHARGONEKAR, P.P., "Stability robustness bounds for linear state-space models with structured uncertainty," *IEEE Trans. on Autom. Contr.*, vol. AC-32, no. 7, pp. 621–623, 1987.

A
MATHEMATICAL BACKGROUND

The aim of this appendix is to provide a quick reference to standard mathematical background material for system and control theory and to trace out a suggested program for a preliminary study or an introductory course.

A.1 SETS, RELATIONS, FUNCTIONS

In this section some basic concepts of algebra are briefly recalled. They include the standard tools for finite-state system analysis, such as binary operations and transformations of sets and partitions.

The word *set* denotes a collection of objects, called *elements* or *members* of the set. Unless a different notation is expressly introduced, sets will be denoted by capital "calligraphic" letters ($\mathcal{X}, \mathcal{Y}, \ldots$), elements of sets or vectors by lower case italic letters ($x, y, \ldots$), numbers and scalars by lower case Greek or italic letters ($\alpha, \beta, \ldots; a, b, \ldots$), linear functions and matrices by capital italic letters ($A, B, \ldots$).

Symbol $\in$ denotes belonging, i.e., $x \in \mathcal{X}$ indicates that x is an element of the set $\mathcal{X}$. Symbol $\notin$ denotes nonbelonging. Particular sets, which will be referred to frequently, are **Z**, **R**, **C**, the sets of all real, integer, complex numbers.

A set is said to be *finite* if the number of its elements is finite. The following definition, called *extension axiom*, provides a connection between the concepts of belonging and equality of sets.

Definition A.1-1. Different Sets

For any two sets $\mathcal{X}$ and $\mathcal{Y}$, $\mathcal{X}$ is equal to $\mathcal{Y}$ ($\mathcal{X} = \mathcal{Y}$) if every element of $\mathcal{X}$ also belongs to $\mathcal{Y}$ and vice versa; if not, the considered sets are said to be *different*.

Notation $\mathcal{X} := \{x_1, x_2, \ldots, x_5\}$ is used to state that a particular set $\mathcal{X}$ is composed of the elements $x_1, x_2, \ldots, x_5$. A set can be specified also by stating a certain number of properties for its elements; in this case the word "class" is often used instead of "set." The corresponding notation is $\mathcal{X} := \{x : p_1(x), p_2(x), \ldots\}$ and is read "$\mathcal{X}$ is the set" (or the class) of all elements x such that statements $p_1(x), p_2(x), \ldots$ are true. For instance

$$\mathcal{X} := \{x : x = 2y,\ y \in \mathbf{Z},\ 0 \le x \le 10\}$$

denotes the set of all even numbers between 0 and 10.

To represent *intervals* defined in the set of real numbers, the shorter notations

$$[\alpha, \beta] := \{x : \alpha \le x \le \beta\}$$
$$(\alpha, \beta] := \{x : \alpha < x \le \beta\}$$
$$[\alpha, \beta) := \{x : \alpha \le x < \beta\}$$
$$(\alpha, \beta) := \{x : \alpha < x < \beta\}$$

are used for the sake of brevity; the first is a closed interval, the second and the third are half-closed intervals, the fourth is an open interval.

In general, symbols $\forall$ (for all), $\exists$ (there exists), $\ni$ (such that) are often used. To denote that the two assertions $p_1(x)$ and $p_2(x)$ are equivalent, i.e., imply each other, notation $p_1(x) \Leftrightarrow p_2(x)$ is used, while to denote that $p_1(x)$ implies $p_2(x)$ we shall write $p_1(x) \Rightarrow p_2(x)$.

Definition A.1-2. Subset

Given two sets $\mathcal{X}$ and $\mathcal{Y}$, $\mathcal{X}$ is said to be a *subset* of $\mathcal{Y}$ if every element of $\mathcal{X}$ is also an element of $\mathcal{Y}$.

In such a case $\mathcal{X}$ is said to be contained in $\mathcal{Y}$ or $\mathcal{Y}$ to contain $\mathcal{X}$, in symbols $\mathcal{X} \subseteq \mathcal{Y}$ or $\mathcal{Y} \supseteq \mathcal{X}$ if equality is not excluded. If, on the contrary, equality is excluded, $\mathcal{X}$ is said to be strictly contained in $\mathcal{Y}$ or $\mathcal{Y}$ to contain strictly $\mathcal{X}$, in symbols $\mathcal{X} \subset \mathcal{Y}$ or $\mathcal{Y} \supset \mathcal{X}$.

The set that contains no elements is said to be the *empty set* and denoted by $\emptyset$: $\emptyset \subseteq \mathcal{X}$ for all $\mathcal{X}$, i.e., the empty set is a subset of every set.

Definition A.1-3. Union of Sets

Given two sets $\mathcal{X}$ and $\mathcal{Y}$, the *union* of $\mathcal{X}$ and $\mathcal{Y}$ (in symbols $\mathcal{X} \cup \mathcal{Y}$) is the set of all elements belonging to $\mathcal{X}$ or to $\mathcal{Y}$, i.e.

$$\mathcal{X} \cup \mathcal{Y} := \{z : z \in \mathcal{X} \text{ or } z \in \mathcal{Y}\}$$

Sect. A.1 Sets, Relations, Functions

Definition A.1-4. Intersection of Sets
Given two sets $\mathcal{X}$ and $\mathcal{Y}$, the *intersection* of $\mathcal{X}$ and $\mathcal{Y}$ (in symbols $\mathcal{X} \cap \mathcal{Y}$) is the set of all elements belonging to $\mathcal{X}$ and to $\mathcal{Y}$, i.e.

$$\mathcal{X} \cap \mathcal{Y} := \{z : z \in \mathcal{X}, \, z \in \mathcal{Y}\}$$

Definition A.1-5. Difference of Sets
Given two sets $\mathcal{X}$ and $\mathcal{Y}$, the *difference* of $\mathcal{X}$ and $\mathcal{Y}$ (in symbols $\mathcal{X} - \mathcal{Y}$) is the set of all elements of $\mathcal{X}$ not belonging to $\mathcal{Y}$, i.e.

$$\mathcal{X} - \mathcal{Y} := \{z : z \in \mathcal{X}, \, z \notin \mathcal{Y}\}$$

Two sets $\mathcal{X}$ and $\mathcal{Y}$ are said to be *disjoint* if $\mathcal{X} \cap \mathcal{Y} = \emptyset$. The *complement* of $\mathcal{X}$ with respect to a given set $\mathcal{E}$ containing $\mathcal{X}$ is $\bar{\mathcal{X}} := \mathcal{E} - \mathcal{X}$.

(a)

(b)

(c)

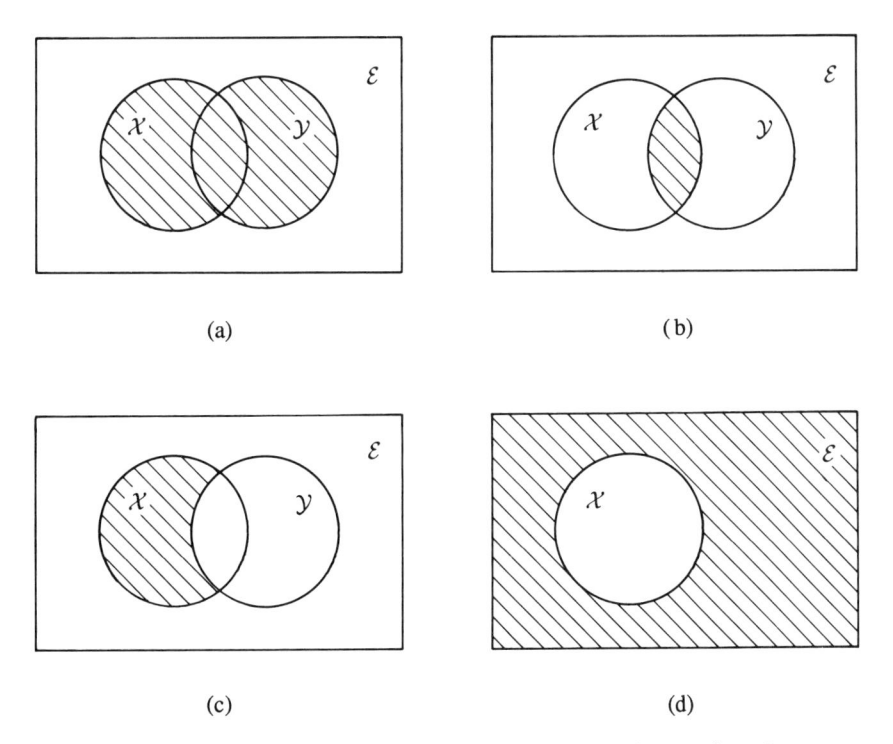

(d)

Figure A-1 Union, intersection, difference and complementation of sets.

A simple and intuitive way of illustrating these concepts is the use of the *Venn diagrams*, shown in Fig. A-1, which refer to sets whose elements are points of a plane, hence susceptible to an immediate graphical representation.

The identities

$$\mathcal{X} \subseteq \mathcal{Y} \text{ e } \mathcal{Y} \subseteq \mathcal{X} \Leftrightarrow \mathcal{X} = \mathcal{Y} \tag{1}$$
$$\mathcal{X} \subseteq \mathcal{Y} \text{ e } \mathcal{Y} \subseteq \mathcal{Z} \Rightarrow \mathcal{X} \subseteq \mathcal{Z} \tag{2}$$
$$\mathcal{X} \subseteq \mathcal{Y} \Leftrightarrow \mathcal{X} \cup \mathcal{Y} = \mathcal{Y} \tag{3}$$
$$\mathcal{X} \subseteq \mathcal{Y} \Leftrightarrow \mathcal{X} \cap \mathcal{Y} = \mathcal{X} \tag{4}$$

and the following properties of sets and operations with sets are easily proved by direct check using Venn diagrams:

1. The commutative laws for union and intersection:

$$\mathcal{X} \cup \mathcal{Y} = \mathcal{Y} \cup \mathcal{X} \tag{5}$$
$$\mathcal{X} \cap \mathcal{Y} = \mathcal{Y} \cap \mathcal{X} \tag{6}$$

2. The associative laws for union and intersection:

$$\mathcal{X} \cup (\mathcal{Y} \cup \mathcal{Z}) = (\mathcal{X} \cup \mathcal{Y}) \cup \mathcal{Z} \tag{7}$$
$$\mathcal{X} \cap (\mathcal{Y} \cap \mathcal{Z}) = (\mathcal{X} \cap \mathcal{Y}) \cap \mathcal{Z} \tag{8}$$

3. The distributivity of union with respect to intersection and intersection with respect to union:

$$\mathcal{X} \cup (\mathcal{Y} \cap \mathcal{Z}) = (\mathcal{X} \cup \mathcal{Y}) \cap (\mathcal{X} \cup \mathcal{Z}) \tag{9}$$
$$\mathcal{X} \cap (\mathcal{Y} \cup \mathcal{Z}) = (\mathcal{X} \cap \mathcal{Y}) \cup (\mathcal{X} \cap \mathcal{Z}) \tag{10}$$

4. The De Morgan laws:

$$\overline{\mathcal{X} \cup \mathcal{Y}} = \bar{\mathcal{X}} \cap \bar{\mathcal{Y}} \tag{11}$$
$$\overline{\mathcal{X} \cap \mathcal{Y}} = \bar{\mathcal{X}} \cup \bar{\mathcal{Y}} \tag{12}$$

and, in the end,

$$\mathcal{X} \subseteq \mathcal{Y} \Leftrightarrow \bar{\mathcal{X}} \supseteq \bar{\mathcal{Y}} \tag{13}$$

By virtue of the associative law, union and intersection can be defined also for a number of sets greater than two: given the sets $\mathcal{X}_1, \mathcal{X}_2, \ldots, \mathcal{X}_n$, their union and their intersection are denoted by

$$\bigcup_{i=1}^{n} \mathcal{X}_i \quad \text{and} \quad \bigcap_{i=1}^{n} \mathcal{X}_i$$

Sect. A.1 Sets, Relations, Functions

More generally, let $\mathcal{J}$ be a set such that $\mathcal{X}_i$ is a well-defined set for all $i \in \mathcal{J}$: it follows that

$$\bigcup_{i \in \mathcal{J}} \mathcal{X}_i := \{x : \exists\, i \in \mathcal{J} \ni x \in \mathcal{X}_i\} \tag{14}$$

$$\bigcap_{i \in \mathcal{J}} \mathcal{X}_i := \{x : x \in \mathcal{X}_i \;\forall\, i \in \mathcal{J}\} \tag{15}$$

Relations (9–12) can be generalized as follows:

$$\mathcal{X} \cup \left(\bigcap_{i \in \mathcal{J}} \mathcal{X}_i\right) = \bigcap_{i \in \mathcal{J}} (\mathcal{X} \cup \mathcal{X}_i) \tag{16}$$

$$\mathcal{X} \cap \left(\bigcup_{i \in \mathcal{J}} \mathcal{X}_i\right) = \bigcup_{i \in \mathcal{J}} (\mathcal{X} \cap \mathcal{X}_i) \tag{17}$$

$$\overline{\bigcup_{i \in \mathcal{J}} \mathcal{X}_i} = \bigcap_{i \in \mathcal{J}} \bar{\mathcal{X}}_i \tag{18}$$

$$\overline{\bigcap_{i \in \mathcal{J}} \mathcal{X}_i} = \bigcup_{i \in \mathcal{J}} \bar{\mathcal{X}}_i \tag{19}$$

Definition A.1-6. Ordered Pair
An *ordered pair* is a set of the type

$$(x, y) := \{\{x\}, \{x, y\}\}$$

where elements x, y are called respectively *first coordinate* and *second coordinate*.

Definition A.1-7. Cartesian Product of Sets
Given two nonvoid sets $\mathcal{X}$ and $\mathcal{Y}$, the *cartesian product* of $\mathcal{X}$ and $\mathcal{Y}$ (in symbols $\mathcal{X} \times \mathcal{Y}$) is the set of all ordered pairs whose first coordinate belongs to $\mathcal{X}$, and the second to $\mathcal{Y}$, i.e.

$$\mathcal{X} \times \mathcal{Y} := \{(x, y) : x \in \mathcal{X},\ y \in \mathcal{Y}\}$$

The cartesian product is distributive with respect to union, intersection, and difference: in other words, referring to the sole union for the sake of simplicity, the following relations hold:

$$\mathcal{X} \times (\mathcal{Y} \cup \mathcal{Z}) = (\mathcal{X} \times \mathcal{Y}) \cup (\mathcal{X} \times \mathcal{Z})$$
$$(\mathcal{X} \cup \mathcal{Y}) \times \mathcal{Z} = (\mathcal{X} \times \mathcal{Z}) \cup (\mathcal{Y} \times \mathcal{Z})$$

The cartesian product can be extended to involve a number of sets greater than two: $\mathcal{X} \times \mathcal{Y} \times \mathcal{Z}$ denotes, for instance, the set of ordered triples whose elements belong respectively to $\mathcal{X}$, $\mathcal{Y}$, and $\mathcal{Z}$. The sets in the product may be equal: $\mathcal{X} \times \mathcal{X}$ or $\mathcal{X}^2$ means the set of all ordered pairs of elements of $\mathcal{X}$, even repeated.

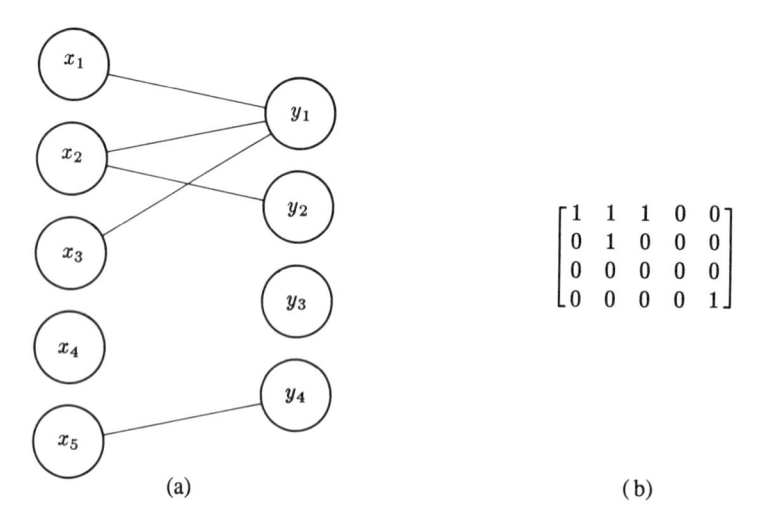

Figure A-2 Graph and adjacency matrix of a relation from $\mathcal{X}$ to $\mathcal{Y}$.

Definition A.1-8. Relation

Given two nonvoid sets $\mathcal{X}$ and $\mathcal{Y}$, any subset r of $\mathcal{X} \times \mathcal{Y}$ is called a (binary) *relation* from $\mathcal{X}$ to $\mathcal{Y}$. If $\mathcal{Y} = \mathcal{X}$, the relation is a subset of $\mathcal{X} \times \mathcal{X}$ and is called *relation in* $\mathcal{X}$.

Example. In a set of people $\mathcal{X}$, relationship is a relation defined by

$$r := \{(x_i, x_j) : x_i \text{ is relative of } x_j\} \tag{20}$$

The *domain* of a relation r from $\mathcal{X}$ to $\mathcal{Y}$ is the set

$$\mathcal{D}(r) := \{x : \exists\, y \ni (x, y) \in r\}$$

while its *codomain* or *range* is

$$\mathcal{C}(r) := \{y : \exists\, x \ni (x, y) \in r\}$$

and the *inverse* of a relation r (denoted by r^{-1}) is defined as

$$r^{-1} := \{(y, x) : (x, y) \in r\}$$

Clearly, $\mathcal{D}(r^{-1}) = \mathcal{C}(r)$, $\mathcal{C}(r^{-1}) = \mathcal{D}(r)$.

The *identity* relation on a set $\mathcal{X}$ is

$$i := \{(x, x) : x \in \mathcal{X}\}$$

Given two relations r and s, the former from $\mathcal{X}$ to $\mathcal{Y}$, the latter from $\mathcal{Y}$ to $\mathcal{Z}$, their *composition* or *product* $s \circ r$ is

$$s \circ r := \{(x, z) : \exists\, y \ni (x, y) \in r, (y, z) \in s\} \tag{21}$$

Sect. A.1 Sets, Relations, Functions

A relation can be defined in several ways. Given the sets $\mathcal{X} = \{x_1, x_2, x_3, x_4, x_5\}$ and $\mathcal{Y} = \{y_1, y_2, y_3, y_4\}$, any relation from $\mathcal{X}$ to $\mathcal{Y}$ can be specified through a *graph*, i.e., a collection of *nodes* or *vertexes* joined to each other by *branches* or *edges* that single out the pairs belonging to the relation [see Fig. A-2(a)], or through a matrix R, called *adjacency matrix*, with as many rows as there are elements of $\mathcal{X}$ and as many columns as there are elements of $\mathcal{Y}$ and such that the generic element r_{ij} is 1 if $(x_j, y_i) \in r$, 0 if $(x_j, y_i) \notin r$ [see Fig. A-2(b)].

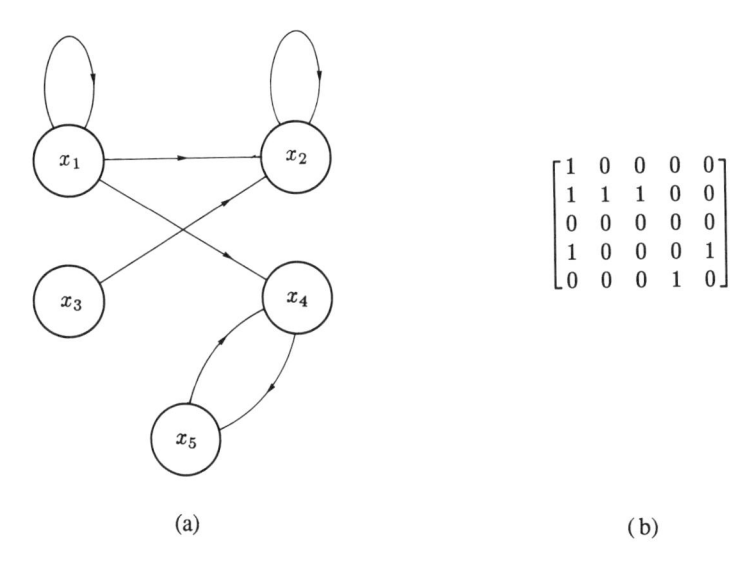

$$\begin{bmatrix} 1 & 0 & 0 & 0 & 0 \\ 1 & 1 & 1 & 0 & 0 \\ 0 & 0 & 0 & 0 & 0 \\ 1 & 0 & 0 & 0 & 1 \\ 0 & 0 & 0 & 1 & 0 \end{bmatrix}$$

(a)   (b)

Figure A-3 Graph and adjacency matrix of a relation in $\mathcal{X}$.

A relation in $\mathcal{X}$ can be represented by means of an *oriented graph* (i.e., a graph whose branches are given a direction by means of an arrow) having as many nodes as there are elements of $\mathcal{X}$, instead of a double number. Such a graph and the corresponding adjacency matrix are shown in Fig. A-3.

Definition A.1-9. Function (Map, Transformation, Operator)

Given two nonvoid sets $\mathcal{X}$ and $\mathcal{Y}$, a *function* (or *map*, *transformation*, *operator*) is a relation f from $\mathcal{X}$ to $\mathcal{Y}$ such that
1. $\mathcal{D}(f) = \mathcal{X}$
2. there are no elements of f with the same first coordinate, i.e. $(x, y_i) \in f, (x, y_j) \in f \Rightarrow y_i = y_j$.

For instance, the relation represented in Fig. A-2(a) is not a function, since $x_4 \notin \mathcal{D}(f)$ and it contains pairs (x_2, y_1) and (x_2, y_2), which have the first coordinates equal and the second different.

If $(x, y) \in f$, y is said to be the *image* of x in f or the *value* of f at x and notation $y = f(x)$ is used, while function as a correspondence between sets is denoted by

$$f : \mathcal{X} \to \mathcal{Y} \quad \text{or} \quad \mathcal{X} \xrightarrow{f} \mathcal{Y}$$

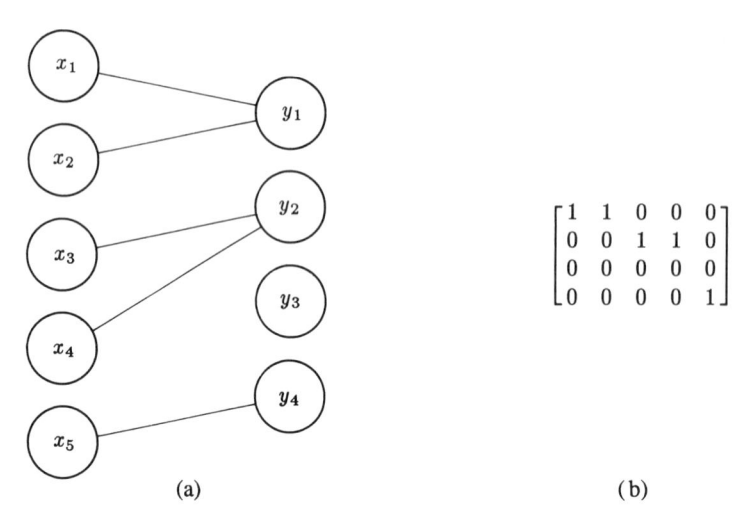

$$\begin{bmatrix} 1 & 1 & 0 & 0 & 0 \\ 0 & 0 & 1 & 1 & 0 \\ 0 & 0 & 0 & 0 & 0 \\ 0 & 0 & 0 & 0 & 1 \end{bmatrix}$$

(a)   (b)

Figure A-4 Graph and matrix of a function from $\mathcal{X}$ to $\mathcal{Y}$.

Referring to the representations shown in Fig. A-2, it can be argued that a relation is a function if and only if no more than one branch leaves any node x_i of the graph or, equivalently, if and only if each column of the adjacency matrix has no more than one element different from zero: this indeed happens in the case referred to in Fig. A-4.

A simpler representation is the so-called *function table* or *correspondence table* shown in Fig. A-5(a). For a function $z = f(x, y)$ whose domain is the cartesian product of two sets $\mathcal{X}$ and $\mathcal{Y}$, a table of the type shown in Fig. A-5(b) can be used.

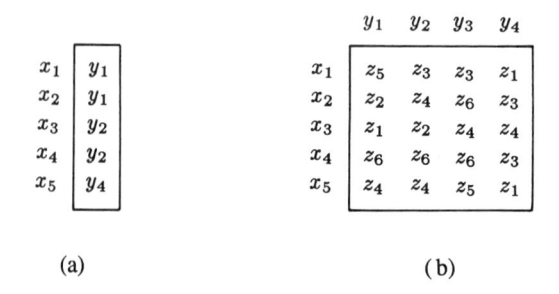

	y_1	y_2	y_3	y_4
x_1	z_5	z_3	z_3	z_1
x_2	z_2	z_4	z_6	z_3
x_3	z_1	z_2	z_4	z_4
x_4	z_6	z_6	z_6	z_3
x_5	z_4	z_4	z_5	z_1

x_1	y_1
x_2	y_1
x_3	y_2
x_4	y_2
x_5	y_4

(a)   (b)

Figure A-5 Tables of functions.

Sect. A.1 Sets, Relations, Functions

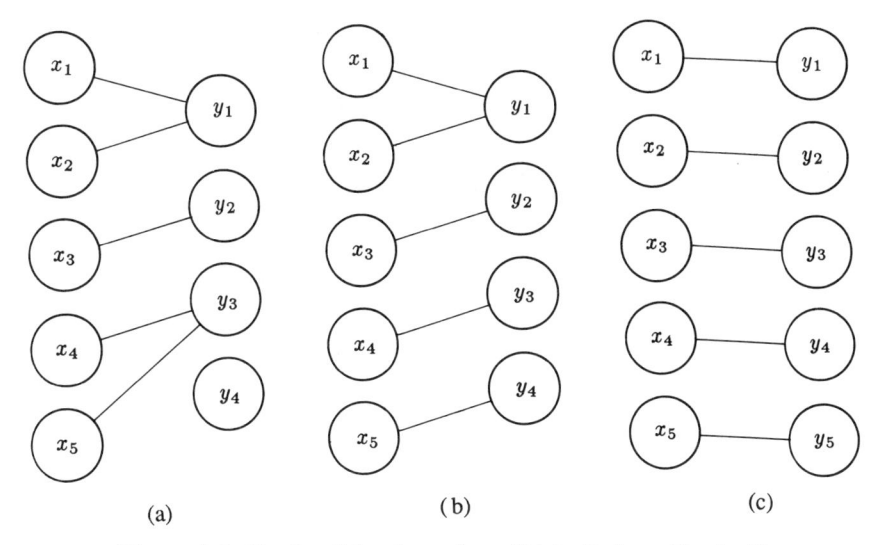

Figure A-6 Graphs of functions: from $\mathcal{X}$ into $\mathcal{Y}$, from $\mathcal{X}$ onto $\mathcal{Y}$, and one-to-one.

For any function $f: \mathcal{X} \to \mathcal{Y}$ by definition $\mathcal{D}(f) = \mathcal{X}$. Given any subset $\mathcal{Z} \in \mathcal{X}$, the *image* of $\mathcal{Z}$ in f is

$$f(\mathcal{Z}) := \{y : y = f(x),\ x \in \mathcal{Z}\}$$

The *image of the function*, im f, is $f(\mathcal{X})$, i.e., the image of its domain. In general $f(\mathcal{X}) \subseteq \mathcal{Y}$ and f is called a function from $\mathcal{X}$ *into* $\mathcal{Y}$ [see Fig. A-6(a)]; if $f(\mathcal{X}) = \mathcal{Y}$, f is called a map of $\mathcal{X}$ *onto* $\mathcal{Y}$ or a *surjection* [see Fig. A-6(b)].

A function $f: \mathcal{X} \to \mathcal{Y}$ is said to be *one-to-one* or an *injection* if $f(x) = f(z)$ implies $x = z$ for all pairs $(x, z) \in \mathcal{X}$ [see Fig. A-6(c)].

A function $f: \mathcal{X} \to \mathcal{Y}$ that is onto and one-to-one is called *invertible*. In fact, it is possible to define its *inverse map* f^{-1} as the unique function such that $y = f(f^{-1}(y))$ for all $y \in \mathcal{Y}$.

Given two functions f and g, their *composed function* $g \circ f$ is defined as the composed relation (21). If $y = f(x)$, $z = g(y)$, notation $z = g(f(x))$ is used. The composed function is invertible if and only if both f and g are invertible.

Given any subset $\mathcal{Z} \subseteq f(\mathcal{X})$, the *inverse image* of $\mathcal{Z}$ in the map f is

$$f^{-1}(\mathcal{Z}) := \{x : y = f(x),\ y \in \mathcal{Z}\}$$

Note that in order to define the inverse image of a set in a map, the map need not be invertible.

The following relations hold.

$$f(\mathcal{X}_1 \cup \mathcal{X}_2) = f(\mathcal{X}_1) \cup f(\mathcal{X}_2) \tag{22}$$

$$f(\mathcal{X}_1 \cap \mathcal{X}_2) \subseteq f(\mathcal{X}_1) \cap f(\mathcal{X}_2) \tag{23}$$
$$f^{-1}(\mathcal{Y}_1 \cup \mathcal{Y}_2) = f^{-1}(\mathcal{Y}_1) \cup f^{-1}(\mathcal{Y}_2) \tag{24}$$
$$f^{-1}(\mathcal{Y}_1 \cap \mathcal{Y}_2) = f^{-1}(\mathcal{Y}_1) \cap f^{-1}(\mathcal{Y}_2) \tag{25}$$

A.1.1 Equivalence Relations and Partitions

Definition A.1-10. **Equivalence Relation**

Given a set $\mathcal{X}$, a relation r in $\mathcal{X}$ is an *equivalence relation* if
1. it is reflexive, i.e. $(x, x) \in r$ for all $x \in \mathcal{X}$;
2. it is symmetric, i.e. $(x, y) \in r \Leftrightarrow (y, x) \in r$;
3. it is transitive, i.e. $(x, y) \in r, (y, z) \in r \Rightarrow (x, z) \in r$.

Example. The relationship in (20) is an equivalence relation.

An equivalence relation is often denoted by the symbol $\equiv$. Thus, instead of $(x, y) \in r$, notation $x \equiv y$ is used.

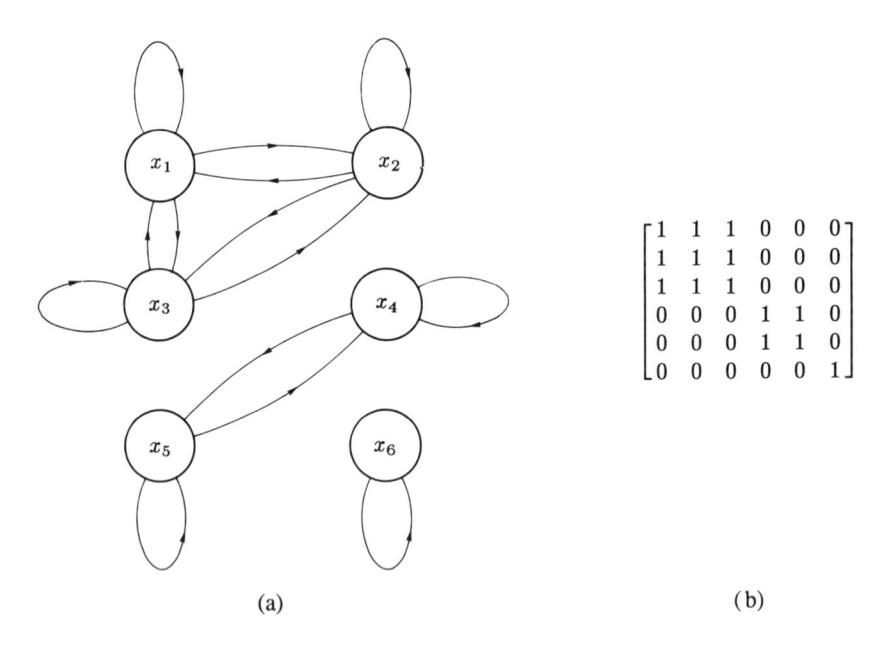

$$\begin{bmatrix} 1 & 1 & 1 & 0 & 0 & 0 \\ 1 & 1 & 1 & 0 & 0 & 0 \\ 1 & 1 & 1 & 0 & 0 & 0 \\ 0 & 0 & 0 & 1 & 1 & 0 \\ 0 & 0 & 0 & 1 & 1 & 0 \\ 0 & 0 & 0 & 0 & 0 & 1 \end{bmatrix}$$

(a)   (b)

Figure A-7 Graph and matrix of an equivalence relation.

The oriented graph of an equivalence relation has the particular shape shown in Fig. A-7(a), where
1. every node is joined to itself by a branch, i.e., every node has a *self-loop*;
2. the presence of any branch implies that of an opposite branch between the same nodes;
3. any two nodes joined to each other must be connected to all nodes connected to them.

Hence, the graph presents disjoint sets of nodes whose elements are connected to each other in all possible ways.

The matrix also has a particular shape. In fact:
1. $r_{ii}=1$ for all i (all the main diagonal elements are 1);
2. $r_{ij}=r_{ji}$ for all i,j (matrix is symmetric);
3. $r_{ij}=r_{jk}$ implies $r_{ik}=r_{ij}$ (or $r_{ik}=r_{jk}$)) for all i,j,k.

By a proper ordering of rows and columns, the matrix of any equivalence relation can be given the structure shown in Fig. A-7(b).

Definition A.1-11. Partition
Given a set $\mathcal{X}$, a *partition* P in $\mathcal{X}$ is a set of nonvoid subsets $\mathcal{X}_1, \mathcal{X}_2, \ldots$ of $\mathcal{X}$ such that
1. $\mathcal{X}_i \cap \mathcal{X}_j = \emptyset$ for $i \neq j$;
2. $\bigcup_i \mathcal{X}_i = \mathcal{X}$.

Sets $\mathcal{X}_1, \mathcal{X}_2, \ldots$ are called the *blocks* of partition P.

Theorem A.1-1.
To any equivalence relation in a given nonvoid set $\mathcal{X}$ there corresponds a partition of $\mathcal{X}$ and vice versa.

Proof. By considering properties of the graph shown in Fig. A-7(a), it is clear that any equivalence relation defines a partition. On the other hand, given any partition $P = \{\mathcal{X}_1, \mathcal{X}_2, \ldots, \mathcal{X}_n\}$, relation

$$r = \{(x,y) : \exists\, i \ni x,y \in \mathcal{X}_i\}$$

is reflexive, symmetric, and transitive, hence an equivalence relation. □

The blocks of the partition induced by an equivalence relation are called *equivalence classes*. An equivalence class is singled out by specifying any element belonging to it.

Graphs of the type shown in Fig. A-7(a) are not used for equivalence relations since it is sufficient to specify their node partitions (i.e., the equivalence classes). Notation $P = \{x_1, x_2, x_3; x_4, x_5; x_6\}$ is more convenient.

A.1.2 Partial Orderings and Lattices

Definition A.1-12. Partial Ordering
Given a set $\mathcal{X}$, a relation r in $\mathcal{X}$ is a *partial ordering* if
1. it is reflexive, i.e. $(x,x) \in r$ for all $x \in \mathcal{X}$;
2. it is antisymmetric, i.e. $(x,y) \in r$, $x \neq y \Rightarrow (y,x) \notin r$;
3. it is transitive, i.e. $(x,y) \in r$, $(y,z) \in r \Rightarrow (x,z) \in r$.

A partial ordering is usually denoted by symbols $\leq$, $\geq$. Thus, instead of $(x,y) \in r$, notation $x \leq y$ or $x \geq y$ is used. A set with a partial ordering is called a *partially ordered set*.

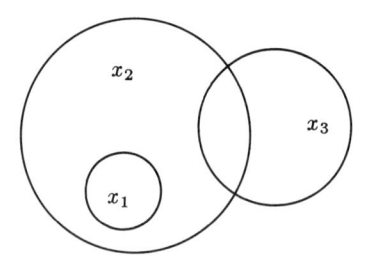

Figure A-8 Partially ordered sets.

Example A.1-1.
In a set of people $\mathcal{X}$, descend is a partial ordering.

Example A.1-2.
Consider a set S of subsets of a given set $\mathcal{X}$: inclusion relation $\subseteq$ is a partial ordering in S. Referring to Fig. A-8, we have $(x_1, x_2) \in r$, i.e., $x_1 \leq x_2$, $(x_2, x_1) \notin r$, $(x_1, x_3) \notin r$, $(x_3, x_1) \notin r$, $(x_2, x_3) \notin r$, $(x_3, x_2) \notin r$.

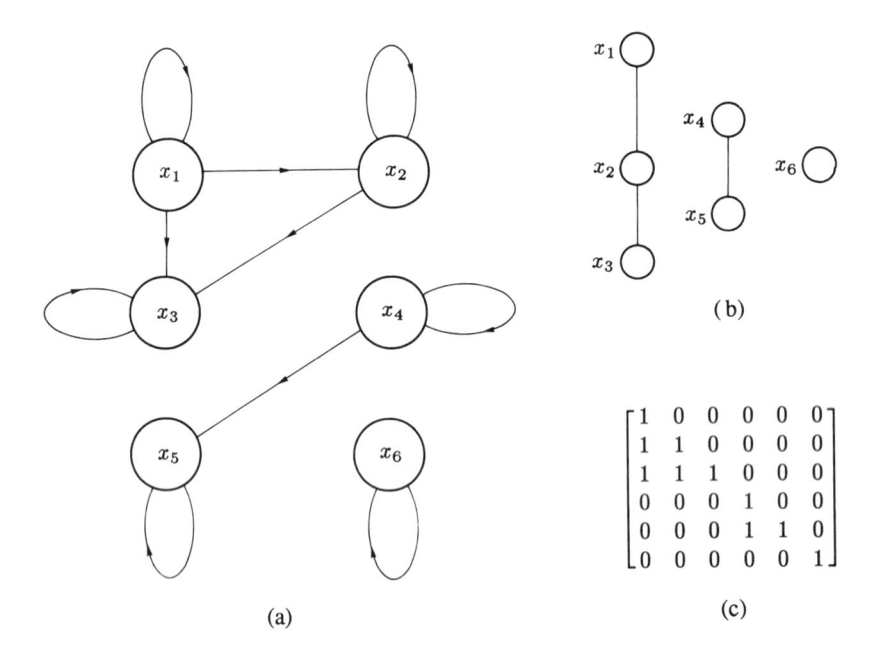

Figure A-9 Graph, Hasse diagram, and matrix of a partial ordering.

The oriented graph of a partial ordering has the particular shape shown in Fig. A-9(a), where

1. every node has a self-loop;
2. the presence of any branch excludes that of an opposite branch between the same nodes;

Sect. A.1 Sets, Relations, Functions

3. any two nodes joined to each other by a *path* (sequence of oriented branches) are also directly connected by a single branch having the same orientation.

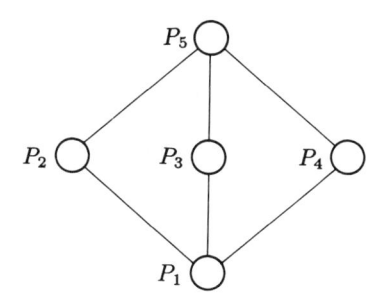

Figure A-10 Hasse diagram of partitions of a set with three elements.

A partial ordering relation is susceptible to a simpler representation through a *Hasse diagram*, where self-loops and connections implied by other connections are not shown and node x_i is represented below x_j if $x_i \leq x_j$. The Hasse diagram represented in Fig. A-9(b) corresponds to the graph shown in Fig. A-9(a).

The matrix of a partial ordering relation [see Fig. A-9(c)] also has a particular shape. In fact:
1. $r_{ii}=1$ for all i;
2. $r_{ij}=1$ implies $r_{ji}=0$ for all i,j;
3. $r_{ij}=r_{jk}=1$ implies $r_{ik}=r_{ij}$ (or $r_{ik}=r_{jk}$) for all i,j,k.

Example. Given a set $\mathcal{X}$, in the set of all partition of $P_1, P_2, \ldots$ of $\mathcal{X}$ a partial ordering can be defined by stating that $P_i \leq P_j$ if every block of P_i is contained in a block of P_j. In particular, let $\mathcal{X} = \{x_1, x_2, x_3\}$: the partitions of $\mathcal{X}$ are $P_1 = \{x_1; x_2; x_3\}$, $P_2 = \{x_1, x_2; x_3\}$, $P_3 = \{x_1; x_2, x_3\}$, $P_4 = \{x_1, x_3; x_2\}$, $P_5 = \{x_1, x_2, x_3\}$: and their Hasse diagram is shown in Fig. A-10.

Definition A.1-13. Lattice

A *lattice* $\mathcal{L}$ is a partially ordered set in which for any pair $x, y \in \mathcal{L}$ there exists a *least upper bound* (l.u.b.), i.e., an $\eta \in \mathcal{L}$ such that $\eta \geq x$, $\eta \geq y$ and $z \geq \eta$ for all $z \in \mathcal{L}$ such that $z \geq x$, $z \geq y$, and a *greatest lower bound* (g.l.b.), i.e., an $\epsilon \in \mathcal{L}$ such that $x \geq \epsilon$, $y \geq \epsilon$ and $\epsilon \geq z$ for all $z \in \mathcal{L}$ such that $z \leq x$, $z \geq y$.

Example A.1-3.

The set of all subsets of a given set $\mathcal{X}$, with the partial ordering relation induced by inclusion $\subseteq$, is a lattice. Its Hasse diagram in the case of a set with three elements is shown in Fig. A-11.

Example A.1-4.

The set of all partitions $P_1, P_2, \ldots$ of a given set $\mathcal{X}$, with the above specified partial ordering, is a lattice.

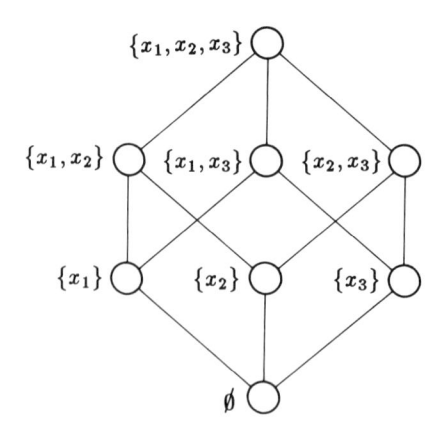

Figure A-11 The lattice of all subsets of a set with three elements.

In a lattice, two binary operations, which will be called addition and multiplication and denoted with symbols + and ·, can be defined through the relations $x+y=\eta$ and $x \cdot y=\epsilon$, i.e., as the operations that associate to any pair of elements their g.l.b. and l.u.b. It is easily shown that + and · satisfy

1. idempotency:

$$x + x = x \quad \forall x \in \mathcal{L}$$
$$x \cdot x = x \quad \forall x \in \mathcal{L}$$

2. commutative laws:

$$x + y = y + x \quad \forall x, y \in \mathcal{L}$$
$$x \cdot y = y \cdot x \quad \forall x, y \in \mathcal{L}$$

3. associative laws:

$$x + (y + z) = (x + y) + z \quad \forall x, y, z \in \mathcal{L}$$
$$x \cdot (y \cdot z) = (x \cdot y) \cdot z \quad \forall x, y, z \in \mathcal{L}$$

4. absorption laws:

$$x + (x \cdot y) = x \quad \forall x, y \in \mathcal{L}$$
$$x \cdot (x + y) = x \quad \forall x, y \in \mathcal{L}$$

A lattice is called *distributive* if distributive laws hold, i.e.

$$x + (y \cdot z) = (x + y) \cdot (x + z) \quad \forall x, y, z \in \mathcal{L}$$
$$x \cdot (y + z) = (x \cdot y) + (x \cdot z) \quad \forall x, y, z \in \mathcal{L}$$

Since + and · are associative, any finite subset of a lattice has a least upper bound and a greatest lower bound; in particular, any finite lattice has a *universal upper bound* or, briefly, a *supremum I* (the sum of all its elements) and an *universal lower bound* or, briefly, an *infimum O* (the product of all its elements), which satisfy

$$I + x = I, \quad O + x = x \quad \forall x \in \mathcal{L}$$
$$I \cdot x = x, \quad O \cdot x = O \quad \forall x \in \mathcal{L}$$

Also a nonfinite lattice may admit universal bounds. In the case of Example A.1-3, the binary operations are union and intersection and the universal bounds are $\mathcal{X}$ and $\emptyset$. In Example A.1-4 the operations are the sum of partitions, defined as the maximal partition (i.e., the partition with the maximum number of blocks) whose blocks are unions of blocks of all partitions to be summed, and the product of partitions, defined as the minimal partition whose blocks are intersections of blocks of all partitions to be multiplied; the lattice is nondistributive and the universal bounds are partition P_M (with $\mathcal{X}$ as the only block) and P_m (with all elements of $\mathcal{X}$ as blocks), called the *maximal partition* and the *minimal partition* respectively.

A.2 FIELDS, VECTOR SPACES, LINEAR FUNCTIONS

The material presented in this section and in the following one is a selection of topics of linear algebra in which the most important concepts and the formalism needed in general system theory are particularly stressed: vector spaces, subspaces, linear transformations, and matrices.

Definition A.2-1. Field
A *field* $\mathcal{F}$ is a set, whose elements are called *scalars*, with two binary operations + (addition) and · (multiplication) characterized by the following properties:

1. commutative laws:

$$\alpha + \beta = \beta + \alpha \quad \forall \alpha, \beta \in \mathcal{F}$$
$$\alpha \cdot \beta = \beta \cdot \alpha \quad \forall \alpha, \beta \in \mathcal{F}$$

2. associative laws:

$$\alpha + (\beta + \gamma) = (\alpha + \beta) + \gamma \quad \forall \alpha, \beta, \gamma \in \mathcal{F}$$
$$\alpha \cdot (\beta \cdot \gamma) = (\alpha \cdot \beta) \cdot \gamma \quad \forall \alpha, \beta, \gamma \in \mathcal{F}$$

3. distributivity of multiplication with respect to addition:

$$\alpha \cdot (\beta + \gamma) = \alpha \cdot \beta + \alpha \cdot \gamma \quad \forall \alpha, \beta, \gamma \in \mathcal{F}$$

4. the existence of a neutral element for addition, i.e., of a unique scalar $0 \in \mathcal{F}$ (called *zero*) such that
$$\alpha + 0 = \alpha \quad \forall \alpha \in \mathcal{F}$$
5. the existence of the opposite of any element: for all $\alpha \in \mathcal{F}$ there exists a unique scalar $-\alpha \in \mathcal{F}$ such that $\alpha + (-\alpha) = 0$
6. the existence of a neutral element for multiplication, i.e., of a unique scalar $1 \in \mathcal{F}$ (called *one*) such that
$$\alpha \cdot 1 = \alpha \quad \forall \alpha \in \mathcal{F}$$
7. the existence of the inverse of any element: for all $\alpha \in \mathcal{F}$, $\alpha \neq 0$ there exists a unique scalar $\alpha^{-1} \in \mathcal{F}$ such that $\alpha \cdot \alpha^{-1} = 1$.

Example A.2-1.
    The set of all real numbers is a field, which is denoted by **R**.
Example A.2-2.
    The set of all complex numbers is a field, which is denoted by **C**.
Example A.2-3.
    The set $\mathbf{B} := \{0, 1\}$, with $+$ and $\cdot$ defined as in the tables shown in Fig. A-12, is also a field, which will be denoted by **B**.

Definition A.2-2. Vector Space
    A *vector space* $\mathcal{V}$ over a field $\mathcal{F}$ is a set, whose elements are called *vectors*, with two binary operations $+$ (addition) and $\cdot$ (multiplication by scalars or external product) characterized by the following properties:

1. commutative law of addition:
$$x + y = y + x \quad \forall x, y \in \mathcal{V}$$
2. associative law of addition:
$$x + (y + z) = (x + y) + z \quad \forall x, y, z \in \mathcal{V}$$
3. the existence of a neutral element for addition, i.e., of a unique vector 0 (called the *origin*) such that
$$x + 0 = x \quad \forall x \in \mathcal{V}$$
4. the existence of the opposite of any element: $\forall x \in \mathcal{V}$ there exists a unique element $-x \in \mathcal{V}$ such that $x + (-x) = 0$
5. associative law of multiplication by scalars:
$$\alpha \cdot (\beta \cdot x) = (\alpha \cdot \beta) \cdot x \quad \forall \alpha, \beta \in \mathcal{F}, \; \forall x \in \mathcal{V}$$
6. the neutrality of the scalar 1 in multiplication by scalars:
$$1 \cdot x = x \quad \forall x \in \mathcal{V}$$
7. distributive law of multiplication by scalars with respect to vector addition:
$$\alpha \cdot (x + y) = \alpha \cdot x + \alpha \cdot y \quad \forall \alpha \in \mathcal{F}, \; \forall x, y \in \mathcal{V}$$
8. distributive law of multiplication by scalars with respect to scalar addition:
$$(\alpha + \beta) \cdot x = \alpha \cdot x + \beta \cdot x \quad \forall \alpha, \beta \in \mathcal{F} \; \forall x \in \mathcal{V}$$

Sect. A.2 Fields, Vector Spaces, Linear Functions

x_1	x_2	$x_1 \oplus x_2$
0	0	0
0	1	1
1	0	1
1	1	0

x_1	x_2	$x_1 \cdot x_2$
0	0	0
0	1	0
1	0	0
1	1	1

Figure A-12 Operations in **B**.

Example A.2-4.

The set of all ordered n-tuples $(\alpha_1, \ldots, \alpha_n)$ of elements of a field $\mathcal{F}$ is a vector space over $\mathcal{F}$. It is denoted by $\mathcal{F}^n$. The sum of vectors and product by a scalar are defined as [1]

$$(\alpha_1, \ldots, \alpha_n) + (\beta_1, \ldots, \beta_n) = (\alpha_1 + \beta_1, \ldots, \alpha_n + \beta_n)$$

$$\alpha(\beta_1, \ldots, \beta_n) = (\alpha \beta_1, \ldots, \alpha \beta_n)$$

The origin is the n-tuple of all zeros. Referring to the examples of fields previously given, we may conclude that $\mathbf{R}^n$, $\mathbf{C}^n$, and $\mathbf{B}^n$ are vector spaces.

Example A.2-5.

The set of all functions $f[t_0, t_1]$ or $f : [t_0, t_1] \to \mathcal{F}^n$, piecewise continuous, i.e., with a finite number of discontinuities in $[t_0, t_1]$, is a vector space over $\mathcal{F}^n$.

Example A.2-6.

The set of all the solutions of the homogeneous differential equation

$$\frac{d^3 x}{dt^3} + 6 \frac{d^2 x}{dt^2} + 11 \frac{dx}{dt} = 0$$

which can be expressed as

$$x(t) = k_1 e^{-t} + k_2 e^{-2t} + k_3 e^{-3t} \quad (k_1, k_2, k_3) \in \mathbf{R}^3 \tag{1}$$

is a vector space over **R**.

Example A.2-7.

For any positive integer n, the set of all polynomials having degree equal or less than $n-1$ and coefficients belonging to **R**, **C**, or **B**, is a vector space over **R**, **C**, or **B** in which the operations are the usual polynomial addition and multiplication by a scalar. The origin is the polynomial with all coefficients equal to zero.

Definition A.2-3. Subspace

Given a vector space $\mathcal{V}$ over a field $\mathcal{F}$, a subset $\mathcal{X}$ of $\mathcal{V}$ is a *subspace* of $\mathcal{V}$ if

$$\alpha x + \beta y \in \mathcal{X} \quad \forall \alpha, \beta \in \mathcal{F}, \quad \forall x, y \in \mathcal{X}$$

Note that any subspace of $\mathcal{V}$ is a vector space over $\mathcal{F}$. The origin is a subspace of $\mathcal{V}$ that is contained in all other subspaces of $\mathcal{V}$; it will be denoted by $\mathcal{O}$.

---

[1] Here and in the following the "dot" symbol is understood, as in standard algebraic notation.

Definition A.2-4. Sum of Subspaces

The *sum* of two subspaces $\mathcal{X}, \mathcal{Y} \in \mathcal{V}$ is the set

$$\mathcal{X} + \mathcal{Y} := \{ z : z = x+y,\ x \in \mathcal{X},\ y \in \mathcal{Y} \}$$

Property A.2-1.

The sum of two subspaces of $\mathcal{V}$ is a subspace of $\mathcal{V}$.

Proof. Let $p, q \in \mathcal{X}+\mathcal{Y}$: by definition there exist two pairs of vectors $x, z \in \mathcal{X}$ and $y, v \in \mathcal{Y}$ such that $p = x+y$, $q = z+v$. Since $\alpha p + \beta q = (\alpha x + \beta z) + (\alpha y + \beta v)$ $\forall \alpha, \beta \in \mathcal{F}$, and $\alpha x + \beta z \in \mathcal{X}$, $\alpha y + \beta v \in \mathcal{Y}$ because $\mathcal{X}$ and $\mathcal{Y}$ are subspaces, it follows that $\alpha p + \beta q \in \mathcal{X} + \mathcal{Y}$, hence $\mathcal{X} + \mathcal{Y}$ is a subspace. □

Definition A.2-5. Intersection of Subspaces

The intersection of two subspaces $\mathcal{X}, \mathcal{Y} \in \mathcal{V}$ is the set

$$\mathcal{X} \cap \mathcal{Y} := \{ z : z \in \mathcal{X},\ z \in \mathcal{Y} \}$$

Property A.2-2.

The intersection of two subspaces of $\mathcal{V}$ is a subspace of $\mathcal{V}$.

Proof. Let $p, q \in \mathcal{X} \cap \mathcal{Y}$, i.e., $p, q \in \mathcal{X}$ and $p, q \in \mathcal{Y}$. Since $\mathcal{X}$ and $\mathcal{Y}$ are subspaces, $\alpha p + \beta q \in \mathcal{X}$ and $\alpha p + \beta q \in \mathcal{Y}$ $\forall \alpha, \beta \in \mathcal{F}$. Thus, $\alpha p + \beta q \in \mathcal{X} \cap \mathcal{Y}$, hence $\mathcal{X} \cap \mathcal{Y}$ is a subspace. □

Definition A.2-6. Direct Sum of Subspaces

Let $\mathcal{X}, \mathcal{Y} \subseteq \mathcal{Z}$ be subspaces of a vector space $\mathcal{V}$ which satisfy

$$\mathcal{X} + \mathcal{Y} = \mathcal{Z} \tag{2}$$

$$\mathcal{X} \cap \mathcal{Y} = \mathcal{O} \tag{3}$$

In such a case $\mathcal{Z}$ is called the *direct sum* of $\mathcal{X}$ and $\mathcal{Y}$. The direct sum is denoted by the symbol $\oplus$; hence the notation $\mathcal{X} \oplus \mathcal{Y} = \mathcal{Z}$ is equivalent to (2,3).

Property A.2-3.

Let $\mathcal{Z} = \mathcal{X} \oplus \mathcal{Y}$. Any vector $z \in \mathcal{Z}$ can be expressed in a unique way as the sum of a vector $x \in \mathcal{X}$ and a vector $y \in \mathcal{Y}$.

Proof. The existence of two vectors x and y such that $z = x+y$ is a consequence of Definition A.2-4. In order to prove uniqueness, assume that $z = x+y = x_1 + y_1$; by difference we obtain $(x-x_1) + (y-y_1) = 0$ or $(x-x_1) = -(y-y_1)$. Since $\mathcal{X} \cap \mathcal{Y} = \mathcal{O}$, the only vector belonging to both subspaces is the origin. Hence $(x-x_1) = (y-y_1) = 0$, i.e., $x = x_1$, $y = y_1$. □

Corollary A.2-1.

Let $\mathcal{Z} = \mathcal{X} + \mathcal{Y}$. All decompositions of any vector $z \in \mathcal{Z}$ in the sum of two vectors belonging respectively to $\mathcal{X}$ and $\mathcal{Y}$ are obtained from any one of them, say $z = x_1 + y_1$, by summing each vector of $\mathcal{X} \cap \mathcal{Y}$ to x_1 and subtracting it from y_1.

Proof. The proof is contained in that of Property A.2-3. □

Definition A.2-7. Linear Variety (Linear Manifold)

Let x_0 be any vector belonging to a vector space $\mathcal{V}$, $\mathcal{X}$ any subspace of $\mathcal{V}$. The set [2]

$$\mathcal{L} = \{ z : z = x_0 + x,\ x \in \mathcal{X} \} = \{x_0\} + \mathcal{X}$$

is called a *linear variety* or *linear manifold* contained in $\mathcal{V}$.

Definition A.2-8. Quotient Space

Let $\mathcal{X}$ be a subspace of a vector space $\mathcal{V}$ over a field $\mathcal{F}$: the set of all linear varieties

$$\mathcal{L} = \{x\} + \mathcal{X},\quad x \in \mathcal{V}$$

is called the *quotient space* of $\mathcal{V}$ over $\mathcal{X}$ and denoted by $\mathcal{V}/\mathcal{X}$.

A quotient space is a vector space over $\mathcal{F}$. Any two elements of a quotient space $\mathcal{L}_1 = \{x_1\} + \mathcal{X}$, $\mathcal{L}_2 = \{x_2\} + \mathcal{X}$ are equal if $x_1 - x_2 \in \mathcal{X}$. The sum of $\mathcal{L}_1$ and $\mathcal{L}_2$ is defined as

$$\mathcal{L}_1 + \mathcal{L}_2 := \{x_1 + x_2\} + \mathcal{X}$$

The external product of $\mathcal{L} = \{x\} + \mathcal{X}$ by a scalar $\alpha \in \mathcal{F}$ is defined as

$$\alpha \mathcal{L} := \{\alpha x\} + \mathcal{X}$$

A.2.1 Bases, Isomorphisms, Linearity

Definition A.2-9. Linearly Independent Set

A set of vectors $\{x_1, \ldots, x_h\}$ belonging to a vector space $\mathcal{V}$ over a field $\mathcal{F}$ is *linearly independent* if

$$\sum_{i=1}^{h} \alpha_i x_i = 0,\quad \alpha_i \in \mathcal{F}\ (i = 1, \ldots, h) \tag{4}$$

implies that all α_i are zero. If, on the other hand, (4) holds with some of the α_i different from zero, the set is *linearly dependent*.

Note that a set in which at least one of the elements is the origin is linearly dependent. If a set is linearly dependent, at least one of the vectors can be expressed as a *linear combination* of the remaining ones. In fact, suppose that (4) holds with one of the coefficients, for instance α_1, different from zero. Hence

$$x_1 = -\sum_{i=2}^{h} \frac{\alpha_i}{\alpha_1} x_i$$

---

[2] Note that the sum of subspaces, as introduced in Definition A.2-4, can be extended to general subsets of a vector space. Of course, the sum of general subsets in general is not a subspace and does not contain the origin, unless both the addends do.

Definition A.2-10. Span of a Set of Vectors
Let $\{x_1,\ldots,x_h\}$ be any set of vectors belonging to a vector space $\mathcal{V}$ over a field $\mathcal{F}$. The *span* of $\{x_1,\ldots,x_h\}$ (denoted by $\mathrm{sp}\{x_1,\ldots,x_h\}$) is the subspace of $\mathcal{X}$

$$\mathrm{sp}\{x_1,\ldots,x_h\} := \Big\{ x : x = \sum_{i=1}^{h} \alpha_i x_i \, , \quad \alpha_i \in \mathcal{F} \ (i=1,\ldots,h) \Big\}$$

Definition A.2-11. Basis
A set of vectors $\{x_1,\ldots,x_h\}$ belonging to a vector space $\mathcal{V}$ over a field $\mathcal{F}$ is a *basis* of $\mathcal{V}$ if it is linearly independent and $\mathrm{sp}\{x_1,\ldots,x_h\} = \mathcal{V}$.

Theorem A.2-1.
Let $(b_1,\ldots,b_n)$ be a basis of a vector space $\mathcal{V}$ over a field $\mathcal{F}$; for any $x \in \mathcal{V}$ there exists a unique n-tuple of scalars $(\alpha_1,\ldots,\alpha_n)$ such that

$$x = \sum_{i=1}^{n} \alpha_i b_i$$

Proof. Existence of $(\alpha_1,\ldots,\alpha_n)$ is a consequence of Definition A.2-11, uniqueness is proved by contradiction. Let $(\alpha_1,\ldots,\alpha_n)$, $(\beta_1,\ldots,\beta_n)$ be two ordered n-tuples of scalars such that

$$x = \sum_{i=1}^{n} \alpha_i b_i = \sum_{i=1}^{n} \beta_i b_i$$

Hence, by difference

$$0 = \sum_{i=1}^{n} (\alpha_i - \beta_i) b_i$$

since the set $\{b_1,\ldots,b_n\}$ is linearly independent, it follows that $\alpha_i = \beta_i$ $(i=1,\ldots,n)$. $\square$

The scalars $(\alpha_1,\ldots,\alpha_n)$ are called the *components* of x in the basis $(b_1,\ldots,b_n)$.

Example A.2-8.
A basis for $\mathcal{F}^n$ is the set of vectors

$$e_1 := (1,0,\ldots,0)$$
$$e_2 := (0,1,\ldots,0)$$
$$\ldots\ldots\ldots$$
$$e_n := (0,0,\ldots,1)$$
(5)

It is called the *main basis* of $\mathcal{F}^n$. Let $x=(x_1,\ldots,x_n)$ be any element of $\mathcal{F}^n$: the i-th component of x with respect to the main basis is clearly x_i.

Example A.2-9.
A basis for the vector space (1), defined in Example A.2-6, is
$$b_1 := e^{-t}, \quad b_2 := e^{-2t}, \quad b_3 := e^{-3t}$$

Example A.2-10.
For the vector space of all piecewise continuous functions $f[t_0, t_1]$ it is not possible to define any basis with a finite number of elements.

Theorem A.2-2.
The number of elements in any basis of a vector space $\mathcal{V}$ is the same as in any other basis of $\mathcal{V}$.

Proof. Let $\{b_1, \ldots, b_n\}$ and $\{c_1, \ldots, c_m\}$ ($m \geq n$) be any two bases of $\mathcal{V}$. Clearly
$$\mathcal{V} = \text{sp}\{c_n, b_1, \ldots, b_n\} \tag{6}$$
Since $c_n \in \mathcal{V}$ it can be expressed as
$$c_n = \sum_{i \in \mathcal{J}_1} \alpha_{1i} b_i, \quad \mathcal{J}_1 := \{1, 2, \ldots, n\}$$
At least one of the α_{1i} ($i \in \mathcal{J}_1$) is different from zero, c_n being different from the origin. Let $\alpha_{1j} \neq 0$, so that b_j can be expressed as a linear combination of $\{c_n, b_1, \ldots, b_{j-i}, b_{j+1}, \ldots, b_n\}$ and can be deleted on the right of (6). By insertion of the new vector c_{n-1} in the set, it follows that
$$\mathcal{V} = \text{sp}\{c_{n-1}, c_n, b_1, \ldots, b_{j-1}, b_{j+1}, \ldots, b_n\} \tag{7}$$
Since $c_{n-1} \in \mathcal{V}$ and $\mathcal{V}$ is the span of the other vectors on the right of (7),
$$c_{n-1} = \beta_{2n} c_n + \sum_{i \in \mathcal{J}_2} \alpha_{2i} b_i, \quad \mathcal{J}_2 := \{1, \ldots, j-1, j+1, \ldots, n\}$$
At least one of the α_{2i} ($i \in \mathcal{J}_2$) is different from zero, c_{n-1} being different from the origin and linearly independent of c_n. Again, on the right of (7) it is possible to delete one of the b_i ($i \in \mathcal{J}_2$). By iteration of the same argument, it follows that
$$\mathcal{V} = \text{sp}\{c_1, \ldots, c_n\}$$
Since set $\{c_1, \ldots, c_n\}$ is linearly independent, we conclude that it is a basis of $\mathcal{V}$, hence $m = n$. □

A vector space $\mathcal{V}$ whose bases contain n elements is called an *n-dimensional vector space*. The integer n is called the *dimension* of $\mathcal{V}$, $n = \dim \mathcal{V}$ in short notation. Example A.2-10 above refers to an *infinite-dimensional vector space*.

Definition A.2-12. Isomorphic Vector Spaces
Two vector spaces V and W over the same field $\mathcal{F}$ are *isomorphic* if there exists a one-to-one correspondence $t : V \to W$ which preserves all linear combinations, i.e.

$$t(\alpha x + \beta y) = \alpha\, t(x) + \beta\, t(y) \quad \forall \alpha, \beta \in \mathcal{F}, \quad \forall x, y \in V \tag{8}$$

To denote that V and W are isomorphic, notation $V \equiv W$ is used.

Function t is called an *isomorphism*. In order to be isomorphic, V and W must have the same dimension; in fact, by virtue of (8) a basis in one of the vector spaces corresponds to a basis in the other. As a consequence of Theorem A.2-2, any n-dimensional vector space V over $\mathcal{F}$ is isomorphic to $\mathcal{F}^n$, hence for any subspace $\mathcal{X} \subseteq V$, $\dim \mathcal{X} \leq \dim V$.

Corollary A.2-2.
Let $\{b_1, \ldots, b_n\}$ be a basis of a vector space V and $\{c_1, \ldots, c_m\}$ a basis of a subspace $\mathcal{X} \in V$. It is possible to extend the set $\{c_1, \ldots, c_m\}$ to a new basis of V by inserting in it a proper choice of $n - m$ of the elements of the old basis $\{b_1, \ldots, b_n\}$.

Proof. The argument developed for the proof of Theorem A.2-2 can be applied in order to substitute m elements of the basis $\{b_1, \ldots, b_n\}$ with $\{c_1, \ldots, c_m\}$. □

Corollary A.2-3.
Let $\mathcal{X}$ be a subspace of a vector space V, with $\dim \mathcal{X} = m$, $\dim V = n$. The dimension of the quotient space $V/\mathcal{X}$ is $n - m$.

Proof. Apply Corollary A.2-2: if $\{b_1, \ldots, b_n\}$ is a basis of V such that $\{b_1, \ldots, b_m\}$ is a basis of $\mathcal{X}$, the linear varieties

$$\{b_i\} + \mathcal{X} \quad (i = m+1, \ldots, n)$$

are clearly a basis of $V/\mathcal{X}$. □

Definition A.2-13. Linear Function (Linear Map)
A function $A : V \to W$, where V and W are vector spaces over the same field $\mathcal{F}$, is a *linear function* or *linear map* or *linear transformation* if

$$A(\alpha x + \beta y) = \alpha A(x) + \beta A(y) \quad \forall \alpha, \beta \in \mathcal{F}, \quad \forall x, y \in V$$

In other words, a linear function is one that preserves all linear combinations. For instance, according to (8) an isomorphism is a linear function. The particular linear function $I : V \to V$ such that $I(x) = x$ for all $x \in V$ is called the *identity function*.

Example. Let V be the vector space of all piecewise continuous functions $x(\cdot) : [t_0, t_1] \to \mathbf{R}$. The relation

$$z = \int_{t_0}^{t_1} e^{-(t_1 - \tau)} x(\tau)\, d\tau$$

defines a linear function $A : V \to \mathbf{R}$.

Definition A.2-14. Image of a Linear Function
Let $A: V \to W$ be a linear function. The set
$$\mathrm{im}A := \{z : z = A(x),\ x \in V\}$$
is called the *image* or *range* of A. The dimension of $\mathrm{im}A$ is called the *rank* of A and denoted by $\rho(A)$.

Definition A.2-15. Kernel (Null Space) of a Linear Function
Let $A: V \to W$ be a linear function. The set
$$\mathrm{ker}A := \{x : x \in V,\ A(x) = 0\}$$
is called the *kernel* or *null space* of A. The dimension of $\mathrm{ker}A$ is called the *nullity* of A and denoted by $\nu(A)$.

Property A.2-4.
Both $\mathrm{im}A$ and $\mathrm{ker}A$ are subspaces (of W and V respectively).

Proof. Let $z, u \in \mathrm{im}A$, so that there exist two vectors $x, y \in V$ such that $z = A(x)$, $u = A(y)$. A being linear
$$\alpha z + \beta u = \alpha A(x) + \beta A(y) = A(\alpha x + \beta y) \quad \forall \alpha, \beta \in \mathcal{F}$$
therefore, provided $\alpha x + \beta y \in V$, it follows that $\alpha z + \beta u \in \mathrm{im}A$. Let $x, y \in \mathrm{ker}A$, so that $A(x) = A(y) = 0$. Therefore
$$A(\alpha x + \beta y) = \alpha A(x) + \beta A(y) = 0 \quad \forall \alpha, \beta \in \mathcal{F}$$
hence, $\alpha x + \beta y \in \mathrm{ker}A$. □

Property A.2-5.
Let $A: V \to W$ be a linear function and $\dim V = n$. The following equality holds:
$$\rho(A) + \nu(A) = n$$

Proof. Let $\{b_1, \ldots, b_n\}$ be a basis of V and $\{b_1, \ldots, b_h\}$, $h \leq n$, a basis of $\mathrm{ker}A$ (see Corollary A.2-2). Clearly
$$\mathrm{sp}\{A(b_1), \ldots, A(b_n)\} = \mathrm{sp}\{A(b_{h+1}), \ldots, A(b_n)\} = \mathrm{im}A$$
thus, in order to prove the property it is sufficient to show that $\{A(b_{h+1}), \ldots, A(b_n)\}$ is a linearly independent set. In fact, the equality
$$\sum_{i=h+1}^{n} \alpha_i A(b_i) = A\left(\sum_{i=h+1}^{n} \alpha_i b_i\right) = 0$$
i.e.
$$\sum_{i=h+1}^{n} \alpha_i b_i \in \mathrm{ker}A$$
implies $\alpha_i = 0$ ($i = h+1, \ldots, n$), because the last $n-h$ components (with respect to the assumed basis) of all vectors belonging to $\mathrm{ker}A$ must be zero. □

Property A.2-6.
A linear function $A: V \to W$ is one-to-one if and only if $\ker A = \mathcal{O}$.

Proof. If. Let $\ker A = \mathcal{O}$: for every pair $x, y \in V$, $x \neq y$, it follows that $A(x-y) \neq 0$, i.e., $A(x) \neq A(y)$.

Only if. By contradiction, suppose $\ker A \neq \mathcal{O}$ and consider a vector x such that $x \in \ker A$, $x \neq 0$. Since $A(x) = 0$, $A(0) = 0$, A is not one-to-one. □

Property A.2-7.
A linear function $A: V \to W$ is one-to-one if and only if it maps linearly independent sets into linearly independent sets.

Proof. If. Let $\{x_1, \ldots, x_h\}$ be a linearly independent set. Suppose, by contradiction, that $\{A(x_1), \ldots, A(x_h)\}$ is linearly dependent, so that there exist h scalars $\alpha_1, \ldots, \alpha_h$ not all zero such that $\alpha_1 A(x_1) + \ldots + \alpha_h A(x_h) = 0$, hence $\alpha_1 x_1 + \ldots + \alpha_h x_h \in \ker A$. Thus, A is not one-to-one by Property A.2-6.

Only if. By virtue of Property A.2-6, $\ker A \neq \mathcal{O}$ if A is not one-to-one; hence, there exists at least one nonzero vector $x \in V$ such that $A(x) = 0$. Any linear independent set that includes x is transformed into a set that is linearly dependent because it includes the origin. □

A.2.2 Projections, Matrices, Similarity

Definition A.2-16. Projection
Let $\mathcal{X}, \mathcal{Y}$ be any pair of subspaces of a vector space V such that $\mathcal{X} \oplus \mathcal{Y} = V$. By virtue of Property A.2-3, for all $z \in V$ there exists a unique pair (x, y) such that $z = x + y$, $x \in \mathcal{X}$, $y \in \mathcal{Y}$. The linear functions $P: V \to \mathcal{X}$ and $Q: V \to \mathcal{Y}$ defined by

$$P(z) = x, \quad Q(z) = y \quad \forall z \in V$$

are called, respectively, the *projection on $\mathcal{X}$ along $\mathcal{Y}$* and the *projection on $\mathcal{Y}$ along $\mathcal{X}$*.

Note that $\operatorname{im} P = \mathcal{X}$, $\ker P = \mathcal{Y}$, $\operatorname{im} Q = \mathcal{Y}$, $\ker Q = \mathcal{X}$.

Definition A.2-17. Canonical Projection
Let $\mathcal{X}$ be any subspace of a vector space V. The linear function $P: V \to V/\mathcal{X}$ defined by $P(x) = \{x\} + \mathcal{X}$ is called the *canonical projection* of V on $V/\mathcal{X}$.

Note that $\operatorname{im} P = V/\mathcal{X}$, $\ker P = \mathcal{X}$.

Definition A.2-18. Invariant Subspace
Let V be a vector space and $A: V \to V$ a linear map. A subspace $\mathcal{X} \subseteq V$ is said to be an *invariant under A*, or an *A-invariant*, if

$$A(\mathcal{X}) \subseteq \mathcal{X}$$

It is easy to prove that the sum and the intersection of two or more A-invariants are A-invariants; hence any subset of the set of all A-invariants contained in V, closed with respect to sum and intersection, is a lattice with respect to $\subseteq, +, \cap$.

Sect. A.2 Fields, Vector Spaces, Linear Functions

The following theorem states a very important connection between linear maps and matrices.

Theorem A.2-3.
Let V and W be finite-dimensional vector spaces over the same field $\mathcal{F}$ and $\{b_1,\ldots,b_n\}$, $\{c_1,\ldots,c_m\}$ bases of V and W respectively. Denote by ξ_i $(i=1,\ldots,n)$ and η_i $(i=1,\ldots,m)$ the components of vectors $x \in V$ and $y \in W$ with respect to these bases. Any linear function $A : V \to W$ can be expressed as

$$\eta_j = \sum_{i=1}^{n} a_{ji}\, \xi_i \quad (j=1,\ldots,m) \tag{9}$$

Proof. By definition of linear function, the following equalities hold:

$$y = A(x) = A\left(\sum_{i=1}^{n} \xi_i\, b_i\right) = \sum_{i=1}^{n} \xi_i\, A(b_i)$$

For each value of i, denote by a_{ji} $(j=1,\ldots,m)$ the components of $A(b_i)$ with respect to the basis $\{c_1,\ldots,c_m\}$, i.e.

$$A(b_i) = \sum_{j=1}^{m} a_{ji}\, c_j$$

By substitution, it follows that

$$y = \sum_{i=1}^{n} \xi_i \left(\sum_{j=1}^{m} a_{ji}\, c_j\right) = \sum_{j=1}^{m}\left(\sum_{i=1}^{n} a_{ji}\, \xi_i\right) c_j$$

which is clearly equivalent to (9). □

Relation (9) can be written in a more compact form as

$$\eta = A\, x \tag{10}$$

where η, A, and ξ are matrices defined as

$$\eta := \begin{bmatrix} \eta_1 \\ \vdots \\ \eta_m \end{bmatrix} \qquad A := \begin{bmatrix} a_{11} & \cdots & a_{1n} \\ \vdots & \ddots & \vdots \\ a_{m1} & \cdots & a_{mn} \end{bmatrix} \qquad \xi := \begin{bmatrix} \xi_1 \\ \vdots \\ \xi_m \end{bmatrix}$$

The following corollaries are direct consequences of the argument developed in the proof of Theorem A.2-3.

Corollary A.2-4.
Let $A: \mathcal{V} \to \mathcal{W}$ be a linear function represented by the matrix A with respect to the bases $\{b_1,\ldots,b_n\}$ of $\mathcal{V}$ and $\{c_1,\ldots,c_m\}$ of $\mathcal{W}$. The columns of A are the components of $A(b_1),\ldots,A(b_n)$ in the basis $\{c_1,\ldots,c_m\}$.

Corollary A.2-5.
Let $A: \mathcal{V} \to \mathcal{V}$ be a linear function represented by the matrix A with respect to the basis $\{b_1,\ldots,b_n\}$ of $\mathcal{V}$. The columns of A are the components of $A(b_1),\ldots,A(b_n)$ in this basis.

Theorem A.2-3 states that a linear function $A: \mathcal{V} \to \mathcal{W}$, where $\mathcal{V}$ and $\mathcal{W}$ are finite-dimensional vector spaces over the same field $\mathcal{F}$, is given by defining a basis for $\mathcal{V}$, a basis for $\mathcal{W}$, and a matrix with elements in $\mathcal{F}$. The special case $\mathcal{V} = \mathcal{F}^n$, $\mathcal{W} = \mathcal{F}^m$ may be a source of confusion: in fact, in this case vectors are m-tuples and n-tuples of scalars, so that apparently only a matrix with elements in $\mathcal{F}$ is needed in order to represent a linear function. This does not contradict Theorem A.2-3, because vectors can be understood to be referred respectively to the main bases of $\mathcal{F}^n$ and $\mathcal{F}^m$ so that components coincide with elements of vectors. In this case it is customary to write (10) directly as

$$z = A x \qquad (11)$$

and call "vectors" the $n \times 1$ and $m \times 1$ matrices representing the n-tuple x and the m-tuple z, so that (11) means "vector z is equal to the product of vector x by matrix A." For the sake of simplicity, in this case a linear function and its representing matrix are denoted with the same symbol; hence, notation $A: \mathcal{F}^n \to \mathcal{F}^m$ means "the linear function from $\mathcal{F}^n$ into $\mathcal{F}^m$ which is represented by the $m \times n$ matrix A with respect to the main bases." Similarly, notations im$A$, ker$A$, $\rho(A)$, $\nu(A)$, A-invariant are referred both to functions and matrices.

Definition A.2-19. Basis Matrix
Any subspace $\mathcal{X} \subseteq \mathcal{F}^n$ can be represented by a matrix X whose columns are a basis of $\mathcal{X}$, so that $\mathcal{X} = \text{im} X$. Such a matrix is called a *basis matrix* of $\mathcal{X}$.

Property A.2-8.
Let $\mathcal{V}$ be an n-dimensional vector space over a field $\mathcal{F}$ ($\mathcal{F} = \mathbf{R}$ or $\mathcal{F} = \mathbf{C}$) and denote by $u, v \in \mathcal{F}^n$ the components of any vector $x \in \mathcal{V}$ with respect to the bases $\{b_1,\ldots,b_n\}$, $\{c_1,\ldots,c_n\}$. These components are related by

$$u = Tv$$

where T is the $n \times n$ matrix having as columns the components of vectors $\{c_1,\ldots,c_n\}$ with respect to the basis $\{b_1,\ldots,b_n\}$.

Proof. Apply Corollary A.2-5 to the identity function I. □

Since the representation of any vector with respect to any basis is unique, matrix T is invertible, so that

$$v = T^{-1}u$$

where T^{-1}, by virtue of Corollary A.2-5, is the matrix having as columns the components of vectors $\{b_1, \ldots, b_n\}$ with respect to the basis $\{c_1, \ldots, c_n\}$.

Changes of basis are very often used in order to study properties of linear transformations by analyzing the structures of their representing matrices with respect to some properly selected bases. Therefore, it is worth knowing how matrices representing the same linear function with respect to different bases are related each other.

Property A.2-9.
Let $A: V \to W$ be a linear function represented by the $m \times n$ matrix A with respect to the bases $\{b_1, \ldots, b_n\}$, $\{c_1, \ldots, c_m\}$. If new bases $\{p_1, \ldots, p_n\}$, $\{q_1, \ldots, q_m\}$ are assumed for V and W, A is represented by the new matrix

$$B = Q^{-1}AP$$

where P and Q are the $n \times n$ and the $m \times m$ matrices whose columns are the components of the new bases with respect to the old ones.

Proof. Denote by $u, v \in \mathcal{F}^n$ and $r, s \in \mathcal{F}^m$ the old and the new components of any two vectors $x \in V$ and $y \in W$ such that y is the image of x in the linear transformation A: by virtue of Property A.2-8

$$u = Pv, \quad r = Qs$$

By substitution into $r = Au$, we obtain

$$s = Q^{-1}APv$$

which directly proves the property. □

In the particular case of a function whose domain and codomain are the same vector space, we may restate the preceding result as follows.

Corollary A.2-6.
Let $A: V \to V$ be a linear function represented by the $n \times n$ matrix A with respect to the basis $\{b_1, \ldots, b_n\}$. In the new basis $\{c_1, \ldots, c_n\}$ function A is represented by the new matrix

$$B = T^{-1}AT \tag{12}$$

where T is the $n \times n$ matrix whose columns are the components of the new basis with respect to the old one.

Let A and B be any two $n \times n$ matrices; if there exists a matrix T such that equality (12) holds, A and B are called *similar matrices* and T a *similarity transformation* or an *automorphism*, i.e., an isomorphism of a vector space with itself.

Note that the same similarity transformations relate powers of B and A: in fact, $B^2 = T^{-1}AT\,T^{-1}AT = T^{-1}A^2T$ and so on.

Theorem A.2-4.
Let $\mathcal{X}$, $\mathcal{Y}$ be subspaces of $\mathcal{F}^n$ such that $\mathcal{X} \oplus \mathcal{Y} = \mathcal{F}^n$ and X, Y basis matrices of $\mathcal{X}$, $\mathcal{Y}$. Projecting matrices on $\mathcal{X}$ along $\mathcal{Y}$ and on $\mathcal{Y}$ along $\mathcal{X}$, i.e., matrices that realize the projections introduced as linear functions in Definition A.2-16, are

$$P = \begin{bmatrix} X & O \end{bmatrix} \begin{bmatrix} X & Y \end{bmatrix}^{-1} \qquad (13)$$

$$Q = \begin{bmatrix} O & Y \end{bmatrix} \begin{bmatrix} X & Y \end{bmatrix}^{-1} \qquad (14)$$

Proof. Since the direct sum of $\mathcal{X}$ and $\mathcal{Y}$ is $\mathcal{F}^n$, matrix $[X Y]$ is nonsingular; therefore, the image of $[X Y]^{-1}$ is $\mathcal{F}^n$, so that $\operatorname{im} P = \mathcal{X}$, $\operatorname{im} Q = \mathcal{Y}$. Since $P+Q=I$, for any $z \in \mathcal{F}^n$ $x := Pz$, $y := Qz$ is the unique pair $x \in \mathcal{X}$, $y \in \mathcal{Y}$ such that $z = x+y$. □

A.2.3 A Brief Survey of Matrix Algebra

In the previous section, matrices were introduced as a means of representing linear maps between finite-dimensional vector spaces. Operations on linear functions, such as addition, multiplication by a scalar, composition, reflect on operations on the corresponding matrices. In the sequel, we will consider only *real matrices* and *complex matrices*, i.e., matrices whose elements are real or complex numbers. A matrix having as many rows as columns is called a *square matrix*. A matrix having only one row is called a *row matrix* and a matrix having only one column is called a *column matrix*. The symbol $(A)_{ij}$ denotes the element of the matrix A belonging to the i-th row and the j-th column and $[a_{ij}]$ the matrix whose general element is a_{ij}.

The main operations on matrices are:

1. *addition of matrices*: given two matrices A and B, both $m \times n$, their sum $C = A+B$ is the $m \times n$ matrix whose elements are the sums of the corresponding elements of A and B, i.e., $c_{ij} = a_{ij} + b_{ij}$ ($i = 1, \ldots, m$; $j = 1, \ldots, n$);

2. *multiplication of a matrix by a scalar*: given a scalar α and an $m \times n$ matrix A, the product $P = \alpha A$ is the $m \times n$ matrix whose elements are the products by α of the elements of A, i.e., $p_{ij} = \alpha a_{ij}$ ($i = 1, \ldots, m$; $j = 1, \ldots, n$);

3. *multiplication of two matrices*:[3] given an $m \times n$ matrix A and a $n \times p$ matrix B, the product $C = AB$ is the $m \times p$ matrix whose elements are defined as

$$c_{ij} = \sum_{k=1}^{n} a_{ik} b_{kj} \quad (i = 1, \ldots, m; \; j = 1, \ldots, p)$$

These operations enjoy the following properties:

---

[3] Of course, the same rules hold for the product $c = A b$, where the "vectors" b and c are simply $n \times 1$ and $m \times 1$ column matrices.

Sect. A.2 Fields, Vector Spaces, Linear Functions

1. commutative law of addition:[4]

$$A + B = B + A$$

2. associative laws:

$$(A + B) + C = A + (B + C)$$
$$(\alpha A) B = \alpha (A B)$$
$$A (B C) = (A B) C$$

3. distributive laws:

$$(\alpha + \beta) A = \alpha A + \beta A$$
$$\alpha (A + B) = \alpha A + \alpha B$$
$$(A + B) C = A C + B C$$
$$A (B + C) = A B + A C$$

A *null matrix* O is a matrix with all elements equal to zero.

An *identity matrix* I is a square matrix with all elements on the main diagonal equal to one and all other elements equal to zero. In particular, the symbol I_n is used for the $n \times n$ identity matrix.

A square matrix A is called *idempotent* if $A^2 = A$, *nilpotent* of index q if $A^{q-1} \neq O$, $A^q = O$.

For any real $m \times n$ matrix A, the symbol A^T denotes the *transpose* of A, i.e., the matrix obtained from A by interchanging rows and columns. Its elements are defined by

$$(A^T)_{ji} := (A)_{ij} \quad (i=1,\ldots,m;\ j=1,\ldots,n)$$

In the complex field, A^* denotes the *conjugate transpose* of A, whose elements are defined by

$$(A^*)_{ji} := (A)^*_{ij} \quad (i=1,\ldots,m;\ j=1,\ldots,n)$$

A real matrix A such that $A^T = A$ is called *symmetric*, while a complex matrix A such that $A^* = A$ is called *hermitian*.

A square matrix A is said to be *invertible* if it represents an invertible linear function; in such a case A^{-1} denotes the *inverse matrix* of A. If A is

---

[4] It is worth noting that in general the multiplication of matrices is not commutative, i.e., in general $AB \neq BA$. Hence, instead of saying "multiply A by B," it is necessary to state "multiply A by B on the left (right)" or "premultiply (postmultiply) A by B."

invertible, the relations $C = AB$ and $B = A^{-1}C$ are equivalent. An invertible matrix and its inverse matrix commute, i.e., $A^{-1}A = AA^{-1} = I$.

In the real field the transpose and the inverse matrices satisfy the following relations:

$$(A^T)^T = A \tag{15}$$
$$(A + B)^T = A^T + B^T \tag{16}$$
$$(AB)^T = B^T A^T \tag{17}$$
$$(A^{-1})^{-1} = A \tag{18}$$
$$(AB)^{-1} = B^{-1} A^{-1} \tag{19}$$
$$(A^{-1})^T = (A^T)^{-1} \tag{20}$$

Note that (17) implies that for any A the matrices AA^T and $A^T A$ are symmetric. In the complex field, relations (15–17, 20) hold for conjugate transpose instead of transpose matrices.

The *trace* of a square matrix A, denoted by tr$A$, is the sum of all the elements on the main diagonal, i.e.

$$\text{tr} A := \sum_{i=1}^{n} a_{ii} \tag{21}$$

Let A be a 2×2 matrix: the *determinant* of A, denoted by det$A$, is defined as

$$\det A := a_{11} a_{22} - a_{12} a_{21}$$

If A is an $n \times n$ matrix, its determinant is defined by any one of the recursion relations

$$\det A := \sum_{i=1}^{n} a_{ij} A_{ij} \quad (j = 1, \ldots, n)$$
$$= \sum_{j=1}^{n} a_{ij} A_{ij} \quad (i = 1, \ldots, n) \tag{22}$$

where A_{ij} denotes the *cofactor* of a_{ij}, which is defined as $(-1)^{i+j}$ times the determinant of the $(n-1) \times (n-1)$ matrix obtained by deleting the i-th row and the j-th column of A.

The transpose (or conjugate transpose) of the matrix of cofactors $[A_{ij}]^T$ (or $[A_{ij}]^*$) is called the *adjoint matrix* of A and denoted by adj$A$.

Any square matrix A such that det$A = 0$ is called *singular*; in the opposite case it is called *nonsingular*.

Sect. A.2 Fields, Vector Spaces, Linear Functions

The main properties of determinants are:
1. in the real field, $\det A = \det A^T$; in the complex field, $\det A = \det A^*$;
2. let B be a matrix obtained from A by interchanging any two rows or columns: $\det B = -\det A$;
3. if any two rows or columns of A are equal, $\det A = 0$;
4. let B be a matrix obtained from A by adding one row or column multiplied by a scalar α to another row or column: $\det B = \det A$;
5. if any row or column of A is a linear combination of other rows or columns, $\det A = 0$;
6. let A, B be square matrices having equal dimensions: $\det(A\,B) = \det A \det B$.

Theorem A.2-5.
Let A be a nonsingular matrix. Its inverse matrix A^{-1} is given by

$$A^{-1} = \frac{\operatorname{adj} A}{\det A} \qquad (23)$$

Proof. Denote by B the matrix on the right of (23): by virtue of property 3 of the determinants

$$\sum_{k=1}^{n} a_{ik} A_{jk} = \begin{cases} \det A & \text{if } i=j \\ 0 & \text{if } i \neq j \end{cases}$$

Then, for any element of the matrix $P := AB$

$$p_{ij} = \frac{1}{\det A} \sum_{k=1}^{n} a_{ik} A_{jk} = \begin{cases} 1 & \text{if } i=j \\ 0 & \text{if } i \neq j \end{cases}$$

Hence, $P = I$, so that $B = A^{-1}$. □

As a consequence of Theorem A.2-5, we may conclude that a square matrix is invertible if and only if it is nonsingular.

Partitioned Matrices. A *partitioned matrix* is one whose elements are matrices, called *submatrices* of the original nonpartitioned matrix. It is easily seen that, if partitioning is congruent, addition and multiplication can be performed by considering each of the submatrices as a single element. To show how partitioning is used, let us consider some examples:

$$\begin{array}{c} \begin{matrix} n_1 & n_2 \end{matrix} \\ \begin{matrix} m_1 \\ m_2 \end{matrix}\begin{bmatrix} A & B \\ C & D \end{bmatrix} \end{array} + \begin{array}{c} \begin{matrix} n_1 & n_2 \end{matrix} \\ \begin{matrix} m_1 \\ m_2 \end{matrix}\begin{bmatrix} E & F \\ G & H \end{bmatrix} \end{array} = \begin{array}{c} \begin{matrix} n_1 & n_2 \end{matrix} \\ \begin{matrix} m_1 \\ m_2 \end{matrix}\begin{bmatrix} A+E & B+F \\ C+G & D+H \end{bmatrix} \end{array}$$

$$\begin{array}{c} \begin{matrix} n_1 & n_2 \end{matrix} \\ \begin{matrix} m_1 \\ m_2 \end{matrix}\begin{bmatrix} A & B \\ C & D \end{bmatrix} \end{array} \cdot \begin{array}{c} \begin{matrix} p_1 & p_2 \end{matrix} \\ \begin{matrix} n_1 \\ n_2 \end{matrix}\begin{bmatrix} E & F \\ G & H \end{bmatrix} \end{array} = \begin{array}{c} \begin{matrix} p_1 & p_2 \end{matrix} \\ \begin{matrix} m_1 \\ m_2 \end{matrix}\begin{bmatrix} AE+BG & AF+BH \\ CE+DG & CF+DH \end{bmatrix} \end{array}$$

$$\begin{matrix} & n_1 & n_2 \\ m_1 \\ m_2 \end{matrix} \begin{bmatrix} A & B \\ C & D \end{bmatrix} \cdot \begin{matrix} & p \\ n_1 \\ n_2 \end{matrix} \begin{bmatrix} E \\ F \end{bmatrix} = \begin{matrix} & p \\ m_1 \\ m_2 \end{matrix} \begin{bmatrix} AE+BF \\ CE+DF \end{bmatrix}$$

Consider a square matrix partitioned into four submatrices as follows:

$$A = \begin{bmatrix} B & C \\ D & E \end{bmatrix} \qquad (24)$$

and assume that B and E are square matrices. It is easy to prove by induction that if one of the off-diagonal matrices C and D is null, the following holds:

$$\det A = \det B \, \det E \qquad (25)$$

If $\det B$ is different from zero, so that B is invertible, by subtracting from the second row of (24) the first multiplied on the left by DB^{-1}, we obtain the matrix

$$\begin{bmatrix} B & C \\ O & E - DB^{-1}C \end{bmatrix}$$

whose determinant is equal to $\det A$ by virtue of the preceding property 4 of the determinants, so that

$$\det A = \det B \, \det(E - DB^{-1}C) \qquad (26)$$

Similarly, if $\det E \neq 0$, by subtracting from the first row of (24) the second multiplied on the left by CE^{-1}, we obtain

$$\det A = \det E \, \det(B - CE^{-1}D) \qquad (27)$$

A.3 INNER PRODUCT, ORTHOGONALITY

Providing a vector space with an inner product is a source of some substantial advantages, particularly when the maximum universality and abstractness of approach is not called for. The most significant of these advantages are: a deeper insight into the geometric meaning of many conditions and properties, a straightforward way to consider and possibly avoid ill-conditioning in computations, a valid foundation for setting duality arguments (depending on properties of adjoint transformations, hence on the introduction of an inner product).

Definition A.3-1. Inner Product
Let $\mathcal{V}$ be a vector space defined over the field $\mathbf{R}$ of real numbers. An *inner product* or *scalar product* is a function $\langle \cdot, \cdot \rangle : \mathcal{V} \times \mathcal{V} \to \mathcal{F}$ that satisfies

1. commutativity:
$$\langle x, y \rangle = \langle y, x \rangle \quad \forall x, y \in \mathcal{V}$$

2. linearity with respect to a left-hand factor:
$$\langle \alpha x + \beta y, z \rangle = \alpha \langle x, z \rangle + \beta \langle y, z \rangle \quad \forall \alpha, \beta \in \mathbf{R}, \quad \forall x, y, z \in \mathcal{V}$$

3. positiveness:
$$\langle x, x \rangle \geq 0 \quad \forall x \in \mathcal{V}$$
$$\langle x, x \rangle = 0 \quad \Leftrightarrow \quad x = 0$$

If $\mathcal{V}$ is defined over the field $\mathbf{C}$ of the complex numbers, the preceding property 3 still holds, while 1 and 2 are replaced respectively by

1. conjugate commutativity:
$$\langle x, y \rangle = \langle y, x \rangle^* \quad \forall x, y \in \mathcal{V}$$

2. conjugate linearity with respect to a left-hand factor:
$$\langle \alpha x + \beta y, z \rangle = \alpha^* \langle x, z \rangle + \beta^* \langle y, z \rangle \quad \forall \alpha, \beta \in \mathbf{C}, \quad \forall x, y, z \in \mathcal{V}$$

A vector space with an inner product is called an *inner product (vector) space*.

Note that in the real field, commutativity and linearity with respect to a left-hand factor imply linearity with respect to a right-hand one, hence *bilinearity*, while in the complex field conjugate commutativity and conjugate linearity with respect to a left-hand factor imply linearity with respect to a right-hand factor. In fact:

$$\begin{aligned} \langle x, \alpha y + \beta z \rangle &= \langle \alpha y + \beta z, x \rangle^* \\ &= \left(\alpha^* \langle y, x \rangle + \beta^* \langle z, x \rangle \right)^* \\ &= \alpha \langle x, y \rangle + \beta \langle x, z \rangle \end{aligned}$$

Example A.3-1.
In $\mathbf{R}^n$ an inner product is

$$\langle x, y \rangle := \sum_{i=1}^{n} x_i y_i \tag{1}$$

and in $\mathbf{C}^n$

$$\langle x, y \rangle := \sum_{i=1}^{n} x_i^* y_i \tag{2}$$

Note that in $\mathbf{R}^n$ and $\mathbf{C}^n$, vectors can be considered as $n \times 1$ matrices, so that the notations $x^T y$ and $x^* y$ may be used instead of $\langle x, y \rangle$.

Example A.3-2.
In any finite-dimensional vector space over **R** or **C** an inner product is defined as in (1) or (2), where $(x_1,\ldots,x_n)$ and $(y_1,\ldots,y_n)$ denote components of vectors with respect to given bases.

Example A.3-3.
In the vector space of all piecewise continuous time functions $f(\cdot) : [t_0, t_1] \to \mathcal{F}^n$ (with $\mathcal{F} = \mathbf{R}$ or $\mathcal{F} = \mathbf{C}$) an inner product is

$$\langle x, y \rangle := \int_{t_0}^{t_1} \langle x(t)\, y(t) \rangle\, dt$$

Definition A.3-2. Euclidean Norm
Let $\mathcal{V}$ be an inner product space and x any vector belonging to $\mathcal{V}$. The real nonnegative number

$$\|x\| := \sqrt{\langle x, x \rangle} \qquad (3)$$

is called the *euclidean norm* of x.

The euclidean norm is a measure of the "length" of x, i.e., of the distance of x from the origin.

Definition A.3-3. Orthogonal Vectors
A pair of vectors x, y belonging to an inner product space are said to be *orthogonal* if $\langle x, y \rangle = 0$.

Definition A.3-4. Orthonormal Set of Vectors
A set of vectors $\{u_1,\ldots,u_n\}$ belonging to an inner product space is *orthonormal* if

$$\langle u_i, u_j \rangle = 0 \quad (i = 1,\ldots,n;\ j = 1,\ldots,i-1,i+1,\ldots,n)$$
$$\langle u_i, u_i \rangle = 1 \quad (i = 1,\ldots,n)$$

From

$$\sum_{i=1}^{n} \alpha_i u_i = 0$$

through the left inner product of both members by $u_1,\ldots,u_n$, we obtain $\alpha_1 = 0,\ldots,\alpha_n = 0$, so we may conclude that an orthonormal set is a linearly independent set.

Definition A.3-5. Orthogonal (Unitary) Matrix
An $n \times n$ real matrix is *orthogonal* if its rows (columns) are orthonormal sets. In the complex field, such a matrix is called *unitary*.

It is easy to check that a necessary and sufficient condition for a square matrix U to be orthogonal (unitary) is

$$U^T U = U U^T = I \quad (U^* U = U U^* = I)$$

Sect. A.3 Inner Product, Orthogonality

Therefore, the inverse of an orthogonal (unitary) matrix can be determined by simply interchanging rows and columns.

Orthogonal (unitary) matrices have the interesting feature of preserving orthogonality of vectors and values of inner products. The product of two or more orthogonal (unitary) matrices is an orthogonal (unitary) matrix. In fact, if A, B are orthogonal:

$$(AB)^T(AB) = B^T(A^T A)B = B^T B = I$$

Similar manipulations can be set for unitary matrices.

Property A.3-1.
Let $\mathcal{V}$ be a finite-dimensional inner product space. The components $(\xi_1, \ldots, \xi_n)$ of any vector $x \in \mathcal{V}$ with respect to an orthonormal basis $(u_1, \ldots, u_n)$ are provided by

$$\xi_i = \langle u_i, x \rangle \quad (i = 1, \ldots, n) \tag{4}$$

Proof. Consider the relation

$$x = \sum_{i=1}^{n} \xi_i u_i$$

and take the left inner product of both members by the orthonormal set $(u_1, \ldots, u_n)$. □

It is possible to derive an orthonormal basis for any finite-dimensional vector space or subspace through the Gram-Schmidt orthonormalization process (see Algorithm B.2-1).

Definition A.3-6. Adjoint of a Linear Map
A linear map B is called *adjoint* to a linear map A if for any two vectors x, y belonging respectively to the domains of A and B the following identity holds:

$$\langle Ax, y \rangle = \langle x, By \rangle$$

Property A.3-2.
Let A be an $m \times n$ real matrix. The inner product (1) satisfies the identity

$$\langle Ax, y \rangle = \langle x, A^T y \rangle \quad \forall x \in \mathbf{R}^n, \forall y \in \mathbf{R}^m \tag{5}$$

while, if A is complex and the inner product is defined as in (2),

$$\langle Ax, y \rangle = \langle x, A^* y \rangle \quad \forall x \in \mathbf{C}^n, \forall y \in \mathbf{C}^m \tag{6}$$

Proof. Equalities (5,6) follow from matrix identity $(Ax)^T y = x^T A^T y$, which is a consequence of (A.2.17), and $(Ax)^* y = x^* A^* y$. □

It follows that if a linear map is represented by a real (a complex) matrix with respect to an orthonormal basis, its adjoint map is represented by the transpose (the conjugate transpose) matrix with respect to the same basis.

Definition A.3-7. Orthogonal Complement of a Subspace
Let $\mathcal{V}$ be an inner product space and $\mathcal{X}$ any subspace of $\mathcal{V}$. The set

$$\mathcal{X}^\perp = \{ y : \langle x, y \rangle = 0, \; x \in \mathcal{X} \} \tag{7}$$

is called the *orthogonal complement* of $\mathcal{X}$.

It is easily seen that $\mathcal{X}^\perp$ is a subspace of $\mathcal{V}$.

Property A.3-3.
Let $\mathcal{V}$ be an inner product space and $\mathcal{X}$ any finite-dimensional subspace of $\mathcal{V}$.[1] Then

$$\mathcal{V} = \mathcal{X} \oplus \mathcal{X}^\perp$$

Proof. First note that $\mathcal{X} \cap \mathcal{X}^\perp = \{0\}$ because $\langle x, x \rangle = 0$ implies $x = 0$. Let $\{u_1, \ldots, u_h\}$ be an orthonormal basis of $\mathcal{X}$. For any $z \in \mathcal{V}$ consider the decomposition $z = x + y$ defined by

$$x = \sum_{i=1}^{h} \xi_i u_i, \quad \xi_i = \langle u_i, z \rangle$$

$$y = z - x = z - \sum_{i=1}^{h} \xi_i u_i$$

Clearly, $x \in \mathcal{X}$. Since $\langle u_i, y \rangle = 0$ $(i = 1, \ldots, h)$, $y \in \mathcal{X}^\perp$. □

All statements reported hereafter in this section will refer to real matrices; their extension to complex matrices simply requires the substitution of transpose matrices with conjugate transpose.

Property A.3-4.
For any real matrix A

$$\ker A^T = (\operatorname{im} A)^\perp \tag{8}$$

Proof. Let $y \in (\operatorname{im} A)^\perp$, so that $\langle y, A A^T y \rangle = 0$ (in fact clearly $A A^T y \in \operatorname{im} A$), i.e. $\langle A^T y, A^T y \rangle = 0$, hence $A^T y = 0$, $y \in \ker A^T$. On the other hand, if $y \in \ker A^T$, $\langle A^T y, x \rangle = 0$ for all x, i.e. $\langle y, Ax \rangle = 0$, hence $y \in (\operatorname{im} A)^\perp$. □

Property A.3-5.
For any real matrix A

$$\rho(A) = \rho(A^T) \tag{9}$$

Proof. Let A be $m \times n$: by virtue of Properties A.3-3 and A.3-4, it is possible to derive a basis of $\mathbf{R}^m$ whose elements belong to $\ker A^T$ and $(\operatorname{im} A)^\perp$: it follows that $\nu(A^T) + \rho(A) = m$; on the other hand, from Property A.2-5, $\rho(A^T) + \nu(A^T) = m$. □

Hence, if A is not a square matrix, $\nu(A) \neq \nu(A^T)$.

---

[1] The finiteness of the dimensions of $\mathcal{X}$ is a more restrictive than necessary assumption: Property A.3-3 applies also in the more general case where $\mathcal{X}$ is any subspace of an infinite-dimensional Hilbert space.

Sect. A.3 Inner Product, Orthogonality

Property A.3-6.
For any real matrix A
$$\text{im} A = \text{im}(A A^T)$$

Proof. Let A be $m \times n$; take a basis of $\mathbf{R}^n$ whose elements are in part a basis of $\text{im} A^T$ and in part a basis of $\text{ker} A$. The vectors of this basis transformed by A span $\text{im} A$ by definition. Since vectors in $\text{ker} A$ are transformed into the origin, it follows that a basis of $\text{im} A^T$ is transformed into a basis of $\text{im} A$. □

A.3.1 Orthogonal Projections, Pseudoinverse of a Linear Map

Definition A.3-8. Orthogonal Projection
Let V be a finite-dimensional inner product space and $\mathcal{X}$ any subspace of V. The *orthogonal projection* of $\mathcal{X}$ is the projection on $\mathcal{X}$ along $\mathcal{X}^\perp$.

Corollary A.3-1.
Let $\mathcal{X}$ be any subspace of $\mathbf{R}^n$ ($\mathbf{C}^n$) and U an orthonormal (unitary) basis matrix of $\mathcal{X}$. The orthogonal projection matrices on $\mathcal{X}$ and $\mathcal{X}^\perp$ are

$$P = U U^T \qquad (P = U U^*) \tag{10}$$
$$Q = I - U U^T \qquad (Q = I - U U^*) \tag{11}$$

Proof. The proof is contained in that of Property A.3-3. □

Lemma A.3-1.
Let V be an inner product vector space and $x, y \in V$ orthogonal vectors. Then
$$\|x+y\|^2 = \|x\|^2 + \|y\|^2$$

Proof.
$$\begin{aligned}\|x+y\|^2 &= \langle (x+y), (x+y) \rangle \\ &= \langle x,x \rangle + \langle x,y \rangle + \langle y,x \rangle + \langle y,y \rangle \\ &= \langle x,x \rangle + \langle y,y \rangle = \|x\|^2 + \|y\|^2 \quad \square\end{aligned}$$

Theorem A.3-1.
Let $\mathcal{X}$ be any subspace of an inner product space and P the orthogonal projection on $\mathcal{X}$. Then
$$\|x - Px\| \leq \|x - y\| \quad \forall y \in \mathcal{X}$$

Proof. Since $(x - Px) \in \mathcal{X}^\perp$, $(Px - y) \in \mathcal{X}$, from Lemma A.3-1 it follows that
$$\|x-y\|^2 = \|x - Px + Px - y\|^2 = \|x - Px\|^2 + \|Px - y\|^2 \quad \square$$

Theorem A.3-1 is called the *orthogonal projection theorem* and states that in an inner product space the orthogonal projection of any vector on a subspace is the vector of this subspace whose distance (in the sense of euclidean norm) from the projected vector is minimal.

Theorem A.3-2.
Let $\mathcal{X}$ be any subspace of $\mathbf{R}^n$ and X a basis matrix of $\mathcal{X}$. Orthogonal projection matrices on $\mathcal{X}$ and $\mathcal{X}^\perp$ are:

$$P = X(X^T X)^{-1} X^T \qquad (12)$$
$$Q = I - X(X^T X)^{-1} X^T \qquad (13)$$

Proof. Let $h := \dim \mathcal{X}$; note that the $n \times n$ matrix $X^T X$ is nonsingular since $\rho(X^T X) = \rho(X^T)$ by virtue of Property A.3-6 and $\rho(X^T) = \rho(X) = h$ by Property A.3-5. Any $x \in \mathcal{X}$ can be expressed as $x = Xa$, $a \in \mathbf{R}^h$, so it is easily checked that $Px = x$. On the other hand, any $y \in \mathcal{X}^\perp$, so that $X^T y = 0$, clearly satisfies $Py = 0$. □

Note that (10, 11) perform the same operations as (12, 13); however, in the latter case the basis referred to has not been assumed to be orthonormal.

Theorem A.3-2 suggests an interesting analysis in geometric terms of some intrinsic properties of linear maps. Consider a linear function $A : \mathbf{R}^n \to \mathbf{R}^m$: as an example, consider the particular case reported in Fig. A-13, where $m = n = 3$, $\rho(A) = 2$.

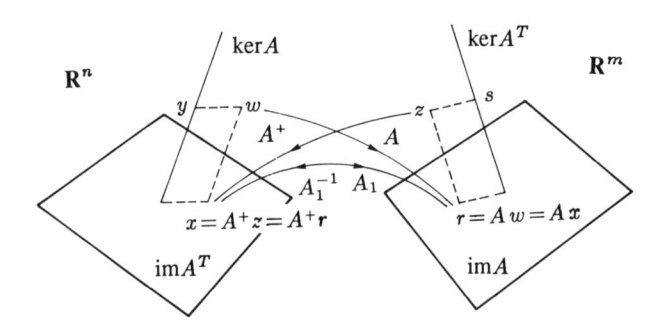

Figure A-13 Characteristic subspaces of a linear map.

Any vector $w \in \mathbf{R}^n$ can be expressed as $w = x + y$, $x \in \operatorname{im} A^T$, $y \in \ker A$, so that $Aw = A(x+y) = Ax = r$. Hence, the linear function A can be considered as the composition of the orthogonal projection on $\operatorname{im} A^T$ and the linear function $A_1 : \operatorname{im} A^T \to \operatorname{im} A$ defined as $A_1(x) = A(x)$ for all $x \in \operatorname{im} A^T$, which is invertible because any basis of $\operatorname{im} A^T$ is transformed by A into a basis of $\operatorname{im} A$ (Property A.3-6).

In order to extend the concept of invertibility of a linear map it is possible to introduce its *pseudoinverse* $A^+ : \mathbf{R}^m \to \mathbf{R}^n$ which works in a similar way: to any vector $z \in \mathbf{R}^m$ it associates the unique vector $x \in \operatorname{im} A^T$ which corresponds in A_1^{-1} to the orthogonal projection r of z on $\operatorname{im} A$. Note that A^+ is unique and that $(A^+)^+ = A$.

Sect. A.3 Inner Product, Orthogonality

Theorem A.3-3.
Let A be any $m \times n$ real matrix. The pseudoinverse of A is

$$A^+ = A^T X (X^T A A^T X)^{-1} X^T \tag{14}$$

where X denotes an arbitrary basis matrix of $\operatorname{im} A$.

Proof. From Property A.3-6 it follows that $A^T X$ is a basis matrix of $\operatorname{im} A^T$, so that according to (12), $A^T X (X^T A A^T X)^{-1} X^T A$ is the matrix that performs the orthogonal projection from $\mathbf{R}^n$ on $\operatorname{im} A^T$. The corresponding linear map can be considered as the composition of $A : \mathbf{R}^n \to \operatorname{im} A$ and $A_1^{-1} : \operatorname{im} A \to \operatorname{im} A^T$, the inverse of one-to-one map A_1 previously defined. Matrix (14) represents A_1^{-1} with respect to the main bases of $\mathbf{R}^n$ and $\mathbf{R}^m$. In fact, let $h := \rho(A) = \rho(A^T)$; $A^T X a$, $a \in \mathbf{R}^h$, is a generic vector belonging to $\operatorname{im} A^T$, and $A A^T X a$ is its image in A or in A_1. By direct substitution

$$A^+ A A^T X a = A^T X (X^T A A^T X)^{-1} X^T A A^T X a = A^T X a$$

Hence, (14) expresses A_1^{-1} when applied to vectors belonging to $\operatorname{im} A$. Furthermore, it maps vectors of $\ker A^T$ into the origin, being $X^T x = 0$ for all $x \in \ker A^T$. $\square$

It is worth investigating the meaning of the pseudoinverse in connection with the matrix linear equation

$$A x = b \tag{15}$$

If (15) admits at least one solution in x, i.e., if $b \in \operatorname{im} A$, $x := A^+ b$ is the solution with the least euclidean norm, i.e., the only solution belonging to $\operatorname{im} A^T$. The set of all solutions is the linear variety

$$\mathcal{X} := \{A^+ b\} + \ker A \tag{16}$$

If, on the other hand, $b \notin \operatorname{im} A$, equation (15) has no solution; in this case the pseudosolution $x := A^+ b$ is the vector having the least euclidean norm transformed by A into a vector whose distance from b is a minimum.

When matrix A has maximum rank, i.e., when its rows or columns are a linearly independent set, the expression of A^+ can be simplified, as stated in the following corollary.

Corollary A.3-2.
Let A be an $m \times n$ real matrix. If $m \leq n$, $\rho(A) = m$, the pseudoinverse of A is

$$A^+ := A^T (A A^T)^{-1} \tag{17}$$

If, on the other hand, $m \geq n$, $\rho(A) = n$, the pseudoinverse of A is

$$A^+ := (A^T A)^{-1} A^T \tag{18}$$

Proof. Referring first to (17), note that $A^T(AA^T)^{-1}A$ is the orthogonal projection matrix from $\mathbf{R}^n$ onto imA^T. Since in this case im$A = \mathbf{R}^m$, there exists for any z at least one w such that $z = Aw$, then $A^T(AA^T)^{-1}z$ is the orthogonal projection of w on imA^T. In order to prove (18), note that $A(A^TA)^{-1}A^T$ is the orthogonal projection matrix from $\mathbf{R}^m$ onto im$A$. Since in this case ker$A = \{0\}$, for any z note that $w := (A^TA)^{-1}A^Tz$ is the unique vector such that Aw coincides with the orthogonal projection of z on im$A$. □

When the pseudoinverse of A can be defined as in (17), its product on the left by A is clearly the $m \times m$ identity matrix, i.e., $AA^+ = I_m$, while in the case where (18) holds, the product on the right of the pseudoinverse by A is the $n \times n$ identity matrix, i.e., $A^+A = I_n$. Hence (17) and (18) are called the *right inverse* and the *left inverse* of matrix A and, in these cases, A is said respectively to be *right-invertible* and *left-invertible*.

Note that relation (18) embodies the least squares method of linear regression, whose geometrical meaning is clarified by the projection theorem; hence, the general pseudoinverse can be advantageously used when a least squares problem admits of more than one solution.

A.4 EIGENVALUES, EIGENVECTORS

Eigenvalues and eigenvectors represent a useful summary of information about the features of linear transformations and are the basic instruments for deriving simplified (canonical) forms of general matrices through similarity transformations. The most important of these forms, the Jordan canonical form, is a powerful means for investigating and classifying linear systems according to their structure. Furthermore, eigenvalues are the protagonists of the paramount problem of control theory, i.e., stability analysis of linear time-invariant systems.

Given any real or complex $n \times n$ matrix A, consider, in the complex field, the equation

$$Ax = \lambda x$$

which can also be written

$$(\lambda I - A)x = 0 \qquad (1)$$

This equation admits nonzero solutions in x if and only if $(\lambda I - A)$ is singular, i.e., if

$$\det(\lambda I - A) = 0 \qquad (2)$$

The left side of equation (2) is a polynomial $p(\lambda)$ of degree n in λ, which is called the *characteristic polynomial* of A: it has real coefficients if A is real. Equation (2) is called the *characteristic equation* of A and admits n roots $\lambda_1, \ldots, \lambda_n$, in general complex, which are called the *eigenvalues* or

Sect. A.4 Eigenvalues, Eigenvectors

characteristic values of A. If A is real, complex eigenvalues are conjugate in pairs. The set of all eigenvalues of A is called the *spectrum* of A.

To each eigenvalue λ_i $(i=1,\ldots,n)$ there corresponds at least one nonzero real or complex vector x_i which satisfies equation (1); this is called an *eigenvector* or *characteristic vector* of A. Since for any eigenvector x_i, αx_i, $\alpha \in \mathbf{R}$, is also an eigenvector, it is convenient to use *normalized eigenvectors*, i.e., eigenvectors with unitary euclidean norm.

Note that, if A is real:

1. the eigenvectors corresponding to complex eigenvalues are complex: in fact $Ax = \lambda x$ cannot be satisfied for λ complex and x real;
2. if λ, x are a complex eigenvalue and a corresponding eigenvector, x^* is an eigenvector corresponding to λ^*: in fact, Ax^* is the conjugate of Ax, $\lambda^* x^*$ the conjugate of λx.

Theorem A.4-1.

Let A be an $n \times n$ real or complex matrix. If the eigenvalues of A are distinct, the corresponding eigenvectors are a linearly independent set.

Proof. Assume, by contradiction, that eigenvectors $x_1,\ldots,x_h$, $h<n$, are linearly independent, while $x_{h+1},\ldots,x_n$ are linearly dependent on them, so that

$$x_j = \sum_{i=1}^{n} \alpha_{ij} x_j \quad (j = h+1,\ldots,n)$$

Since x_i is an eigenvector, it follows that

$$Ax_j = \lambda_j x_j = \sum_{i=1}^{h} \alpha_{ij} x_i \quad (j = h+1\ldots,n)$$

$$Ax_j = A\left(\sum_{i=1}^{h} \alpha_{ij} x_i\right) = \sum_{i=1}^{h} \alpha_{ij} A x_i$$

$$= \sum_{i=1}^{h} \alpha_{ij} \lambda_i x_i \quad (j = h+1,\ldots,n)$$

By difference, we obtain in the end

$$0 = \sum_{i=1}^{n} \alpha_{ij}(\lambda_j - \lambda_i) x_i \quad (j = h+1,\ldots,n)$$

Since the scalars α_{ij} cannot all be zero and the set $\{x_1,\ldots,x_h\}$ is linearly independent, it follows that $\lambda_i = \lambda_j$ for at least one pair of indexes i, j. □

Similarity transformations of matrices were introduced in Subsection A.2.2. A fundamental property of similar matrices is the following.

Property A.4-1.
Similar matrices have the same eigenvalues.

Proof. Let A, B be similar, so that $B = T^{-1}AT$. Hence

$$\det(\lambda I - T^{-1}AT) = \det(\lambda T^{-1}IT - T^{-1}AT)$$
$$= \det\left(T^{-1}(\lambda I - A)T\right)$$
$$= \det T^{-1} \det(\lambda I - A) \det T$$

Since $\det T^{-1}$ and $\det T$ are different from zero, any λ such that $\det(\lambda I - A) = 0$ also satisfies $\det(\lambda I - B) = 0$. □

Any $n \times n$ real or complex matrix A is called *diagonalizable* if it is similar to a diagonal matrix Λ, i.e., if there exists a similarity transformation T such that $\Lambda = T^{-1}AT$. In such a case, Λ is called the *diagonal form* of A.

Theorem A.4-2.
An $n \times n$ real or complex matrix A is diagonalizable if and only if it admits a linearly independent set of n eigenvectors.

Proof. If. Let $\{t_1, \ldots, t_n\}$ be a linearly independent set of eigenvectors, so that

$$A t_i = \lambda_i t_i \quad (i = 1, \ldots, n) \tag{3}$$

Note that in (3) $\lambda_1, \ldots, \lambda_n$ are not necessarily distinct. Equation (3) can be compacted as

$$AT = T\Lambda \tag{4}$$

where T is nonsingular and Λ is diagonal.

Only if. If A is diagonalizable, there exists a diagonal matrix Λ and a nonsingular matrix T such that (4) and, consequently, (3) hold. Hence, the columns of T must be eigenvectors of A. □

Note that by virtue of Theorem A.4-2 any square matrix A whose eigenvalues are distinct admits a diagonal form in which the elements on the main diagonal are the eigenvalues of A. On the other hand, if A has multiple eigenvalues, it is still diagonalizable only if every multiple eigenvalue corresponds to as many linearly independent eigenvectors as its degree of multiplicity.

In general, the diagonal form is complex also when A is real; this is a difficulty in computations, so that it may be preferable, when A is diagonalizable, to derive a real matrix, similar to A, having a 2×2 submatrix on the main diagonal for each pair of conjugate complex eigenvalues. The corresponding similarity transformation is a consequence of the following lemma.

Sect. A.4 Eigenvalues, Eigenvectors

Lemma A.4-1.
Let $\{u_1+jv_1,\ldots,u_h+jv_h, u_1-jv_1,\ldots,u_h-jv_h\}$ be a linearly independent set in the complex field. Then the set $\{u_1,\ldots,u_h,v_1,\ldots,v_h\}$ is linearly independent in the real field.

Proof. Consider the identity

$$\sum_{i=1}^{h}(\alpha_i+j\beta_i)(u_i+jv_i) + \sum_{i=1}^{h}(\gamma_i+j\delta_i)(u_i-jv_i)$$

$$= \sum_{i=1}^{h}\left((\alpha_i+\gamma_i)u_i + (\delta_i-\beta_i)v_i\right) +$$

$$j\sum_{i=1}^{h}\left((\delta_i+\beta_i)u_i + (\alpha_i-\gamma_i)v_i\right) \tag{5}$$

By contradiction: since, for each value of i, $(\alpha_i+\gamma_i)$, $(\alpha_i-\gamma_i)$, $(\delta_i+\beta_i)$, $(\delta_i-\beta_i)$ are four arbitrary numbers, if set $\{u_1,\ldots,u_h\}$ is linearly dependent it is possible to null the linear combination (5) with coefficients α_i, β_i, γ_i, δ_i ($i=1,\ldots,n$) not all zero. $\square$

Theorem A.4-3.
Let A be a diagonalizable real matrix, λ_i ($i=i,\ldots,h$) its real eigenvalues, $\sigma_i+j\omega_i$, $\sigma_i-j\omega_i$ ($i=1,\ldots,k$), $k=(n-h)/2$, its complex eigenvalues. There exists a similarity transformation T such that

$$B = T^{-1}AT$$

$$= \begin{bmatrix} \lambda_1 & 0 & \cdots & 0 & 0 & 0 & \cdots & 0 & 0 \\ 0 & \lambda_2 & \cdots & 0 & 0 & 0 & \cdots & 0 & 0 \\ \vdots & \vdots & \ddots & \vdots & \vdots & \vdots & \ddots & \vdots & \vdots \\ 0 & 0 & \cdots & \lambda_h & 0 & 0 & \cdots & 0 & 0 \\ 0 & 0 & \cdots & 0 & \sigma_1 & \omega_1 & \cdots & 0 & 0 \\ 0 & 0 & \cdots & 0 & -\omega_1 & \sigma_1 & \cdots & 0 & 0 \\ \vdots & \vdots & \ddots & \vdots & \vdots & \vdots & \ddots & \vdots & \vdots \\ 0 & 0 & \cdots & 0 & 0 & 0 & \cdots & \sigma_k & \omega_k \\ 0 & 0 & \cdots & 0 & 0 & 0 & \cdots & -\omega_k & \sigma_k \end{bmatrix} \tag{6}$$

Proof. Since A is diagonalizable, it admits a set of n linearly independent eigenvectors: let T be the matrix whose columns are the real eigenvectors and the real and imaginary part of each pair of complex eigenvectors. For a general pair of conjugate complex eigenvalues $\sigma\pm j\omega$ and the corresponding eigenvectors $u\pm jv$, relation (1) becomes

$$Au \pm jAv = \sigma u - \omega v \pm j(\sigma v + \omega u)$$

or, by splitting the real and imaginary parts

$$Au = \sigma u - \omega v$$
$$Av = \omega u + \sigma v$$

which imply for B the structure shown in (6), since u and v are columns of the transformation matrix T. □

Unfortunately, not all square matrices are diagonalizable, so that we are faced with the problem of defining a canonical form[1] which is as close as possible to the diagonal form and similar to any $n \times n$ matrix. This is the Jordan form, of paramount importance for getting a deep insight into the structure of linear transformations. On the other hand, unfortunately the Jordan form turns out to be rather critical from the computational standpoint, since the transformed matrix may be ill-conditioned and very sensitive to small parameter variations and rounding errors.

A.4.1 The Schur Decomposition

A much less critical similarity transformation, which can be used for any real or complex square matrix, is considered in the following theorem.

Theorem A.4-4. The Schur Decomposition
Let A be any real or complex $n \times n$ matrix. There exists a unitary similarity transformation U that takes A into the upper-triangular form

$$B = U^*AU = \begin{bmatrix} \lambda_1 & b_{12} & \cdots & b_{1n} \\ 0 & \lambda_2 & \cdots & b_{2n} \\ \vdots & \vdots & \ddots & \vdots \\ 0 & 0 & \cdots & \lambda_n \end{bmatrix} \tag{7}$$

Proof. The argument is by induction. First assume $n=2$ and denote by λ_1 and λ_2 the eigenvalues of A; by u_1 a normalized eigenvector corresponding to λ_1; by u_2 any vector orthogonal to u_1 with unitary euclidean norm. Define W as the 2×2 matrix having u_1, u_2 as columns so that, provided $Au_1 = \lambda_1 u_1$

$$W^*AW = \begin{bmatrix} \langle u_1, Au_1 \rangle & \langle u_1, Au_2 \rangle \\ \langle u_2, Au_1 \rangle & \langle u_2, Au_2 \rangle \end{bmatrix}$$

$$= \begin{bmatrix} \lambda_1 & b_{12} \\ 0 & b_{22} \end{bmatrix} = \begin{bmatrix} \lambda_1 & b_{12} \\ 0 & \lambda_2 \end{bmatrix}$$

---

[1] In a given class of matrices (e.g., general real or complex matrices, or square matrices or idempotent matrices) a canonical form is one that induces a partition under the property of similarity, in the sense that in every subclass only one matrix is in canonical form.

where the last equality is a consequence of Property A.4-1.

Assume that the theorem is true for $(n-1) \times (n-1)$ matrices. Let λ_1 be an eigenvalue of A, u_1 a corresponding normalized eigenvector, and $u_2, \ldots, u_n$ any orthonormal set orthogonal to u_1. By a procedure similar to the previous one, it is proved that the unitary matrix W whose columns are $u_1, \ldots, u_n$ is such that

$$W^*AW = \begin{bmatrix} \lambda_1 & b_{12} & \cdots & b_{1n} \\ 0 & b_{22} & \cdots & b_{2n} \\ \vdots & \vdots & \ddots & \vdots \\ 0 & b_{n2} & \cdots & b_{nn} \end{bmatrix} = \begin{bmatrix} \lambda_1 & b_0 \\ O & B_1 \end{bmatrix}$$

where b_0 denotes a $1 \times (n-1)$ matrix and B_1 an $(n-1) \times (n-1)$ matrix. By the induction hypothesis, there exists an $(n-1) \times (n-1)$ unitary matrix U_1 such that $U_1^* B_1 U_1$ is an upper-triangular matrix. Let

$$V := \begin{bmatrix} 1 & O \\ O & U_1 \end{bmatrix}$$

It follows that

$$V^*(W^*AW)V = \begin{bmatrix} 1 & O \\ O & U_1^* \end{bmatrix} \begin{bmatrix} \lambda_1 & b_0 \\ O & B_1 \end{bmatrix} \begin{bmatrix} 1 & O \\ O & U_1 \end{bmatrix}$$

$$= \begin{bmatrix} \lambda_1 & b \\ O & U_1^* B_1 U_1 \end{bmatrix} \quad \square$$

A.4.2 The Jordan Canonical Form. Part I

An important role in deriving the Jordan canonical form is played by the properties of nilpotent linear maps.

Lemma A.4-2.
A linear map $A : \mathcal{F}^n \to \mathcal{F}^n$ is nilpotent of index $q \leq n$ if and only if all its eigenvalues are zero.

Proof. Only if. Let A be nilpotent of order q. Assume that λ is a nonzero eigenvalue, x a corresponding eigenvector. From $Ax = \lambda x$, it follows that $A^2 x = \lambda Ax, \ldots, A^q x = \lambda A^{q-1} x$. Since λ is nonzero, $A^q x = 0$ implies $A^{q-1} x = A^{q-2} x = \ldots = Ax = 0$, hence $x = 0$, which contradicts the assumption that x is an eigenvector.

If. Let B be an upper-triangular form of A. If all the eigenvalues of A are zero, the main diagonal of B is zero. It is easily seen that in B^2, B^3, ... successive diagonals above the main diagonal vanish, so that at most B^n, and consequently A^n, is a null matrix. $\square$

Theorem A.4-5.
Let $A : \mathcal{F}^n \to \mathcal{F}^n$ be a linear map nilpotent of index q. There exists a similarity transformation T that takes A into the canonical form:

$$B = T^{-1}AT = \begin{bmatrix} B_1 & O & \cdots & O \\ O & B_2 & \cdots & O \\ \vdots & \vdots & \ddots & \vdots \\ O & O & \cdots & B_r \end{bmatrix} \tag{8}$$

where the B_i ($i=1,\ldots,n$) denote $m_i \times m_i$ matrices with $m_1 = q$ and $m_i \leq m_{i-1}$ ($i=2,\ldots,r$) having the following structure:

$$B_i = \begin{bmatrix} 0 & 1 & 0 & \cdots & 0 \\ 0 & 0 & 1 & \cdots & 0 \\ 0 & 0 & 0 & \cdots & 0 \\ \vdots & \vdots & \vdots & \ddots & \vdots \\ 0 & 0 & 0 & \cdots & 0 \end{bmatrix} \quad (i=1,\ldots,r)$$

Described in words, the nilpotent canonical form is a form in which every element that is not on the diagonal just above the main diagonal is zero and the elements of this diagonal are sets of 1's separated by single 0's.

Proof. Consider the sequence of subspaces

$$\mathcal{X}_0 = \{0\}$$
$$\mathcal{X}_i = \ker A^i \quad (i=1,\ldots,q)$$

which can be obtained by means of the recursion algorithm:

$$\mathcal{X}_0 = \{0\}$$
$$\mathcal{X}_i = A^{-1}\mathcal{X}_{i-1} \quad (i=1,\ldots,q)$$

Clearly, $\mathcal{X}_{i-1} = A\mathcal{X}_i$ ($i=1,\ldots,q$) and $\mathcal{X}_0 \subset \mathcal{X}_1 \subset \mathcal{X}_2 \subset \ldots \subset \mathcal{X}_q$, $\mathcal{X}_q = \mathcal{F}^n$. Note that $\dim \mathcal{X}_i = n - \dim \mathcal{Y}_i$, where $\mathcal{Y}_i := \mathrm{im}\, A^i$ (Property A.2-5). Since, for any subspace $\mathcal{Y}$,

$$\dim(A\mathcal{Y}) = \dim \mathcal{Y} - \dim\left(\mathcal{Y} \cap \ker A\right)$$

it follows that

$$\dim \mathcal{Y}_i - \dim \mathcal{Y}_{i-1} = -\dim\left(\mathcal{Y}_{i-1} \cap \ker A\right)$$

so that, denoting the variation of dimension at each step by

$$\delta_i := \dim \mathcal{X}_i - \dim \mathcal{X}_{i-1} = \dim \mathcal{Y}_{i-1} - \dim \mathcal{Y}_i \quad (i=0,\ldots,q)$$

it follows that $\delta_1 \geq \delta_2 \geq \ldots \geq \delta_q$. To show how the canonical form is derived, consider a particular case in which $q=7$ and in the sequence $\mathcal{X}_0, \ldots, \mathcal{X}_7$ the dimensional variations $\delta_1, \ldots, \delta_7$ are (3, 3, 2, 2, 1, 1, 1). Since $\mathcal{X}_0 = 0$, $\mathcal{X}_7 = \mathcal{F}^n$, their sum must be n, hence $n = 13$. Take any vector x belonging to $\mathcal{X}_7$ but not to $\mathcal{X}_6$, so that $A^q x = 0$, $A^{q-1} x \neq 0$. Note that Ax belongs to $\mathcal{X}_6$ but not to $\mathcal{X}_5$ (because $A^{q-1} Ax = 0$, $A^{q-2} Ax \neq 0$) and is linearly independent of x since x does not belong to $\mathcal{X}_6$; by iteration, $A^2 x$ belongs to $\mathcal{X}_5$ but not to $\mathcal{X}_4$ and is linearly independent of $\{x, Ax\}$ because $\{x, Ax\}$ does not belong to $\mathcal{X}_5$, $A^3 x$ belongs to $\mathcal{X}_4$ but not to $\mathcal{X}_3$ and is linearly independent of $\{x, Ax, A^2 x\}$; since $\delta_4 = 2$, it is possible to take another vector y in $\mathcal{X}_4$, linearly independent of $A^3 x$ and which does not belong to $\mathcal{X}_3$, so that $\{x, Ax, A^2 x, A^3 x, y\}$ is a linearly independent set; $A^4 x$ and Ay belong to $\mathcal{X}_3$ but not to $\mathcal{X}_2$, are linearly independent of the previous set, which does not belong to $\mathcal{X}_3$ and are a linearly independent pair since $\delta_3 = \dim \mathcal{X}_3 - \dim \mathcal{X}_2 = 2$. For the same

Sect. A.4 Eigenvalues, Eigenvectors

reason the transformed vectors A^5x, A^2y, which belong to $\mathcal{X}_2$ but not to $\mathcal{X}_1$, form a linearly independent set with the previously considered vectors and, in addition, a further linearly independent vector z can be selected on $\mathcal{X}_2$ since $\delta_2 = 3$. By a similar argument it may be stated that A^6x, A^3y and Az are a linearly independent set and are also linearly independent of the previously considered vectors. In conclusion, the chains

$$x, Ax, A^2x, \ldots, A^6x$$
$$y, Ay, A^2y, A^3y \tag{9}$$
$$z, Az$$

are a basis for $\mathcal{F}^n$. In order to obtain the canonical structure (8) this basis must be rearranged as

$$p_1 = A^6x, \quad p_2 = A^5x, \ldots, \quad p_7 = x$$
$$p_8 = A^3y, \quad p_9 = A^2y, \ldots, \quad p_{11} = y$$
$$p_{12} = Az, \quad p_{13} = z$$

The 1's in the canonical form are due to the fact that, while the first vectors of the above sequences (p_1, p_8 and p_{12}) are transformed by A into the origin (i.e., into the subspace $\mathcal{X}_0$ of the previously considered sequence), and are therefore eigenvectors of A, subsequent vectors are each transformed in the previous one. □

Corollary A.4-1.

All vectors (9) are a linearly independent set if the last vectors of the chains are linearly independent.

Proof. Denote the last vectors of the chains (9) by v_1, v_2, v_3 and the last vectors but one by u_1, u_2, u_3. Assume

$$\alpha_1 u_1 + \alpha_2 u_2 + \alpha_3 u_3 + \beta_1 v_1 + \beta_2 v_2 + \beta_3 v_3 = 0 \tag{10}$$

by multiplying both members by A, it follows that

$$\alpha_1 v_1 + \alpha_2 v_2 + \alpha_3 v_3 = 0$$

hence, v_1, v_2, v_3 being a linearly independent set, $\alpha_1 = \alpha_2 = \alpha_3 = 0$. By substituting in (10), the equality

$$\beta_1 v_1 + \beta_2 v_2 + \beta_3 v_3 = 0$$

is obtained, which implies $\beta_1 = \beta_2 = \beta_3 = 0$, hence vectors in (10) are a linearly independent set. This argument can be easily extended to prove that all vectors (9) are a linearly independent set: by multiplying a linear combination of them by A^6, then by A^5 and so on it is proved that the linear combination is equal to zero only if all the coefficients are zero. □

Theorem A.4-6.
Let $A: \mathcal{F}^n \to \mathcal{F}^n$ be a linear map. There exists a pair of A-invariant subspaces $\mathcal{X}, \mathcal{Y}$ such that $\mathcal{X} \oplus \mathcal{Y} = \mathcal{F}^n$ and, moreover, $A|_\mathcal{X}$ is nilpotent, $A|_\mathcal{Y}$ invertible.

Proof. Let $\mathcal{X}_i := \ker A^i$, so that, as in the proof of the previous theorem,
$$\mathcal{X}_0 \subset \mathcal{X}_1 \subset \mathcal{X}_2 \subset \ldots \subset \mathcal{X}_q$$
where q is the least integer such that $\ker A^q = \ker A^{q-1}$.
Assume $\mathcal{X} := \ker A^q$, $\mathcal{Y} := \operatorname{im} A^q$. By Property A.2-5, $\dim \mathcal{X} + \dim \mathcal{Y} = n$ so that in order to prove that $\mathcal{X} \oplus \mathcal{Y} = \mathcal{F}^n$ it is sufficient to prove that $\mathcal{X} \cap \mathcal{Y} = \{0\}$. Assume, by contradiction, $x \in \mathcal{X}$, $x \in \mathcal{Y}$, $x \neq 0$. From $x \in \mathcal{X}$ it follows that $A^q x = 0$, from $x \in \mathcal{Y}$ it follows that there exists a vector $y \neq 0$ such that $x = A^q y$. Consequently $A^{2q} y = 0$, hence $y \in \ker A^{2q}$: since $\ker A^{2q} = \ker A^q$, it follows that $y \in \ker A^q$, so that $x = A^q y = 0$. $A|_\mathcal{X}$ is nilpotent of index q because $A^q x = 0$ for all $x \in \mathcal{X}$, while $A^{q-1} x$ is different from zero for some $x \in \mathcal{X}$ since $\ker A^{q-1} \subset \mathcal{X}$.
$A|_\mathcal{Y}$ is invertible because $Ax = 0$, $x \in \mathcal{Y}$ implies $x = 0$. In fact, let $x \in \mathcal{Y}$ and $x = A^q y$ for some y. Since $Ax = A^{q+1} y$, $Ax = 0$ implies $y \in \ker A^{q+1}$; from $\ker A^{q+1} = \ker A^q$ it follows that $y \in \ker A^q$, hence $x = A^q y = 0$. □

Theorem A.4-7. The Jordan Canonical Form
Let $A: \mathcal{F}^n \to \mathcal{F}^n$ be a linear map. There exists a similarity transformation T which takes A into the canonical form, called *Jordan form*:

$$J = T^{-1} A T = \begin{bmatrix} B_1 & O & \cdots & O \\ O & B_2 & \cdots & O \\ \vdots & \vdots & \ddots & \vdots \\ O & O & \cdots & B_h \end{bmatrix} \tag{11}$$

where matrices B_i $(i = 1, \ldots, h)$ are as many as the number of distinct eigenvalues of A, and are block-diagonal:

$$B_i = \begin{bmatrix} B_{i1} & O & \cdots & O \\ O & B_{i2} & \cdots & O \\ \vdots & \vdots & \ddots & \vdots \\ O & O & \cdots & B_{i,k_i} \end{bmatrix} \quad (i = 1, \ldots, h)$$

while matrices B_{ij}, which are called *Jordan blocks*, have the following structure:

$$B_{ij} = \begin{bmatrix} \lambda_i & 1 & 0 & \cdots & 0 \\ 0 & \lambda_i & 1 & \cdots & 0 \\ 0 & 0 & \lambda_i & \cdots & 0 \\ \vdots & \vdots & \vdots & \ddots & \vdots \\ 0 & 0 & 0 & \cdots & \lambda_i \end{bmatrix} \quad (i = 1, \ldots, h;\ j = 1, \ldots, k_i)$$

Described in words, the Jordan canonical form is a block-diagonal form such that in every block all elements on the main diagonal are equal to one of the distinct eigenvalues of A, the elements on the diagonal just above the main diagonal are sets of 1's separated by single 0's and all other elements are zero. The dimension of each

Sect. A.4 Eigenvalues, Eigenvectors

block is equal to the multiplicity of the corresponding eigenvalue as a root of the characteristic equation.

Proof. Consider the matrix $(A-\lambda_1 I)$. By virtue of Theorem A.4-6 there exists a pair of $(A-\lambda_1 I)$-invariants $\mathcal{X}$ and $\mathcal{Y}$ such that $\mathcal{X} \oplus \mathcal{Y} = \mathcal{F}^n$ and $(A-\lambda_1 I)|_{\mathcal{X}}$ is nilpotent, while $(A-\lambda_1 I)|_{\mathcal{Y}}$ is invertible. Furthermore, $\mathcal{X} = \ker(A-\lambda_1 I)^{m_1}$, where m_1 is the least integer such that $\ker(A-\lambda_1 I)^{m_1} = \ker(A-\lambda_1 I)^{m_1+1}$.

Note that $\mathcal{X}$ and $\mathcal{Y}$ are not only $(A-\lambda_1 I)$-invariants, but also A-invariants because $(A-\lambda_1 I)x \in \mathcal{X}$ for all $x \in \mathcal{X}$ implies $Ax \in \mathcal{X}$ for all $x \in \mathcal{X}$, since clearly $\lambda_1 x \in \mathcal{X}$. Because of Theorems A.4-5, A.4-6, there exists a similarity transformation T_1 such that

$$C_1 = T_1^{-1}(A-\lambda_1 I)T_1 = \begin{bmatrix} C_{11} & O & \cdots & O & \\ O & C_{12} & \cdots & O & \\ \vdots & \vdots & \ddots & \vdots & O \\ O & O & \cdots & C_{1,k_1} & \\ & & O & & D_1 \end{bmatrix}$$

$$C_{1j} = \begin{bmatrix} 0 & 1 & 0 & \cdots & 0 \\ 0 & 0 & 1 & \cdots & 0 \\ 0 & 0 & 0 & \cdots & 0 \\ \vdots & \vdots & \vdots & \ddots & \vdots \\ 0 & 0 & 0 & \cdots & 0 \end{bmatrix} \quad (j=1,\ldots,k_1)$$

where D_1 is nonsingular. Since $T_1^{-1}(\lambda_1 I)T_1 = \lambda_1 I$, it follows that

$$B_1 = T_1^{-1} A T_1 = \begin{bmatrix} B_{11} & O & \cdots & O & \\ O & B_{12} & \cdots & O & \\ \vdots & \vdots & \ddots & \vdots & O \\ O & O & \cdots & B_{1,k_1} & \\ & & O & & D_1+\lambda_1 I \end{bmatrix}$$

where matrices B_{1j} $(j=1,\ldots,k_1)$ are defined as in the statement. Since λ_1 cannot be an eigenvalue of $D_1+\lambda_1 I$, the sum of the dimensions of the Jordan blocks B_{1i} $(i=i,\ldots,k_1)$ is equal to n_1, the multiplicity of λ_1 in the characteristic polynomial. The same procedure can be applied to $A|_{\mathcal{Y}}$ in connection with eigenvalue λ_2 and so on, until the canonical form (11) is obtained. □

A.4.3 Some Properties of Polynomials

The characteristic polynomial of a matrix represents a significant link between polynomials and matrices; indeed, many important properties of matrices and their functions are related to eigenvalues and eigenvectors.

A deeper insight into the structure of a general linear map can be achieved by studying properties of the minimal polynomial of the corresponding matrix, which will be defined in the next section.

First of all, we briefly recall some general properties of polynomials. Although we are mostly interested in real polynomials, for the sake of generality we will refer to complex polynomials, which can arise from factorizations of real ones in some computational procedures.

A polynomial $p(\lambda)$ is called *monic* if the coefficient of the highest power of λ is one. Note that the characteristic polynomial of a square matrix, as defined at the beginning of this section, is monic.

Given any two polynomials $p(\lambda)$ and $\psi(\lambda)$ with degrees n and m, $m \leq n$, it is well known that there exist two polynomials $q(\lambda)$ with degree $n-m$ and $r(\lambda)$ with degree $\leq m-1$ such that

$$p(\lambda) = \psi(\lambda) q(\lambda) + r(\lambda) \tag{12}$$

$q(\lambda)$ and $r(\lambda)$ are called the *quotient* and the *remainder* of the division of $p(\lambda)$ by $\psi(\lambda)$: their computation is straightforward, according to the polynomial division algorithm, which can be easily performed by means of a tableau and implemented on computers. Note that when $p(\lambda)$ and $\psi(\lambda)$ are monic, $q(\lambda)$ is monic but $r(\lambda)$ in general is not.

When in (12) $r(\lambda)$ is equal to zero, $p(\lambda)$ is said to be *divisible* by $\psi(\lambda)$ and $\psi(\lambda)$ to *divide* $p(\lambda)$ or be a *divisor* of $p(\lambda)$. When, on the other hand, $p(\lambda)$ is not divisible by $\psi(\lambda)$, $p(\lambda)$ and $\psi(\lambda)$ admit a *greatest common divisor* (g.c.d.) which can be computed by means of the euclidean process of the successive divisions: consider the iteration scheme

$$\begin{aligned} p(\lambda) &= \psi(\lambda) q_1(\lambda) + r_1(\lambda) \\ \psi(\lambda) &= r_1(\lambda) q_2(\lambda) + r_2(\lambda) \\ r_1(\lambda) &= r_2(\lambda) q_3(\lambda) + r_3(\lambda) \\ &\cdots\cdots \\ r_{k-2}(\lambda) &= r_{k-1}(\lambda) q_k(\lambda) + r_k(\lambda) \end{aligned} \tag{13}$$

which converges to a division with zero remainder. In fact, the degrees of the remainders are reduced by at least one at every step. The last nonzero remainder is the g.c.d. of $p(\lambda)$ and $\psi(\lambda)$. In fact, let $r_k(\lambda)$ be the last nonzero remainder, which divides $r_{k-1}(\lambda)$ because in the next relation of the sequence the remainder is zero: for the last of (13), it also divides $r_{k-2}(\lambda)$, by virtue of the previous one, it divides $r_{k-3}(\lambda)$, and repeating throughout it is inferred that it also divides $\psi(\lambda)$ and $p(\lambda)$. Hence, $r_k(\lambda)$ is a common divisor of $p(\lambda)$ and $\psi(\lambda)$: it still has to be proved that it is the greatest common divisor, i.e., that $p(\lambda)/r_k(\lambda)$ and $\psi(\lambda)/r_k(\lambda)$ have no common divisor other than a constant. Divide all (13) by $r_k(\lambda)$: it is clear that any other common divisor is also a divisor of $r_1(\lambda)/r_k(\lambda)$, $r_2(\lambda)/r_k(\lambda)$, ..., $r_k(\lambda)/r_k(\lambda) = 1$, so it is a constant.

This procedure can be extended to computation of the g.c.d. of any finite number of polynomials. Let $p_1(\lambda), \ldots, p_h(\lambda)$ be the given polynomials, ordered by nonincreasing degrees: first compute the g.c.d. of $p_1(\lambda)$ and $p_2(\lambda)$ and denote it by $\alpha_1(\lambda)$: then, compute the g.c.d. of $\alpha_1(\lambda)$ and $p_3(\lambda)$ and denote it by $\alpha_2(\lambda)$, and so on. At the last step the g.c.d. of all polynomials is obtained as $\alpha_{h-1}(\lambda)$.

Two polynomials whose g.c.d. is a constant are called *coprime*.

Lemma A.4-3.

Let $p(\lambda)$, $\psi(\lambda)$ be coprime polynomials. There exist two polynomials $\alpha(\lambda)$ and $\beta(\lambda)$ such that

$$\alpha(\lambda) p(\lambda) + \beta(\lambda) \psi(\lambda) = 1 \tag{14}$$

Proof. Derive $r_1(\lambda)$ from the first of (13), then substitute it in the second, derive $r_2(\lambda)$ from the second and substitute in the third, and so on, until the g.c.d. $r_k(\lambda)$ is derived as

$$r_k(\lambda) = \mu(\lambda) p(\lambda) + \nu(\lambda) \psi(\lambda)$$

By dividing both members by $r_k(\lambda)$, which is a constant since $p(\lambda)$ and $\psi(\lambda)$ are coprime, (14) follows. □

A straightforward extension of Lemma A.4-3 is formulated as follows: let $p_1(\lambda), \ldots, p_h(\lambda)$ be any finite number of pairwise coprime polynomials. Then there exist polynomials $\psi_1(\lambda), \ldots, \psi_h(\lambda)$ such that

$$\psi_1(\lambda) p_1(\lambda) + \ldots + \psi_h(\lambda) p_h(\lambda) = 1 \tag{15}$$

Computation of the least common multiple (l.c.m.) of two polynomials $p(\lambda)$ and $\psi(\lambda)$ reduces to that of their g.c.d. In fact, let $\alpha(\lambda)$ be the g.c.d. of $p(\lambda)$ and $\psi(\lambda)$, so that $p(\lambda) = \alpha(\lambda) \gamma(\lambda)$ and $\psi(\lambda) = \alpha(\lambda) \delta(\lambda)$, where $\gamma(\lambda)$, $\delta(\lambda)$ are coprime. The l.c.m. $\beta(\lambda)$ is expressed by

$$\beta(\lambda) = \alpha(\lambda) \gamma(\lambda) \delta(\lambda) = \big(p(\lambda) \psi(\lambda)\big) / \alpha(\lambda) \tag{16}$$

Computation of the l.c.m. of any finite number of polynomials can be performed by steps, like the computation of the g.c.d. described earlier.

A.4.4 Cyclic Invariant Subspaces, Minimal Polynomial

Let us now consider the definition of the minimal polynomial of a linear map. First, we define the minimal polynomial of a vector with respect to a linear map.

Let $A : \mathcal{F}^n \to \mathcal{F}^n$ be a linear map and x any vector in $\mathcal{F}^n$. Consider the sequence of vectors

$$x, \; Ax, \; A^2 x, \; \ldots, \; A^k x, \; \ldots \; : \tag{17}$$

there exists an integer k such that vectors $\{x, Ax, \ldots, A^{k-1}x\}$ are a linearly independent set, while $A^k x$ can be expressed as a linear combination of them, i.e.

$$A^k x = -\sum_{i=0}^{k-1} \alpha_i A^i x \qquad (18)$$

The span of vectors (17) is clearly an A-invariant subspace; it is called a *cyclic invariant subspace of A generated by x*. Let $p(\lambda)$ be the monic polynomial

$$p(\lambda) := \lambda^k + \alpha_{k-1} \lambda^{k-1} + \ldots + \alpha_0$$

so that (18) can be written as

$$p(A) x = 0$$

$p(A)$ is called the *minimal annihilating polynomial* of x (with respect to A). It is easily seen that every annihilating polynomial of x, i.e., any polynomial $\psi(\lambda)$ such that $\psi(A) x = 0$, is divisible by $p(\lambda)$: from

$$\psi(\lambda) = p(\lambda) q(\lambda) + r(\lambda)$$

or

$$\psi(A) x = p(A) q(A) x + r(A) x$$

with $\psi(A) = 0$ by assumption and $p(A) x = 0$ since $p(\lambda)$ is the minimal annihilating polynomial of x, it follows that $r(A) x = 0$, which is clearly a contradiction because $r(\lambda)$ has a lower degree than the minimal annihilating polynomial $p(\lambda)$.

Let $p(\lambda)$ and $\psi(\lambda)$ be the minimal annihilating polynomials of any two vectors x and y: it follows that the l.c.m. of $p(\lambda)$ and $\psi(\lambda)$ is the minimal polynomial that annihilates all the linear combinations of x and y.

The *minimal polynomial of the linear map A* is the minimal polynomial which annihilates any vector $x \in \mathcal{F}^n$ and can be obtained as the l.c.m. of the minimal annihilating polynomials of the vectors $\{e_1, \ldots, e_n\}$ of any basis of $\mathcal{F}^n$ (for instance the main basis). Hence, the minimal polynomial $m(A)$ is the minimal annihilating polynomial of the whole space, i.e., the polynomial with minimal degree such that

$$\ker(m(A)) = \mathcal{F}^n \qquad (19)$$

Lemma A.4-4.
For any polynomial $p(\lambda)$, $\ker(p(A))$ is an A-invariant subspace.

Proof. Let $x \in \ker(p(A))$, so that $p(A) x = 0$. Since A and $p(A)$ commute, it follows that $p(A) A x = A p(A) x = 0$, so that $A x \in \mathcal{X}$. □

Theorem A.4-8.

Let $m(\lambda)$ be the minimal polynomial of A and $p(\lambda)$, $\psi(\lambda)$ coprime polynomials that factorize $m(\lambda)$, i.e., such that $m(\lambda) = p(\lambda)\psi(\lambda)$; define $\mathcal{X} := \ker(p(A))$, $\mathcal{Y} := \ker(\psi(A))$. Then

$$\mathcal{X} \oplus \mathcal{Y} = \mathcal{F}^n$$

and $p(\lambda)$, $\psi(\lambda)$ are the minimal polynomials of the restrictions $A|_\mathcal{X}$ and $A|_\mathcal{Y}$.

Proof. Since $p(\lambda)$ and $\psi(\lambda)$ are coprime, by virtue of Lemma A.4-3 there exist two polynomials $\alpha(\lambda)$, $\beta(\lambda)$ such that

$$\alpha(\lambda)p(\lambda) + \beta(\lambda)\psi(\lambda) = 1$$

hence

$$\alpha(A)p(A)z + \beta(A)\psi(A)z = z \quad \forall z \in \mathcal{F}^n \tag{20}$$

Relation (20) can be rewritten as $z = x + y$, with $x := \beta(A)\psi(A)z$, $y := \alpha(A)p(A)z$. It is easily seen that $x \in \mathcal{X}$, $y \in \mathcal{Y}$, since $p(A)x = \beta(A)m(A)z = 0$, $\psi(A)y = \alpha(A)m(A)z = 0$. Let z be any vector belonging both to $\mathcal{X}$ and $\mathcal{Y}$, so that $p(A)z = \psi(A)z = 0$: in (20) the right side member is zero, hence $z = 0$.

In order to prove that $p(\lambda)$ is the minimal polynomial for the restriction of A to $\mathcal{X}$, let $\mu(\lambda)$ be any annihilating polynomial of this restriction; in the relation

$$\mu(A)\psi(A)z = \psi(A)\mu(A)x + \mu(A)\psi(A)y$$

the first term on the right side is zero because $\mu(A)$ annihilates x, the second term is zero because $\psi(A)y = 0$, hence $\mu(A)\psi(A)z = 0$ and, since z is arbitrary, $\mu(\lambda)\psi(\lambda)$ is divisible by the minimal polynomial $m(\lambda) = p(\lambda)\psi(\lambda)$, so that $\mu(\lambda)$ is divisible by $p(\lambda)$. It follows that $p(\lambda)$ is an annihilating polynomial of the restriction of A to $\mathcal{X}$ which divides any other annihilating polynomial, hence it is the minimal polynomial of this restriction. By a similar argument we may conclude that $\psi(\lambda)$ is the minimal polynomial of the restriction of A to $\mathcal{Y}$. $\square$

Theorem A.4-8 can be extended to any factorization of the minimal polynomial $m(\lambda)$ into a product of pairwise coprime polynomials. Let $\lambda_1, \ldots, \lambda_h$ be the roots of $m(\lambda)$, $m_1, \ldots, m_h$ their multiplicities, so that

$$m(\lambda) = (\lambda - \lambda_1)^{m_1}(\lambda - \lambda_2)^{m_2} \ldots (\lambda - \lambda_h)^{m_h} : \tag{21}$$

the linear map A can be decomposed according to the direct sum of A-invariants

$$\mathcal{X}_1 \oplus \mathcal{X}_2 \oplus \ldots \oplus \mathcal{X}_h = \mathcal{F}^n \tag{22}$$

defined by

$$\mathcal{X}_i := \ker(A - \lambda_i I)^{m_i} \quad (i = 1, \ldots, h) \tag{23}$$

Furthermore, the minimal polynomial of the restriction $A|_{\mathcal{X}_i}$ is

$$(\lambda - \lambda_i)^{m_i} \quad (i = i, \ldots, h)$$

The A-invariant $\mathcal{X}_i$ is called the *eigenspace corresponding to the eigenvalue* λ_i. Clearly, the concept of eigenspace is an extension of that of eigenvector.

Corollary A.4-2.
The minimal polynomial $m(\lambda)$ of a linear map A has the same roots as its characteristic polynomial $p(\lambda)$, each with nongreater multiplicity, so that $m(\lambda)$ is a divisor of $p(\lambda)$.

Proof. Since the minimal polynomial of $A|_{\mathcal{X}_i}$ is $(\lambda-\lambda_i)^{m_i}$ and $\mathcal{X}_i$ is a $(A-\lambda_i I)$-invariant, the restriction $(A-\lambda_i I)|_{\mathcal{X}_i}$ is nilpotent of index m_i, so that by virtue of Lemma A.4-2 all its eigenvalues are zero and consequently the only eigenvalue of $A|_{\mathcal{X}_i}$ is λ_i. Since any decomposition of a linear map induces a partition of the roots of its characteristic polynomial, it follows that $\dim \mathcal{X}_i = n_i$ (the multiplicity of λ_i in the characteristic polynomial of A) and the characteristic polynomial of $A|_{\mathcal{X}_i}$ is $(\lambda-\lambda_i)^{n_i}$. □

As a consequence of Corollary A.4-2, we immediately derive the *Cayley-Hamilton theorem*: the characteristic polynomial $p(\lambda)$ of any linear map A is an annihilator for the whole space, i.e., $p(A)=O$.

A.4.5 The Jordan Canonical Form. Part II

Theorem A.4-7 is the most important result concerning linear maps, so that, besides the proof reported in Subsection A.4.2, which is essentially of a "geometric" type, it is interesting to consider also a proof based on the aforementioned properties of polynomials.

Proof of Theorem A.4-7.
Consider the decomposition (22,23). Since the restriction $(A-\lambda_i I)|_{\mathcal{X}_i}$ is nilpotent of index m_i, according to Theorem A.4-5 for every root λ_i of the minimal polynomial there exist k_i chains of vectors

$$x_1, (A-\lambda_i)x_1, (A-\lambda_i)^2 x_1, \ldots, (A-\lambda_i)^{m_{i1}-1} x_1$$
$$x_2, (A-\lambda_i)x_2, (A-\lambda_i)^2 x_2, \ldots, (A-\lambda_i)^{m_{i2}-1} x_2$$
$$\ldots\ldots\ldots$$

which form a basis for $\mathcal{X}_i$. The length of the first chain is $m_{i1}=m_i$, the multiplicity of λ_i as a root of the minimal polynomial. Let $v_{ij\ell}$ ($i=1,\ldots,h$; $j=1,\ldots,k_i$; $\ell=1,\ldots,m_{ij}$) be the vectors of the above chains considered in reverse order, so that

$$\begin{aligned} v_{ij,\ell-1} &= (A-\lambda_i I) v_{ij\ell} \quad \text{hence} \quad A v_{ij\ell} = \lambda_i v_{ij\ell} + v_{ij\ell-1} \\ & (i=1,\ldots,h;\ j=1,\ldots,k_i;\ \ell=2,\ldots,m_{ij}) \\ (A-\lambda_i I) v_{ij1} &= 0 \quad \text{hence} \quad A v_{ij1} = \lambda_i v_{ij1} \\ & (i=1,\ldots,h;\ j=1,\ldots,k_i) \end{aligned} \quad (24)$$

The Jordan canonical form is obtained as follows: first, by means of decomposition (22,23), obtain a block diagonal matrix such that each block has one eigenvalue of A as its only eigenvalue: since the product of all characteristic polynomials is equal to the characteristic polynomial of A, the dimension of each block is equal to the

Sect. A.4 Eigenvalues, Eigenvectors

multiplicity of the corresponding eigenvalue as a root of the characteristic equation of A; then, for each block, consider a further change of coordinates, assuming the set of vectors defined previously as the new basis: it is easily seen that every chain corresponds to a Jordan block, so that for each eigenvalue the maximal dimension of the corresponding Jordan blocks is equal to its multiplicity as a root of the minimal polynomial of A. $\square$

Chains (24), which terminate with an eigenvector, are called *chains of generalized eigenvectors* (corresponding to the eigenvalue λ_i) and generate cyclic invariant subspaces that are called *cyclic eigenspaces* (corresponding to the eigenvalue λ_i). They satisfy the following property, which is an immediate consequence of Corollary A.4-1.

Property A.4-2.

Chains of generalized eigenvectors corresponding to the same eigenvalue are a linearly independent set if their last elements are a linearly independent set.

The characteristic polynomial of a general Jordan block B_{ij} can be written as

$$(\lambda - \lambda_i)^{m_{ij}} \quad (i = 1, \ldots, h; \; j = 1, \ldots, k_i) \tag{25}$$

It is called an *elementary divisor* of A and clearly coincides with the minimal polynomial of B_{ij}. The product of all the elementary divisors of A is the characteristic polynomial of A, while the product of all the elementary divisors of maximal degree among those corresponding to the same eigenvalue is the minimal polynomial of A.

A.4.6 The Real Jordan Form

As the diagonal form, the Jordan canonical form may be complex also when A is real, but we may derive a "real" Jordan form by a procedure similar to that developed in the proof of Theorem A.4-3 for the diagonal form, based on the fact that the generalized eigenvectors of a real matrix are conjugate by pairs as the eigenvectors.

In order to briefly describe the procedure, consider a particular case: suppose that $\lambda = \sigma + j\omega$, $\lambda^* = \sigma - j\omega$ are a pair of complex eigenvalues, $p_i = u_i + jv_i$, $p_i^* = u_i - jv_i$ ($i = 1, \ldots, 4$) a pair of chains of generalized eigenvectors corresponding to a pair of 4×4 complex Jordan blocks. Assume the set $(u_1, v_1, u_2, v_2, u_3, v_3, u_4, v_4)$ instead of $(p_1, p_2, p_3, p_4, p_1^*, p_2^*, p_3^*, p_4^*)$ as columns of the transforming matrix T.

Instead of a pair of complex conjugate Jordan blocks a single "real" Jordan block, but with double dimension, is obtained, since structures of blocks change as follows:

$$\begin{bmatrix} \lambda & 1 & 0 & 0 & & & & & \\ 0 & \lambda & 1 & 0 & & & & & \\ 0 & 0 & \lambda & 1 & & & O & & \\ 0 & 0 & 0 & \lambda & & & & & \\ & & & & \lambda^* & 1 & 0 & 0 \\ & & O & & 0 & \lambda^* & 1 & 0 \\ & & & & 0 & 0 & \lambda^* & 1 \\ & & & & 0 & 0 & 0 & \lambda^* \end{bmatrix} \rightarrow$$

$$\begin{bmatrix} \sigma & \omega & 1 & 0 & & & & & \\ -\omega & \sigma & 0 & 1 & & O & & O & \\ & & \sigma & \omega & 1 & 0 & & & \\ O & & -\omega & \sigma & 0 & 1 & & O & \\ & & & & \sigma & \omega & 1 & 0 & \\ O & & O & & -\omega & \sigma & 0 & 1 & \\ & & & & & & \sigma & \omega & \\ O & & O & & O & & -\omega & \sigma \end{bmatrix} \quad (26)$$

In fact, from

$$A p_1 = \lambda p_1$$
$$A p_i = p_{i-1} + \lambda p_i \quad (i = 2, 3, 4)$$
$$A p_1^* = \lambda^* p_1^*$$
$$A p_i^* = p_{i-1}^* + \lambda^* p_i^* \quad (i = 2, 3, 4)$$

which imply the particular structure of the former matrix, by substitution it is possible to derive the equivalent relations

$$A u_1 = \sigma u_1 - \omega v_1$$
$$A v_1 = \omega u_1 + \sigma v_1$$
$$A u_i = u_{i-1} + \sigma u_i - \omega v_i \quad (i = 2, 3, 4)$$
$$A v_i = v_{i-1} + \omega u_i + \sigma v_i \quad (i = 2, 3, 4)$$

which imply the structure of the latter matrix.

A.4.7 Computation of the Characteristic and Minimal Polynomial

Computer-oriented methods to derive the coefficients of the characteristic and minimal polynomial can provide very useful hints in connection with structure and stability analysis of linear systems. Although some specific algorithms to directly compute the eigenvalues of generic square matrices are more accurate, when matrices are sparse (as happens in most system theory problems) it may be more convenient to derive first the coefficients of the characteristic polynomial and then use the standard software for the roots of polynomial equations.

Sect. A.4 Eigenvalues, Eigenvectors

Consider the characteristic polynomial of an $n \times n$ real matrix in the form

$$p(\lambda) = \lambda^n + a_1 \lambda^{n-1} + \ldots + a_{n-1} \lambda + a_n := \det(\lambda I - A) \qquad (27)$$

where coefficients a_i ($i = 1, \ldots, n$) are functions of matrix A. Recall the well known relations between the coefficients and the roots of a polynomial equation

$$
\begin{aligned}
-a_1 &= \sum_i \lambda_i \\
a_2 &= \sum_{i \neq j} \lambda_i \lambda_j \\
&\ldots \\
(-1)^n a_n &= \lambda_1 \lambda_2 \ldots \lambda_n
\end{aligned}
\qquad (28)
$$

On the other hand, by expanding the determinant on the right of (27) it immediately turns out that

$$a_1 = -\operatorname{tr} A = -(a_{11} + a_{22} + \ldots + a_{nn}) \qquad (29)$$

while setting $\lambda = 0$ yields

$$a_n = \det(-A) = (-1)^n \det A \qquad (30)$$

The elements of matrix $\lambda I - A$ are polynomials in λ with real coefficients. Theorem A.2-5 allows us to set the equality

$$
\begin{aligned}
(\lambda I - A)^{-1} &= \frac{\operatorname{adj}(\lambda I - A)}{\det(\lambda I - A)} \\
&= \frac{\lambda^{n-1} B_0 + \lambda^{n-2} B_1 + \ldots + \lambda B_{n-2} + B_{n-1}}{\det(\lambda I - A)}
\end{aligned}
\qquad (31)
$$

where B_i ($i = 0, \ldots, n-1$) denote real $n \times n$ matrices. It follows from (31) that the inverse of $\lambda I - A$ is an $n \times n$ matrix whose elements are ratios of polynomials in λ with real coefficients such that the degree of the numerator is, at most, $n-1$ and that of the denominator n.

Algorithm A.4-1. (Souriau-Leverrier)

Joint computation of matrices B_i ($i = 0, \ldots, n-1$) of relation (31) and coefficients a_i ($i = 1, \ldots, n$) of the characteristic polynomial (27) is performed by means of the recursion formulae:[2]

---

[2] The last formula on the left of (32) is not strictly necessary and can be used simply as a check of computational precision.

$$B_0 = I \qquad\qquad a_1 = -\operatorname{tr} A$$
$$B_1 = A B_0 + a_1 I \qquad\qquad a_2 = -(1/2)\operatorname{tr}(A B_1)$$
$$\ldots \qquad\qquad \ldots$$
$$B_i = A B_{i-1} + a_i I \qquad\qquad a_{i+1} = -(1/(i+1))\operatorname{tr}(A B_i) \qquad (32)$$
$$\ldots \qquad\qquad \ldots$$
$$B_{n-1} = A B_{n-2} + a_{n-1} I \qquad\qquad a_n = -(1/n)\operatorname{tr}(A B_{n-1})$$
$$O = A B_{n-1} + a_n I$$

Proof. From (27, 31) it follows that

$$(\lambda I - A)(\lambda^{n-1} B_0 + \ldots + \lambda B_{n-2} + B_{n-1})$$
$$= (\lambda^n + a_1 \lambda^{n-1} + \ldots + a_n) I \qquad (33)$$

Formulae for the B_i's are immediately derived by equating the corresponding coefficients on the left and right sides of (33).

Let s_i be the sum of the i-th powers of the roots of the characteristic equation; consider the following well known Newton formulae:

$$a_1 = -s_1$$
$$2 a_2 = -(s_2 + a_1 s_1)$$
$$3 a_3 = -(s_3 + a_1 s_2 + a_2 s_1) \qquad (34)$$
$$\ldots$$
$$n a_n = -(s_n + a_1 s_{n-1} + \ldots + a_{n-1} s_1)$$

From (29) it follows that $s_1 = \operatorname{tr} A$. Since the eigenvalues of A^i are the i-th powers of the eigenvalues of A, clearly

$$s_i = \operatorname{tr} A^i \quad (i = 1, \ldots, n)$$

so that the i-th of (34) can be written as

$$i a_i = -\operatorname{tr}(A^i + a_1 A^{i-1} + \ldots + a_{i-1} A) \quad (i = 1, \ldots, n) \qquad (35)$$

On the other hand, matrices B_i provided by the algorithm are such that

$$B_{i-1} = A^{i-1} + a_1 A^{i-2} + \ldots + a_{i-2} A + a_{i-1} I \quad (i = 1, \ldots, n) \qquad (36)$$

Formulae (32) for the a_i's are immediately derived from (35, 36). □

The Souriau-Leverrier algorithm allows us to develop arguments to prove the main properties of the characteristic and minimal polynomial alternative to those presented in Subsection A.4-4. For instance, the Cayley-Hamilton theorem, already derived as a consequence of Corollary A.4-2, can be stated and proved as follows.

Sect. A.4 Eigenvalues, Eigenvectors

Theorem A.4-9. (Cayley-Hamilton)
Every square matrix satisfies its characteristic equation.

Proof. By eliminating matrices $B_{n-1}, B_{n-2}, \ldots, B_0$ one after the other, proceeding from the last to the first of the formulae on the left of (32), it follows that

$$O = A^n + a_1 A^{n-1} + \ldots + a_n I \qquad \square$$

We recall that the minimal polynomial of A is the polynomial $m(\lambda)$ with minimal degree such that $m(A) = O$. Of course, to assume the minimal polynomial to be monic does not affect the generality. The minimal polynomial is unique: in fact, the difference of any two monic polynomials with the same degree that are annihilated by A, is also annihilated by A and its degree is less by at least one. Furthermore, the minimal polynomial is a divisor of every polynomial $p(\lambda)$ such that $p(A) = O$: in fact, by the division rule of polynomials

$$p(\lambda) = m(\lambda) + r(\lambda)$$

[where it is known that the degree of the remainder $r(\lambda)$ is always at least one less than that of the divisor $m(\lambda)$] equalities $p(A) = O$, $m(A) = O$ imply $r(A) = O$, which contradicts the minimality of $m(\lambda)$ unless $r(\lambda)$ is zero. An algorithm to derive the minimal polynomial is provided by the following theorem.

Theorem A.4-10.
The minimal polynomial of an $n \times n$ matrix A can be derived as

$$m(\lambda) = \frac{\det(\lambda I - A)}{b(\lambda)}$$

where $b(\lambda)$ denotes the greatest common divisor monic of all the minors of order $n-1$ of matrix $\lambda I - A$, i.e., of the elements of $\mathrm{adj}(\lambda I - A)$.

Proof. By definition, $b(\lambda)$ satisfies the relation

$$\mathrm{adj}(\lambda I - A) = b(\lambda) B(\lambda)$$

where $B(\lambda)$ is a polynomial matrix whose elements are coprime (i.e., have a greatest common divisor monic equal to one). Let $p(\lambda) := \det(\lambda I - A)$: from (31) it follows that

$$p(\lambda) I = b(\lambda) (\lambda I - A) B(\lambda) \qquad (37)$$

which means that $b(\lambda)$ is a divisor of $p(\lambda)$. Let

$$\varphi(\lambda) := \frac{p(\lambda)}{b(\lambda)}$$

so that (37) can be written as

$$\varphi(\lambda) = (\lambda I - A) B(\lambda) \qquad (38)$$

which, by an argument similar to that developed in the proof of Theorem A.4-9, implies $\varphi(A) = O$. Hence, the minimal polynomial $m(\lambda)$ must be a divisor of $\varphi(\lambda)$, so there exists a polynomial $\psi(\lambda)$ such that

$$\varphi(\lambda) = m(\lambda) \psi(\lambda) \qquad (39)$$

Since

$$\lambda^i I - A^i = (\lambda I - A)(\lambda^{i-1} I - \lambda^{i-2} A + \ldots + A^{i-1})$$

by simple manipulations we obtain

$$m(\lambda I) - m(A) = (\lambda I - A) C(\lambda)$$

where $C(\lambda)$ denotes a proper polynomial matrix and, since $m(A) = O$

$$m(\lambda) I = (\lambda I - A) C(\lambda)$$

From the preceding relation it follows that

$$\varphi(\lambda) I = m(\lambda) \psi(\lambda) I = \psi(\lambda) (\lambda I - A) C(\lambda)$$

hence, by (38)

$$B(\lambda) = \psi(\lambda) C(\lambda)$$

Since the g.c.d. of the elements of $B(\lambda)$ is a constant, $\psi(\lambda)$ must be a constant, say k. Relation (39) becomes $\varphi(\lambda) = k m(\lambda)$; recalling that $\varphi(\lambda)$ and $m(\lambda)$ are monic, we finally get $\varphi(\lambda) = m(\lambda)$. □

A.5 HERMITIAN MATRICES, QUADRATIC FORMS

It will be shown in this section that the eigenvalues and the eigenvectors of real symmetric or, more generally, hermitian matrices, have special features that result in much easier computability.

Theorem A.5-1.
   The eigenvalues of any hermitian matrix are real.

   Proof. Let A be a hermitian square matrix, λ an eigenvalue of A, and x a corresponding normalized eigenvector, so that

$$A x = \lambda x$$

Taking the left inner product by x yields

$$\langle x, Ax \rangle = \langle x, \lambda x \rangle = \lambda \langle x, x \rangle = \lambda$$

On the other hand, since A is hermitian, it follows that $\langle x, Ax \rangle = \langle Ax, x \rangle = \langle x, Ax \rangle^*$, i.e., the right side of the relation is real, hence λ is real. □

Sect. A.5 Hermitian Matrices, Quadratic Forms

Property A.5-1.
Any hermitian matrix is diagonalizable by means of a unitary transformation.

Proof. Consider the Schur decomposition

$$A = URU^*$$

From $A = A^*$ it follows that $URU^* = UR^*U^*$, hence $R = R^*$ so that R, as a hermitian upper triangular matrix, must be diagonal. □

The following corollaries are consequences of previous statements.

Corollary A.5-1.
Any real symmetric matrix is diagonalizable by means of an orthogonal transformation.

Corollary A.5-2.
Any hermitian matrix admits a set of n orthonormal eigenvectors.

Symmetric and hermitian matrices are often used in connection with quadratic forms, which are defined as follows.

Definition A.5-1. Quadratic Form
A *quadratic form* is a function $q : \mathbf{R}^n \to \mathbf{R}$ or $q : \mathbf{C}^n \to \mathbf{R}$ expressed by

$$q(x) = \langle x, Ax \rangle = x^T A x = \sum_{i=1}^{n} \sum_{j=1}^{n} a_{ij} x_i x_j \tag{1}$$

or

$$q(x) = \langle x, Ax \rangle = x^* A x = \sum_{i=1}^{n} \sum_{j=1}^{n} a_{ij} x_i^* x_j \tag{2}$$

with A symmetric in (1) and hermitian in (2).

A quadratic form is said to be *positive (negative) definite* if $\langle x, Ax \rangle > 0$ (< 0) for all $x \neq 0$, *positive (negative) semidefinite* if $\langle x, Ax \rangle \geq 0$ (≤ 0) for all $x \neq 0$.

In the proofs of the following theorems we will consider only the more general case of A being hermitian. When A is symmetric, unitary transformations are replaced with orthogonal transformations.

Theorem A.5-2.
A quadratic form $\langle x, Ax \rangle$ is positive (negative) definite if and only if all the eigenvalues of A are positive (negative); it is positive (negative) semidefinite if and only if all the eigenvalues of A are nonnegative (nonpositive).

Proof. By virtue of Property A.5-1, the following equalities can be set:

$$\langle x, Ax \rangle = \langle x, U\Lambda U^* x \rangle = \langle U^* x, \Lambda U^* x \rangle$$

Let $z := U^* x$, so that

$$\langle x, Ax \rangle = \langle z, \Lambda z \rangle = \sum_{i=1}^{n} \lambda_i z_i^2$$

which proves both the if and the only if parts of the statement, since the correspondence between x and z is one-to-one. □

By extension, a symmetric or hermitian matrix is said to be *positive (negative) definite* if all its eigenvalues are positive (negative), *positive (negative) semidefinite* if all its eigenvalues are nonnegative (nonpositive). The following theorem states a useful criterion, called the *Sylvester criterion*, to test positive definiteness without any eigenvalue computation.

Theorem A.5-3. (Sylvester)
Let A be a symmetric matrix and A_i ($i = 1, \ldots, n$) be the successive leading principal submatrices of A, i.e., submatrices on the main diagonal of A whose elements belong to the first i rows and i columns of A. The quadratic form $\langle x, Ax \rangle$ is positive definite if and only if $\det A_i > 0$ ($i = 1, \ldots, n$), i.e., if and only if the n successive leading principal minors of A are positive.

Proof. Only if. Note that $\det A$ is positive since all the eigenvalues of A are positive by virtue of Theorem A.5-2. Positive definite "reduced" quadratic forms are clearly obtained by setting one or more of the variables x_i equal to zero: since their matrices coincide with principal submatrices of A, all the principal minors, in particular those considered in the statement, are positive.

If. By induction, we suppose that $\langle x, A_{k-1} x \rangle$ is positive definite and that $\det A_{k-1} > 0$, $\det A_k > 0$ ($2 \le k \le n$) and we prove that $\langle x, A_k x \rangle$ is positive definite. The stated property clearly holds for $k=1$. Consider

$$A_k = \begin{bmatrix} A_{k-1} & a \\ a^T & a_{kk} \end{bmatrix}$$

By virtue of relation (A.2.25) it follows that

$$\det A_k = \det A_{k-1} \cdot \left(a_{kk} - \langle a, A_{k-1}^{-1} a \rangle \right)$$

hence, $a_{kk} - \langle a, A_{k-1}^{-1} a \rangle > 0$.

Consider the quadratic form $\langle x, A_k x \rangle$ and assume $x = Pz$, where P is the nonsingular matrix defined by

$$P := \begin{bmatrix} I & -A_{k-1}^{-1} \\ O & 1 \end{bmatrix}$$

Clearly, $\langle x, A_k x \rangle = \langle z, P^T A_k P z \rangle$ is positive definite, since

$$P^T A_k P = \begin{bmatrix} A_{k-1} & O \\ O & a_{kk} - \langle a, A_{k-1}^{-1} a \rangle \end{bmatrix} \quad \square$$

Corollary A.5-3.

The quadratic form $\langle x, Ax \rangle$ is negative definite if and only if $(-1)^i \det A_i > 0$ ($i = 1, \ldots, n$), i.e., if and only if the n successive leading principal minors of A are alternatively negative and positive.

Proof. Apply Theorem A.5-3 to the quadratic form $\langle x, -Ax \rangle$, which clearly is definite positive. □

The Sylvester criterion allows the positive (negative) definiteness of a quadratic form to be checked by considering the determinants of n symmetric matrices, each obtained by bordering the previous one. On the other hand, it is easily shown that a quadratic form is positive (negative) semidefinite if and only if *all* its principal minors are nonnegative (nonpositive).[1]

Theorem A.5-4.

Let the quadratic form $\langle x, Ax \rangle$ be positive (negative) semidefinite. Then

$$\langle x_1, Ax_1 \rangle = 0 \quad \Leftrightarrow \quad x_1 \in \ker A$$

Proof. Since $\ker A = \ker(-A)$, we can assume that A is positive semidefinite without any loss of generality. By Property A.5-1 and Theorem A.5-2, $A = U\Lambda U^*$, where Λ is a diagonal matrix of nonnegative real numbers (the eigenvalues of A). Let $B := U\sqrt{\Lambda}U^*$: clearly $A = BB$, with B symmetric, positive semidefinite and such that $\operatorname{im} B = \operatorname{im}(BB) = \operatorname{im} A$ (by virtue of Property A.3-6); hence, $\ker B = \ker A$. Therefore, the expression $\langle x_1, Ax_1 \rangle = 0$ is equivalent to $\langle Bx_1, Bx_1 \rangle = 0$, which implies $x_1 \in \ker B$, or $x_1 \in \ker A$. □

A.6 METRIC AND NORMED SPACES, NORMS

The introduction of a metric in a vector space becomes necessary when approaching problems requiring a quantitative characterization of the elements of such spaces. In connection with dynamic system analysis, the need of a metric in the state space arises when convergence of trajectories must be considered, as, for instance, in the study of stability from the most general standpoint. In this section, besides introducing basic concepts relating to metrics and norms, a constructive proof is presented of the main existence and uniqueness theorem

---

[1] Swamy [19] reports the following example. The eigenvalues of matrix

$$A := \begin{bmatrix} 1 & 1 & 1 \\ 1 & 1 & 1 \\ 1 & 1 & 0 \end{bmatrix}$$

are $0, 1+\sqrt{3}, 1-\sqrt{3}$ so that A is not positive semidefinite, even if its three successive principal minors are nonnegative $(1, 0, 0)$.

of differential equations solutions, which is very important in connection with the definition itself and analysis of the basic properties of dynamic systems.

The concept of metric space is as primitive and general as the concept of set, or rather is one of the simplest specializations of the concept of set: a metric space is a set with a criterion for evaluating distances. In axiomatic form it is defined as follows.

Definition A.6-1. Metric Space, Metric

A *metric space* is a set $\mathcal{M}$ with a function $\delta(\cdot,\cdot) : \mathcal{M} \times \mathcal{M} \to \mathbf{R}$, called *metric*, which satisfies

1. positiveness:
$$\delta(x,y) \geq 0 \quad \forall x,y \in \mathcal{M}$$
$$\delta(x,y) = 0 \iff x = y$$

2. symmetry:
$$\delta(x,y) = \delta(y,x) \quad \forall x,y \in \mathcal{M}$$

3. triangle inequality:
$$\delta(x,z) \leq \delta(x,y) + \delta(y,z) \quad \forall x,y,z \in \mathcal{M}$$

In Section A.3 the euclidean norm has been defined as a consequence of the inner product. The following axiomatic definition of a norm is not related to an inner product and leads to a concept of normed space that is completely independent of that of inner product space.

Definition A.6-2. Normed Space, Norm

A vector space $\mathcal{V}$ over a field $\mathcal{F}$ (with $\mathcal{F}=\mathbf{R}$ or $\mathcal{F}=\mathbf{C}$) is called a *normed space* if there exists a function $\|\cdot\| : \mathcal{V} \to \mathbf{R}$, called *norm*, which satisfies

1. positiveness:
$$\|x\| \geq 0 \quad \forall x \in \mathcal{V}$$
$$\|x\| = 0 \iff x = 0$$

2. commutativity with product by scalars:
$$\|\alpha x\| = |\alpha| \, \|x\| \quad \forall \alpha \in \mathcal{F}, \; \forall x \in \mathcal{V}$$

3. triangle inequality:
$$\|x+y\| \leq \|x\| + \|y\| \quad \forall x,y \in \mathcal{V}$$

Note that in the field of scalars $\mathcal{F}$, a norm is represented by the "absolute value" or "modulus." Every normed space is also a metric space: in fact it is possible to assume $\delta(x,y) := \|x-y\|$. In the sequel, only the symbol $\|x-y\|$ will be used to denote a distance, i.e., only metrics induced by norms will be considered, since greater generality is not necessary.

More than one norm can be defined in the same vector space, as the following examples clarify.

Sect. A.6 Metric and Normed Spaces, Norms

Example A.6-1.

In $\mathbf{R}^n$ or $\mathbf{C}^n$ the following norms are used:

$$\|x\|_1 := \sum_{i=1}^{n} |x_i| \tag{1}$$

$$\|x\|_2 := \sqrt{\sum_{i=1}^{n} |x_i|^2} \tag{2}$$

$$\|x\|_\infty := \sup_{1 \leq i \leq n} |x_i| \tag{3}$$

Their geometrical meaning in space $\mathbf{R}^2$ is illustrated in Fig. A-14, where the shapes of some constant-norm loci are represented.

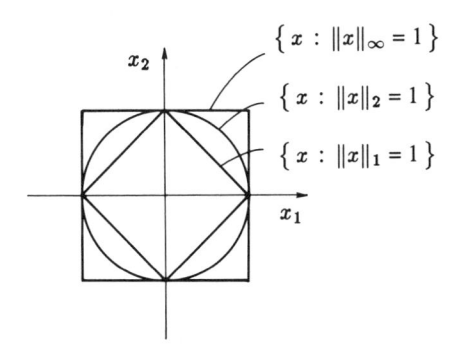

Figure A-14 Constant norm loci in $\mathbf{R}^2$.

Example A.6-2.

In the vector space of infinite sequences $s(\cdot) : \mathbf{Z} \to \mathbf{R}^n$ or $s(\cdot) : \mathbf{Z} \to \mathbf{C}^n$ norms similar to the previous ones are defined as:

$$\|s(\cdot)\|_1 := \sum_{i=1}^{\infty} \|s(i)\|_1 \tag{4}$$

$$\|s(\cdot)\|_2 := \sqrt{\sum_{i=1}^{\infty} \|s(i)\|_2^2} \tag{5}$$

$$\|s(\cdot)\|_\infty := \sup_{1 \leq i < \infty} \|s(i)\|_\infty \tag{6}$$

Example A.6-3.

In the vector space of functions $f(\cdot) : T \to \mathbf{R}^n$ or $s(\cdot) : T \to \mathbf{C}^n$, where T denotes the set of all nonnegative real numbers, the most common norms are:

$$\|f(\cdot)\|_1 := \int_0^\infty \|f(t)\|_1 \, dt \tag{7}$$

$$\|f(\cdot)\|_2 := \sqrt{\int_0^\infty \|f(t)\|_2^2 \, dt} \tag{8}$$

$$\|f(\cdot)\|_\infty := \sup_{t \in T} \|f(t)\|_\infty \tag{9}$$

The norm $\|x\|_2$ is the euclidean norm already introduced in Section A.3 as that induced by the inner product; in the sequel the symbol $\|\cdot\|$, without any subscript, will be referred to a generic norm.

Clearly, all the previously defined norms satisfy axioms 1 and 2 of Definition A.6-2. In the case of euclidean norms the triangle inequality is directly related to the following result.

Theorem A.6-1.

Let V be an inner product space. The euclidean norm satisfies the *Schwarz inequality*:

$$|\langle x, y \rangle| \leq \|x\|_2 \|y\|_2 \quad \forall x, y \in V \tag{10}$$

Proof. From

$$0 \leq \langle x+\alpha y, x+\alpha y \rangle = \langle x, x \rangle + \alpha \langle x, y \rangle + \alpha^* \langle y, x \rangle + \alpha \alpha^* \langle y, y \rangle$$

assuming $y \neq 0$ (if y is zero (10) clearly holds) and $\alpha := -\langle y, x \rangle / \langle x, x \rangle$, it follows that

$$\langle x, x \rangle \langle y, y \rangle \geq \langle x, y \rangle \langle y, x \rangle = |\langle x, y \rangle|^2$$

which leads to (10) by extraction of the square root. □

In order to prove that euclidean norms satisfy triangle inequality, consider the following manipulations:

$$\begin{aligned}
\|x+y\|_2^2 &= \langle x+y, x+y \rangle \\
&= \langle x, x \rangle + \langle x, y \rangle + \langle y, x \rangle + \langle y, y \rangle \\
&\leq \|x\|_2^2 + 2|\langle x, y \rangle| + \|y\|_2^2
\end{aligned}$$

then use (10) to obtain

$$\|x+y\|_2^2 \leq \|x\|_2^2 + 2\|x\|_2 \|y\|_2 + \|y\|_2^2 = \left(\|x\|_2 + \|y\|_2\right)^2$$

Sect. A.6 Metric and Normed Spaces, Norms

which is the square of the triangle inequality.

The norms considered in the previous examples are particular cases of the more general norms, called *p-norms*

$$\|x\|_p := \left(\sum_{i=1}^{n} |x_i|^p\right)^{\frac{1}{p}} \quad (1 \leq p \leq \infty) \tag{11}$$

$$\|s(\cdot)\|_p := \left(\sum_{i=1}^{\infty} \|s(i)\|_p^p\right)^{\frac{1}{p}} \quad (1 \leq p \leq \infty) \tag{12}$$

$$\|f(\cdot)\|_p := \left(\int_0^{\infty} \|f(t)\|_p^p \, dt\right)^{\frac{1}{p}} \quad (1 \leq p \leq \infty) \tag{13}$$

in particular, the norms subscripted with ∞ are the limits of (11–13) as p approaches infinity.

It is customary to denote with l_p the space of sequences $s(\cdot)$ measurable according to the norm (12), i.e., such that the corresponding infinite series converges, and with $l_p(n)$ the spaces of sequences of n terms; clearly l_p is infinite-dimensional, while $l_p(n)$ is finite-dimensional.

Similarly, L_p denotes the space of functions $f(\cdot)$ measurable according to (13), i.e., such that the corresponding improper integral exists, while $L_p[t_0, t_1]$ denotes the space of functions defined in the finite interval $[t_0, t_1]$, for which, in norm (13), the integration interval is changed accordingly.

For the most general norms the proof of the triangle inequality is based on the following *Hölder inequalities*:

$$\sum_{i=1}^{n} |x_i y_i| \leq \left(\sum_{i=1}^{n} |x_i|^p\right)^{\frac{1}{p}} \left(\sum_{i=1}^{n} |y_i|^q\right)^{\frac{1}{q}} \quad \left(\frac{1}{p} + \frac{1}{q} = 1\right) \tag{14}$$

$$\int_a^b |x(t) y(t)| \, dt \leq \left(\int_a^b |x(t)|^p \, dt\right)^{\frac{1}{p}} \left(\int_a^b |y(t)|^q \, dt\right)^{\frac{1}{q}} \quad \left(\frac{1}{p} + \frac{1}{q} = 1\right) \tag{15}$$

which generalize Schwarz inequality (10). The proofs of Hölder inequalities and consequent triangle inequality are omitted here. Also without proof we report the following property, which clearly holds in the particular cases considered in Fig. A-14.

Property A.6-1.
Norms (11–13) satisfy the inequalities

$$\|\cdot\|_i \geq \|\cdot\|_j \quad \text{for } i \leq j \tag{16}$$

Norms of transformations are often used in conjunction with norms of vectors to characterize the "maximum variation of length" of the image of a vector with respect to the vector itself. In the particular case of a linear transformation, the norm is defined as follows.

A.6.1 Matrix Norms

Definition A.6-3. Norm of a Linear Map
Let $\mathcal{V}$ and $\mathcal{W}$ be two normed vector spaces over the same field $\mathcal{F}$ and $\mathcal{L}$ the set of all the linear maps from $\mathcal{V}$ to $\mathcal{W}$. A *norm of the linear map* $A \in \mathcal{L}$ is a function $\|\cdot\| : \mathcal{L} \to \mathbf{R}$ which satisfies

$$\|A\| \geq \sup_{\|x\|=1} \|Ax\| \tag{17}$$

$$\|\alpha A\| = |\alpha| \|A\| \tag{18}$$

If $\mathcal{V} = \mathcal{F}^n$ and $\mathcal{W} = \mathcal{F}^m$, the linear map is represented by an $m \times n$ matrix and the same definition applies for the *norm of the matrix* A.

For the sake of simplicity, only norms of matrices will be considered in the sequel. From (17) and (18) it follows that

$$\|Ax\| \leq \|A\| \|x\| \quad \forall x \in \mathcal{V} \tag{19}$$

Norms of matrices satisfy the following fundamental properties, called the *triangle inequality* and the *submultiplicative property*

$$\|A + B\| \leq \|A\| + \|B\| \tag{20}$$
$$\|AB\| \leq \|A\| \|B\| \tag{21}$$

which are consequences of the manipulations

$$\|(A + B)x\| \leq \|Ax\| + \|Bx\|$$
$$\leq \|A\| \|x\| + \|B\| \|x\| = (\|A\| + \|B\|) \|x\|$$

$$\|ABx\| \leq \|A\| \|Bx\| \leq \|A\| \|B\| \|x\|$$

The most frequently used matrix norms are defined as consequences of the corresponding vector norms, simply by taking the equality sign in relation

(17). Referring to the vector norms defined by (1–3), the corresponding norms of an $m \times n$ matrix A are

$$\|A\|_1 = \sup_{1 \le j \le n} \sum_{i=1}^{m} |a_{ij}| \qquad (22)$$

$$\|A\|_2 = \sqrt{\lambda_M} \qquad (23)$$

$$\|A\|_\infty = \sup_{1 \le i \le m} \sum_{j=1}^{n} |a_{ij}| \qquad (24)$$

In (23) λ_M denotes the greatest eigenvalue of $A^T A$ if A is real, or $A^* A$ if A is complex.

In all three cases relation (19) holds with equality sign for at least one vector x having unitary norm. This property can easily be checked in the cases of norms (22) and (24), while for (23) it is proved by solving the problem of maximizing the quadratic form $\langle Ax, Ax \rangle$ under the constraint $\langle x, x \rangle = 1$. Recall that the quadratic form can be written also $\langle x, A^T A x \rangle$ if A is real, or $\langle x, A^* A x \rangle$ if A is complex.

Refer to the case of A being real and take into account the constraint by means of a Lagrange multiplier λ: the problem is solved by equating to zero the partial derivatives with respect to the components of x of the function

$$f(x) = \langle x, A^T A x \rangle - \lambda(\langle x, x \rangle - 1)$$

i.e.,

$$\operatorname{grad} f(x) = A^T A x - \lambda x = (A^T A - \lambda I)x = 0$$

It follows that at a maximum of $\|Ax\|_2$ vector x is an eigenvector of $A^T A$. On the other hand, if x is an eigenvector of $A^T A$

$$\|y\|_2 = \sqrt{\langle Ax, Ax \rangle} = \sqrt{\lambda \langle x, x \rangle} = \sqrt{\lambda}$$

hence the eigenvector that solves the maximization problem corresponds to the greatest eigenvalue of $A^T A$ and norm $\|A\|_2$ is its square root. The preceding argument extends to the case of A being complex by simply substituting A^T with A^*.

Another interesting matrix norm is the *Frobenius norm*

$$\|A\|_F := \sqrt{\sum_{i=1}^{m} \sum_{j=1}^{n} |a_{ij}|^2} \qquad (25)$$

Property A.6-2.
Norms of matrices (22–25) satisfy the inequality

$$\|A\| \leq \sum_{i=1}^{m} \sum_{j=1}^{n} |a_{ij}| \qquad (26)$$

Proof. In the cases of norms (22) and (24) the validity of (26) is clear, while for (23) it is proved as follows: let x_0 be a normalized eigenvector of $A^T A$ or $A^* A$ corresponding to the greatest eigenvalue λ_0; the components of x_0 have absolute value less than or equal to one, hence the components of the transformed vector $y := A x_0$ are such that

$$|y_i| \leq \sum_{j=1}^{n} |a_{ij}| \quad (i = 1, \ldots, m)$$

This relation, joined to

$$\|A\|_2 = \|y\|_2 = \sqrt{\sum_{i=1}^{m} |y_i|^2} \leq \sum_{i=1}^{m} |y_i|$$

proves (26). The inequality in the latter relation is due to the square of a sum of positive numbers being greater than or equal to the sum of their squares (in fact the expansion of the square also includes the double-products). In the case of norm (25) this inequality directly proves the result. □

Property A.6-3.
The 2-norm (23) and the F-norm (25) are invariant under orthogonal transformations (in the real field) or unitary transformations (in the complex field).

Proof. Refer to the real case and let U $(m \times m)$ and V $(n \times n)$ be orthogonal matrices. Assume $B := U^T A V$: since

$$B^T B - \lambda I = V^T A^T A V - \lambda I = V^T (A^T A - \lambda I) V$$

$B^T B$ and $A^T A$ have the same eigenvalues. Hence, the 2-norm and the F-norm, which are related to these eigenvalues by

$$\|A\|_2 = \sqrt{\sup_i \lambda_i} \qquad \|A\|_F = \sqrt{\operatorname{tr}(A^T A)} = \sqrt{\sum_{i=1}^{n} \lambda_i}$$

are clearly equal. □

Expressing the 2-norm and the F-norm in terms of the eigenvalues of $A^T A$ immediately yields

$$\|A\|_2 \leq \|A\|_F \leq \sqrt{n} \|A\|_2$$

Sect. A.6 Metric and Normed Spaces, Norms

The following definitions are specific of metric spaces.

Definition A.6-4. Sphere
Let V be a normed space. Given $x_0 \in V$, $r \in \mathbf{R}$, the set

$$\mathcal{O}(x_0, r) = \left\{ x : \|x - x_0\| < r \right\}$$

is called an *open sphere* with center at x_0 and radius r. If equality is allowed, i.e., $\|x - x_0\| \le r$, the set is a *closed sphere*.

The open sphere $\mathcal{O}(x_0, \epsilon)$ is also called an ϵ-neighborhood of x_0.

Definition A.6-5. Interior Point of a Set
Let $\mathcal{X}$ be a set in a normed space V. A vector $x \in \mathcal{X}$ is called an *interior point* of $\mathcal{X}$ if there exists a real number $\epsilon > 0$ such that $\mathcal{O}(x, \epsilon) \subset \mathcal{X}$.

Definition A.6-6. Limit Point of a Set
Let $\mathcal{X}$ be a set in a normed space V. A vector x not necessarily belonging to $\mathcal{X}$ is called a *limit point* or *accumulation point* of $\mathcal{X}$ if for any real $\epsilon > 0$ there exists an $y \in \mathcal{X}$ such that $y \in \mathcal{O}(x, \epsilon)$.

Definition A.6-7. Isolated Point of a Set
Let $\mathcal{X}$ be a set in a normed space V. A vector $x \in \mathcal{X}$ is called an *isolated point* of $\mathcal{X}$ if there exists an $\epsilon > 0$ such that $\mathcal{O}(x, \epsilon)$ does not contain any other point of $\mathcal{X}$.

The set of all interior points of $\mathcal{X}$ is called the *interior* of $\mathcal{X}$ and denoted by int$\mathcal{X}$, the set of all limit points is called the *closure* of $\mathcal{X}$ and denoted by clo$\mathcal{X}$. Since in every neighborhood of a point of $\mathcal{X}$ there is a point of $\mathcal{X}$ (the point itself), any set $\mathcal{X}$ is contained in its closure. A *boundary point* of $\mathcal{X}$ is a point of clo$\mathcal{X}$ which is not an interior point of $\mathcal{X}$; the set of all boundary points of $\mathcal{X}$ is called the *boundary* of $\mathcal{X}$.

An *open set* is a set whose points are all interior, a *closed set* is a set that contains all its boundary points or that coincides with its closure.

A typical use of norms is related to the concept of convergence and limit.

Definition A.6-8. Limit of a Sequence of Vectors
A sequence of vectors $\{x_i\}$ ($i = 1, 2, \ldots$) belonging to a normed space V is said to *converge* to x_0 if for any real $\epsilon > 0$ there exists a natural number N_ϵ such that $\|x_n - x_0\| < \epsilon$ for all $n \ge N_\epsilon$; x_0 is called the *limit* of $\{x_i\}$.

The convergence defined previously clearly depends on the particular norm referred to. However, it is possible to prove that when $\mathcal{X}$ is finite-dimensional all norms are *equivalent*, in the sense that convergence with respect to any norm implies convergence with respect to all the other norms and that the limit of any converging sequence is unique.

Theorem A.6-2.
A set $\mathcal{X}$ in a normed space V is closed if and only if the limit of any converging sequence with elements in $\mathcal{X}$ belongs to $\mathcal{X}$.

Proof. If. By virtue of Definition A.6-8, if the limit belongs to $\mathcal{X}$, it is necessarily a limit point of $\mathcal{X}$.

Only if. Suppose that $\mathcal{X}$ is not closed and that x_0 is a limit point of $\mathcal{X}$ not belonging to $\mathcal{X}$. Again by virtue of Definition A.6-8 for any value of the integer i it is possible to select in $\mathcal{X}$ a vector $x_i \in \mathcal{O}(x_0, 1/i)$ and in this way to obtain a sequence converging to x_0. □

A.6.2 Banach and Hilbert Spaces

Definition A.6-9. Continuous Map

Let $\mathcal{V}$ and $\mathcal{W}$ be normed spaces. A map $T: \mathcal{V} \to \mathcal{W}$ is said to be *continuous* at $x_0 \in \mathcal{V}$ if for any real $\epsilon > 0$ there exists a real $\delta > 0$ such that

$$\|x - x_0\| < \delta \quad \Rightarrow \quad \|T(x) - T(x_0)\| < \epsilon$$

In other words T is continuous at x_0 if for any $\epsilon > 0$ there exists a $\delta > 0$ such that the image of $\mathcal{O}(x_0, \delta)$ is contained in $\mathcal{O}(T(x_0), \epsilon)$.

Theorem A.6-3.

A map $T: \mathcal{V} \to \mathcal{W}$ is continuous at $x_0 \in \mathcal{V}$ if and only if

$$\lim_{i \to \infty} x_i = x_0 \quad \Rightarrow \quad \lim_{i \to \infty} T(x_i) = T(x_0)$$

Proof. If. Suppose that there exists in $\mathcal{V}$ a sequence converging to x_0 that is transformed by T into a sequence belonging to $\mathcal{W}$ that does not converge to $T(x_0)$; then clearly T cannot be continuous according to Definition A.6-9.

Only if. Suppose that T is noncontinuous, so that there exists a real $\epsilon > 0$ such that for any $\delta > 0$, $\mathcal{O}(x_0, \delta)$ contains vectors whose images do not all belong to $\mathcal{O}(T(x_0), \epsilon)$. Hence, it is possible to select in every $\mathcal{O}(x_0, 1/n)$ a vector x_n such that $T(x_0) \notin \mathcal{O}(T(x_0), \epsilon)$, so that

$$\lim_{n \to \infty} x_n = x_0$$

while $\{T(x_n)\}$ does not converge. □

An important criterion for testing convergence of a sequence without any knowledge of its limit is based on Cauchy sequences, which are defined as follows.

Definition A.6-10. Fundamental (Cauchy) Sequence

A sequence of vectors $\{x_i\}$ belonging to a normed space $\mathcal{V}$ is said to be a *fundamental sequence* or a *Cauchy sequence* if for any real $\epsilon > 0$ there exists an N_ϵ such that

$$\|x_n - x_m\| < \epsilon \quad \forall m, n \geq N_\epsilon$$

It is well known that in the field **R** of reals every converging sequence is a Cauchy sequence and, conversely, every Cauchy sequence converges. This property does not hold in all normed spaces, since every converging sequence

Sect. A.6 Metric and Normed Spaces, Norms

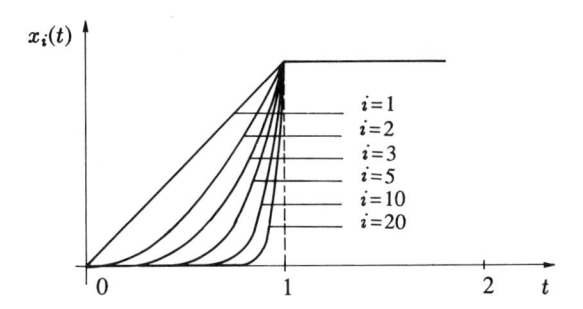

Figure A-15 Continuous functions converging to a discontinuous function.

is a Cauchy sequence but, in general, the contrary is not true. The direct assertion is a consequence of the triangle inequality: in fact, let $\{x_i\}$ converge to x_0, so that for any real $\epsilon > 0$ there exists an N_ϵ such that $\|x_k - x_0\| < \epsilon/2$ for all $k \geq N_\epsilon$, hence $\|x_n - x_m\| \leq \|x_m - x_0\| + \|x_n - x_0\| < \epsilon$ for all $m, n \geq N_\epsilon$.

The following definition is basic for most functional analysis developments.

Definition A.6-11. Banach and Hilbert Spaces

A normed space V is said to be *complete* if every Cauchy sequence with elements in V converges to a limit belonging to V. A complete normed space is also called a *Banach space*. An inner product space that is complete with respect to the norm induced by the inner product is called a *Hilbert space*.

Example A.6-4.

As an example of a noncomplete normed space, consider the space of real-valued continuous functions defined in the interval $[0, 2]$, with the norm

$$\|x(\cdot)\|_1 := \int_0^2 |x(t)|\, dt \tag{27}$$

In this space the sequence

$$x_i(t) = \begin{cases} t^i & (0 \leq t < 1) \\ 1 & (1 \leq t \leq 2) \end{cases} \quad (i = 1, 2, \ldots) \tag{28}$$

(some elements of which are shown in Fig. A-15) is a Cauchy sequence with respect to the norm (27) since for any real $\epsilon > 0$ it is possible to select an N_ϵ such that

$$\|x_n(\cdot) - x_m(\cdot)\|_1 = \frac{1}{n+1} - \frac{1}{m+1} < \epsilon \quad \forall m, n \geq N_\epsilon$$

but converges to the function

$$x(t) = \begin{cases} 0 & (0 \leq t < 1) \\ 1 & (1 \leq t \leq 2) \end{cases}$$

which does not belong to the considered space since it is not continuous.

Example A.6-5.
R is a Banach space, since every Cauchy sequence of real numbers converges.

Example A.6-6.
$\mathbf{R}^n$ and $\mathbf{C}^n$ are Banach spaces. In fact, suppose that $\{x_i\}$ ($i = 1, 2, \ldots$) is a Cauchy sequence, so that for any $\epsilon > 0$ there exists an N_ϵ such that

$$\|x_p - x_q\| \leq \epsilon \quad \forall p, q \geq N_\epsilon$$

Denote by x_{pi}, x_{qi} the i-th elements of x_p, x_q: by virtue of Property A.6-1 it follows that

$$|x_{pi} - x_{qi}| \leq \|x_p - x_q\|_\infty \leq \|x_p - x_q\| < \epsilon \quad (i=1,\ldots,n)$$

hence $\{x_{pi}\}$ ($i=1,2,\ldots$) is a Cauchy sequence of real numbers. Let

$$\bar{x}_i := \lim_{p \to \infty} \{x_{pi}\} \quad (i = 1, \ldots, n)$$

and

$$\bar{x} := (\bar{x}_1, \ldots, \bar{x}_n)$$

Clearly

$$\lim_{p \to \infty} x_p = \bar{x}, \quad \text{with } \bar{x} \in \mathbf{R}^n \text{ or } \bar{x} \in \mathbf{C}^n$$

Example A.6-7.
The space $C[t_0, t_1]$ of real-valued continuous functions, with the norm

$$\|x(\cdot)\|_\infty = \sup_{t_0 \leq t \leq t_1} |x(t)| \tag{29}$$

is a Banach space. In fact, let $\{x_i(\cdot)\}$ ($i=1,2,\ldots$) be a Cauchy sequence in $C[t_0, t_1]$, so that for any $\epsilon > 0$ there exists an N_ϵ such that

$$\sup_{t_0 \leq t \leq t_1} |x_n(t) - x_m(t)| < \epsilon \quad \forall m, n \geq N_\epsilon$$

hence

$$|x_n(t) - x_m(t)| < \epsilon \quad \forall m, n \geq N_\epsilon, \ \forall t \in [t_0, t_1]$$

Therefore $\{x_i(t)\}$ ($i=1,2,\ldots$) is a Cauchy sequence of real numbers for all $t \in [t_0, t_1]$.
Let

$$\bar{x}(t) := \lim_{i \to \infty} x_i(t) \quad \forall t \in [t_0, t_1]$$

It will be proved that the function so defined is the limit in $C[t_0, t_1]$, i.e., with respect to the norm (29), of the sequence $\{x_i\}$ and that $\bar{x}(\cdot) \in C[t_0, t_1]$. For any $\epsilon > 0$ there exists an N_ϵ such that

$$|x_n(t) - \bar{x}(t)| < \frac{\epsilon}{2} \quad \forall n \geq N_\epsilon, \ \forall t \in [t_0, t_1]$$

Sect. A.6 Metric and Normed Spaces, Norms

i.e.,
$$\sup_{t_0 \leq t \leq t_1} |x_n(t) - \bar{x}(t)| < \epsilon \quad \forall n \geq N_\epsilon$$

Clearly this means that
$$\lim_{i \to \infty} x_i(\cdot) = \bar{x}(\cdot)$$

By virtue of the triangle inequality, it follows that
$$|\bar{x}(t) - \bar{x}(\tau)| \leq |\bar{x}(t) - x_n(\tau)| + |x_n(\tau) - x_n(t)| + |x_n(t) - \bar{x}(\tau)|$$
$$\forall n, \; \forall t, \tau \in [t_0, t_1]$$

since it has been proved that convergence of $\{x_i(t)\}$ ($i = 1, 2, \ldots$) to $\bar{x}(t)$ is uniform with respect to t, for any $\epsilon > 0$ there exists an N_ϵ such that for $n \geq N_\epsilon$ the first and the third terms on the right side of the above relation are less than $\epsilon/3$, so that

$$|\bar{x}(t) - \bar{x}(\tau)| \leq \frac{2\epsilon}{3} + |x_n(\tau) - x_n(t)| \quad \forall n \geq N_\epsilon, \; \forall t, \tau \in [t_0, t_1]$$

Since $x_n(\cdot)$ is a continuous function, there exists a real $\delta > 0$ such that

$$|\tau - t| < \delta \quad \Rightarrow \quad |x_n(\tau) - x_n(t)| < \frac{\epsilon}{2}$$

hence
$$|\bar{x}(\tau) - \bar{x}(t)| < \epsilon$$

This means that $\bar{x}(\cdot)$ is continuous at t; since t is arbitrary, it is continuous in $[t_0, t_1]$. It is remarkable that sequence (28), which is Cauchy with respect to the norm (27), is not Cauchy with respect to (29).

Example A.6-8.
The space l_p ($1 \leq p \leq \infty$) is a Banach space.

Example A.6-9.
The space $L_p[a, b]$ is a Banach space.

A.6.3 The Main Existence and Uniqueness Theorem

Proof of existence and uniqueness of solutions of differential equations is a typical application of normed spaces. Consider the vector differential equation
$$\dot{x}(t) = f(t, x(t)) \tag{30}$$

with the aim of determining a class of functions $f : \mathbf{R} \times \mathcal{F}^n \to \mathcal{F}^n$ such that (30) has a unique solution for any given initial state and initial instant of time, i.e., there exists a unique function $x(\cdot)$ satisfying (30) and such that $x(t_0) = x_0$ for any $x_0 \in \mathcal{F}^n$ and for any real t_0. Here, as before, $\mathcal{F} := \mathbf{R}$ or $\mathcal{F} := \mathbf{C}$. Only a set of sufficient conditions is sought, so that it is important that this class is large enough to include all cases of practical interest.

Theorem A.6-4.

The differential equation (30) admits a unique solution $x(\cdot)$ which satisfies the initial condition $x(t_0)=x_0$ for any given real t_0 and any given $x_0 \in \mathcal{F}^n$ if

1. for all $x \in \mathcal{F}^n$ function $f(\cdot, x)$ is piecewise continuous for $t \geq t_0$;
2. for all $t \geq t_0$ which are not discontinuity points of $f(\cdot, x)$ and for any pair of vectors $u, v \in \mathcal{F}^n$ the following *Lipschitz condition* [1] is satisfied:

$$\|f(t, u) - f(t, v)\| \leq k(t)\|u - v\| \tag{31}$$

where $k(t)$ is a bounded and piecewise continuous real-valued function, $\|\cdot\|$ any norm in $\mathcal{F}^n$.

Proof. *Existence.* By using the Peano-Picard successive approximations method, set the sequence of functions

$$\begin{aligned} x_0(t) &:= x_0 \\ x_i(t) &:= x_0 + \int_{t_0}^{t} f\bigl(\tau, x_{i-1}(\tau)\bigr) d\tau \quad (i = 1, 2, \dots) \end{aligned} \tag{32}$$

It will be proved that this sequence, for $t \in [t_0, t_1]$ with t_1 arbitrary, converges uniformly to a function $x(t)$, which is an integral of (30). In fact, by taking the limit of the sequence under the integral sign, it follows that

$$x(t) = x_0 + \int_{t_0}^{t} f\bigl(\tau, x(\tau)\bigr) d\tau$$

In order to prove the uniform convergence of (32), consider the series

$$s(t) := \sum_{i=1}^{\infty} \bigl(x_i(t) - x_{i-1}(t)\bigr)$$

and note that its n-th partial sum is

$$s_n(t) = \sum_{i=1}^{n} \bigl(x_i(t) - x_{i-1}(t)\bigr) = x_n(t) - x_0(t)$$

Therefore, the series converges uniformly if and only if the sequence converges uniformly. The series converges uniformly in norm in the interval $[t_0, t_1]$ if the series with scalar elements

$$\sigma(t) := \sum_{i=1}^{\infty} \|x_i(t) - x_{i-1}(t)\| \tag{33}$$

---

[1] Note that set $\mathcal{M}_k$ of all functions that satisfy the Lipschitz condition at t is closed, it being the closure of the set of all differentiable functions that satisfy $\|\text{grad}_x f(t, x)\| \leq k(t)$, where

$$\text{grad}_x f(t, x) := \left(\frac{\partial f(t, x)}{\partial x_1}, \dots, \frac{\partial f(t, x)}{\partial x_n}\right)$$

Sect. A.6 Metric and Normed Spaces, Norms

converges uniformly in $[t_0, t_1]$; in fact the sum of $\sigma(t)$ is clearly greater than or equal to the norm of the sum of $s(t)$.

From (32) it follows that

$$x_{i+1}(t) - x_i(t) = \int_{t_0}^t \Big(f\big(\tau, x_i(\tau)\big) - f\big(\tau, x_{i-1}(\tau)\big)\Big) d\tau$$

hence

$$\|x_{i+1}(t) - x_i(t)\| \leq \int_{t_0}^t \|f\big(\tau, x_i(\tau)\big) - f\big(\tau, x_{i-1}(\tau)\big)\| d\tau$$

By virtue of hypothesis 2 of the statement

$$\|x_{i+1}(t) - x_i(t)\| \leq k_1 \int_{t_0}^t \|x_i(\tau) - x_{i-1}(\tau)\| d\tau$$

where

$$k_1 := \sup_{t_0 \leq t \leq t_1} k(t)$$

Since, in particular

$$\|x_1(t) - x_0(t)\| \leq \int_{t_0}^t \|f\big(\tau, x_0(\tau)\big)\| d\tau \leq k_1 \|x_0\| (t - t_0)$$

by recursive substitution it follows that

$$\|x_i(t) - x_{i-1}(t)\| \leq \|x_0\| \frac{k_1^i (t-t_0)^i}{i!} \quad (i = 1, 2, \ldots)$$

which assures uniform convergence of series (33) by the comparison test: in fact, the right side of the above relation is the general element of the exponential series which converges uniformly to

$$\|x_0\| e^{k_1 (t-t_0)}$$

Uniqueness. Let $y(t)$ be another solution of differential equation (30) with the same initial condition, so that

$$y(t) = x_0 + \int_{t_0}^t f\big(\tau, y(\tau)\big) d\tau$$

By subtracting (32), it follows that

$$y(t) - x_i(t) = \int_{t_0}^t \Big(f\big(\tau, y(\tau)\big) - f\big(\tau, x_{i-1}(\tau)\big)\Big) d\tau \quad (i = 1, 2, \ldots)$$

hence, by condition 2 of the statement

$$\|y(t) - x_i(t)\| \le \int_{t_0}^{t} k(\tau) \|y(\tau) - x_{i-1}(\tau)\| \, d\tau$$

since

$$\|y(t) - x_0\| \le \|y(t)\| + \|x_0\| \le k_2 + \|x_0\|$$

where

$$k_2 := \sup_{t_0 \le t \le t_1} \|y(t)\|$$

by recursive substitution we obtain

$$\|y(t) - x_i(t)\| \le \left(k_2 + \|x_0\|\right) \frac{k_1^i (t-t_0)^i}{i!} \quad (i = 1, 2, \ldots)$$

This shows that $\{x_i(t)\}$ converges uniformly in norm to $y(t)$. □

Note that condition (31) implies the continuity of function f with respect to x for all t that are not discontinuity points while, on the contrary, continuity of f with respect to x does not imply (31). For example, the differential equation

$$\dot{x}(t) = 2\sqrt{x(t)}$$

with the initial condition $x(0)=0$, admits two solutions: $x(t)=0$ and $x(t)=t^2$; in this case the function at the right side is continuous but does not meet the Lipschitz condition at $x=0$.

Corollary A.6-1.
Any solution of differential equation (30) is continuous.

Proof. Clearly all functions of sequence (32) are continuous. It has been proved that this sequence converges with respect to the norm $\|\cdot\|_\infty$ for vector-valued functions, so that single components of the elements of sequence converge with respect to the norm $\|\cdot\|_\infty$ for real-valued functions, hence are Cauchy sequences. Since $C[t_0, t_1]$ is a complete space (see Example A.6-7 of discussion on Banach spaces), the limits of these sequences are continuous functions. □

REFERENCES

1. BELLMAN, R., *Introduction to Matrix Analysis*, McGraw-Hill, New York, 1960.
2. BIRKHOFF, G., and MACLANE, S., *A Survey of Modern Algebra*, Macmillan, New York, 1965.
3. BOULLION, T.L., and ODELL, P.L., *Generalized Inverse Matrices*, Wiley-Interscience, New York, 1971.
4. CODDINGTON, E.A., and LEVINSON, N., *Theory of Ordinary Differential Equations*, McGraw-Hill, New York, 1955.
5. DESOER, C.A., *Notes for a Second Course on Linear Systems*, Van Nostrand Reinhold, New York, 1970.
6. DURAND, E., *Solutions Numériques des Équations Algébriques, Tome I et II*, Masson, Paris, 1961.
7. FADDEEV, D.K., and FADDEEVA, V.N., *Computational Methods of Linear Algebra*, Freeman & Co, San Francisco and London, 1963.
8. GANTMACHER, F.R., *The Theory of Matrices, Vol. 1*, Chelsea, New York, 1959.
9. GREVILLE, T.N.E., "The pseudoinverse of a rectangular or singular matrix and its application to the solution of systems of linear equations," *SIAM Newsletter*, vol. 5, pp. 3–6, 1957.
10. HALMOS, P.R., *Measure Theory*, Van Nostrand, Princeton, 1950.
11. —, *Finite-Dimensional Vector Spaces*, Van Nostrand, Princeton, 1958.
12. KAUFMAN, I., "The inverse of the Vandermonde matrix and the transformation to the Jordan canonical form," *IEEE Trans. on Aut. Control*, vol. AC-14, pp. 774–777, 1969.
13. LANCZOS, C., *Applied Analysis*, Prentice Hall, Englewood Cliffs, N.J., 1956.
14. LIPSCHUTZ, S., *Linear Algebra*, Schaum's outline series, McGraw-Hill, New York, 1968.
15. PEASE, M.C. III, *Methods of Matrix Algebra*, Academic, New York, 1965.
16. PENROSE, R., "A generalized inverse for matrices," *Proc. Cambridge Phil. Soc.*, vol. 51, pp. 406–413, 1955.
17. POLAK, E., and WONG, E., *Notes for a First Course on Linear Systems*, Van Nostrand Reinhold, New York, 1970.
18. PORTER, W.A., *Modern Foundations of Systems Engineering*, Macmillan, New York, 1966.
19. SWAMY, K.N., "On Sylvester criterion for positive-semidefinite matrices," *IEEE Trans. Autom. Control*, vol. AC-18, no. 3, p. 306, 1973.
20. WIBERG, D.M., *State Space and Linear Systems*, Schaum's outline series, McGraw-Hill, New York, 1971.

B
COMPUTATIONAL BACKGROUND

In this appendix some widely used algorithms, particularly suitable to set a computational support for practical implementation of the geometric approach techniques, are briefly presented from a strictly didactic standpoint.

B.1 THE GAUSS-JORDAN ELIMINATION METHOD AND THE LU FACTORIZATION

In its most diffused form, the Gauss-Jordan elimination method is used to invert nonsingular square matrices and is derived from the Gaussian elimination (pivotal condensation) method, which provides the solution of a set of n linear algebraic equations in n unknowns by subtracting multiples of the first equation from the others in order to eliminate the first unknown from them, and so on. In this way the last equation will be in a single unknown, which is immediately determined, while the others are subsequently derived by backward recursive substitution of the previously determined unknowns in the other reduced equations. The Gauss-Jordan method is presented herein in a general form that is oriented toward numerical handling of subspaces rather than strict matrix inversion.[1]

Definition B.1-1. Elementary Row and Column Operations

The following operations on matrices are called *elementary row (column) operations*:

1. permutation of row (column) i with row (column) j;
2. multiplication of row (column) i by a scalar α;
3. addition of row (column) j multiplied by scalar α to row (column) i.

---

[1] This extension is due to Desoer [A,5].

Sect. B.1 The Gauss-Jordan Elimination Method and the LU Factorization

The elementary row operations can be performed by premultiplying the considered matrix by matrices P_1, P_2, P_3, obtained by executing the same operations on the identity matrix, i.e.

$$P_1 = \begin{matrix} \\ \\ i \\ \\ j \\ \\ \end{matrix} \begin{bmatrix} 1 & & \overset{i}{} & & \overset{j}{} & & \\ & 1 & & & & & \\ & & 0 & & 1 & & \\ & & & 1 & & & \\ & & 1 & & 0 & & \\ & & & & & & 1 \end{bmatrix}, \quad P_2 = \begin{matrix} \\ \\ \\ i \\ \\ \\ \end{matrix} \begin{bmatrix} 1 & & \overset{i}{} & & & \\ & 1 & & & & \\ & & 1 & & & \\ & & & \alpha & & \\ & & & & 1 & \\ & & & & & 1 \end{bmatrix},$$

$$P_3 = \begin{matrix} \\ \\ i \\ \\ j \\ \\ \end{matrix} \begin{bmatrix} 1 & & \overset{i}{} & \overset{j}{} & & \\ & 1 & & & & \\ & & 1 & \alpha & & \\ & & & 1 & & \\ & & & & 1 & \\ & & & & & 1 \end{bmatrix} \qquad (1)$$

where all the omitted elements are understood to be zero. Similarly, the elementary column operations can be performed by postmultiplying by matrices Q_1, Q_2, Q_3, obtained by means of the same operations on the identity matrix. Note that the matrices that perform the elementary row (column) operations are nonsingular since, clearly, $\det P_1 = -1$, $\det P_2 = \alpha$, $\det P_3 = 1$. Furthermore, note that $P_1 P_1^T = I$ (in fact P_1 is orthogonal) and that $Q_1 = P_1^T$, hence $P_1 Q_1 = Q_1 P_1 = I$.

The elementary row and column operations are very useful for matrix computations by means of digital computers. For instance, they are used in the following basic algorithm which, for a general matrix A, allows us to derive $\rho(A)$ and $\nu(A)$, basis matrices for $\text{im}\,A$ and $\ker A$ and the inverse matrix A^{-1} if A is invertible.

Algorithm B.1-1. (Gauss-Jordan)

Let A be an $m \times n$ matrix; denote by B the matrix, also $m \times n$, on which the operations of the algorithm are from time to time performed and by i the current iteration number of the algorithm:

1. Initialize: $i \leftarrow 1$, $B \leftarrow A$;
2. Consider the elements of B with row and column indices equal to or greater than i and select that (or any of those) having the greatest absolute value. If all the considered elements are zero, stop;

3. Let b_{pq} be the element selected at the previous step: interchange rows i and p, columns i and q, so that $b_{ii} \neq 0$:

$$b_{ik} \leftrightarrow b_{pk} \quad (k=1,\ldots,n), \quad b_{ki} \leftrightarrow b_{kq} \quad (k=1,\ldots,m)$$

4. Add row i multiplied by $-b_{ji}/b_{ii}$ to every row j with $j \neq i$:

$$b_{jk} \leftarrow b_{jk} - b_{ik}\frac{b_{ji}}{b_{ii}} \quad (k=1,\ldots,n;\ j=1,\ldots,i-1,i+1,\ldots,m)$$

5. Multiply row i by $1/b_{ii}$:

$$b_{ik} \leftarrow \frac{b_{ik}}{b_{ii}} \quad (k=1,\ldots,n)$$

6. Increment i: $i \leftarrow i+1$; then, if $i < m+1$, $i < n+1$ go to step 2. If $i = m+1$ or $i = n+1$, stop.

The element of greatest absolute value selected at step 2 and brought to the position corresponding to b_{ii} at step 3 is called the *pivot* for the i-th iteration.

As an example, consider a 5×8 matrix A and suppose $\rho(A) = 3$. By using the above algorithm we obtain a matrix B with the following structure:

$$B = \begin{bmatrix} 1 & 0 & 0 & b_{14} & b_{15} & b_{16} & b_{17} & b_{18} \\ 0 & 1 & 0 & b_{24} & b_{25} & b_{26} & b_{27} & b_{28} \\ 0 & 0 & 1 & b_{34} & b_{35} & b_{36} & b_{37} & b_{38} \\ 0 & 0 & 0 & 0 & 0 & 0 & 0 & 0 \\ 0 & 0 & 0 & 0 & 0 & 0 & 0 & 0 \end{bmatrix} = PAQ \qquad (2)$$

in which the elements b_{ij} ($i = 1,\ldots,3;\ j = 4,\ldots,8$) are in general different from zero; in (2) P and Q denote the products of the matrices corresponding to the elementary row and column operations performed during application of the algorithm. Matrix B can also be represented in partitioned form as

$$B = \begin{bmatrix} I_r & B_{12} \\ O & O \end{bmatrix} \qquad (3)$$

in which the zero rows are present only if the algorithm stops at step 2 rather than step 6.

At step 2 we are faced with the problem of setting a threshold for machine zeros. It is very common to assume a small real number, related to a matrix norm and to the numerical precision of the digital processor: denoting by ϵ the "machine zero" (for instance, $\epsilon = 10^{-16}$), a possible expression for threshold is

$$t = k\epsilon \|A\|_F \qquad (4)$$

where k denotes a suitable power of 10 (for instance 100 or 1000), introduced in order to get a certain distance from machine zeros so that results of numerical computations still have significance.

The Gauss-Jordan Algorithm, provided with the preceding linear independence test based on a threshold, solves the following standard computations of matrix analysis, related to numerical handling of subspaces.

Rank and Nullity. The rank of A is equal to r, the number of nonzero rows of B. In fact, by virtue of Property A.2-7 $\rho(A)=\rho(PAQ)$, P and Q being nonsingular; hence $\rho(A)=\rho(B)$, which, due to the particular structure of B, is equal to the number of its nonzero rows. By virtue of Property A.2-5, the nullity of A, $\nu(A)$, is immediately derived as $n-r$.

Image. A basis matrix R for im$A$ is provided by the first r columns of matrix AQ. In fact, since Q is nonsingular, im$A$ = im(AQ). Let R be the matrix formed by the first r columns of AQ; clearly im$R \subseteq$ im$A$, but, the columns of PR being the first r columns of the $m \times m$ identity matrix, it follows that $\rho(R)=r$, hence, im$R$ = im$A$.[2]

Kernel. A basis matrix N for ker$A$ is given by $N = QX$, where X is the $n \times (n-r)$ matrix
$$X := \begin{bmatrix} -B_{12} \\ I_{n-r} \end{bmatrix}$$
In fact, since $\nu(B) = n-r$, matrix X, whose columns are clearly a linearly independent set such that $BX = O$, is a basis matrix for ker$B$. Hence, $BX = PAQX = PAN = O$: since P is nonsingular, $AN = O$, so that im$N \subseteq$ ker$A$. But $\rho(N) = n-r$ because of the nonsingularity of Q and, since $\nu(A) = n-r$ too, it follows that im$N$ = ker$A$.

Inverse. If A is square nonsingular, its inverse A^{-1} is QP, where Q and P are the matrices obtained by applying Algorithm B.1-1 to A. In fact, in this case relation (4) becomes
$$I = PAQ$$
or
$$P = (AQ)^{-1} = Q^{-1}A^{-1}$$
(we recall that Q is orthogonal); then
$$A^{-1} = QP$$

Hence, A^{-1} can be computed by performing the same elementary row operations on the identity matrix I as were performed on B while applying Algorithm B.1-1 and then executing in reverse order the permutations, to which the columns of B had been subjected, on rows of the obtained matrix.

The following well known result is easily derived as a consequence of the Gauss-Jordan algorithm.

---

[2] Note that Algorithm B.1-1 realizes the *direct selection* of a linearly independent subset with the maximum number of elements among the vectors of any set given in $\mathcal{F}^n$: in fact the columns of matrix R are related to those of A in this way, since Q performs only column permutations.

Theorem B.1-1. The LU Factorization
Let A be a nonsingular $n \times n$ real or complex matrix. There exist both a lower triangular matrix L and an upper triangular matrix U such that $A = LU$.

Proof. Applying Algorithm B.1-1 without steps 2 and 3 and executing the summation in step 4 only for $j > i$ we obtain a matrix B such that

$$B = PA$$

which is upper triangular with ones on the main diagonal; on the other hand, P is lower triangular, being the product of elementary row operation matrices of types P_2 and P_3 in (1), which are respectively diagonal and lower triangular in this case. Assume $U := B$. $L = P^{-1}$ is lower triangular as the inverse of a lower triangular matrix. The easily derivable recursion relations

$$\ell_{ii} = \frac{1}{p_{ii}} \quad (i = 1, \ldots, n)$$

$$\ell_{ji} = -\ell_{ii} \sum_{k=i+1}^{j} p_{ki} \ell_{jk} \quad (i = n-1, \ldots, 1; j = n, \ldots, i+1) \tag{5}$$

can be used to compute its nonzero elements. □

B.2 THE GRAM-SCHMIDT ORTHONORMALIZATION PROCESS AND THE QR FACTORIZATION

An algorithm that turns out to be very useful in numerical computations related to the geometric approach is the Gram-Schmidt orthonormalization process. In its basic formulation it solves the following problem: given a linearly independent set in an inner product space, determine an orthonormal set with the same span. The corresponding computational algorithm can be provided with a linear independence test in order to process a general set of vectors (not necessarily linearly independent) and in this modified version it becomes the basic algorithm to perform all the fundamental operations on subspaces such as sum, intersection, orthogonal complementation, direct and inverse linear transformations, and so on.

Algorithm B.2-1. (Gram-Schmidt)
Let $\mathcal{V}$ be an inner product space and $\{a_1, \ldots, a_h\}$ a linearly independent set in $\mathcal{V}$. An orthonormal set $\{q_1, \ldots, q_h\}$ such that $\text{sp}(q_1, \ldots, q_h) = \text{sp}(a_1, \ldots, a_h)$ is determined by the following process:
  1. Initialize:

$$v_1 \leftarrow a_1$$
$$q_1 \leftarrow \|v_1\|^{-1} v_1 \tag{1}$$

Sect. B.2 The Gram-Schmidt Orthonormalization Process and the QR Factorization

2. Apply the recursion relations:

$$v_i \leftarrow a_i - \sum_{j=1}^{i-1} \langle q_j, a_i \rangle q_j$$

$$q_i \leftarrow \|v_i\|^{-1} v_i \quad (i = 2, \ldots, h) \tag{2}$$

Proof. By means of an induction argument it is easy to check that $\langle q_j, v_i \rangle = 0$ ($j = 1, \ldots, i-1$), hence $\langle q_j, q_i \rangle = 0$ ($j = 1, \ldots, i-1$). Thus, every vector determined by applying the process, whose euclidean norm is clearly one, is orthogonal to all the previous ones. Furthermore, $\text{sp}(q_1, \ldots, q_h) = \text{sp}(a_1, \ldots, a_h)$, since $\{q_1, \ldots, q_h\}$ is a linearly independent set whose elements are linear combinations of those of $\{a_1, \ldots, a_h\}$. □

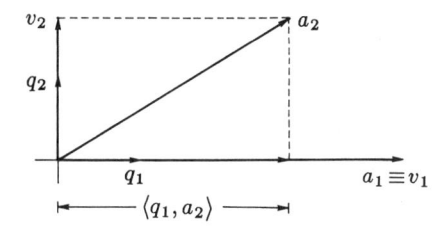

Figure B-1 The Gram-Schmidt process in $\mathbf{R}^2$.

As an example, Fig. B-1 shows the elements of the Gram-Schmidt process in $\mathbf{R}^2$. The geometric meaning of the process is the following: at the i-th iteration the orthogonal projection of vector a_i on the span of the previous vectors is subtracted from a_i itself, thus obtaining a vector v_i that is orthogonal to all the previous ones (in fact it is the orthogonal projection on the orthogonal complement of their span); then q_i is obtained by simply normalizing v_i. Note that, in order to apply the Gram-Schmidt orthonormalization process, $\mathcal{V}$ need not be finite dimensional.

As a direct consequence of the Gram-Schmidt orthonormalization process, we derive the following theorem.

Theorem B.2.1. The QR Factorization

Let A be a nonsingular $n \times n$ real or complex matrix. There exist both an orthogonal or unitary matrix Q and an upper triangular matrix R such that $A = QR$.

Proof. Denote by $(a_1, \ldots, a_n)$ the ordered set of all columns of A, which is linearly independent by assumption. Similarly, denote by Q the $n \times n$ matrix with vectors $(q_1, \ldots, q_n)$ as columns, obtained by applying the orthonormalization process to $(a_1, \ldots, a_n)$. From $(1, 2)$ we can easily derive the equalities

$$a_1 = \|v_1\| q_1$$

$$a_i = \|v_i\| \, q_i + \sum_{j=1}^{i-1} \langle q_j, a_i \rangle \, q_j \quad (i=2,\ldots,n) \qquad (3)$$

which can be written in compact form as

$$A = QR \qquad (4)$$

where R is the $n \times n$ upper triangular matrix whose nonzero elements are

$$r_{ii} = \|v_i\| \quad (i=1,\ldots,n)$$
$$r_{ij} = \langle q_i, a_j \rangle \quad (i=1,\ldots,n; \, j=i+1,\ldots,n) \qquad \square \qquad (5)$$

Note that, according to (5), a matrix R is derived with all the main diagonal elements positive. However, this property is not guaranteed in all the QR factorization algorithms available for computers but, if not satisfied, it is possible to change the sign of all elements in the rows of R corresponding to negative diagonal elements and in the corresponding columns of Q: this is equivalent to postmultiplying Q and premultiplying R by the same diagonal matrix composed of 1's and -1's, which is clearly orthogonal.

Matrix R can be derived while executing the computations of Algorithm B.1-1 or, at the end of such computations, by using

$$U = Q^T A \qquad (U = A^* A) \qquad (6)$$

In particular, at the i-th step, the nonzero subvector of u_i (i.e., the column vector containing the first i elements of u_i) is provided by

$$u'_i = Q_i^T a_i \qquad (u'_i = Q_i^* a_i)$$

where Q_i denotes the submatrix composed of the first i columns of Q. Moreover, since $|\det Q| = 1$, it follows that

$$|\det A| = |\det R| = \prod_{i=1}^{n} |r_{ii}| \qquad (7)$$

i.e., the absolute value of $\det A$ is equal to the product of the euclidean norms of the orthogonal projections of columns (rows) of A on the orthogonal complement of the subspace spanned by the previous columns (rows).

B.2.1 The QR Factorization for Singular Matrices

Consider an $m \times n$ real or complex matrix A and apply Algorithm B.1-1 to its columns in sequence: if $v_i = 0$ for a certain i, a_i can be expressed as a linear combination of $a_1, \ldots, a_{i-1}$ and is omitted. At the end of the process an orthonormal set $\{q_1, \ldots, q_h\}$ is obtained whose span is equal to that of the original set.

A very significant drawback of this procedure when it is practically implemented on a digital computer is that, due to the rounding errors, machine zeros appear instead of true zeros when, with a reasonable approximation, a_i is linearly dependent on the previous vectors, so it is necessary to decide whether to include it or not according to an appropriate selection criterion. Usually a threshold t similar to (4) of the previous section is introduced for selection in order to avoid loss of orthogonality of the computed vectors. Furthermore, a significant improvement in precision of the linear dependence test is obtained if at each step the vector with projection having maximal euclidean norm is processed first.

These considerations lead to the following result, which extends Theorem B.2-1 to generic matrices.

Theorem B.2-2. The Extended QR Factorization

Let A be an $m \times n$ real or complex matrix. There exist an $m \times m$ orthogonal or unitary matrix Q, an $m \times n$ upper triangular matrix R with nonnegative nonincreasing diagonal elements, and an $n \times n$ permutation matrix P such that

$$AP = QR \qquad (8)$$

To be more precise, let $r := \operatorname{rank} A$; matrix R has the form

$$R = \begin{bmatrix} R_{11} & R_{12} \\ O & O \end{bmatrix} \qquad (9)$$

where R_{11} is $r \times r$ upper triangular with positive nonincreasing diagonal elements.

Proof. By applying the Gram-Schmidt algorithm, determine an $m \times r$ matrix Q_1, an $r \times r$ matrix R_{11}, and a permutation matrix P such that columns of $Q_1 R_{11}$ are equal to the first r columns of AP. Let $[M_1 \ M_2] := AP$, with M_1, M_2 respectively $m \times r$ and $m \times (n-r)$. Thus, $Q_1 R_{11} = M_1$. Since columns of M_2 are linear combinations of those of M_1, hence of Q_1, a matrix R_{12} exists such that $M_2 = Q_1 R_{12}$, or $R_{12} = Q_1^T M_2$ ($R_{12} = Q_1^* M_2$). Then determine Q_2 such that $Q := [Q_1 \ Q_2]$ is orthogonal or unitary. It follows that

$$AP = [M_1 \ M_2] = [Q_1 \ Q_2] \begin{bmatrix} R_{11} & R_{12} \\ O & O \end{bmatrix} \quad \square \qquad (10)$$

The extended QR factorization solves the following standard computations of matrix analysis, related to numerical handling of subspaces.

Rank and Nullity. The rank of A, $\rho(A)$, is equal to r, the number of nonzero diagonal elements of R and the nullity of A, $\nu(A)$, is equal to $n - r$.

Image. A basis matrix for im$A$ is Q_1, formed by the first r columns of Q.

Kernel. A basis matrix for ker$A$ is

$$P \begin{bmatrix} -R_{11}^{-1} R_{12} \\ I_{n-r} \end{bmatrix}$$

as can easily be checked by using (10).

Inverse. If A is square nonsingular, its inverse is

$$P R^{-1} Q^T \qquad (P R^{-1} Q^*)$$

Note that, R being upper triangular, its inverse is easily computed by means of relations dual to (5) of the previous section, reported herein for the sake of completeness. Let $U := R^{-1}$: then

$$u_{ii} = \frac{1}{r_{ii}} \quad (i = 1, \ldots, n)$$

$$u_{ij} = -u_{ii} \sum_{k=i+1}^{j} r_{ik} u_{kj} \quad (i=n-1, \ldots, 1; j=n, \ldots, i+1) \qquad (11)$$

In conclusion, the Gram-Schmidt orthonormalization process when used for numerical handling of subspaces solves the same problems as the Gauss-Jordan algorithm. The advantages of the Gauss-Jordan method are simpler and faster computations and preservation of the span of any subset of the given vectors, while the advantages of the Gram-Schmidt method are continuous correction of possible ill-conditioning effects through reorthonormalization of basis matrices and a more efficient linear independence test.

B.3 THE SINGULAR VALUE DECOMPOSITION

The *singular value decomposition* (SVD) is a very efficient tool to perform matrix computations and to handle, in particular, singular matrices. The *singular values* of a general $m \times n$ real or complex matrix A are defined as the square roots of the eigenvalues of $A^T A$ in the real case or $A^* A$ in the complex case. In the real case they have an interesting geometric meaning, being the euclidean norms of the principal axes of the ellipsoid in $\mathbf{R}^m$ into which the unitary sphere in $\mathbf{R}^n$ is transformed by A. The existence of the SVD is constructively proved in the following theorem.

Sect. B.3 The Singular Value Decomposition

Theorem B.3-1. The Singular Value Decomposition

Let A be an $m \times n$ real or complex matrix. There exists an $m \times m$ orthogonal or unitary matrix U, an $m \times n$ diagonal matrix S with nonnegative nonincreasing elements, and an $n \times n$ orthogonal or unitary matrix P such that

$$A = USV^T \qquad (A = USV^*) \qquad (1)$$

Proof. Only the real case is considered in the proof since the extension to the complex case is straightforward. Apply the Schur decomposition to $A^T A$:

$$A^T A = V \Sigma V^T$$

Since $A^T A$ is symmetric semidefinite positive, Σ is diagonal with nonnegative elements, which can be assumed to be nonincreasing without any loss of generality: in fact, if not, consider a permutation matrix P such that $\Sigma_p := P^T \Sigma P$ has this feature, and redefine $V \leftarrow VP$, $\Sigma \leftarrow \Sigma_p$.

Let r be the number of nonzero elements of Σ: clearly $r \leq \inf(m, n)$; denote by V_1 the $n \times r$ matrix formed by the first r columns of V and by Σ_1 the $r \times r$ diagonal matrix made with the nonzero elements of Σ. Then

$$A^T A = V_1 \Sigma_1 V_1^T \qquad (2)$$

Let $S_1 := \sqrt{\Sigma_1}$ and define the $m \times r$ matrix U_1 as

$$U_1 := A V_1 S_1^{-1} \qquad (3)$$

It is easy to check that columns of U_1 are an orthogonal set. In fact

$$U_1^T U_1 = S_1^{-1} V_1^T A^T A V_1 S_1^{-1} = I_r$$

since, from (2)

$$\Sigma_1 = S_1^2 = V_1^T A^T A V_1$$

The SVD "in reduced form" directly follows from (3):

$$A = U_1 S_1 V_1^T$$

To obtain the standard SVD, let $U := [U_1 \; U_2]$, $V := [V_1 \; V_2]$, with U_2, V_2 such that U, V are orthogonal, and define the $m \times n$ matrix S as

$$S := \begin{bmatrix} S_1 & O \\ O & O \end{bmatrix} \qquad \square$$

All the standard computations of matrix analysis considered in the previous sections are easily performed with the SVD.

Rank and Nullity. The rank of A, $\rho(A)$, is equal to r (the number of nonzero elements of S) and the nullity of A, $\nu(A)$, is equal to $n - r$.

Image. A basis matrix for $\operatorname{im} A$ is U_1, formed by the first r columns of U.

Kernel. A basis matrix for $\ker A$ is V_2, formed by the last $n-r$ columns of V.

Pseudoinverse. Let A be a general $m \times n$ real matrix; then

$$A^+ = V_1 S_1^{-1} U_1^T \quad \text{or} \quad A^+ = V S^+ U^T$$

where S^+ is the $n \times m$ matrix defined by

$$S^+ := \begin{bmatrix} S_1^{-1} & O \\ O & O \end{bmatrix}$$

In fact, consider relation (A.3.14) and assume $X := U_1$. It follows that

$$\begin{aligned}
A^+ &= A^T X (X^T A A^T X)^{-1} X^T \\
&= V_1 S_1 U_1^T U_1 (U_1^T U_1 S_1 V_1^T V_1 S_1 S_1 U_1^T U_1)^{-1} U_1^T \\
&= V_1 S_1 S_1^{-2} U_1^T = V_1 S_1^{-1} U_1^T
\end{aligned}$$

Condition number. Let A be an $n \times n$ real or complex matrix. The *condition number* of A is a "demerit figure" about the nonsingularity of A, and is defined as the ratio s_1/s_n (the greatest over the smallest singular value of A). It ranges from one to infinity: it is very high when A is badly conditioned, infinity if it is singular, and one for orthogonal and unitary matrices.

B.4 SOME MATLAB SUBROUTINES FOR GEOMETRIC APPROACH COMPUTATIONS

This section reports the lists of Matlab subroutines (m-files) for the most common computational problems of the geometric approach, based on the extended QR decomposition.[1] The first and second of them (`ima` and `ortc`) are the basic tools: they provide respectively orthonormal bases for $\operatorname{im} A$ and $(\operatorname{im} A)^\perp$. A flag is provided in `ima` in order to avoid random permutation of already computed orthonormal vectors when it is used in recursion algorithms. The subroutines require both the general "\matlab" and the specific "\matlab\control" computational environment and may be located in the special subdirectory "\matlab\ga". A comment at the beginning of each routine briefly explains its aim and features: it can be displayed by means of Matlab's "help" command.

---

[1] Matlab is a package known worldwide for matrix computations developed by The MathWorks Inc., 21 Eliot Street, South Natick, MA 01760.

Sect. B.4 Some Matlab Subroutines for Geometric Approach Computations

```matlab
function Q = ima(A,p)
% IMA    Orthogonalization. Q=ima(A,p) is an orthonormal basis for imA .
%        If p=1 permutations are allowed, while if p=0 they are not .

%        Basile and Marro 4-20-90

tol = eps*norm(A,'fro')*10^4;
[ma,na] = size(A);
if p == 1
[Q,R,E] = qr(A);
  if (na == 1)|(ma == 1)
    d = R(1,1);
  else
    d = diag(R);
  end
  d = (abs(d))';
  nul = find(d > tol);
  r=length(nul > 0);
  if r > 0
    Q = Q(:,nul);
  else
    Q = zeros(ma,1);
  end
else
  ki = 1;
  A1=A;
  while ki == 1
    [ma,na] = size(A1);
    punt = 1:na;
    [Q,R] = qr(A1);
    if (na == 1)|(ma == 1)
      d = R(1,1);
    else
      d = diag(R);
    end
    d = (abs(d))';
    nul = find(d <= tol);
    n = min(nul);
    punt = find(punt ~= n);
    if (length(nul) == 0)|(length(punt) == 0)
      ki = 0;
    else
      A1 = A1(:,punt);
    end
  end
  if (length(n) == 1)&(length(punt) == 0)
    Q = zeros(ma,1);
  else
    r = length(d);
    if r > 0
      Q=Q(:,1:r);
      Q = -Q;
      Q(:,r) = -Q(:,r);
    end
  end
end
% --- last line of ima ---

function Q = ortc(A)
% ORTC   Complementary orthogonalization. Q=ortc(A) is an orthonormal
%        basis for the orthogonal complement of imA .

%        Basile and Marro 4-20-90 (from Matlab's null)

   [Q,R,E] = qr(A);
tol = eps*norm(A,'fro')*10^4;
```

```
   [ma,na] = size(A);
if na == 1
   d = R(1,1);
else
   d = diag(R);
end
   d = (abs(d))';
   dd = ma-na;
if dd >= 1
   d=[d zeros(1,dd)];
end
nul = find(d <= tol);
   r=length(nul > 0);
if r > 0
   Q=Q(:,nul);
else
   Q = zeros(ma,1);
end
% --- last line of ortc ---

function Q = sums(A,B)
% SUMS   Sum of subspaces Q = sums(A,B) is an orthonormal basis for
%        the subspace im[A B] = imA + imB .
%
%        Basile and Marro 4-20-90

   Q = ima([A B],1);
% --- last line of sums ---

function Q = ints(A,B)
% INTS   Intersection of subspaces Q = ints(A,B) is an orthonormal basis
%        for the subspace (imA) intersection (imB) .
%
%        Basile and Marro 4-20-90

   Q = ortc(sums(ortc(A),ortc(B)));
% --- last line of ints ---

function Q = invt(A,X)
% INVT   Inverse transform. Q=invt(A,p) is an orthonormal basis for the
%        inverse map of imX in A.
%
%        Basile and Marro 4-20-90

Q=ortc(A'*ortc(X))
% --- last line of invt ---

function Q = ker(A)
% KER    Kernel. Q=ker(A) is an orthonormal basis for kerA.
%
%        Basile and Marro 6-20-90

Q = ortc(A');
% --- last line of ker ---

function Q = mininv(A,B)
% MININV Q = mininv(A,B) is an orthonormal basis for the minimal
%        A-invariant containing imB .
%        The routine implements Algorithm 3.2-1 of "Controlled and
%        Conditioned Invariant Subspaces in Linear System Theory"
%
%        Basile and Marro 4-20-90

   nv = 0;
   B1 = ima(B,0);
   Q=B1;
```

Sect. B.4 Some Matlab Subroutines for Geometric Approach Computations

```
  [mq,nq] = size(Q);
  n = 0;
while (nq-nv) > 0
  nv = nq;
  Q = ima([B1 A*Q],0);
  [mq,nq] = size(Q);
% Monitoring iterations
%  n = n+1;
%  disp('   iterations in progress'), n, Q
end
% --- last line of mininv

function Q = maxinv(A,X)
% MAXINV Q = maxinv(A,X) is an orthonormal basis for the maximal
%          A-invariant contained in imX .
%          The routine implements Relation 3.2.7 of "Controlled and
%          Conditioned Invariant Subspaces in Linear System Theory"

%          Basile and Marro 4-20-90

  Q = ortc(mininv(A',ortc(X)));
% --- last line of maxinv ---

function Q = miinco(A,C,X)
% MIINCO Q = miinco(A,C,X) is an orthonormal basis for the minimal
%          (A,imC)-conditioned invariant containing imX .
%          The routine implements Algorithm 4.1-1 of "Controlled and
%          Conditioned Invariant Subspaces in Linear System Theory"

%          Basile and Marro 4-20-90

  nv = 0;
  X1 = ima(X,0);
  Q=X1;
  [mq,nq] = size(Q);
  n = 0;
while (nq-nv) > 0
  nv = nq;
  Q = ima([X1 A*(ints(Q,C))],0);
  [mq,nq] = size(Q);
% Monitoring iterations
%  n = n+1;
%  disp('   iterations in progress'), n, Q
end
% --- last line of miinco ---

function Q = mainco(A,B,X)
% MAINCO Q = mainco(A,B,X) is an orthonormal basis for the maximal
%          (A,imB)-controlled invariant contained in imX .
%          The routine implements Relation 4.1.7 of "Controlled and
%          Conditioned Invariant Subspaces in Linear System Theory"

%          Basile and Marro 4-20-90

  Q = ortc(miinco(A',ortc(B),ortc(X)));
% --- last line of mainco ---

function [P,Q] = stabi(A,X)
% STABI  [P,Q] = stabi(A,X) provides as P and Q the critical matrices
%          for the internal and external stability of an A-invariant X .

%          Basile and Marro 4-20-90

% Checks and messages
  tol=norm(A,'fro')*eps*10^4;
  [mx,nx] = size(X);
```

```
         no = norm(X,'fro');
    if (nx == 1)&(no < tol)
         nx = 0;
    end
         T=maxinv(A,X);
         [my,ny] = size(T);
         no = norm(T,'fro');
    if (ny == 1)&(no < tol)
         ny = 0;
    end
    if (ny ~= nx)
    error('   X is not an A-invariant in STABI')
    end
    %
      X=T;
      [ma,na] = size(A);
      T = ima(X,1);
      [m,n] = size(T);
    no = norm(T,'fro');
    if (n == 1)&(no < tol)
      n = 0;
      T = eye(m);
    else
      T = ima([T ortc(T)],0);
    end
      A1 = inv(T)*A*T;
      P = A1(1:n,1:n);
      n1 = n+1;
      Q = A1(n1:na,n1:na);
    % --- last line of stabi ---

    function [P,Q] = stabv(A,B,X)
    % STABV   [P,Q] = stabv(A,B,X) provides as P and Q the critical matrices
    %         for the internal and external stabilizability of an (A,B)-
    %         controlled invariant X .

    %         Basile and Marro 4-20-90

    % Check and message
      [mx,nx] = size(X);
      T=mainco(A,B,X);
      [my,ny] = size(T);
    if (ny ~= nx)
       error('   X is not a controlled invariant in STABV')
    end
    %
      X = T;
      tol = norm(A,'fro')*eps*10^4;
      [ma,na] = size(A);
      P = miinco(A,X,B);
      R = ints(X,P);
      [m1,n1] = size(R);
      no = norm(R,'fro');
    if (n1 == 1)&(no < tol)
      n1 = 0;
    end
    if n1 == 0
      T = ima(X,0);
      Q = P;
    else
      T = ima([R X],0);
      Q = ima([R P],0);
    end
      [m2,n2] = size(T);
      [mx,nx] = size(Q);
      Q = Q(:,(n1+1):nx);
```

Sect. B.4 Some Matlab Subroutines for Geometric Approach Computations

```
      no = norm(T,'fro');
   if (n2 == 1)&(no < tol)
      n2 = 0;
   end
   if n2 == 0
      T = ima([P ortc(P)],0);
   else
      T = ima([T ortc(T)],0);
      T(:,(n2+1):(n2+nx-n1)) = Q;
   end
      [m3,n3] = size(T);
   if (n3 == 1)&(no < tol)
      n3 = 0;
      T = eye(m3);
   end
      A1 = inv(T)*A*T;
      ni = n1+1;
      nx = n2-n1;
      P = A1(ni:nx,ni:nx);
      [T Q] = stabi(A,sums(X,mininv(A,B)));
   % --- last line of stabv ---

   function F = effe(A,B,X)
   % EFFE    F = effe(A,B,X) is a state-to-input feedback matrix such that
   %         the (A,imB)-controlled invariant X is an (A+B*F)-invariant .
   %         The routine implements Algorithm 4.1-3 of "Controlled and
   %         Conditioned Invariant Subspaces in Linear System Theory"
   %
   %         Basile and Marro 4-20-90

   % Checks and messages
      [mx,nx] = size(X);
      [my,ny] = size(mainco(A,B,X));
   if ny ~= nx
   error('   X is not a controlled invariant in EFFE')
   end
      [mb,nb] = size(B);
      [my,ny] = size(ima(B,1));
   if ny ~= nb
   error('   B is not full rank in EFFE')
   end
   %
      tol = norm(A,'fro')*eps*10^4;
      [ma,na] = size(A);
      T = ints(X,B);
      T = ima(T,1);
      [m1,n1] = size(T);
      no = norm(T,'fro');
   if (n1 == 1)&(no < tol)
      n1 = 0;
   end
   if n1 == 0
      T = ima(X,1);
   else
      T = ima([T X],0);
   end
      [m2,n2] = size(T);
      no = norm(T,'fro');
   if (n2 == 1)&(no < tol)
      n2 = 0;
   end
   if n2 == 0
      T = ima(B,1);
   else
      T = ima([T B],0);
   end
```

```
      [m3,n3] = size(T);
      no = norm(T,'fro');
   if (n3 == 1)&(no < tol)
      n3 =0;
      T = eye(m3);
   else
      T = ima([T ortc(T)],0);
   end
      c1 = n1+1;
      c2 = n3+1;
      T = [B T(:,c1:n2) T(:,c2:na)];
      A1 = inv(T)*A*T;
      c1 = na - n3;
      F = [-A1(1:nb,1:n3) zeros(nb,c1)]*inv(T);
   % --- last line of effe ---

   function z = gazero(A,B,C,D)
   % GAZERO Invariant zeros.
   %          z = GAZERO(A,B,C) or z = GAZERO(A,B,C,D) returns in a column vector
   %          the invariant zeros of the state-space system
   %                  .
   %                  x = Ax + Bu
   %                  y = Cx (+ Du)
   %
   %          The algorithm is based on the standard geometric approach definition
   %          of invariant zeros. A different algorithm is used in the program
   %          tzero of the standard Matlab Control System Toolbox .

   %          Basile and Marro 4-20-90

   tol = norm(A,'fro')*eps*100;
   error(nargchk(3,4,nargin));
   [ma,na] = size(A);
   [mb,nb] = size(B);
   [mc,nc] = size(C);
   k1 = 1;
   nargs = nargin;
   if nargs == 3
     k1 = 0;
     D=zeros(mc,nb);
   end
   if norm(D,'fro') < tol
     k1 = 0;
   end
   error(abcdchk(A,B,C,D));
   if(k1 == 1)
     A=[A zeros(na,mc); C zeros(mc)];
     B=[B; D];
     C=[zeros(mc,na) eye(mc)];
   end
   E = ortc(C');
   J = mainco(A,B,E);
   [P,Q] = stabv(A,B,J);
   z = eig(P);
   % --- last line of gazero ---

   function F=place1(A,B,P)
   % PLACE1 Eigenvalue assignment
   %          F = PLACE1(A,B,P) returns as F the state feedback matrix
   %          such that the h assignable eigenvalues of A+B*F are those
   %          specified in the first h rows of the column vector P .
   %          F = PLACE1(A,B) assigns these eigenvalues in interactive mode .
   %          To enter complex eigenvalues use symbol j for the imaginary unit .

   %          Basile and Marro 4-20-90
```

Sect. B.4 Some Matlab Subroutines for Geometric Approach Computations

```matlab
j=sqrt(-1);
[ma,na]=size(A);
[mb,nb]=size(B);
T=mininv(A,B);
[m,n]=size(T);
nargs = nargin;
if n ~= 0
  if nargs == 2
    disp(   'Enter vector P of eigenvalues to be located: number of components')
      n
      ni=1;
      P=zeros(n,1);
      while (ni <= n)
        P(ni,1)=input(   'Enter an eigenvalue: ')
        ni=ni+1;
      end
  else
    if length(P) < n
      error('    not enough eigenvalues transmitted to PLACE1');
    end
    P=P(1:n,:);
  end
  disp('    thinking')
  disp(' ')
  if (n < na)
    T=ima([T ortc(T)],0);
    A1=inv(T)*A*T;
    B1=inv(T)*B;
    A1=A1(1:n,1:n);
    B1=B1(1:n,:);
    F1=place(A1,B1,P);
    F1=[F1 zeros(nb,(na-n))];
    F=-F1*T;
  else
    F=place(A,B,P);
    F=-F;
  end
else
  disp('    no eigenvalues can be assigned in PLACE1')
end
% --- last line of place1 ---

function [B,C,D] = setze(A,B,C,D,P)
% SETZE  Set zeros.
%          [B,C,D] = SETZE(A,B,C,D) or [B,C,D] = SETZE(A,B,C,D,P) . If called
%          with D = 0 it sets C and D (if nb <= mc) or B and D (if nb > mc)
%          such that the invariant zeros of the system
%                      .
%                     x = Ax + Bu
%                     y = Cx + Du
%
%          are those communicated in interactive mode or specified in p .
%          If called with D ~= 0, it maintains the communicated value for D .
%          The routine implements Algorithm 4.5-1 of "Controlled and
%          Conditioned Invariant Subspaces in Linear System Theory"

%          Basile and Marro 4-20-90

nargs = nargin;
if (nargs ~= 4)&(nargs ~= 5)
  error('    wrong number of arguments in SETZE')
end
[t,na] = size(A);
[t,nb] = size(B);
[mc,t] = size(C);
s = 1;
```

```
if nb <= mc
  s = 0;
end
if s == 0
  A1 = A;
  B1 = B;
  C1 = C;
  D1 = D;
else
  A1 = A';
  B1 = C';
  C1 = B';
  D1 = D';
  r = nb;
  nb = mc;
  mc = r;
end
if nargs == 4
  F1 = place1(A1,B1);
else
  F1 = place1(A1,B1,P);
end
if norm(D,'fro') == 0
  D1 = zeros(mc,nb);
  j = 1;
  for i = 1:mc
    D1(i,j) = 1;
    if j == nb
      j = j - nb;
    end
    j = j+1;
  end
end
if s == 0
  C = -D1*F1;
  D = D1;
else
  B = -F1'*D1';
  D = D1';
end
% --- last line of setze ---

function R = mlocus(A,B,C,D,k)
%RLOCUS Multivariable root locus.
%        R = MLOCUS(A,B,C,D,k) finds the root locus for the multivariable
%        system:
%                    .
%                    x = Ax + Bu     u = -k*y
%                    y = Cx + Du
%
%        The number of outputs must be equal to that of inputs.
%        MLOCUS returns a matrix R with LENGTH(k) rows and as many columns
%        as the state space dimension is (dimension of the square matrix A).
%        Each row of the matrix corresponds to a gain from vector k.
%        The root locus may be plotted with  PLOT(R,'x').
%        The program is a generalized version of Matlab's rlocus.m

%        Basile and Marro 4-20-90

[t,na] = size(A);
[t,nb] = size(B);
[mc,t] = size(C);
if nb ~= mc
  error('    different number of inputs and outputs in MLOCUS');
end
error(abcdchk(A,B,C,D));
```

```
nk = length(k);
R  = sqrt(-1) * ones(na,nk);
% Find eigenvalues of:  A - B*inv(I+k*D)*k*C:
Id = eye(nb);
for i=1:nk
  R(:,i) = eig(A-k(i)*B*inv(Id+k(i)*D)*C);
end
R = R.';
% --- last line of mlocus ---
```

REFERENCES

1. FORSYTHE, G.E., MALCOLM, A., and MOLER, C.B., *Computer Methods for Mathematical Computations*, Prentice Hall, Englewood Cliffs, N.J., 1977.
2. FORSYTHE, G.E., and MOLER, C., *Computer Solutions of Linear Algebraic Systems*, Prentice Hall, Englewood Cliffs, N.J., 1967.
3. GOLUB, H., and VAN LOAN, F., *Matrix Computations*, Second Edition, John Hopkins University Press, Baltimore and London, 1989.
4. RALSTON, A., and WILF, H.S., *Mathematical Methods for Digital Computers Vols. I and II*, Wiley, New York, 1960.
5. STEWART, G.W., *Introduction to Matrix Computations*, Academic, New York, 1973.
6. STOER, J., and BULIRSCH, R., *Introduction to Numerical Analysis*, Springer-Verlag, New York, 1980.
7. WESTLAKE, J.R., *A Handbook of Numerical Matrix Inversion and Solution of Linear Equations*, Wiley, New York, 1968.
8. WILKINSON, J.H., *The Algebraic Eigenvalue Problem*, Clarendon Press, Oxford, 1965.

INDEX

Accessible disturbance, 299
Accessible disturbance localization by dynamic feedback, 299
Accumulation point, 425
Adjacency matrix, 361
Adjoint
  map, 389
  system, 59, 61, 183
  variable, 183
Algebraic feedback, 172
Algebraic Riccati equation, 195
Algorithm
  for state feedback matrix, 210
  for the maximal controlled invariant, 210
  for the maximal robust controlled invariant, 345
  for the minimal conditioned invariant, 209
  for the minimal robust self-bounded controlled invariant, 349
  to compute state feedback matrix F, 210
Analysis problems, 32
Asymptotic estimation in presence of disturbances, 228
Asymptotic estimator
  unknown-input, nonpurely dynamic, 229, 244
  unknown-input, purely dynamic, 229, 244
Asymptotic observer, 168
Asymptotic stability of linear systems, 102
Automaton, 36
Automorphism, 381

Autonomous regulator, 316

Backward rectangular approximation, 88
Banach space, 427
Bang-bang control function, 185
Basis, 374
Basis matrix, 112, 380
BIBO stability of linear systems, 104
BIBS stability of linear systems, 103
Binary powering method, 72
Binary relation, 360
Block-companion form, 96, 167
Block diagrams, 25
Blocking structure, 241
Blocking zero, 241
Bound in magnitude, 176
Boundary of a set, 425
Boundary point, 425
Branch, 27, 361
Branching point, 26

Canonical forms
  MIMO, 155
  relative to input, 150, 160
  relative to output, 152, 160
  SISO, 150
Canonical projection, 378
Canonical realizations, 150
Cascade connection, 29
Cauchy sequence, 426
Causality, 16
Cause, 2
Cayley-Hamilton theorem, 408
Change of basis, 129
Characteristic equation, 394

455

Characteristic *(cont.)*
 polynomial, 394
 value, 395
 vector, 395
Class, 356
Clock signal, 36
Closed half-space, 179
Closed set, 425
Closed sphere, 425
Closure of a set, 425
Codomain of a relation, 360
Cofactor, 384
Companion form, 151
Compensator
 dual observer-based, 289
 full-order, 289
 full-order, dual observer-based, 290
 full-order, observer-based, 290
 observed-based, 289
 reduced-order, 296
 reduced-order, dual observer-based, 298
 reduced-order, observer-based, 297
Complement
 of a set, 357
 of an invariant, 133
Complementability of an invariant, 133
Complete space, 427
Completely controllable system, 121
Completely observable system, 123
Component, 374
Composition, 16
Composition of two relations, 360
Concave function, 180
Condition number, 444
Conditioned invariant, 205
 complementable, 223
 input-containing, 247
 maximal self-hidden, 215
 self-hidden, 215
Cone, 180
Connection
 closed-loop, 34
 feedback, 34
 feedforward, 33
 open-loop, 33
Consistency, 16
Continuous systems, 87

Continuous map, 426
Control
 adaptive, 35
 between two given states, 25, 42, 114
 closed-loop, 34
 feedback, 34, 287
 feedforward, 33, 288
 for a given output function, 25
 for a given output sequence, 43
 input synthesis, 33
 open-loop, 33
 optimal, 35
 self-tuning, 35
 to a given output, 25, 43
 to the origin from a known initial state, 175
 to the origin from an unknown initial state, 175
 tracking, 35
Controllability, 21, 140
 after sampling, 149
 analysis, 32
 canonical form, 150
 canonical realization, 152
 index, 156
 of finite-state systems, 40
 of linear discrete systems, 118, 123
 of linear systems, 111
 referring to the Jordan form, 148
 set, 121
 subspace, 121
Controllable pair, 121
Controllable set, 22
 to the origin, 111
Controlled invariant, 204
 complementable, 222
 externally stabilizable, 218
 internally stabilizable, 217
 minimal self-bounded, 213
 output-nulling, 246
 self-bounded, 211
Controlled output, 311
Controlled system, 252
Controller, 32, 252
 canonical form, 151
 canonical realization, 153
Convex function, 180
Convex set, 179

Convolution integral, 77
Correspondence table, 362
Cost, 176
Cover, 53
Cyclic eigenspace, 409
Cyclic invariant subspace, 406
Cyclic map, 150

De Morgan, 358
Detectability, 165
Determinant, 384
 of a graph, 28
 of a partial graph, 28
Diagnosis, 24, 25, 47
Diagonal form, 396
Differentiators, 231
Dimension
 of a vector space, 375
 of a convex set, 179
Dirac impulse, 77
Direct sum, 372
Directionally convex set, 186
Discrete systems, 87
Disturbance, 4, 33, 253
 input, 35, 311
Disturbance localization
 by dynamic compensator, 273
 with d accessible, 225, 244
 with d unaccessible, 224, 244
 with stability, 225
Divisible polynomials, 404
Domain of a relation, 360
Dual observer, 173
Dual-lattices, 266
Dynamic feedback, 172
Dynamic precompensator, 173

Edge, 361
Effect, 2
Eigenspace, 407
Eigenvalue, 394
Eigenvalues
 of hermitian matrices, 414
 of symmetric matrices, 414
Eigenvector, 395
 generalized, 409
Electric circuit, 4
Electric motor, 5

Elementary divisor, 409
Empty set, 356
Environment, 2
Equilibrium point, 19
Equilibrium state, 19
 temporary, 19
Equivalence
 classes, 365
 partition, 45
 relation, 364
Equivalent norms, 425
Equivalent systems, 45
Error variable, 311
Estimate error, 169
Euclidean norm, 388
Euclidean process, 404
Event, 17
Exogenous modes, 82
Exogenous variable, 4
Exosystem, 81, 82, 254
Experiment
 adaptive, 47, 49
 multiple, 47
 preset, 47, 49
 simple, 47
Extended
 plant, 275, 311, 318
 state, 82, 255
 system, 317
Extension axiom, 355

Feedback, 160
Feedback connection, 30
Field, 369
Final state set, 177
Finite-state machine, 36
Finite-state system, 36
First-order hold approximation, 88
Five-map system, 252
Forced
 motion, 21
 response, 21
 system, 4
Forcing action, 135
Forcing actions subspace, 121
Forward rectangular approximation, 88
Four-map system (A, B, C, D), 134
Francis robust synthesis algorithm, 336

Free motion, 21
Free response, 21
Frequency response, 239
Frobenius norm, 423
Function, 361
　invertible, 363
　of a matrix, 62
　table, 362
　value, 361
Functional controllability, 236
Functional controller, 237
　stable, 238
Fundamental
　lattices, 266
　lemma of the geometric approach, 135
　sequence, 426
　theorem on the autonomous regulator, 321

Gain, 27
Gain constant, 340
Gauss-Jordan elimination method, 434
Generalized frequency response
　of the output, 240
　of the state, 240
Generating vector, 150
Gram-Schmidt orthonormalization process, 438
Gramian matrix, 112
Graph, 361
　oriented, 361
Greatest common divisor, 404

Hamiltonian
　function, 183
　matrix, 195
　system, 183
Hasse diagram, 367
Hilbert space, 427
Hold device, 32, 87
Hölder inequalities, 421
Homing, 25, 49
Hyper-robust
　disturbance localization problem, 350
　regulation, 344, 351
Hyper-robustness, 344
Hyperplane, 178

Identifiability analysis, 33
Identification, 34
Identity
　function, 376
　observer, 169
　relation, 360
Image, 361, 377
　of a set in a function, 363
Impulse response, 77
　of a discrete-time system, 79
Inaccessible states subspace, 122
Induced map on a quotient space, 132
Infimum, 369
Informative output, 35, 253, 311
Initial state set, 177
Injection, 363
Inner product, 387
Inner product space, 387
Input, 2
　distribution matrix, 135
　function, 4
　function set, 11
　set, 11
　signal, 4
　structural indices, 156
Input-output model, 78
Input-output representation, 89
　of a continuous-time system, 89
　of a discrete-time system, 90
Input-state-output model, 78
Integral depending on a parameter, 5
Integrator, 98
Interconnected systems, 25
Interior of a set, 425
Interior point, 425
Internal model, 288
Internal model principle, 313, 327
Internal modes, 82
Interpolating polynomial method, 62, 67, 73
Intersection, 357
　of two controlled invariants, 208
　of two self-bounded controlled invariants, 212
　of two subspaces, 127, 129
Invariant, 129, 378
　complementable, 133
　externally stable, 137

Invariant *(cont.)*
  internally stable, 137
  zero structure, 239
Invariant zeros, 92, 236, 238, 290
Inverse
  function, 363
  linear transform of a subspace, 129
  linear transformation of a subspace, 127
  map, 363
  of a relation, 360
  system, 237
  system stable, 237
Invertibility, 236
  unknown-state, unknown-input, 231
  zero-state, unknown-input, 231
IO model, 78
IO representation, 89
ISO model, 78
Isochronous surface, 189
Isocost hyperplane, 192
Isolated point, 425
Isomorphism, 376

Jordan form, 64, 67, 73, 399, 408
Jordan realization, 96

Kalman canonical decomposition, 143
Kalman regulator, 193
Kernel, 377
Kernel of a convolution integral, 77
Kleinman algorithm, 198

Lattice, 367
  distributive, 368
  of invariants, 131
  of self-bounded controlled invariants, 212
Liapunov equation, 107
Liapunov function, 107
Limit of a sequence, 425
Limit point, 425
Line segment, 179
Linear
  combination, 373
  dependence, 373
  function, 376
  independence, 373

Linear *(cont.)*
  manifold, 373
  map, 376
  system time-invariant, 79
  transform of a subspace, 129
  transformation, 376
  transformation of a subspace, 127
  variety, 178, 373
Lipschitz condition, 430
Loop, 28
LQR problem, 193
LU factorization, 438

Maclaurin expansion, 64
Main basis, 374
Manipulable input, 33, 253, 311
Map, 361
Mason formula, 28
Mathematical model, 2
Matrix
  adjoint, 384
  column, 382
  companion, 151
  complex, 382
  conjugate transpose, 383
  diagonalizable, 396
  exponential, 66
  exponential integral, 81, 84
  hermitian, 383
  idempotent, 383
  identity, 383
  inverse, 383
  left-invertible, 394
  nilpotent, 383
  nonsingular, 384
  null, 383
  orthogonal, 388
  partitioned, 385
  pseudoinverse, 392
  real, 382
  right-invertible, 394
  row, 382
  singular, 384
  square, 382
  stable, 136
  symmetric, 383
  transpose, 383
  unitary, 388

Maximal (A,B)-controlled invariant contained in $\mathcal{E}$, 205
Maximal A-invariant contained in $\mathcal{C}$, 131
Maximal robust $(A(q), B(q))$-controlled invariant contained in $\mathcal{E}$, 345
Maximum condition, 183
Mealy model, 36
Measurable attribute, 1
Memory of a finite-state system, 52
Metric, 418
Metric space, 418
Minimal $(A,\mathcal{C})$-conditioned invariant containing $\mathcal{D}$, 205
Minimal A-invariant containing $\mathcal{B}$, 131
Minimal polynomial, 406
Minimal realization, 94, 144
Minimal-dimension resolvent, 321
Minimal-order robust synthesis algorithm, 339
Minimum-energy control, 190
Minimum-phase system, 314
Minimum-time control, 188
Model, 168
Model-following control, 301
Modeling, 32
Modes, 69, 74
Moore model, 36
Motion, 4, 17
Motion analysis, 32

Newton formulae, 412
Next-state function, 10, 12
Nilpotent canonical form, 399
Node, 27, 361
  dependent, 27
  independent, 27
  input, 27
Noninteracting controller, 303
Nonmanipulable input, 33, 253
Nontouching loops, 28
Nontouching paths, 28
Norm, 418
  of a linear map, 422
  of a matrix, 422
Normed space, 418
Null space, 377
Nullity, 377

Observability, 23, 140
  after sampling, 149
  analysis, 32
  canonical form, 152
  canonical realization, 153
  index, 160
  of linear discrete systems, 118, 123
  of linear systems, 111
  referring to the Jordan form, 148
Observable pair, 123
Observation, 34
  of the initial state, 117
  problem, 49
Observer
  canonical form, 152
  canonical realization, 153
  reduced-order, 295
Open half-space, 179
Open set, 425
Open sphere, 425
Operations on subspaces, 127
Operations with sets, 358
Operator, 361
Optimal control, 176
Optimization, 176
Order of an input-output representation, 89
Ordered pair, 359
Oriented branch, 27
Orthogonal
  complement, 390
  complement of a subspace, 129
  complementation of a subspace, 127
  projection, 391
  projection matrix, 392
  projection theorem, 391
  vectors, 388
Orthonormal set, 388
Output, 2, 33
  distribution matrix, 135
  feedback, 173
  function, 4, 10, 12
  injection, 160
  map, 12
  set, 11
  signal, 4
  structural indices, 160
  table, 38

Output *(cont.)*
  trajectory, 18
Overall system, 311, 316

Pairwise diagnosis experiment, 46
Parallel connection, 30
Parallel realization, 101, 145
Partial ordering, 365
Partialization, 50
Partition, 365
  maximal, 369
  minimal, 369
Path, 28, 367
Peano-Baker sequence, 59
Peano-Picard successive approximations method, 430
Performance index, 34, 176
Physical realizability condition, 91
Plant, 254, 318
  extended, 275, 311, 318
Polar cone of a convex set, 180
Pole, 92
Pole assignment
  in MIMO systems, 162
  in SISO systems, 162
Pontryagin maximum principle, 182
Power of a matrix, 72
Product of two relations, 360
Projection, 378
Projection in the extended state space, 258

QR factorization, 439
  extended, 441
Quadratic form, 415
Quadruple, 134, 245
Quantizer, 32
Quotient space, 373

Range, 377
Range of a relation, 360
Rank, 377
Reachable set, 22
  from the origin, 111
  in infinite time with bounded energy, 200
  in infinite time with bounded quadratic cost, 199

Reachable set *(cont.)*
  on a given subspace, 216
  with bounded energy, 192
Realization problem, 94
Reconstructability, 23
  unknown-state, unknown-input, 231
  zero-state, unknown-input, 231
Reduction, 54
Reduction to the minimal form, 43
Reference input, 33, 253, 311
Regulated output, 35, 253
Regulated plant, 317
Regulation requirement, 274
Regulator, 32
  dual observer-based, 289
  full-order, 289
  full-order, dual observer-based, 290
  full-order, observer-based, 290
  observer-based, 289
  problem, 273
  reduced-order, 296
  reduced-order, dual observer-based, 299
  reduced-order, observer-based, 299
Relative
  complementability of an invariant, 138
  interior of a convex set, 179
  stability of an invariant, 137
Resolvent pair, 280
Response, 4
  analysis, 32
Response function, 5, 18, 77
  zero-input, 21
  zero-state, 21
Restriction of a linear map, 132
Riccati equation, 195
Robust
  controlled invariant, 344
  regulator, 310
  regulator synthesis algorithm, 332
  self-bounded controlled invariant, 344, 348
Robustness, 286

Sampled data, 9
Sampler, 31
Scalar, 369

Scalar product, 387
Schur decomposition, 110, 398
Schur form, 67, 74
Schwarz inequality, 420
Search for resolvents, 267
Self-loop, 28, 364
Separation property, 172
　of regulation, 324
Sequence detector, 39
Sequence of samples, 80
Set, 355
　finite, 355
　partially ordered, 365
Sets
　disjoint, 357
　equal, 355
Shifted input function, 19
Signal exogenous, 4
Signal-flow graphs, 25
Similar matrices, 381
Similarity transformation, 381
Singular value decomposition, 442
Singular values, 442
Sinusoid, 84
Souriau-Leverrier algorithm, 411
Span, 374
Spectrum, 395
Stability, 101
　analysis, 32
　BIBO, 148
　BIBS, 146
　in the sense of Liapunov, 102
　external, of an invariant, 137
　internal, of an invariant, 137
　of linear systems, 102
　of linear time-invariant discrete systems, 107
Stability requirement, 273, 274
Stabilizability, 164
Stabilizing unit, 324
State, 3, 15
　feedback, 160
　observation, 25, 47
　observer, 52
　reconstruction, 25, 49, 51
　set, 11, 15
State transition function, 5, 16, 76
　zero-input, 21

State transition function (cont.)
　zero-state, 21
State transition matrix, 57, 76
State-to-input feedback, 160
State-velocity function, 4, 12
States
　equivalent, 18, 45
　indistinguishable, 18
　indistinguishable in k steps, 43
Steady condition of the state, 239
Stimulus, 4
Straight line, 178
Strict regulator, 324
Strictly convex function, 180
Structure requirement, 273, 274
Submatrix, 385
Submultiplicative property, 422
Subset, 356
Subspace, 178, 371
Subspaces
　intersection of, 372
　sum of, 372
Sum of two conditioned invariants, 208
Sum of two self-hidden conditioned
　　invariants, 215
Sum of two subspaces, 127, 129
Summing junction, 26
Superposition of the effects, 20
Support hyperplane, 180
Supremum, 369
Surge tank installation, 7
SVD (singular value decomposition),
　442
Sylvester criterion, 416
Sylvester equation, 109, 111, 133
Synchronizing event, 36
Synthesis
　of a state observer, 33
　of an automatic control apparatus, 33
　of an identifier, 33
　problems, 32
System, 1
　autonomous, 4
　causal, 15
　completely observable, 24
　completely reconstructable, 24
　connected, 23
　constant, 13

System *(cont.)*
  continuous-time, 11, 12
  discrete-time, 9, 12
  distributed-parameter, 8
  dynamic, 3
  electromechanical, 5
  finite-dimensional, 15
  finite-memory, 52
  finite-state, 10, 15
  forced, 4
  free, 4
  hydraulic, 7
  in minimal form, 18, 45
  infinite-dimensional, 15
  invertibility, 231
  linear, 13
  matrix, 135
  memoryless, 3, 12, 37
  minimal, 18, 45
  nonanticipative, 15
  nonlinear, 13
  observable by a suitable experiment, 24
  oriented, 2
  purely algebraic, 3, 11, 37
  purely combinatorial, 37
  purely dynamic, 6, 12
  reconstructable by a suitable experiment, 24
  sampled data, 9
  theory, 2
  time-invariant, 13
  time-varying, 13
  with memory, 3
Systems equivalent, 19

Test signal, 315
Three-map system (A, B, C), 134
Time set, 11
Time-invariant LQR problem, 195
Time-orientation, 16
Time-shifting of causes and effects, 19
Trace, 58, 384
Tracking, 35
Trajectory, 17
Transfer function, 91
Transfer matrix, 92
Transformation, 361

Transient condition of the state, 240
Transition graph, 38
Transition table, 38
Transmission zeros, 238
Transmittance, 27
  of a loop, 28
  of a path, 28
Trapezoidal approximation, 88
Triangle inequality, 422
Triple, 134
Two-point boundary value problem, 188

Unaccessible disturbance, 299
Unassignable
  external eigenvalues of a conditioned invariant, 220
  internal eigenvalues of a controlled invariant, 218
Uncertainty domain, 310
Union, 356
Unit delay, 36, 98
Unit ramp, 84
Unit step, 83
Universal bound, 369
Unobservability
  set, 122
  subspace, 122, 232
Unobservable set containing a given subspace, 216
Unreconstructability subspace, 232
Upper-triangular form, 398

Value admissible, 11
Variable, 2
  exogenous, 4
  manipulable, 4
  nonmanipulable, 4
Vector, 370
Vector space, 370
Venn diagrams, 357
Vertex, 361
Vertex of a cone, 180

Zero, 92
  blocking, 241
  invariant, 92, 236, 238, 291
Zero assignment, 248
Zero-order hold approximation, 88

List of Matlab Geometric Approach Subroutines

(For more information, use the help command)

Q = **ima(A,p)** Orthonormalization.

Q = **ortc(A)** Complementary orthogonalization.

Q = **sums(A,B)** Sum of subspaces.

Q = **ints(A,B)** Intersection of subspaces.

Q = **invt(A,X)** Inverse transform of a subspace.

Q = **ker(A)** Kernel of a matrix.

Q = **mininv(A,B)** Minimal A-invariant containing im$B$.

Q = **maxinv(A,X)** Maximal A-invariant contained in im$X$.

Q = **miinco(A,C,X)** Minimal (A,C)-conditioned invariant containing im$B$.

Q = **mainco(A,B,X)** Maximal (A,B)-controlled invariant contained in im$X$.

[P,Q] = **stabi(A,X)** Matrices for the internal and external stablity of the A-invariant im$X$.

[P,Q] = **stabv(A,B,X)** Matrices for the internal and external stablity of the (A,B)-controlled invariant im$X$.

F = **effe(A,B,X)** State feedback matrix such that the (A,B)-controlled invariant im$X$ is an $(A+BF)$-invariant.

z = **gazero(A,B,C,D)** Invariant zeros of (A,B,C) or (A,B,C,D).

F = **place1(A,B,P)** Assigns all possible eigenvalues of $A+BF$ (specified in column vector P or in interactive mode).

[B,C,D] = **setze(A,B,C,D,P)** Assigns all possible invariant zeros of quadruple (A,B,C,D) (specified in column vector P or in interactive mode).

[B,C,D] = **setze1(A,B,C,D,P)** Like setze, with different behavior in the square case.

R = **mlocus(A,B,C,D,k)** Multivariable root locus.

V = **compl(A,B,E,P1,in)** Complementation algorithm for multivariable robust regulator design.

F = **stabf(A,B,V.P1,in,P)** State feedback matrix for multivariable robust regulator design.

B = **randma(A,coe,ik)** Random perturbation of a matrix for robustness check in multivariable robust regulator design.

reg1 Interactive multivariable robust regulator design with Algorithm 6.2-3.

reg2 Interactive multivariable robust regulator design with Algorithm 6.3-3.

reg3 Interactive multivariable robust regulator design with Algorithm 6.3-4.

examp1, examp2, ... Files of numerical data for reg1, reg2 and reg3.